Wildlife and Wind Farms, Conflicts and Solutions

Dedication

This work is dedicated to my family: my wife Eleanor, with whom I share a vision of a better future for our planet; my children Merlin & Phoenix who are still young enough to wonder, and Morgan & Rowan, who took their place in society as women somewhere along the way; and my Mum and Dad. My Mum was taken from us before these books were completed and is acutely missed.

Wildlife and Wind Farms, Conflicts and Solutions

Volume 4
Offshore: Monitoring and Mitigation

Edited by
Martin R. Perrow

Pelagic Publishing | www.pelagicpublishing.com

Published by Pelagic Publishing
www.pelagicpublishing.com
PO Box 874, Exeter, EX3 9BR, UK

Wildlife and Wind Farms, Conflicts and Solutions
Volume 4 Offshore: Monitoring and Mitigation

ISBN 978-1-78427-131-2 (Pbk)
ISBN 978-1-78427-132-9 (ePub)
ISBN 978-1-78427-134-3 (PDF)

This book should be cited as: Perrow, M.R. (ed) (2019) Wildlife and Wind Farms, Conflicts and Solutions. Volume 4 Offshore: Monitoring and Mitigation. Pelagic Publishing, Exeter, UK.

A catalogue record for this book is available from the British Library.

Colour reproduction of this book was made possible thanks to sponsorship by Vattenfall Wind Power Limited™. For more information visit https://corporate.vattenfall.com/

Cover images:

Top: Use of a double big bubble curtain in combination with a hydro-sound damper system around the pile during installation of one of the turbine foundations at Wikinger offshore wind farm in the Baltic Sea in 2016. (Hydrotechnik Lübeck GmbH)

Left: Radio-tagged Little Tern *Sternula albifrons* foraging at sea around Scroby Sands (UK); thought to be the first seabird species to be tagged in relation to wind-farm studies (2003–2006). (Martin Perrow)

Middle: A multi-sensor bird detection system combining radar technology (left) with a visual and thermal camera (right) as installed at Thanet offshore wind farm (UK) during the Offshore Renewables Joint Industry Programme Bird Collision Avoidance Study. (Thomas W Johansen)

Right: Image from an aerial survey of the seal haul-out on the main sandbank of Scroby Sands as part of the monitoring programme of the effects of the nearby Scroby Sands wind farm (UK) upon Harbour Seals *Phoca vitulina* and Grey Seals *Halichoerus grypus*. (Air Images Ltd)

Contents

Contributors

Sophy Allen Specialist Services and Programmes Team, Chief Scientist Directorate, Natural England, Ground Floor, Sterling House, Dix's Field, Exeter EX1 1QA, UK

Lothar Bach Freilandforschung, zoologische Gutachten, Hamfhofsweg 125 b, D-28357 Bremen, Germany

Richard J. Berridge ECON Ecological Consultancy Ltd, Unit 7 Octagon Business Park, Little Plumstead, Norwich NR13 5FH, UK

Juliane Biehl Berlin Institute of Technology (TU Berlin), Environmental Assessment & Planning Research Group, Sekr. EB 5, Straße des 17. Juni 145, D-10623 Berlin, Germany

Richard Caldow Birds Team – Specialist Services & Programmes, Natural England, W4, Dorset Council, County Hall, Colliton Park, Dorchester, Dorset DT1 1XJ, UK

Filipe Canário STRIX, Environment and Innovation, Rua Roberto Ivens 1314 esc. 1.14–15, 4450-251 Matosinhos, Portugal

Timothy Coppack APEM Ltd, Riverview, A17 Embankment Business Park, Heaton Mersey, Stockport SK4 3GN, UK

Aonghais S. C. P. Cook British Trust for Ornithology, The Nunnery, Thetford IP24 2PU, UK

Thomas G. Dahlgren NORCE, Postboks 22 Nygårdstangen, N-5838 Bergen, Norway and University of Gothenburg, Department of Marine Sciences, Box 461, SE 405 30 Gothenburg, Sweden

Marie Dahmen Berlin Institute of Technology (TU Berlin), Environmental Assessment & Planning Research Group, Sekr. EB 5, Straße des 17. Juni 145, D-10623 Berlin, Germany

Tobias Dittmann IfAÖ Institut für Angewandte Ökosystemforschung GmbH, Carl-Hopp-Str. 4a, D-18069 Rostock, Germany

Gesa Geissler Berlin Institute of Technology (TU Berlin), Environmental Assessment & Planning Research Group, Sekr. EB 5, Straße des 17. Juni 145, D-10623 Berlin, Germany

Larry Griffin Wildfowl & Wetlands Trust, Slimbridge, Gloucester GL2 7BT, UK

Linus Hammar Chalmers University of Technology, Department of Technology Management and Economics, Division of Environmental Systems Analysis, SE-412 96 Gothenburg, Sweden

Andrew J.P. Harwood ECON Ecological Consultancy Ltd, Unit 7 Octagon Business Park, Little Plumstead, Norwich NR13 5FH, UK

Stefan Heinänen DHI A/S, Agern Allé 5, DK-2970 Hørsholm, Denmark

Reinhold Hill Avitec Research, Sachsenring 11, D-27711 Osterholz-Scharmbeck, Germany

Baz Hughes Wildfowl & Wetlands Trust, Slimbridge, Gloucester GL2 7BT, UK

Ommo Hüppop Institute of Avian Research 'Vogelwarte Helgoland', An der Vogelwarte 21, D-26386 Wilhelmshaven, Germany

Sue King Sue King Consulting Ltd, The Coach House, Hampsfell Road, Grange-over-Sands LA11 6BG, UK

Johann Köppel Berlin Institute of Technology (TU Berlin), Environmental Assessment & Planning Research Group, Sekr. EB 5, Straße des 17. Juni 145, D-10623 Berlin, Germany

Olivia Langhamer Chalmers University of Technology, Department of Technology Management and Economics, Division of Environmental Systems Analysis, SE-412 96 Gothenburg, Sweden

Aulay Mackenzie Wrexham Glyndŵr University, Mold Road, Wrexham LL11 2AW, UK

Elizabeth A. Masden Environmental Research Institute, North Highland College-UHI, University of the Highlands and Islands, Ormlie Road, Thurso KW14 7EE, UK

Roel May Norwegian Institute for Nature Research (NINA), PO Box 5685 Sluppen, NO-7485, Trondheim, Norway

Sara Méndez-Roldán Niras Consulting Ltd, St Giles Court, 24 Castle Street, Cambridge CB3 0AJ, UK

Markus Molis Avitec Research, Sachsenring 11, D-27711 Osterholz-Scharmbeck, Germany

Georg Nehls BioConsult SH GmbH & Co. KG, Schobüller Str. 36, 25813 Husum, Germany

Tim Norman Niras Consulting Ltd, St Giles Court, 24 Castle Street, Cambridge CB3 0AJ, UK

Steve Pelletier Stantec Consulting Services, Inc., 30 Park Drive, Topsham, ME 04086, USA

Martin R. Perrow ECON Ecological Consultancy Ltd, Unit 7 Octagon Business Park, Little Plumstead, Norwich NR13 5FH, UK

Lindsay Porter SMRU Hong Kong, The University of St. Andrews, 17/F Tower 1, Lippo Centre, 89 Queensway, Hong Kong SAR

Michelle E. Portman Segoe Building 502, Faculty of Architecture & Town Planning, Technion-Israel Institute of Technology, Haifa 32000, Israel

Eileen Rees Wildfowl & Wetlands Trust, Slimbridge, Gloucester GL2 7BT, UK

Jan T. Reubens Flanders Marine Institute, Wandelaarkaai 7, 8400 Ostend, Belgium

Viola H. Ross-Smith British Trust for Ornithology, The Nunnery, Thetford, IP24 2PU, UK

Meike Scheidat Wageningen Marine Research, PO Box 68, 1970AB IJmuiden, The Netherlands

Axel Schulz IfAÖ Institut für Angewandte Ökosystemforschung GmbH, Carl-Hopp-Str. 4a, D-18069 Rostock, Germany

Eleanor R. Skeate ECON Ecological Consultancy Ltd, Unit 7 Octagon Business Park, Little Plumstead, Norwich NR13 5FH, UK

Henrik Skov DHI A/S, Agern Allé 5, DK-2970 Hørsholm, Denmark

Chris B. Thaxter British Trust for Ornithology, The Nunnery, Thetford, IP24 2PU, UK

Frank Thomsen DHI A/S, Agern Allé 5, DK-2970 Hørsholm, Denmark

Ricardo Tomé STRIX, Environment and Innovation, Rua Roberto Ivens 1314 esc. 1.14–15, 4450-251 Matosinhos, Portugal

Mark L. Tomlinson ECON Ecological Consultancy, Unit 7 Octagon Business Park, Little Plumstead, Norwich NR13 5FH, UK

Inge van der Knaap Marine Biology Research Group, Ghent University, Krijgslaan 281/S8, 9000 Ghent, Belgium

Tobias Verfuß The Carbon Trust, CBC House, 24 Canning Street, Edinburgh EH3 8EG, UK

Robin Ward Niras Consulting Ltd, St Giles Court, 24 Castle Street, Cambridge CB3 0AJ, UK

Andy Webb HiDef Aerial Surveying Ltd, Phoenix Court, Earl Street, Cleator Moor, Cumbria, CA25 5AU, UK

Preface

Wind farms are seen to be an essential component of global renewable energy policy and the action to limit the effects of climate change. There is, however, considerable concern over the effects of wind farms on wildlife, especially on birds on bats onshore, and seabirds and marine mammals offshore. On a positive note, there is increasing optimism and evidence that by operating as reefs and by limiting commercial fishing activity, offshore wind farms may become valuable in conservation terms, perhaps even as marine protected areas.

With respect to any negative effects, Environmental Impact Assessment (EIA) adopted in many countries should, in theory, reduce any impacts to an acceptable level, particularly as this should also incorporate any mitigation required. Although a wide range of monitoring and research studies have been undertaken, only a small body of that work appears to be represented in the peer-reviewed literature. The latter is burgeoning, however, concomitant with the interest in the interactions between wind energy and wildlife as expressed by the continuing Conference on Wind Energy and Wildlife Impacts (CWW) series of international conferences on the topic. A total of 342 participants from 29 countries attended CWW 2017 in Estoril, maintaining similar numbers on both counts to that achieved at CWW 2015 in Berlin. Let us hope that CWW 2019 in Stirling (Scotland) has a similar audience. Recent considerable interest within individual countries suggests this should be the case. For example, the Wind Energy and Wildlife seminar (Eolien et biodiversité) including both onshore and offshore wind, attracted over 400 participants to Artigues-près-Bordeaux, France on 21–22 November 2017.

Even with specific knowledge of the literature as well as participation in meetings, I reached the conclusion (some time ago now) that there was a clear need for a coherent overarching review of potential and actual effects of wind farms and, perhaps even more importantly, how impacts could be successfully avoided or mitigated. Understanding the tools available to conduct meaningful research is also clearly fundamental to any research undertaken. A meeting with Nigel Massen of Pelagic Publishing in late 2012 at the Chartered Institute of Ecology & Environmental Management Renewable Energy and Biodiversity Impacts conference in Cardiff, UK crystallised the notion of a current treatise and the opportunity to bring it to reality. Even then, the project could not have been undertaken without the significant financial support of ECON Ecological Consultancy Ltd. expressed as my time.

At the outset of the project I did not imagine the original concept (one volume for each of the onshore and offshore disciplines) would morph into a four-volume series; with onshore and offshore each having a volume dedicated to (i) documenting current knowledge of the effects – the conflicts with wildlife; and (ii) providing a state-of-the-science guide to the available tools for monitoring and assessment and the means of avoiding, minimising and mitigating potential impacts – the solutions. I also did not imagine that the gestation time to produce the volumes would be nearly seven years, or that the offshore volumes would run two years behind the onshore volumes. The offshore industry has developed rapidly in the last few years and this meant many potential authors were swamped by their workloads within various roles within the industry. Perhaps inevitably, several authors fell by the wayside, which caused delays and some stop and start in the process as replacements were found. However, I believe this has meant that the books have ultimately benefited by being able to document the rapid progress that has occurred in the last few years and by having a particularly active team of authors at the top of their game.

In this Volume 4, documenting monitoring and mitigation as the solutions to any conflicts offshore, the concept was to cover the main groups that have been the focus of monitoring efforts, namely birds, especially seabirds, marine mammals, fish and invertebrate communities. As relatively little work has yet to be conducted on fish, these are included with invertebrates in the opening chapter *Monitoring invertebrates and fish*, before subsequent chapters on *Monitoring marine mammals* and *Surveying seabirds.* Both the *Monitoring invertebrates and fish* and *Monitoring marine mammals* chapters include information on telemetry, with the latter perhaps providing the most relevant information on any telemetry of mega-fish such as sharks, or even sea turtles required in future wind-farm studies. Such has been the rapid advance in, and popularity of, the application of telemetry in relation to birds, including both seabirds and migrant terrestrial species such as waterfowl; a separate chapter on *Telemetry and tracking of birds* is warranted. This concludes the monitoring section of the volume, before a switch in focus to *Modelling collision risk and predicting population-level consequences.* The recent work offshore expands that conducted onshore and documented in Volume 2 of this series, notably in trying to tackle the greater degree of 'unknowns' offshore, where the rate of collision cannot be confirmed through carcass collection, as it may be onshore. This has stimulated the technological advance of remote methods to detect collisions in the challenging conditions at sea that are thoroughly reviewed in *Measuring bird and bat collision and avoidance,* and that provides the means of populating the collision risk models covered in the previous chapter. The chapter also includes exciting work on the telemetry of small migratory birds such as passerines, that have similar traits to bats, to complement the earlier *Telemetry and tracking of birds* chapter.

The volume then moves into mitigation, beginning with *Mitigating the effects of noise* that reviews all current technological options in reducing construction noise in particular, especially where this is undertaken with pile-driving, which is widely thought to be one of the principal threats of offshore wind-farm development to wildlife such as marine mammals and fish. Mirroring the mitigation hierarchy approach adopted in the equivalent onshore chapter in Volume 2, *Mitigation for birds* outlines some of the mostly theoretical options for reducing collision and displacement of birds, several of which also stem from research onshore. The subtitle '*with implications for bats*' nods to some of the additional information provided in a stand-alone box. The final chapter of the nine, *Perspectives on marine spatial planning*, provides a history of how plans to install offshore wind farms have been a key driver of marine spatial planning and provides a series of recommendations in relation to future planning requirements that are a crucial first step in avoiding potential impacts of wind farms.

To promote coherence within and across volumes, a consistent style was adopted for all chapters, with seven sub-headings: Summary, Introduction, Scope, Themes, Concluding remarks, Acknowledgements and References. For ease of reference, the latter are reproduced after each chapter. The carefully selected sub-headings break from standard academic structure (i.e. some derivative of Abstract, Introduction, Methods, Results and Conclusions) in order to provide flexibility for the range of chapters over all four volumes, many of which are reviews of information, whilst others provide more prescriptive recommendations or even original research. Some sub-headings require a little explanation. For example, the Summary provides an up to ~300-word overview of the entire chapter, whilst the Concluding remarks provide both conclusions and any recommendations in a section of generally ~500 words. The Scope sets the objectives of the chapter, and for the benefit of the reader describes what is, and what is not, included. Any methods are also incorporated therein. The Themes provide the main body of the text, and are generally divided into as few sub-heading levels as possible. Division of material between wind-farm phases (e.g. construction, operation and decommissioning) was generally avoided as this increased the number of sub-headings and led to an unwieldy structure. Rather, any clear differences between wind-farm phases, for example where specific monitoring methods should be applied, were incorporated into specific sub-headings.

As well as being liberally decorated with tables, figures and especially photographs, which are reproduced in colour courtesy of sponsorship by Vattenfall, most chapters also contain Boxes. These were designed to be provide important examples of a particular point or case or suffice as an all-round exemplar; and thus 'stand-alone' from the text. In a few cases, these have been written by an invited author(s) on the principle that it is better to see the words from the hands of those involved rather than paraphrase published studies. My sincere thanks go to all 26 chapter authors and further 21 box authors (excluding myself in both cases) for their contributions. I take any deficiencies in the scope and content in this and its sister volume to be my responsibility, particularly as both closely align to my original vision, and many authors have patiently tolerated and incorporated my sometimes extensive editorial changes to initial outlines and draft manuscripts.

Finally, it needs to be stated that with a current epicentre in northwestern Europe in the North and Baltic Seas, the coverage of this book could not exactly be global. However, as the offshore wind industry develops at almost breakneck speed in a great range of countries, I hope the information and experiences gleaned from the pages of this book can be applied in a global context; with the proviso that applying any lessons learned to marine systems elsewhere on the planet would need to be accompanied by specific research to account for any inevitable differences in ecological structure and functioning of those systems. Hopefully, this book is a further step towards the sustainable development of wind farms and the ultimate goal of a win–win[1] scenario for renewable energy and wildlife.

Martin R. Perrow
ECON Ecological Consultancy Ltd.
21 June 2019

1 Kiesecker, J.M., Evans, J.S., Fargione, J., Doherty, K., Foresman, K.R., Kunz, T.H., Naugle, D., Nibbelink, N.P. & Niemuth, N.D. (2011) Win–win for wind and wildlife: a vision to facilitate sustainable development. *PLoS ONE* 6: e17566.

CHAPTER 1

Monitoring invertebrates and fish

THOMAS G. DAHLGREN, LINUS HAMMAR and OLIVIA LANGHAMER

Summary

Current knowledge is often sufficient to predict and prevent degradation of ecosystems, through careful planning, technology choice, good practice, and mitigation and restoration measures. Baseline studies and environmental monitoring programmes must be carefully planned and employ rigorous data collection if the analyses are to be capable of showing significant results for both anticipated and unforeseen effects. Since baseline studies of wind farms are performed at a fine scale, monitoring data may also serve as fundamental research on faunal distribution and behaviour, as well as playing a crucial role in the consenting and permitting processes. The continuously growing offshore wind industry requires certainty about effects before, during and after installation, to minimise or even eliminate any potential detrimental impacts on marine habitats and ecosystems. This chapter reviews monitoring methods, technologies and scientific tools that are commonly applied to monitoring fish and invertebrates at operational sites. Seabed communities are defined as the interacting assemblage of organisms in the sediment (infauna) and on sediment and structures (epifauna and algae), as well as demersal and pelagic fish. Case studies from scientifically developed monitoring programmes and studies are provided as examples of rigorous data gathering capable of showing environmental change. A successful baseline study and monitoring programme is likely to include a combination of traditional tools, such as fyke-net fishing, and modern and developing technologies, such as sonar and the use of telemetry to track the response of individual fish to wind-farm construction and operation. To further our understanding of long-term and cumulative impacts, data from monitoring programmes should ideally be collected, analysed and reported in a way that allows for future additional analyses. Examples of such data include changes in benthic community structure that may affect shelf ecosystem services such as carbon mineralisation and/or burial, or changes in fish diversity and abundance.

Introduction

Environmental issues play a central role in the consenting process of offshore wind projects. As evidenced by Volume 3 in this series (Perrow 2019), the already substantial volume of science-based information on the environmental impact of offshore wind farms (OWFs) is still growing, most importantly with regard to long-term and cumulative impacts. In combination with an increasingly comprehensive palette of strategies to avoid short-term impact and long-term degradation of ecosystems, the arguments to stop a wind farm based on the precautionary principle alone are becoming weaker. After more than 15 years of environmental monitoring surveys and scientific research at OWFs, much has been learned regarding the effects on marine ecosystems. The strongest and best understood environmental stressors are associated with the construction phase, although the operational phase likely affects oceanographic processes (Broström *et al.* 2019) and the distribution and abundance of animals and therefore ecosystem functioning (Perrow 2019). In comparison to wind power on land, the use of large turbines in OWFs is facilitated by a combination of low landscape impact, eased logistics as a result of ship-borne transport and assembly of turbines, and remote operation and maintenance. However, OWFs are relatively sparsely packed, sometimes with 1 km or more between turbines. Hence, the direct areal impact on marine habitats is very minor in relation to installed capacity. Typically, the foundation footprint, and thus the habitat required to be offset, is small compared to the total occupied area. Nevertheless, physical habitat change is one of the most evident environmental issues for seabed communities, comprising the interacting assemblage of organisms in the sediment (infauna) and on sediment and structures (epifauna and algae), and benthic fish (Dannheim *et al.* 2019). Foundations, scour protection and cable trenches typically induce habitat change, expressed as both introductions of new 'hard-surface' and intertidal habitats (habitat gain), and loss of habitats, such as particular sediments due to extraction, burial or changed granulometry caused by altered current patterns.

Introduction of large structures in the sea causes a potentially beneficial 'reef effect', which contributes towards increased biodiversity by attracting both filter-feeding epifauna and larger mobile scavengers, such as crabs and benthic fish (Dannheim *et al.* 2019). Demersal and benthopelagic fish use the structures as fish-aggregation devices, attracted by biofouling organisms (Stenberg *et al.* 2015; Gill & Wilhelmsson 2019). The species involved include large predators, such as Atlantic Cod *Gadus morhua* (Reubens *et al.* 2013; 2014a; 2014b). Top predators, including marine mammals (Nehls *et al.* 2019), especially seals (Russell *et al.* 2014), and some seabirds, such as Great Cormorant *Phalacrocorax carbo* and Great Black-backed Gull *Larus marinus* (Vanermen & Stienen 2019), may also be attracted, resulting in a more patchy distribution of these groups. An additional effect of the increased availability of hard structures is a significant increase in filter-feeding bivalves (*Mytilus* spp.), with the potential to alter the area's benthic productivity rate and water clearance (Krone *et al.* 2013). Altogether, redistribution of marine organisms potentially changes ecosystem functions and predator–prey dynamics, affecting sediment-dwelling, pelagic and sessile organisms. The introduction of new habitat types allows for the colonisation of new, possibly invasive, species that benefit from artificial structures in open water. As OWFs are often sited far from a natural shore, the artificial intertidal habitat can potentially be used as a stepping stone for splash-zone and shallow-water dependent propagules that would otherwise perish in open water. One example is the Pacific Oyster *Crassostrea gigas* (Gutow *et al.* 2014; de Mesel *et al.* 2015).

During construction, dredging and trenching also cause short-term increased turbidity that can harm both filter-feeding organisms (Rosenberg 1977) and fish eggs and larvae (Westerberg *et al.* 1996). However, in most circumstances the sediment spill from construction works causes high particle concentrations only over a very localised area, often not exceeding naturally occurring levels farther than 100 m from the source (Bergström *et al.* 2012). Other pollutants that may cause harm to the environment include the fuels and lubricants that could be accidentally spilled from the vessels present throughout the life of the wind farm, but especially in the construction phase (Rees & Judd 2019). Similarly, the cooling agents and lubricants needed for wind-turbine transmission gear and generators within an operational wind farm may be harmful even if released in very small quantities. A more site-specific problem is the possible release of old pollutants previously dumped or accumulated in the sediment, if these are suspended during cable trenching or seabed preparation (e.g. gravity foundations require dredging). This specific risk of secondary exposure due to the potential release of toxic sediment was a reason for halting the planning of a wind-power project prospected by Vindplats Göteborg in 2014, at an old dredge dump site in the mouth of the Göta älv River on the Swedish west coast.

Electromagnetism is emitted from cables within the wind farm and, to a larger extent, from power cables connecting with land. The electric component of the electromagnetic field (EMF) produced is effectively shielded by cable armour, but the magnetic component persists and, in turn, gives rise to a secondary induced electric field. Elasmobranchs use electric fields to detect prey buried in the sediment and can therefore be disturbed by cables of the kind used in wind farms (Gill 2005; Gill & Wilhelmsson 2019). It has also been shown that migrating fish, such as eels, can be disturbed by strong magnetic fields, slightly diverting them from their track (Westerberg & Lagenfelt 2008). To date, field experiments conducted at wind farms have not been able to show any significant effects on either benthos or fish, but it is reasonable to believe that wind-power cables with strong EMFs can have some level of effect on sensitive animals (Lagenfelt *et al.* 2012).

In contrast, in theory, the noise and vibration energies from pile driving are a more severe threat to the ecosystem at least in the short-term. Pile driving is required to install the more common monopole turbine designs as well as the pin piles associated with jacket and tripod designs (Jameson *et al.* 2019). The assumed acute effects on fish caused by pile-driving noise during construction include behavioural changes, such as escape, avoidance, relocation and change of community structure. The physiological effects include acute tissue or barotrauma damage induced by seismic pressure waves to animals with air-filled cavities, often followed by low survival rates (Nedwell *et al.* 2003a; Müller 2007; Popper & Hastings 2009; Bailey *et al.* 2010). During operation of a wind farm, noise emissions are low, but they increase with turbine and wind-farm size (Nedwell *et al.* 2003a). It has been shown that some fish species can be stressed by operational noise as a low-intensity stressor (Engås *et al.* 1995; Kastelein *et al.* 2008; Caiger *et al.* 2012) by having, for example, an increased respiration rate (Wikström & Granmo 2008). Continuous noise has also been shown to impair reproductive success in Atlantic Cod (Sierra-Flores *et al.* 2015). Still, habituation can occur when fish are continuously exposed to noise levels comparable to those in harbours and areas of intense shipping activity. Indirect effects of noise on fish include the masking of natural sounds or bioacoustics (Wahlberg & Westerberg 2005; Simpson *et al.* 2015). In many fish species, the use of sound is common for finding prey, avoiding predators, interspecies communication, finding a mate, and even for orientation and navigation. Noise has also been shown to act as a settling cue for fish and decapod species (Montgomery *et al.* 2006). Consequently, species abundance and community dynamics can be affected negatively in operational wind farms. Elasmobranchs, for example, have well-developed ears and sound plays an important role in their lives, although there are as yet no studies on the impact of OWF construction and operational noise on elasmobranchs.

To date, noise effects on marine invertebrates have also not been studied. However, the response to noise in invertebrates is rather species-specific and cannot be generalised. Crustaceans, such as shrimps, krills, crabs and lobsters, respond with a high variation in behaviour, feeding rate, growth, reproduction, metabolic level and mortality (Lagardère 1982; Carroll *et al.* 2017). Hitherto, detrimental effects of turbine noise have been shown only under laboratory conditions, while no corresponding effects have been detected in monitoring at actual wind farms.

A range of relatively general actions may be taken to mitigate the effects of operation and especially construction of OWFs on invertebrates and fish detailed below (Box 1.1). Fish, in particular, potentially benefit greatly from the mitigation of pile-driving noise typically directed at marine mammals, such as the use of bubble curtains and casings, as well as adjustment of piling energy, pulse prolongation and advances in piling technology (see Thomsen & Verfuß, Chapter 7). Irrespective of whether mitigation is attempted, wind-farm projects are typically obliged to involve environmental monitoring programmes. The purposes of monitoring include verification that any effect was indeed within acceptable impact boundaries, with or without mitigation, and detecting any unforeseen environmental impacts. Such monitoring also contributes to the general understanding of how OWFs affect the local and regional environment. A properly designed baseline study and monitoring programme may also be used to test the validity of future suggestions or even accusations that the wind farm had a role in any observed ecological degradation, when other stressors, such as climate change or increased shipping activities, could have been the cause. Hence, it is of crucial importance that monitoring programmes use appropriate methods and are based on a rigorous statistical framework accounting for site-specific spatial and temporal physical and biological variations.

Box 1.1 Avoiding and mitigating wind-farm impacts upon invertebrates and fish

Here, best practice to avoid and mitigate the impact of wind farms, during siting, construction, operation and decommissioning, upon infaunal, epifaunal and demersal and pelagic fish assemblages is described following a traditional mitigation hierarchy of avoid, reduce, compensate and restore (Cuperus *et al.* 1996; 1999; Vaissière *et al.* 2015). The mitigation of noise impacts is described in more detail by Thomsen & Verfuß (Chapter 7).

Siting

Considerations over the siting of wind farms and turbines to reduce any effects on seabed integrity and vulnerable species and habitats will help to avoid an overall impact on the ecosystem. The micro-siting of individual turbines within a wind farm may minimise any impact on more valuable habitats in a patchy mosaic, such as seagrass (*Zostera*) meadows, mussel banks, or patches of coarse sand and gravel in an otherwise silty area. Micro-siting should also be considered to avoid the need for dredging in a heterogeneous seabed with alternating soft and hard substrates and to avoid patches of contaminants on the seabed.

The macro-siting of the entire wind farm will inevitably be subject to a large number of factors, of which avoiding potential habitat degradation is only one. Such macro-siting considerations include avoiding unique vulnerable environments or areas that maintain sensitive and important ecological functions, such as particular fish spawning sites and wintering sites for seabirds. The avoidance of specific sites can also be considered in order to minimise the risk of their acting as stepping stones for invasive species. Siting considerations must, however, be based on well-documented anticipated impact in a risk-analysis framework, otherwise there is a risk that various avoidance arguments are used as a pretext to protect unrelated interests. In addition, siting of a project in a disturbed area can be preferable to a development in a pristine environment (Inger *et al.* 2009). The reason for this is two-fold. First, already degraded environments typically represent a benthic community that is more resilient to disturbance compared to pristine environments. Secondly, positive effects of wind farms, such as exclusion of fishing and increased diversity and productivity due to the reef effect, are likely to be more valuable in a degraded ecosystem. An offshore wind farm may actually help to improve the ecological status in a previously degraded environment (Bergström *et al.* 2014).

Wind-farm design

The technical design of a wind farm has an influence on how individual turbine foundations are positioned in relation to benthic habitats. If turbines are located in seagrass beds, mussel beds or other biotopes of particular ecological value, the benthic footprint will be more detrimental than if turbines are positioned on less vulnerable biotopes, such as mud and sand bottoms. It is, however, generally difficult to determine the exact positioning of turbines in a wind farm at the consent stage as a result of the rapid technological development of the type and size of foundations and turbines. Wind farms are therefore often planned according to the 'box model', meaning that the outer borders of the wind farm are determined and the design of the wind farm inside this 'box' is only indicated according to a 'design envelope'. In cases where vulnerable or particularly valuable biotopes are present inside the wind-farm area, it can be difficult to ensure that these sites are avoided during the final design. A suggested solution to this is that the consenting authorities apply legal conditions, stating that the final positioning of foundations and cable trenches must avoid ecologically valuable sites during the micro-siting phase.

The design of a wind farm may also influence how hydrodynamics are affected, with a relatively dense wind farm having more potential impact than a more sparsely designed wind farm with a lot of space between turbines. Moreover, it has been postulated that a wind farm that captures a large portion of the incoming wind may induce an artificial upwelling in the wake caused on the lee side (Broström 2008), and field evidence supporting this theory has been gathered (Broström *et al.* 2019).

Timing

Timing of construction events can be an important element of risk mitigation (Bergström *et al.* 2012; Hammar *et al.* 2012). It is well known that pile driving generates hazardous high-intensity sound and that dredging under certain conditions can cause locally harmful levels of dissolved particles. Seasonal events to avoid such impacts are typically related to ecological functions such as reproduction and migration. Regarding the benthic community, the importance of such time-dependent considerations is more relevant for fish than for invertebrates (Bergström *et al.* 2012). In one case, the construction of a wind farm during the spawning period of a local population of Atlantic Herring *Clupea harengus* was thought to be responsible for a significant reduction in the strength of recruitment, with a significant knock-on effect on the feeding conditions of a breeding seabird, the Little Tern *Sternula albifrons*, at what was then its most important colony in the UK (Perrow *et al.* 2011).

In Sweden, it is common practice to allow potentially harmful construction events only within a limited time window. This applies both to offshore wind power and to other activities consented under the Swedish Environmental Code. For large projects, narrow time windows for construction can increase construction time considerably, especially when combined with windows of favourable weather. The question of time restrictions during the construction phase is a balance between shorter, intense disturbance and prolonged but lower disturbance. Piling and dredging at a single foundation take hours and days, respectively. A large wind farm with hundreds of turbines may thus take years to construct, even without time restrictions. Long-term disturbance is typically considered more significant than short-term disturbance because a more inclusive part of the population will be exposed and since no time will be given for recovery. Yet such temporally prolonged disturbance can have little ecological impact if important functions are not affected (Hammar *et al.* 2014). If the wind farm is large enough, the construction activity can be alternated between opposing sites within the area, relieving the benthic community on one side of the farm at a time.

Mitigation through technology choice

Other than siting and timing, discussed above, the material, type and design of the foundations are important features that can be optimised in an environmental sense based on general and local circumstances. For instance, there is a range of foundation types that minimise the use of heavy piling. In areas sensitive to piling noise, consideration should be given to avoiding the use of monopile foundations (heavy piling) and instead relying on gravity foundations, drilled foundations or foundations based on small-diameter piles (Hammar *et al.* 2008), although these are more expensive.

Compared to steel foundations, concrete-based gravity foundations trigger fouling organisms to settle at a significantly higher rate (Degraer 2012) and also emit less turbine vibration during operation (Hammar *et al.* 2008). However, concrete foundations have the disadvantage of being significantly more expensive than steel.

Jacket foundations have complex structures that enhance the reef effect and generate weaker vibrations than monopile foundations, albeit more than gravity foundations. Therefore, the foundation type, material and coating can be specifically chosen to influence the levels of settling larvae, the extent of the reef effect and the level of noise (vibration) disturbance (Table 1.1). In general, monopile foundations emit most noise but have least impact on the seabed community in terms of fouling, reef effect and space occupation. Conversely, gravity foundations and jacket foundations emit lower noise but impose a higher change on the benthic community.

It should be noted that, regarding the effects of vibrations and noise, only limited data are available, covering only a small range of turbine and foundation types. More research and monitoring are needed to understand how particular types of tower, turbine, gear and generator design affect the level of noise transmitted into the water. Earlier types of tower-to-foundation connectors (grouting) have been replaced with

Table 1.1 Relative magnitudes of effects on seabed communities from different types of foundations (note that all effects are not necessarily negative at all sites)

Effects	Foundation type		
	Gravity	Jacket	Monopile
Temporary effects			
Sediment spill	+++	+	+
Installation noise	+	++	+++
Permanent effects			
Space occupation	++	++	+
Fouling rate	++	+	+
Reef effect	++	+++	+
Noise emissions	+	++	+++

Modified from Hammar *et al.* (2008).

heavy rubber pads that more effectively absorb vibrations. While this is done to prolong the lifetime of the turbine, it has the additional benefit of reducing the noise energy transmitted from the nacelle to the foundation and surrounding water.

As the offshore wind industry develops and occupies more space in the ocean, it becomes increasingly important to implement technical adjustments in order to minimise stressors that generate subtle effects, such as underwater noise, so that long-term ecological effects are prohibited (Slabbekoorn *et al.* 2010). Underwater noise is included among the 11 descriptors for good environmental status in the European Union Marine Strategy Framework Directive (EU 2008), and as such, should be limited.

Compensation and restoration

Compensation to stakeholders to offset the impacts on ecosystem services may include a range of measures (Elliott *et al.* 2014). A paper based on data collected from a French project organises ecosystem services into three categories: (a) provisioning, (b) regulating and (c) cultural (which is not within the scope of this chapter) (Kermagoret *et al.* 2014). For provisioning ecosystem services (a), two major impacts were perceived from stakeholders: loss of fishing opportunities and gain of clean energy. The loss of fishing opportunities is expected from the restriction in exploitable area and the disturbance during the development phase. However, there are examples in the UK where lucrative pot fisheries for European Lobster *Homarus gammarus* and Edible Crab *Cancer pagurus* have developed as a result of the presence of wind-farm structures. Some comments from stakeholders highlighted the potential for increased production of offspring within the wind farm, which effectively acts as a marine protected area (Dannheim *et al.* 2019). This may increase the future catch per effort outside the wind farm (Bergström *et al.* 2014). As a compensation for the loss of fishing opportunity, a seeding programme of the most important catch, Great Scallop *Pecten maximus*, was agreed in the French project. In this case study, fisheries were also to be compensated by receiving 35% of the annual tax based on the electricity production that the project will pay the community (€14,000 per MW and year). The money was to be used to fund projects promoting sustainable fisheries and is thus the most important perceived impact on regulatory ecosystem service (b), coupled with a contribution to global climate regulation, which is inherent in all wind-farm projects. At a local scale, a reef effect was to be expected at each turbine foundation, which is positive in some respects (higher diversity and higher productivity), although the net impact was not possible to judge and no further compensation was suggested.

Irrespective of foundation type, the footprint of a turbine removes one type of habitat and replaces it with another. To compensate, a concrete gravity foundation can be cast in order to enhance the surface shape to maximise the availability of habitats for decapods such as crabs and lobsters, for example (Langhamer & Wilhelmsson 2009). In cases where a specific habitat, such as *Zostera* meadows, has been replaced or damaged, compensatory introduction of new *Zostera* meadows could be undertaken in a suitable area nearby.

Scope

This chapter is based on both monitoring reports and peer-reviewed scientific literature, with the intention of reviewing and outlining the status and applications of several different monitoring methods, focusing on fish and marine invertebrates. The main contributions are reported from the regulatory and science-driven environmental monitoring programmes in north-west European waters, including Belgium (Degraer 2012), the Netherlands (Lindeboom *et al.* 2011; 2015), Sweden (Bergström *et al.* 2013a, 2013b), Denmark (Stenberg *et al.* 2011), Germany (BSH & BMU 2014) and the UK (Huddleston 2010; MMO 2014). These countries have the longest experience in environmental impact assessments and monitoring of offshore wind-power installations. Academic research on ecological and environmental effects related to offshore wind power has so far mostly been carried out by Denmark, Germany, the UK and Sweden. Nevertheless, this research reflects the state of the art for wind-power monitoring in general and includes theoretical considerations for offshore wind development in developing markets in other parts of the world, such as the USA and Japan (Jameson *et al.* 2019).

Unfortunately, many of the monitoring programmes developed at existing OWFs have not been planned adequately. Some surveys may be limited in the development of survey methods while others may have focused on single-species systems. In a review of monitoring programmes for offshore wind projects in Europe, only 33% of the programmes had conducted a power analysis and in only 10% was the study based on a random sampling design (Enhus *et al.* 2017). Therefore, they may not have been able to deliver results as valuable as was intended, such as on effects on marine communities and/or ecosystems, or data that can be used in a wider context (MMO 2014). Nevertheless, there have been many studies on many different species and possible wind-farm induced stressors in the Baltic and North Sea region (reviewed in Bergström *et al.* 2014). The accumulated understanding of the main environmental impacts on fish and marine invertebrates, including the benthic communities, is substantial, although not yet entirely satisfactory with respect to long-term ecological effects (Dannheim *et al.* 2019; Gill & Wilhelmsson 2019).

Thus, it remains critical to carefully design wind-farm studies, and this forms the first theme of this chapter. This touches on the principles of what and when to sample, which is also covered to some extent in Box 1.1 in relation to avoiding impacts. The rest of the themes are focused on monitoring the different parts of the faunal assemblage, partly according to their location and including infauna, bottom coverage, epifouling of turbine bases and scour protection, benthic and demersal fish, and benthopelagic and pelagic fish. The basis of monitoring ecosystem functioning is then considered before the Concluding remarks, which are particularly focused on future studies.

Themes

Study design

To promote the sustainable development of offshore wind power as a primary source of renewable energy, it is essential to assess its effects and impacts. The study design is of great importance to be able to detect changes that can be related to wind farms. Taking a sufficient (depending on variation) number of samples and applying a before–after control-impact (BACI) design are two fundamental ways of strengthening a monitoring programme (Underwood 1994). Baseline studies typically have to be initiated 2–3 years before construction works begin to allow for sampling of temporal variation in biological parameters, including species' composition, populations, biomass and coverage.

The major problem for monitoring programmes is the frequent, if not constant, lack of sufficient statistical power. This can lead to serious consequences, since a potential negative impact may continue without being detected. Statistical power describes how well a conducted study covers the risk of not detecting an actual effect (change) due to having too few replicates. In biological and ecological science, it is standard to consider P=0.05 as the threshold for significant results. This means that it has to be at least 95% certain that an effect indicated in a study reflects an actual effect. Since biological data are variable in nature and many marine organisms are particularly variable in abundance, many replicates are typically needed to detect an effect. If there is an effect that is smaller than can be detected by the number of sampled replicates, due to natural variation, the conclusion of the study will still be that no effect was found. Wrongly, this conclusion is often interpreted as there was no effect. In order to draw the conclusion that effects are not likely, the statistical power needs to be high. Typically, the statistical power should be 0.8 for a significant change, meaning that the study has an 80% chance of detecting an effect size of interest (typically 25–50% change) (Antcliffe 1999). The consequence of the low statistical power in many monitoring programmes is that even if many topics have been studied and few effects have been established, it is not possible to conclude that the studies demonstrate a lack of effect. The possible effects may just be smaller than the applied monitoring efforts are designed to detect. The statistical power may, in some cases, not be presented in the monitoring reports, but it appears likely that only a few of the hitherto presented monitoring studies on the effects of wind power on benthic communities have a statistical power close to the conventional level of 0.8 (Lindeboom *et al.* 2011; Enhus *et al.* 2017). To move the field forward and make statistically rigorous conclusions on non-significant effects, monitoring programmes should focus on a few questions and perhaps key indicator species that are likely to be sensitive to change (Box 1.2), and apply a high sampling rate to resolve them with high statistical power. An example of high statistical power, but showing no significant effect, can be found in the quantitative experiment of Langhamer *et al.* (2016), capturing, marking and recapturing Common Shore Crab *Carcinus maenas* at the Lillgrund OWF. About 4,000 crabs were marked and monitored, but a very low recapture rate was observed. Furthermore, no wind-farm effects, either positive or negative, could be shown on the distribution of sex or colour morphs or body condition of the crabs, in comparison to nearby reference areas.

An optimal experimental design would be to monitor the same species over a longer period and to conduct comparable experiments in other OWFs to observe the effects of time and locality, especially when there is a large time and spatial variation.

When selecting among sampling methods, it is important to strive for minimising variation in data and to make sure that the method is feasible for quantitative data collection under harsh offshore conditions. A good example of targeted methodology is provided by the study on Viviparous Eelpout *Zoarces viviparus* at Lillgrund OWF (Box 1.2).

Box 1.2 Monitoring Viviparous Eelpout *Zoarces viviparus*, a key indicator of change, at Lillgrund offshore wind farm in Sweden

About 4 years after the installation of the Lillgrund offshore wind farm (OWF) in Sweden (Figure 1.1), a study on local populations of Viviparous Eelpout *Zoarces viviparus* was initiated, comparing them with a similar reference site (Langhamer *et al.* 2018). Eelpout is a benthic rather stationary species that has been used as a key indicator organism in the Baltic and North Seas for monitoring anthropogenic effects. Since the fish is an ovoviviparous species, reproductive success and fry development can be linked to each individual pregnant female, including their health status (Figure 1.2). During field sessions in autumn 2011 and 2012, Viviparous Eelpout was captured with two or three linked and baited double fyke nets at ten randomised locations in the Lillgrund OWF, and at a natural reference site 8 km south of Lillgrund, deployed from a small fishing vessel (Figure 1.3). Thus, fyke nets were set at 40–60 stations on one day and collected and emptied on the next day.

Figure 1.1 Lillgrund offshore wind farm in the Baltic Strait, Sweden, established in 2007 and consisting of 48 turbines. (Thomas Dahlgren)

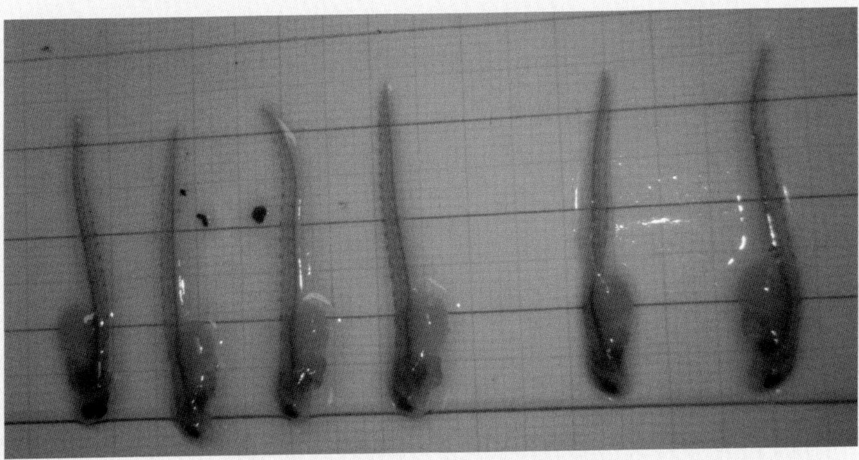

Figure 1.2 Size and condition status analysis of fry of the Viviparous Eelpout *Zoarces viviparus*. (Olivia Langhamer)

In all, 50 and 17 females were captured in 2011 and 2012, respectively. Biometrics, condition and reproductive success were assessed, comparing the wind-farm site with natural reference sites in the same area. Analysis in the laboratory showed that the operational wind farm had no impact on either the condition on the brood development of female Viviparous Eelpout. In supplementary capture–mark–recapture studies, 716 individuals were marked in Lillgrund and in two control sites for the assessment of population dynamics. The results indicated that Viviparous Eelpout neither specifically aggregated in, nor avoided the wind farm. Furthermore, no clear reef effect attracting the species to the foundations and scour protection was observed. Since there were no differences in the parameters of Viviparous Eelpout measured 5 years after the installation of the wind farm in comparison with the natural surroundings, these studies indicate good environmental status and no impact (either positive or negative) of the Lillgrund wind farm.

Figure 1.3 Small fishing vessel set up for Viviparous Eelpout *Zoarces viviparus* survey fishing using fyke nets. Note the tank on the port side for live transport of captured fish. (Thomas Dahlgren)

Monitoring infauna

It is believed that sediment structure and predation pressures may change in soft-sediment habitats surrounding turbine foundations, in turn affecting the infauna community structure (Miller *et al.* 2013). Coastal marine soft bottoms are ecologically important for the aquatic ecosystem owing to high biological activity and production, and are used by economically important fish and shellfish species as nurseries (Pihl & Wennhage 2002). The addition of hard structures can be stabilising and hence protect a variety of benthic organisms against wave action, ice scouring and predation, or against human activities, such as trawling. Monitoring the magnitude and range of these effects, including changes in benthic infaunal community structure and function, can be of high scientific interest. A variety of grabs, for example the Van Veen grab or the larger Smith–McIntyre grab, or Day grab samplers which are commonly used in the UK as standard samplers, are typically used for macrofauna sampling, whereas cores are used for meiofauna, sediment structure and content analysis (Figure 1.4, Figure 1.5). These allow quantitative and qualitative studies of the benthic infauna and are suitable for BACI monitoring. A sampling strategy of an area typically involves a randomised or stratified sampling programme with four or five replicated grabs at each sample station. However, both the sampling density and the number of replicates required vary with the variation in the parameter studied. The confidence interval is low (statistical power is high) when a larger sample is taken and when the variation is low. As a role of thumb, a sample of 30 when the average of a parameter is 100 will give a confidence interval of 90 and 99 when the standard error is 100 and 10, respectively. Corresponding numbers for a sample size of 10 are 70 and 97. When sampling for temporal changes, care must be taken to repeat the sampling at the same time of the year. The grab samples are carefully sieved directly on retrieval, using a 1 mm screen, and the samples fixed for taxonomic work back in the laboratory. Meiofauna, sediment structure and chemical content analysis may be best sampled using a multicorer that will take up to 12 replicate cores in one sampling effort. These cores can, in turn, be sectioned for specific investigations such as the distribution of meiofauna at different

Figure 1.4 Sampling instruments for sediment seabeds. (a) The megacorer is a development of the multicorer. It takes 12 cores from the seabed, leaving a virtually undisturbed surface, and is useful for infauna and meiofauna samples. (b) A Smith McIntyre Grab, generally used for macrofauna samples. (Thomas Dahlgren)

Figure 1.5 (a) A Van Veen Grab on a sieving table installed on a catamaran workboat. (b) Retrieval of a dredge for kelp to sample epifauna from the same vessel. (Uni Research, Bergen)

sediment depths. Meiofauna are typically extracted from fixed sediment core samples in the laboratory using a suspension decantation strategy (Pfannkuche & Thiel 1988).

The sediment characteristics often determine the type of grab or core sampler that can be used. For coarse sediments, the Haps Corer (0.014 m^2 with a depth of 50 cm) has been used for soft-bottom studies on the Swedish west coast (Langhamer 2010), since other kinds of grabs were not able to penetrate the seafloor. The sample surface of both the Van Veen and Day grabs is 0.1 m^2, giving a larger sampled area than the Haps Corer and a higher chance of sampling macrofaunal organisms. The Vibrocorer, sampling 0.01–0.007 m^2, is mostly used for geological purposes and is more appropriate for coarse sediments.

Monitoring bottom coverage

In some situations and/or localities, it may be interesting to monitor changes in coverage of different bottom types and biotopes inside a wind farm. For instance, where wind farms are located in shallow areas with seagrass and kelp forests, it may be useful to map and monitor the impacts of these valuable habitats (Shaskov *et al.* 2014; Schläppy *et al.* 2014).

Since detailed observations are less important than areal coverage when studying bottom types and habitat distribution, remotely operated vehicles (ROVs) or drop cameras can be appropriate. In a study at the Danish Middelgrunden wind farm, the distributions of Eelgrass *Zostera marina* and Blue Mussel *Mytilus edulis* were monitored through scuba diving, showing a strong increase in Eelgrass coverage and a corresponding decline in coverage of mussel beds (Lynge 2004).

At high-energy sites with mixed surface types and a faceted seascape comprising various habitat types, a mix of different gear and sampling approaches may be used. A baseline study was conducted at a planned OWF off the Norwegian coast (Dahlgren *et al.* 2014). The site is unique in that the seabed had a complex bathymetry dominated by rocky outcrops with patches of boulders, coarse sand and gravel. Dense forests of a kelp, Tangle *Laminaria hyperborea*, dominated at depths ranging from 5 to 20 m. Deeper sites had bare rocks, stones, pebble or coarse gravel covered by encrusting red algae such as *Lithothamnion* spp. This type of habitat is notoriously hard to sample, and quantitative sampling gears such as grabs are useless. It was further exposed to full ocean wave action, and strong currents limited the possibility of conducting transect studies using scuba divers. Instead, a combination of grabs (at deeper, >80 m, sedimented sites within the area), rock dredges and video collected with ROVs was used (Figure 1.6). To facilitate data extraction from the video transects, randomised 'samples' of 10 m seafloor images were constructed using mosaics from the raw video data (Figure 1.7) (Šaškov *et al.* 2015). Subsequently, these mosaic images were assessed for area cover using various features defined by different pixel characteristics, such as colour. Benthic fish were also monitored using a full range of passive gear such as crab pots, fyke nets, bottom-set longlines and gill nets (see *Monitoring benthic and demersal fish*, below). The feature and species distribution data from the benthic habitat mapping were further used to model the distribution of certain species and features such as kelp, sea urchins (e.g. *Echinus* spp.) and sea stars (*Asterias asterias* and *Marthasterias glacialis*) (Schläppy *et al.* 2014). Based on the foreseeable change in a set of descriptors, these models can be used to hypothesise on the future impact of the wind farm (Schläppy *et al.* 2014; Šaškov *et al.* 2015).

Figure 1.6 Example of a small remotely operated vehicle (ROV) that is easily operated from the back deck of a vessel. This ROV is powered by an external source of electricity, but other types carry batteries and hence require less infrastructure on the platform. (Thomas Dahlgren)

Figure 1.7 Mosaic seafloor image based on remotely operated vehicle video imagery used to analyse feature coverage such as the red algae *Lithothamnion* spp. (Uni Research, Bergen)

Monitoring epifouling on the turbine foundations and scour protection

The epifouling community on turbine foundations and scour protection have been studied extensively (Andersson *et al.* 2009; Andersson & Öhman 2010; Gutow *et al.* 2014; de Mesel *et al.* 2015). Non-destructive underwater photography with a fixed frame and video transects are 'classic' methods used by marine biologists and are particularly valuable where the sampling efficiency of other physical sampling methods is low or where the succession of fouling organisms over time is to be monitored. Scientific scuba diving is one of the more traditional multi-purpose survey methods, although at some sites where it is impossible to conduct diving surveys because of strong currents for example, ROVs may be used to obtain photographs or videos (Figure 1.7). Both methods deliver high-quality images that enable the identification of much of the epifauna present. These images cover a small area of the structures (or seafloor) and do not provide information on the overall distribution of faunal communities. It is also important to note that smaller ROVs are restricted by their limited capability to operate in currents in excess of 1.5 knots, as well as by the water depth and the length of umbilical connection.

For better qualitative and quantitative monitoring of biofouling succession and colonisation, settlement panels and scraping may be applied. Here, recruitment and growth, and even settlement of non-invasive species, can be detected and analysed. However, in overall terms, it is important to choose a suitable combination of monitoring methodologies for epifouling in order to obtain a meaningful synthesis of the different epifaunal communities. Studies on epifouling communities have been conducted with the focus on habitat gain, showing a succession of colonisation of local species, higher biomass and diversity on wind-turbine structures and adjacent boulders (Wilhelmsson & Malm 2008). In the brackish Baltic Sea, Blue Mussel usually dominates offshore wind structures (Dannheim *et al.* 2019) and its biomass has been shown to exceed that of the surrounding mussel beds by an order of magnitude (Wilhelmsson & Malm 2008; Andersson & Öhman 2010; Krone *et al.* 2013) and may have regional effects on the ecosystem (van der Molen *et al.* 2014). Typically, habitat gain has been associated with positive effects and with habitat improvement (Gutow *et al.* 2014). Still, negative effects may appear when submerged structures, such as foundations and scour protection, attract invasive and non-indigenous species (e.g. Coolen *et al.* 2016). In monitoring programmes in the North Sea and in the Baltic, some new species for these waters have been found, suggesting that OWFs could be stepping stones to facilitate further spread (Leonhard & Pedersen 2006; Brodin & Andersson 2009; Degraer & Brabant 2009).

Rapid developments in computing and molecular biology have also opened up the use of DNA in biodiversity assessments or identification of specific species, such as invasive species (e.g. Deiner *et al.* 2017). While more research and development work is required for consistent use in the analysis of community structure using a metabarcoding approach, where a small sample of sediment or water is used to estimate the presence of all species at a site, the detection of specific species is more straightforward. In this application, a species-specific molecular essay is used to detect the presence and concentration of a target DNA molecule, such as from an invasive species, in a water sample using the droplet digital polymerase chain reaction (Nathan *et al.* 2014).

Monitoring benthic and demersal fish

One example of extensive monitoring of the benthic fish community in a wind farm was conducted using fyke nets for both monitoring and mark–recapture studies at the Lillgrund OWF in Sweden (Bergström *et al.* 2013; Langhamer *et al.* 2018) (Box 1.2). Many replicates (36 stations in each of the impact areas and two reference areas) and long time series (4 years before and 3 years after) provided a good understanding of which species were attracted by the foundations, such as Atlantic Cod, and which were not, such as Viviparous Eelpout (Box 1.2). Moreover, the use of fykes allowed a detailed investigation of the biometrics, reproductive output (broods) and population structure of the latter species (Langhamer *et al.* 2018) (Box 1.2). In the study at Lillgrund, bait was used to attract individuals and obtain high catch rates. An advantage of using fyke nets is that fish are mostly caught uninjured and in generally good condition, unlike using gill nets or actively towed gear. Furthermore, fish can be safely released after capture, which makes this method suitable for tagging experiments on living fish. Fyke nets are, however, rather selective and do not cover all species, but mainly target smaller demersal and benthic fish species/individuals, as well as macro-crustaceans.

Gill nets and survey nets (gill nets with a range of mesh sizes) have also been used in wind-farm baseline and monitoring programmes to sample benthic fish in some circumstances (e.g. Dahlgren *et al.* 2014). Gill nets are very target specific but have the advantage of being long (hundreds of metres) and are thus able to survey a larger area, thereby reducing variation. Survey nets are short (tens of metres) but have various mesh sizes in panels per net, thus covering many species, but typically at the cost of high variation in the data delivered, leading to low statistical power, unless a large number of nets are used. In areas where large crustaceans, such as Green Crab *Carcinus maenas*, Edible Crab *Cancer pagurus* and hermit crabs *Pagurus* spp., are common, the use of nets becomes expensive as entangled crustaceans are hard to remove without destroying the mesh.

Cage or pot fishing is another traditional monitoring method with a long history that can be used for population studies, including telemetry of macro-crustaceans and fish as well as impact assessments. Nedwell *et al.* (2003b) used cage fishing for experimental noise studies in which they investigated behavioural changes in Brown Trout *Salmo trutta* exposed to pile driving and construction noise of a harbour terminal. A similar methodology can be applied to behavioural studies around offshore wind power constructions. The use of fishing cages to monitor macro-crustacean and fish species and communities is an effective, non-destructive method and can be used on most kinds of seabeds. One disadvantage is the limited sizes of the cages, since they are designed to be used for commercial fishing, thus limiting the diversity of sizes and species captured.

Small beam trawls (i.e. 2 m width) are often used to semi-quantitatively survey epifauna species' abundance and diversity (Frauenheim *et al.* 1989). Catching efficiency is dependent on sediment characteristics but also largely on the behaviour of the species, such as burying capacity and swimming speed. In a study where three trawls were towed after each other, the catching efficiency varied from 51% in the Hermit Crab *Pagurus bernhardus* to 9% in the both burying and swimming crab *Liocarcinus holsatus* (Reiss *et al.* 2006).

An alternative method for sampling benthic fish alongside epifauna is to use video observations (Glover *et al.* 2010). Offshore data collection using ROVs (see *Monitoring epifouling on the turbine foundations and scour protection*, above) (Figure 1.6) may be expensive, and stationary video, which has been used extensively in reef monitoring in Australia (Harvey *et al.* 2013), may be used instead. With video observations, it is possible to sample not only relative abundance but also behavioural traits, since observations sustained over time can be made. As turbines emit low-frequency noise during operation, this could initiate changes in behaviour among different animal groups that are particularly interesting to monitor (Hammar 2014). With the use of stereo-video methodology, where two synchronised cameras simultaneously record the same object, video observations can generate accurate quantitative measurements of size, swimming speed and distance to fixed objects such as the foundations (Harvey *et al.* 2001; Watson *et al.* 2005; Hammar *et al.* 2015). Stereo-video methodology has hitherto not been used in wind-farm monitoring, but experiences from tidal power impact studies indicate that the method has great potential (Hammar *et al.* 2013). To obtain non-biased data, the camera systems should be used remotely (i.e. not carried by divers). The drawback with video-based surveys is the extensive time required for analysis.

Monitoring benthopelagic and pelagic fish

Gill-net fishing, a method commonly applied in commercial fisheries, may be used to monitor fish in the water column (see *Monitoring benthic and demersal fish*, above). Multimesh nets encompassing all sizes of fish are typically used to detail fish abundance, fish communities and diversity before installation and during operation of the wind farm. Identical methods and sampling efforts should be applied over several years both before and after construction. At Horns Rev in Denmark, additional surveys were conducted by scuba divers around the turbines to measure fish abundance in relation to distance from the turbine (Leonhard *et al.* 2011). The long-term surveys using gill nets recorded a higher diversity of fish in the wind farm compared to a control site after construction, with Whiting *Merlangius merlangus*, Dab *Limanda limanda* and sandeels *Ammodytes* spp. being the most abundant species. Fish community patterns did, however, show a high spatial and temporal variability in distribution and occurrence. Despite the use of multiple mesh sizes, gill-net fishing is known for its selectivity towards larger individuals, thus leading to an underestimation of small individuals.

Pelagic fish abundance has also been monitored using hydroacoustic methods in combination with pelagic trawls for species determination and ground proofing (Krägefsky 2014). At the German offshore wind energy test site Alpha Ventus, such methods indicated lower abundances of pelagic fish during the construction of the wind farm, but no significant impact on the pelagic fish community during the operational phase (Krägefsky 2014). Monitoring in the operational phase also indicates and helps researchers to evaluate the effects of no-take zones, as commercial trawl fisheries may be prevented or at restricted in their operations after the construction of the site, particularly from the perspective of safety over the risk of snagging cables or other structures.

Although hydroacoustics have been in use for some time and are relatively sophisticated, the limitations in species determination still make it difficult to observe pelagic fish and burrowing species, and to estimate fish length and shape in schools. Other difficulties include reflections from other marine organisms such as jellyfish and squid *Loligo* spp., which can be misinterpreted, as well as backscatter from the seafloor and wind-farm structures potentially masking aggregating fish. Nonetheless, there have been developments in the technology, particularly for use in highly energetic environments where weather windows are rare (Hvidt *et al.* 2006).

Fixed sonar has also been used for OWF monitoring in relation to day/night distributions of fish, fish abundance in control versus impact sites, and seasonal variations in fish abundance and density.

Finally, individual telemetry has been used to investigate the behaviour of larger benthopelagic species such as Atlantic Cod (Reubens *et al.* 2013; 2014a; 2014b; 2019), showing seasonal patterns of presence but also detailed behavioural traits such as habitat selectivity at a small spatial scale. Such studies have been extended to investigate the response of fish to construction noise (Box 1.3).

Box 1.3 Tracking wild Atlantic Cod *Gadus morhua* within offshore wind farms and the use of acoustic telemetry

Inge van der Knaap & Jan T. Reubens

Previous research on the effects of offshore wind farms (OWFs) has shown that once the turbines are constructed, the scour bed, that is the hard substrate surrounding the base of each turbine, has a strong positive influence on the presence of Atlantic Cod *Gadus morhua* in the area (Reubens *et al.* 2013). Atlantic Cod inhabit the offshore artificial structures during summer and autumn and use the scour beds as feeding grounds (Reubens *et al.* 2014a). The presence of offshore turbines therefore has an effect on the number and distribution of Atlantic Cod in the North Sea. However, how the construction phase of offshore turbines affects marine fish has been little studied. Here, we use acoustic telemetry to investigate the effects of wind-farm construction in the North Sea on Atlantic Cod.

By 2020, European Union member states are expected to obtain 43 GW of electric energy from offshore wind (Seanergy 2020 2012), which is 2.7 times the total installed capacity in 2018 of 15.78 GW (WindEurope 2018), implying that construction activities in the North Sea will be intensive and ongoing for some years to come. Most turbines are constructed using pile driving, hammering the foundation piles into the seabed. This method elicits a low-frequency sound, with peaks between 20 and 100 Hz, at a high decibel level (~250 dB re 1 μPa at source). Fish that possess organs to detect sound pressure are able to hear sounds in the range of 10–500 Hz (Slabbekoorn *et al.* 2010) and are therefore potentially affected by construction noise. In extreme cases, high exposure levels of sound can lead to the death of the fish through rupture of the swimbladder and/or other organs or internal bleeding. However, the most widespread effects of noise will most likely be behavioural and physiological changes.

Acoustic telemetry is a well-known method to study fish movement, behaviour and physiology over a longer period (Hussey *et al.* 2015). This technique uses tags, carried by the free-swimming fish (Figure 1.8), that transmit an acoustic signal that can be detected by strategically placed receiver stations. Each tag emits a unique ID code, resulting in presence/absence information on an individual level. Advances in electronic tagging are ongoing and apart from presence/absence tags, electronic tags including temperature, pressure and acceleration sensors are now becoming widely available, offering a whole new range of monitoring possibilities. Using accelerometer sensors, it has become possible to study the activity of an individual. Furthermore, using indoor experiments, links can be made between acceleration and oxygen consumption, which is a proxy for metabolic activity (Metcalfe *et al.* 2016), allowing us to estimate the metabolic costs of the activity levels that fish are exhibiting in the wild.

To investigate the effects of the construction phase on Atlantic Cod, we tagged 14 and 27 individuals (Figure 1.8) in 2016 and 2017, respectively, at Belwind, an operational (construction was completed in 2010) wind farm in the Belgian part of the North Sea, while construction activities at the surrounding wind farms of Nobelwind and Rentel were still ongoing (Figure 1.9). A network of acoustic receivers was set up around the base of a turbine within Belwind (2017 set-up shown in Figure 1.10), with Atlantic Cod tagged with acoustic tags at that same turbine. This made it possible to track the individual fish that inhabited the turbine before, during and after construction activities. The first 14 fish were tagged with acoustic Vemco tags without sensors that provide information on presence and position, whereas the second batch of 27 individuals were tagged with Vemco tags containing both pressure and accelerometer sensors. These sensors provide information on depth usage as well as activity levels of the individual fish, resulting in an even more detailed picture of the horizontal and vertical movement and behaviour. Furthermore, we performed indoor experiments during which a correlation was made between acceleration and oxygen consumption, which provides information on the energetic costs of observed activity levels.

Even though telemetry samples only a small portion of the entire population, information density per individual is very high. Analysis of the results is still

Figure 1.8 Example of the size of Atlantic Cod *Gadus morhua* captured for tagging studies.

ongoing, although a detailed fine-scale picture of the movement behaviour and physiology of Atlantic Cod in response to OWF construction activities, as well as other environmental factors, is anticipated.

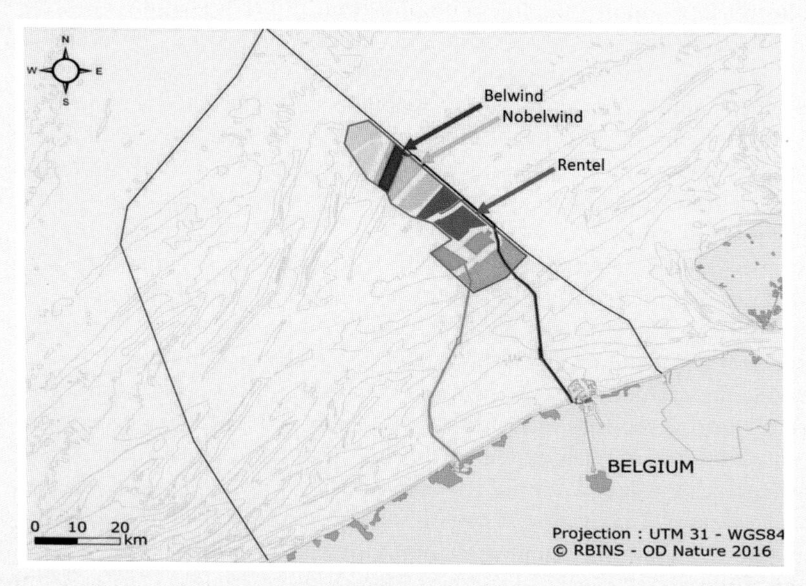

Figure 1.9 Map of the area designated for wind-farm construction within the Belgian Part of the North Sea (BPNS). Different colours indicate different wind-park areas, including Belwind (red), Nobelwind (light blue) and Rentel (dark blue).

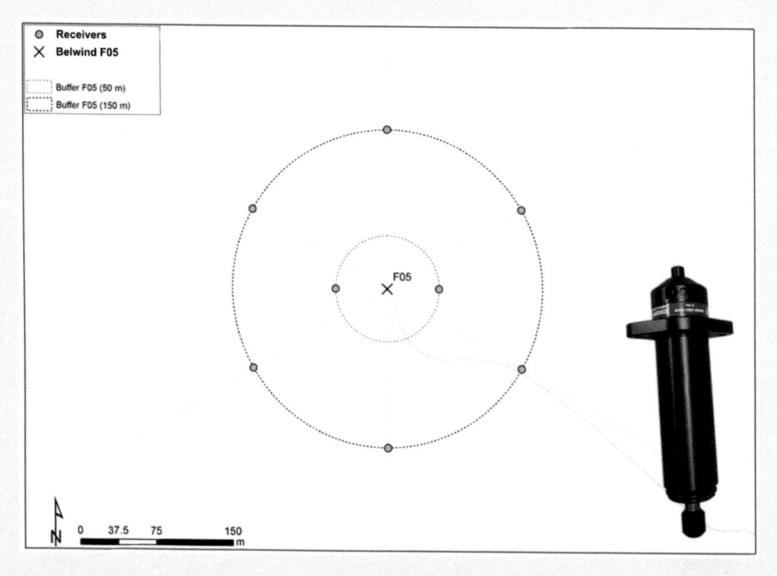

Figure 1.10 Example of the network of bottom-moored receivers that was placed around a turbine in Belwind wind farm in 2017. The eight green dots indicate the position of the receivers around the turbine F05 (X), covering an area with a radius of 150 m. A Vemco receiver is illustrated in the bottom right of the picture (not to scale).

Ecosystem functioning

The development of the offshore wind industry is an example of how anthropogenic activities are changing the marine environment and ecology. When habitats are modified, this affects species distribution and connectivity, including population dynamics and source-sink relationships. Methods for monitoring ecosystem functioning, such as infauna respiration levels, include the deployment of respiration chambers where gas levels are measured *in situ* (Sørensen *et al.* 1979). Other methods include change in faunal functional groups identified from community structure analyses or accumulation in the sediment of nutrients such as organic carbon or phosphorus.

The use of acoustic sonars as a remote sensing technique for the detection of ecosystem functioning, in terms of small-scale seabed features, has been shown to be very effective. Beds of seagrass and Horse Mussel *Modiolus modiolus*, and the reef-building Ross Worm *Sabellaria spinulosa* have been monitored and evaluated by sonar technology in planned and operating wind farms (Pearce *et al.* 2014). Quantitative and qualitative habitat mapping, which included species distributions and densities over large areas, has been taken a step farther following the incorporation of biological data gathered for 'ground-truthing' to provide a higher level of confidence. In the Thanet OWF in the UK, acoustic data sampling identified biogenic reef habitats formed by *Sabellaria spinulosa* which, when combined with images from associated macrofauna, allowed for early-stage micro-siting of the turbines. Ecosystem functioning of the site was then integrated in the project to cause minimal damage and habitat degradation during construction (Pearce *et al.* 2014). In this example, the industry worked closely with marine scientists and conservation agencies to ensure a sustainable construction process. The results of a before–after survey showed an increase in biogenic reefs and their associated macrofauna in the wind-farm area, indicating a positive influence of the wind farm on ecosystem functioning. This is an important step not only for enhancing ecosystem monitoring methodologies, but also for future management of sensitive marine habitats in an emerging offshore wind-power industry.

In another study, OWFs and ecosystem functioning have been reviewed and modelled with a focus on the conservation of valuable species and groups of fish and invertebrates (Raoux *et al.* 2017). An ecosystem model described the before-and-after scenarios of ecosystem functioning and food-web structures. The models showed a general increase in total ecosystem activity, a positive response of higher trophic-level species and a change in keystone groups in response to the reef effect. Predictive models are already used as a complement to monitoring and field surveys, and the future use of ecosystem modelling seems highly likely to increase with the expansion of the offshore wind industry.

Concluding remarks

Although there will always be a temporary 'benthic footprint', there are numerous ways of avoiding, mitigating and compensating for negative impacts of offshore wind power on the seabed community. With respect to benthic organisms in particular, considerate micro-siting of each foundation is probably the most effective means of reducing ecologically significant impact. In addition, the choice of foundation type plays a prominent role in determining the kind of benthic community that will develop around the new structures.

Technological advances, in terms of habitat mapping, video monitoring and image recognition, will allow for more effective data collection. An increased monitoring effort in coastal areas will also render a better understanding of spatial and temporal variation that will allow researchers to plan and conduct monitoring programmes with a higher possibility of detecting any changes caused by OWFs. While there is a compelling lack of evidence for any direct negative effect from wind turbines, statistical power must improve within monitoring programmes, coupled with further investigation of the long-term effects from low-intensity noise, possibly through stress and impaired learning (Ebbesson & Braithwaite 2012). Further studies are also needed at a regional scale on the (cumulative) ecosystem effects from future large-scale offshore wind developments, possibly using seascape risk analysis methods (Copping *et al.* 2014; Hammar 2014; Damian & Merck 2014). The relaxed fishing effort inside most wind farms, in particular the reduction or cessation of trawling, will probably have a large impact on the ecosystem, both at a local benthic scale and through reduced turbidity at larger scales where intense trawling is replaced by wind farms. However, since trawled seabeds are slow to recover (e.g. Olsgard *et al.* 2008) and little is understood about how a general increase in turbidity levels has affected the marine ecosystem, these effects will probably take some time to detect. But what is gained from relaxed benthic trawling can be lost through increased scour caused by turbine foundations. The increased turbidity caused by wakes from foundations in some seafloor sediment types and water depths has the potential to cause ecosystem-scale long-term and cumulative impacts. In this regard, satellite-derived image data have been used to monitor the increased suspended sediments at two OWFs in the Thames estuary (Vanhellemont & Ruddick 2014).

Efforts have also been made to model specific cumulative effects based on forecasts of future offshore wind developments. One such study focused on the regional ecosystem-scale effect of the increased populations of filter-feeding mussels growing on wind-farm turbine foundations (Slavik *et al.* 2018). This study concluded that the increase in Blue Mussel populations in a fully developed southern North Sea would have non-negligible effects upon regional annual primary production of up to a few per cent, and cause larger changes (±10%) in the phytoplankton stock and thus upon water clarity during blooms. In other words, at least at short timescales, there could be a significant effect on pelagic primary production and nutrient availability. Models were also used in another project with the aim of understanding the regional-scale dispersal of an invasive species (Coolen *et al.* 2016). The authors concluded that a native species would not overlap with the invasive species in habitat preference, and that the two species would also be likely be able to coexist after a large expansion of wind farms in the North Sea. Another potential effect of upscaling at a regional scale is the potential increase in primary production caused by upwelling. Based on models, Broström (2008) suggested that the reduction in wind speed downwind from a wind farm caused local upwelling (and downwelling). Fine-resolution measurements confirm these models, and upwelling can potentially increase primary production in the North Sea by several per cent (Broström *et al.* 2019).

Acknowledgements

We wish to thank the editor of this volume, Martin Perrow, for encouragement and insightful comments and suggestions that greatly improved this chapter.

References

Andersson, M.H. & Öhman, M.C. (2010) Fish and sessile assemblages associated with wind-turbine constructions in the Baltic Sea. *Marine and Freshwater Research* 61: 642–650.

Andersson, M., Berggren, M., Wilhelmsson, D. & Öhman, M. (2009) Epibenthic colonization of concrete and steel pilings in a cold-temperate embayment: a field experiment. *Helgoland Marine Research* 63: 249–260.

Antcliffe, B.L. (1999) Environmental impact assessment and monitoring: the role of statistical power analysis. *Impact Assessment and Project Appraisal* 17: 33–43.

Bailey, H., Senior, B., Simmons, D., Rusin, J., Picken, G. & Thompson, P.M. (2010) Assessing underwater noise levels during pile-driving at an offshore windfarm and its potential effects on marine mammals. *Marine Pollution Bulletin* 60: 888–897.

Bergström, L., Kautsky, L., Malm, T., Ohlsson, H., Wahlberg, M., Rosenberg, R. & Capetillo, N. (2012) The effects of wind power on marine life. A synthesis. Stockholm: Swedish Environmental Protection Agency. Report No. 6512. Retrieved 1 December 2018 from www.naturvardsverket.se/publikationer

Bergström L, Lagenfelt I, Sundqvist F, Andersson M, Sigray P (2013a) Fiskundersökningar vid Lillgrund vindkraftpark. Havs- och vattenmyndighetens rapport 2013, p. 18. Retrieved 22 March 2019 from http://space.hgo.se/wpcvi/wp-content/uploads/import/pdf/Kunskapsdatabas%20miljo/Flora%20och%20fauna/marina%20organismer/rapport-2013-18-lillgrund-140127.pdf

Bergström, L., Sundqvist, F. & Bergström, U. (2013b) Effects of an offshore wind farm on temporal and spatial patterns in the demersal fish community. *Marine Ecology Progress Series* 485: 199–210.

Bergström, L., Kautsky, L., Malm, T., Rosenberg, R., Wahlberg, M., Åstrand Capetillo, N. & Wilhelmsson, D. (2014) Effects of offshore wind farms on marine wildlife – a generalized impact assessment. *Environmental Research Letters* 9: 034012.

Brodin, Y. & Andersson, M.H. (2009) The marine splash midge *Telmatogon japonicus* (Diptera; Chironomidae) – extreme and alien? *Biological Invasions* 11: 1311–1317.

Broström, G. (2008) On the influence of large wind farms on the upper ocean circulation. *Journal of Marine Systems* 74: 585–591.

Broström, G., Ludewig, E., Schneehorst, A. & Pohlmann, T. (2019) Atmosphere and ocean dynamics. In Perrow, M.R. (ed.) *Wildlife and Wind Farms, Conflicts and Solutions. Volume 3. Offshore: Potential Effects*. Exeter: Pelagic Publishing. pp. 47–63.

BSH & BMU (2014) *Ecological Research at the Offshore Windfarm Alpha Ventus: Challenges, results and perspectives*. Federal Maritime and Hydrographic Society of Germany (BSH) and Federal Ministry for the Environment, Nature Conservation and Nuclear Safety (BMU). Weisbaden: Springer Spektrum.

Caiger, P.E., Montgomery, J.C. & Radford, C.A. (2012) Chronic low-intensity noise exposure affects the hearing thresholds of juvenile snapper. *Marine Ecology Progress Series* 466: 225–232.

Carroll, A.G., Przeslawski, R., Duncan, A., Gunning, M. & Bruce, B. (2017) A critical review of the potential impacts of marine seismic surveys on fish & invertebrates. *Marine Pollution Bulletin* 114: 9–24.

Coolen, J., Lengkeek, W., Degraer, S., Kerckhof, F., Kirkwood, R. & Lindeboom, H. (2016) Distribution of the invasive *Caprella mutica* Schurin, 1935 and native *Caprella linearis* (Linnaeus, 1767) on artificial hard substrates in the North Sea: separation by habitat. *Aquatic Invasions* 11: 437–449.

Copping, A., Hanna, L., van Cleve, B., Blake, K. & Anderson, R.M. (2014) Environmental risk evaluation system – an approach to ranking risk of ocean energy development on coastal and estuarine environments. *Estuaries and Coasts* 38: 287–302.

Cuperus, R., Canters, K.J. & Piepers, A. (1996) Ecological compensation of the impacts of a road. Preliminary method for the A50 road link (Eindhoven-Oss, The Netherlands). *Ecological Engineering* 7: 327–349.

Cuperus, R., Canters, K.J., de Haes, H. & Friedman, D.S. (1999) Guidelines for ecological compensation associated with highways. *Biological Conservation* 90: 41–51.

Dahlgren, T.G., Schläppy, M.-L., Shashkov, A., Andersson, M., Rzhanov, Y. & Fer, I. (2014) Assessing the impact of windfarms in subtidal, exposed marine areas. In Shields, M.A. & Payne, A.I.L. (eds) *Marine Renewable Energy Technology and Environmental Interactions. Humanity and the sea.* Dordrecht: Springer Science and Business Media. pp. 39–48.

Damian, H.-P. & Merck, T. (2014) Cumulative impacts of offshore windfarms. In BSH & BMU (eds) *Ecological Research at the Offshore Windfarm Alpha Ventus: Challenges, results and perspectives.* Weisbaden: Springer Spektrum. pp. 193–198.

Dannheim, J., Degraer, S., Elliott, M., Smyth, K. & Wilson, J.C. (2019) Seabed communities. In Perrow, M.R. (ed.) *Wildlife and Wind Farms, Conflicts and Solutions. Volume 3. Offshore: Potential effects.* Exeter: Pelagic Publishing. pp. 64–85.

de Mesel, I., Kerckhof, F., Norro, A., Rumes, B. & Degraer, S. (2015) Succession and seasonal dynamics of the epifauna community on offshore wind farm foundations and their role as stepping stones for non-indigenous species. *Hydrobiologia* 756: 37–50.

Degræer, S. (2012) Offshore wind farms in the Belgian part of the North Sea. Royal Belgian Institute of Natural Sciences, Management Unit of the North Sea Mathematical Models, Marine Ecosystem Management Unit. Retrieved 1 December 2018 from www.mumm.ac.be

Degraer, S. & Brabant, R. (eds) (2009) Offshore wind farms in the Belgian part of the North Sea: state of the art after two years of environmental monitoring. Report by Ghent University, Institute for Agricultural and Fisheries Research (ILVO), Management Unit of the North Sea Mathematical Models (MUMM), Research Institute for Nature and Forest (INBO), and Royal Belgian Institute of Natural Sciences (RBINS). Brussels: MUMM. Retrieved from https://odnature.naturalsciences.be/downloads/mumm/windfarms/monitoring_windmills_2009_final.pdf

Deiner, K., Bik, H.M., Mächler, E., Seymour, M., Lacoursière-Roussel, A. & Altermatt, F. (2017) Environmental DNA metabarcoding: transforming how we survey animal and plant communities. *Molecular Ecology* 26: 5872–5895.

Ebbesson, L.O.E. & Braithwaite, V.A. (2012) Environmental effects on fish neural plasticity and cognition. *Journal of Fish Biology* 81: 2151–2174.

Elliott, M., Burdon, D., Hemingway, K.L. & Apitz, S.E. (2007) Estuarine, coastal and marine ecosystem restoration: confusing management and science – a revision of concepts. *Estuarine Coastal and Shelf Science* 74: 349–366.

Engås, A., Misund, O.A., Soldal, A.V., Horvei, B. & Solstad, A. (1995) Reactions of penned herring and cod to playback of original, frequency-filtered and time-smoothed vessel sound. *Fisheries Research* 22: 243–254.

Enhus, C., Bergström, H., Müller, R., Ogonowski, M. & Isæus, M. (2017) Kontrollprogram för vindkraft i vatten. – Sammanställning och granskning, samt förslag till rekommendationer för utformning av kontrollprogram. Report No. 6741. Stockholm: Swedish Environmental Protection Agency. Retrieved 1 December 2018 from https://www.naturvardsverket.se/Documents/publikationer6400/978-91-620-6741-0.pdf?pid=19705

European Union (EU) (2008) The Marine Strategy Framework Directive of the European Union. Brussels: European Commission. Retrieved on 22 March 2019 from http://ec.europa.eu/environment/marine/eu-coast-and-marine-policy/marine-strategy-framework-directive/index_en.htm

Frauenheim, K., Neumann, V., Thiel, H. & Türkay, M. (1989) The distribution of the larger epifauna during summer and winter in the North Sea and its suitability for environmental monitoring. *Senckenbergiana Maritima* 20: 101e118.

Gill, A.B. (2005) Offshore renewable energy: ecological implications of generating electricity in the coastal zone. *Journal of Applied Ecology* 42: 605–615.

Gill, A.B. & Wilhelmsson, D. (2019) Fish. In Perrow, M.R. (ed.) *Wildlife and Wind Farms, Conflicts and Solutions. Volume 3. Offshore: Potential effects.* Exeter: Pelagic Publishing. pp. 86–111.

Glover, A.G., Higgs, N.D., Bagley, P.M., Carlsson, R., Davies, A.J., Kemp, K.M., Last, K.S., Norling, K., Rosenberg, R., Wallin, K.-A., Källström, B. & Dahlgren, T.G. (2010) A live video observatory reveals temporal processes at a shelf-depth whale-fall. *Cahiers de Biologie Marine* 51: 375–381.

Gutow, L., Teschke, K., Schmidt, A., Dannheim, J., Krone, R. & Gusky, M. (2014) Rapid increase of benthic structural and functional diversity at the Alpha Ventus offshore test site. In BSH & BMU (eds) *Ecological Research at the Offshore Windfarm Alpha Ventus: Challenges, results and perspectives.* Weisbaden: Springer Spektrum. pp. 67–81.

Hammar, L. (2014) Power from the brave new ocean: marine renewable energy and ecological risks. PhD thesis, Chalmers University of Technology, Gothenburg. Retrieved 1 December 2018 from http://publications.lib.chalmers.se/publication/196091-power-from-the-brave-new-ocean-marine-renewable-energy-and-ecological-risks

Hammar, L., Andersson, S. & Rosenberg, R. (2008) Adapting offshore wind power foundations to local environment. Vindval Report No. 6367. Stockholm: Swedish Environmental Protection Agency. Retrieved 1 December 2018 from http://naturvardsverket.se/Documents/publikationer/978-91-620-6367-2.pdf

Hammar, L., Andersson, S., Eggertsen, L., Haglund, J., Gullström, M., Ehnberg, J. & Molander, S. (2013) Hydrokinetic turbine effects on fish swimming behaviour. *PLoS ONE* 8(12): e84141. doi: 10.1371/journal.pone.0084141.

Hammar, L., Wikström, A. & Molander, S. (2014) Assessing ecological risks of offshore wind power on Kattegat cod. *Renewable Energy* 66: 414–424.

Hammar, L., Eggertsen, L., Andersson, S., Ehnberg, J., Rickard, A., Gullström, M. & Molander, S. (2015) A probabilistic model for hydrokinetic turbine collision risks: exploring impacts on fish. *PLoS ONE* 10(3): e0117756. doi: 10.1371/journal.pone.0117756.

Harvey, E., Fletcher, D. & Shortis, M. (2001) A comparison of the precision and accuracy of estimates of reef-fish lengths determined visually by divers with estimates produced by a stereo-video system. *Fishery Bulletin* 99: 63–71.

Harvey, E.S., McLean, D.L., Frusher, S., Haywood, M.D.D., Newman, S.J. & Williams, A. (2013) The use of BRUVs as a tool for assessing marine fisheries and ecosystems: a review of the hurdles and potential. Fisheries Research and Development Corporation and The University of Western Australia. Retrieved 1 December 2018 from https://www.researchgate.net/profile/Euan_Harvey/publication/260122217_The_use_of_BRUVs_as_a_tool_for_assessing_marine_fisheries_and_ecosystems_a_review_of_the_hurdles_and_potential/links/0f31752fa13c201fbe000000/The-use-of-BRUVs-as-a-tool-for-assessing-marine-fisheries-and-ecosystems-a-review-of-the-hurdles-and-potential.pdf

Huddleston, J. (ed.) (2010) *Understanding the Environmental Impacts of Offshore Windfarms*. Oxford: COWRIE.

Hussey, N.E., Kessel, S.T., Aarestrup, K., Cooke, S.J., Cowley, P.D., Fisk, A.T., Harcourt, R.G., Holland, K.N., Iverson, S.J., Kocik, J.F. & Flemming, J.E.M. (2015) Aquatic animal telemetry: a panoramic window into the underwater world. *Science* 348: 1255642.

Hvidt, C.B., Leonhard, S.B., Klaustrup, M. & Pedersen, J. (2006) Hydroacoustic monitoring of fish communities at offshore wind farms. Horns Rev Offshore Wind Farm, Annual Report 2005. Document No. 2624-03-003 Rev2.doc. Fredericia: Vattenfall. Retrieved 1 December 2018 from https://corporate.vattenfall.dk/globalassets/danmark/om_os/horns_rev/status-report-2005.pdf

Inger, R., Attrill, M.J., Bearhop, S., Broderick, A.C., James Grecian, W., Hodgson, D.J., Mills, C., Sheehan, E., Votier, S.C., Witt, M.J. & Godley, B.J. (2009) Marine renewable energy: potential benefits to biodiversity? An urgent call for research. *Journal of Applied Ecology* 46: 1145–1153.

Jameson, H., Reeve, E., Laubek, B. & Sittel, H. (2019) The nature of offshore wind farms. In Perrow, M.R. (ed.) *Wildlife and Wind Farms, Conflicts and Solutions. Volume 3. Offshore: Potential effects*. Exeter: Pelagic Publishing. pp. 1–29.

Kastelein, R.A., van der Heul, S., Verboom, W.C., Jennings, N., van der Veen, J. & Haan, D. (2008) Startle response of captive North Sea fish species to underwater tones between 0.1 and 64 kHz. *Marine Environmental Research* 65: 369–377.

Kermagoret, C., Levrel, H. & Carlier, A. (2014) The impact and compensation of offshore wind farm development: analysing the institutional discourse from a French case study. *Scottish Geographical Journal* 130: 188–206.

Krägefsky, S. (2014) Effects of the Alpha Ventus offshore test site on pelagic fish. In BSH & BMU (eds) *Ecological Research at the Offshore Windfarm Alpha Ventus: Challenges, results and perspectives*. Weisbaden: Springer Spektrum. pp. 83–94.

Krone, R., Gutow, L., Joschko, T.J. & Schröder, A. (2013) Epifauna dynamics at an offshore foundation – implications of future wind power farming in the North Sea. *Marine Environmental Research* 85: 1–12.

Lagardère, J.P. (1982) Effects of noise on growth and reproduction of *Crangon crangon* in rearing tanks. *Marine Biology* 71: 177–185.

Lagenfelt, I., Andersson, I. & Westerberg, H. (2012) Blankålsvandring, vindkraft och växelströmsfält, 2011. Vindval. Report No. 6479. Stockholm: Swedish Environmental Protection Agency. Retrieved 1 December 2018 from https://www.naturvardsverket.se/Documents/publikationer6400/978-91-620-6479-2.pdf?pid=3787

Langhamer, O. (2010) Effects of wave energy converters on the surrounding soft-bottom macrofauna (west coast of Sweden). *Marine Environmental Research* 69: 374–381.

Langhamer, O. & Wilhelmsson, D. (2009) Colonisation of fish and crabs of wave energy foundations and the effects of manufactured holes – a field experiment. *Marine Environmental Research* 68: 151–157.

Langhamer, O., Holand, H. & Rosenqvist, G. (2016) Effects of an offshore wind farm (OWF) on the common shore crab *Carcinus maenas*: tagging pilot experiments in the Lillgrund Offshore Wind Farm (Sweden). *PLoS ONE* 11(10): e0165096. doi: 10.1371/journal.pone.0165096.

Langhamer, O., Dahlgren, T.G. & Rosenqvist, G. (2018) Effect of an offshore wind farm on the viviparous eelpout: biometrics, brood development and population studies in Lillgrund, Sweden. *Ecological Indicators* 84: 1–6.

Leonhard, S.B. & Pedersen, J. (2006) Benthic communities at Horns Rev before, during and after construction of Horns Rev offshore wind farm. Final Report/Annual Report 2005. Fredericia: Vattenfall. Retrieved 1 December 2018 from https://corporate.vattenfall.dk/globalassets/danmark/om_os/horns_rev/status-report-2005.pdf

Leonard, S.B., Stenberg, C. & Støttrup, J. (eds) (2011) Effect of the Horns Rev 1 offshore wind farm on fish communities: follow-up seven years after construction. Report No. 246-2011. Charlottenlund: DTU Aqua, Institut for Akvatiske Ressourcer. Retrieved 1 December 2018 from https://www.aqua.dtu.dk:443//-/media/Institutter/Aqua/Publikationer/Forskningsrapporter_201_250/246_2011_effect_of_the_horns_rev_1_offshore_wind_farm_on_fish_communities.ashx

Lindeboom, H.J., Kouwenhoven, H.J., Bergman, M.J.N., Bouma, S., Brasseur, S.M.J.M., Daan, R., Fijn, R.C., de Haan, D., Dirksen, S., van Hal, R., Lambers, R.H.R., ter Hofstede, R., Krijgsveld, K.L., Leopold, M. & Scheidat, M. (2011) Short-term ecological effects of an offshore wind farm in the Dutch coastal zone; a compilation. *Environmental Research Letters* 6(3): 035101.

Lindeboom, H., Degraer, S., Dannheim, J., Gill, A.B. & Wilhelmsson, D. (2015) Offshore wind park monitoring programmes, lessons learned and recommendations for the future. *Hydrobiologia* 756: 169–180.

Lynge, H. (2004) *Middelgrunden: Biologisk undersøgelse ved vindmølleparken på Middelgrunden ved København, efterår 2003*. Roskilde: Miljø- og Energi.

Metcalfe, J.D., Wright, S., Tudorache, C. & Wilson, R.P. (2016) Recent advances in telemetry for estimating the energy metabolism of wild fishes. *Journal of Fish Biology* 88: 284–297.

Miller, R.G., Hutchison, Z.L., Macleod, A.K., Burrows, M.T., Cook, E.J., Last, K.S. & Wilson, B. (2013) Marine renewable energy development: assessing the benthic footprint at multiple scales. *Frontiers in Ecology and the Environment* 11: 433–440.

Marine Management Organisation (MMO) (2014) Review of post-consent offshore wind farm monitoring data associated with licence conditions. MMO Project No. 1031. Report produced for the Marine Management Organisation. Retrieved 1 December 2018 from https://assets.publishing.service.gov.uk/government/uploads/system/uploads/attachment_data/file/317787/1031.pdf

Montgomery, J.C., Jeffs, A., Simpson, S.D., Meekan, M. & Tindle, C. (2006) Sound as an orientation cue for the pelagic larvae of reef fishes and decapod crustaceans. *Advances in Marine Biology* 51: 143–196.

Müller, C. (2007) Behavioural reactions of cod (*Gadus morhua*) and plaice (*Pleuronecta platessa*) to sound resembling offshore wind turbine noise. PhD thesis, Humboldt University, Berlin. Retrieved 1 December 2018 from https://edoc.hu-berlin.de/handle/18452/16377

Nathan, L.M., Simmons, M., Wegleitner, B.J., Jerde, C.L. & Mahon, A.R. (2014) Quantifying environmental DNA signals for aquatic invasive species across multiple detection platforms. *Environmental Science & Technology* 48: 12800–12806.

Nedwell, J., Langworthy, J. & Howell, D. (2003a) Assessment of sub-sea acoustic noise and vibration from offshore wind turbines and its impact on marine wildlife; initial measurements of underwater noise during construction of offshore wind farms, and comparison with background noise. COWRIE Report No. 544 R 0424. Bishop's Waltham: Subacoustech Ltd. Retrieved 1 December 2018 from https://tethys.pnnl.gov/sites/default/files/publications/Noise_and_Vibration_from_Offshore_Wind_Turbines_on_Marine_Wildlife.pdf

Nedwell, J., Turnpenny, A., Langworthy, J. & Edwards, B. (2003b) Measurements of underwater noise during piling at the Red Funnel Terminal, Southampton, and observations of its effect on caged fish. Report No. 558 R 0207. Bishop's Waltham: Subacoustech Ltd. Retrieved 1 December 2018 from http://underwaternoise.org.uk/information/downloads/558R0207.pdf

Nehls, G., Harwood, A.J.P & Perrow, M.R. (2019) Marine mammals. In Perrow, M.R. (ed.) *Wildlife and Wind Farms, Conflicts and Solutions. Volume 3. Offshore: Potential effects*. Exeter: Pelagic Publishing. pp. 112–141.

Olsgard, F., Schaanning, M.T., Widdicombe, S., Kendall, M.A. & Austen, M.C. (2008) Effects of bottom trawling on ecosystem functioning. *Journal of Experimental Marine Biology and Ecology* 366: 123–133.

Pearce, B., Fariñas-Franco, J.M., Wilson, C., Pitts, J., de Burgh, A. & Somerfield, P.J. (2014) Repeated mapping of reefs constructed by *Sabellaria spinulosa* Leuckart, 1849 at an offshore wind farm site. *Continental Shelf Research* 83: 3–13.

Perrow, M.R. (2019) A synthesis of effects and impacts. In Perrow, M.R. (ed.) *Wildlife and Wind Farms, Conflicts and Solutions. Volume 3. Offshore: Potential effects*. Exeter: Pelagic Publishing. pp. 235–277.

Perrow, M.R., Gilroy, J.J., Skeate, E.R. & Tomlinson, M.L. (2011) Effects of the construction of Scroby Sands offshore wind farm on the prey base of little tern *Sternula albifrons* at its most important UK colony. *Marine Pollution Bulletin* 62: 1661–1670.

Pfannkuche, O. & Thiel, H. (1988) Sample processing. In Higgins, R.P. & Thiel, H. (eds) *Introduction to the Study of Meiofauna*. Washington, DC: Smithsonian Institution Press. pp. 134–145.

Pihl, L. & Wennhage, H. (2002) Structure and diversity of fish assemblages on rocky and soft bottom shores on the Swedish west coast. *Journal of Fish Biology* 61: 148–166.

Popper, A.N. & Hastings, M.C. (2009) The effects of anthropogenic sources of sound on fishes. *Journal of Fish Biology* 75: 455–489.

Raoux, A., Tecchio, S., Pezy, J.P., Lassalle, G., Degraer, S., Wilhelmsson, D., Cachera, M., Ernande, B., Le Guen, C., Haraldsson, M., Grangere, K., Le Loc'h, F., Dauvin, J.C. & Niquil, N. (2017) Benthic and fish aggregation inside an offshore wind farm: which effects on the trophic web functioning? *Ecological Indicators* 72: 33–46.

Rees, J.M. & Judd, A.D. (2019) Physical and chemical effects. In Perrow, M.R. (ed.) *Wildlife and Wind Farms, Conflicts and Solutions. Volume 3. Offshore: Potential effects*. Exeter: Pelagic Publishing. pp. 30–46.

Reiss, H., Kroncke, I. & Ehrich, S. (2006) Estimating the catching efficiency of a 2-m beam trawl for sampling epifauna by removal experiments. *ICES Journal of Marine Science* 63: 1453–1464.

Reubens, J.T., Pasotti, F., Degraer, S. & Vincx, M. (2013) Residency, site fidelity and habitat use of Atlantic cod (*Gadus morhua*) at an offshore wind farm using acoustic telemetry. *Marine Environmental Research* 90: 128–135.

Reubens, J.T., Degraer, S. & Vincx, M (2014a) The ecology of benthopelagic fishes at offshore wind farms: a synthesis of 4 years of research. *Hydrobiologia* 727: 121–136.

Reubens, J.T., De Rijcke, M., Degraer, S. & Vincx, M (2014b) Diel variation in feeding and movement patterns of juvenile Atlantic cod at offshore wind farms. *J Sea Research* 85: 214–221.

Reubens, J.T., Degraer, S. & Vincx, M. (2019) Box 5.1: Acoustic telemetry to investigate the presence and movement behaviour of Atlantic Cod *Gadus morhua* in a Belgian offshore wind farm. In Perrow, M.R. (ed.) *Wildlife and Wind Farms, Conflicts and Solutions. Volume 3. Offshore: Potential effects*. Exeter: Pelagic Publishing. pp. 98–100.

Rosenberg, R. (1977) Effects of dredging operations on estuarine benthic macrofauna. *Marine Pollution Bulletin* 8: 102–104.

Russell, D.J.F., Brasseur, S.M.J.M., Thompson, D., Hastie, G.D., Janik, V.M., Aarts, G., McClintock, B.T., Matthiopoulos, J., Moss, S.E.W. & McConnell, B. (2014) Marine mammals trace anthropogenic structures at sea. *Current Biology* 24: R638–R639.

Schläppy, M.-L., Shashkov, A. & Dahlgren, T.G. (2014) Impact hypothesis for offshore wind farms: explanatory models for species distribution at extremely exposed rocky areas. *Continental Shelf Research* 83: 14–23

Seanergy 2020 (2012) Delivering offshore electricity to the EU: spatial planning of offshore renewable energies and electricity grid infrastructures in an integrated EU maritime policy. Brussels: European Wind Energy Association. Retrieved 25 October 2019 from http://www.ewea.org/fileadmin/files/library/publications/reports/Seanergy_2020.pdf

Šaškov A, Dahlgren TG, Rzhanov Y, Schlappy M-L (2015) Comparison of manual and semiautomatic underwater imagery analyses for monitoring of benthic hard-bottom organisms at offshore renewable energy installations. *Hydrobiologia* 756:139–153.

Sierra-Flores, R., Atack, T., Migaud, H. & Davie, A. (2015) Stress response to anthropogenic noise in Atlantic cod *Gadus morhua* L. *Aquacultural Engineering* 67: 67–76.

Simpson, S.D., Purser, J. & Radford, A.N. (2015) Anthropogenic noise compromises antipredator behaviour in European eels. *Global Change Biology* 21: 586–593.

Slabbekoorn, H., Bouton, N., van Opzeeland, I., Coers, A., ten Cate, C. & Popper, A.N. (2010)

A noisy spring: the impact of globally rising underwater sound levels on fish. *Trends in Ecology and Evolution* 25: 419–42, 7.

Slavik, K., Lemmen, C., Zhang, W., Kerimoglu, O., Klingbeil, K. & Wirtz, K. (2018) The large scale impact of offshore windfarm structures on pelagic primary production in the southern North Sea. *Hydrobiologia*. doi: 10.1007/s10750-018-3653-5.

Sørensen, J., Jørgensen, B.B. & Revsbech, N.P. (1979) A comparison of oxygen, nitrate, and sulfate respiration in coastal marine sediments. *Microbial Ecology* 5: 105–115.

Stenberg, C, Støttrup, J.G., van Deurs, M., Berg, C.W., Dinesen, G.E., Mosegaard, H., Grome, T.M. & Leonhard, SB (2015) Long-term effects of an offshore wind farm in the North Sea on fish communities. Marine Ecology Progress Series 528: 257–265.

Underwood, A. (1994) On beyond BACI – sampling designs that might reliably detect environmental disturbances. *Ecological Applications* 4: 3–15.

Vaissière, A.-C. & Levrel, H. (2015) Biodiversity offset markets: what are they really? An empirical approach to wetland mitigation banking. *Ecological Economics* 110: 81–88.

van der Molen, J., Smith, H.C.M., Lepper, P., Limpenny, S. & Rees, J. (2014) Predicting the large-scale consequences of offshore wind turbine array development on a North Sea ecosystem. *Continental Shelf Research* 85: 60–72.

Vanermen, N. & Stienen, E.W.M. (2019) Seabirds: displacement. In Perrow, M.R. (ed.) *Wildlife and Wind Farms, Conflicts and Solutions. Volume 3. Offshore: Potential effects*. Exeter: Pelagic Publishing. pp. 174–205.

Vanhellemont, Q. & Ruddick, K. (2014) Turbid wakes associated with offshore wind turbines observed with Landsat 8. *Remote Sensing of Environment* 145(C): 105–115.

Wahlberg, M. & Westerberg, H. (2005) Hearing in fish and their reactions to sounds from offshore wind farms. *Marine Ecology Progress Series* 288: 295–309.

Watson, D., Harvey, E., Anderson, M. & Kendrick, G. (2005) A comparison of temperate reef fish assemblages recorded by three underwater stereo-video techniques. *Marine Biology* 148: 415–425.

Westerberg, H. & Lagenfelt, I. (2008) Sub-sea power cables and the migration behaviour of the European eel. *Fisheries Management and Ecology* 15: 369–375.

Westerberg, H., Rönnbäck, P. & Frimansson, V. (1996) Effects on suspended sediments on cod egg and larvae and on the behaviour of adult herring and cod. ICES CM 1996/E:26. Copenhagen: International Council for the Exploration of the Sea. Retrieved 1 December 2018 from http://www.ices.dk/sites/pub/CM%20Doccuments/1996/E/1996_E26.pdf

Wikström, A. & Granmo, Å. (2008) En studie om hur bottenlevande fauna påverkas av ljud från vindkraftverk till havs. Report No. 5856. Bromma: Swedish Environmental Protection Agency. Retrieved 1 December 2018 from https://www.naturvardsverket.se/Om-Naturvardsverket/Publikationer/ISBN/5800/978-91-620-5856-2/.

Wilhelmsson, D. & Malm, T. (2008) Fouling assemblages on offshore wind power plants and adjacent substrata. *Estuarine Coastal and Shelf Science* 79: 459–466.

WindEurope (2018) Offshore wind in Europe: key trends and statistics 2017. February 2018. Brussels: WindEurope. Retrieved 27 July 2018 from https://windeurope.org/wp-content/uploads/files/about-wind/statistics/WindEurope-Annual-Offshore-Statistics-2017.pdf

CHAPTER 2

Monitoring marine mammals

MEIKE SCHEIDAT and LINDSAY PORTER

Summary

Wind farms can potentially affect marine mammals at any stage of project development. This chapter outlines the range of monitoring methods that can be used to investigate changes in marine mammal abundance, distribution and behaviour, over long- and short-term periods, in order to evaluate both wind-farm impacts and the effectiveness of mitigation strategies put in place during construction activities. Data on population abundance and distribution over a larger spatial scale are typically obtained from systematic surveys from vessels or aircraft limited to periods of good weather and daylight hours. Such visual surveys provide good spatial coverage in a relatively short time, thus presenting a 'snapshot' of marine mammal occurrence and distribution. Towed passive acoustic monitoring (PAM) arrays that record marine mammal vocalisations can also be used in poor-visibility conditions at night or during rough seas. Detection of fine-scale changes in habitat use or behaviour, however, is best undertaken through continuous monitoring with static or fixed PAM, provided the species under study is acoustically distinctive and vocalises regularly. Such devices generally archive data for later retrieval, although new 'real-time' data transmission devices are becoming available. Acoustic monitoring can also provide information on behaviour, such as foraging, and may be the only feasible method where populations occur at low density or are cryptic. Ideally, a combination of both visual and acoustic monitoring methodologies provides the best spatial and temporal coverage. Telemetry can be used to great effect, particularly on pinnipeds. Other tools, such as unmanned aerial vehicles, also show promise. Regardless of the technique used, the overriding challenge is that sufficient data must be collected so that robust statistical analyses can be performed. Therefore, well-designed surveys and the evaluation of the likely statistical power of any anticipated analyses are of paramount importance when selecting survey methods.

Introduction

Offshore wind-farm (OWF) development began in European waters in the early 2000s and has been concentrated in the North and Baltic Seas (Figure 2.1). Starting in 2015, the number of developments increased dramatically, and total capacity had quadrupled by the end of 2017. At the beginning of 2018, more than 4,000 turbines were operational, comprising 92 wind farms located across the jurisdiction of 11 countries. The first installation of a floating wind farm was completed in Europe in 2018. China was the next economy to invest heavily in this industry, and its first wind farm came online in 2010. Today, China has some 260 turbines located across eight wind farms. The USA's first operational OWF came online at the end of 2016, with several additional large-scale developments proposed for the US east coast to be initiated in 2019 (Massachusetts Clean Energy Center 2018). Taiwan will also see major wind-farm development in the next decade, with its first demonstration projects coming online by 2019/2020 and an additional 11 projects having been approved in 2018. Globally, wind farms are located within 50 km of the shore and within the 50 m depth contour, although most are closer or in shallower water. As technology improves, however, locations farther offshore will become more feasible; for example, developments proposed for European waters are some 200 km from land. Most wind farms comprise monopile structures, the installation of which, through pile driving, can cause considerable increases in underwater noise levels and therefore has the potential to detrimentally affect marine species. Marine mammals are particularly sensitive to underwater noise and have been a primary focus of Environmental Impact Assessments during wind-farm construction. Initially, when development was small scale, potential impacts were considered localised in nature; however, this is no longer the case. With the proliferation of OWFs in European waters, as well as the intention to develop widely in other areas, impacts could be large scale and far reaching, and the potential cumulative impact of multiple developments has yet to be assessed.

In comparison to other taxa, marine mammal populations can be extremely challenging to monitor. As outlined by Nehls *et al.* (2019), the term 'marine mammals' includes a diverse group, ranging from species that spend their entire life in the water, such as cetacea

Figure 2.1 View of the Dutch Princess Amalia Wind Park (PAWP) and the Offshore Wind Park Egmond aan Zee (OWEZ) in the North Sea during an aerial survey for marine mammals. Wind farms are increasingly being installed in clusters, potentially affecting large areas of habitat. (Steve Geelhoed)

and sirenia, to some that spend part of their life cycle on land or ice, such as pinnipeds. Individuals are generally only visible when they come to the surface to breathe, or in the case of pinnipeds when they haul out on coasts, ice floes or sandbars. Perhaps most challenging of all, some species are distributed throughout large areas and migrate over long distances, requiring any comprehensive study to commission substantial survey resources and obtain the cooperation of multiple national jurisdictions. In addition, in the offshore areas typically utilised for wind-farm development, strong winds and heavy seas prevail. This presents significant challenges for the physical study of marine mammals as surveys are most effective during calm sea conditions. As such, there may be limited windows of opportunity to access the study area, by either boat or aircraft, to collect sufficient data to detect any significant change in marine mammal behaviour and distribution. This is further compounded by the short duration of wind-farm construction, which again limits the time available to collect adequate data. Ideally, absolute abundance estimates with associated measurements of uncertainty should be obtained from monitoring programmes; however, relative abundance indices can also be used if the main aim of monitoring is to document changes compared to the baseline.

In terms of potential interactions between marine mammals and wind farms, two groups have been considered thus far: small cetaceans and pinnipeds. This chapter will focus on monitoring methods suitable for these groups. The aim of monitoring the abundance, distribution and behaviour of small cetaceans and pinnipeds in relation to wind-farm developments is to detect changes over time and to relate any perceived change to natural variation, conservation action or anthropogenic activities. When the first wind farms were installed, only discrete areas were affected, and at that time, it was thought unlikely that entire populations of typically wide-ranging species would be affected. As such, initial studies focused on the individuals that inhabited the waters in close proximity to the development. However, there are some cases where the potential impact from wind-farm development can cover entire subpopulations of small cetaceans, for example the Moray Firth Bottlenose Dolphin *Tursiops truncatus* in Scotland and the Eastern Taiwan Strait Humpback Dolphin *Sousa chinensis* in Taiwan. As wind farms proliferate throughout European waters, larger swathes of both coastal and offshore areas are being developed. Since development is proposed to increase rapidly throughout the next decade, there is a growing concern that the cumulative impacts of multiple and concurrent wind-farm development could be significant. There is emerging literature that indicates that wind farms may already be having ecosystem-scale effects (Perrow 2019) such as the potential for shifting primary productivity regimes in the southern North Sea (Slavik *et al.* 2018). Globally, as new wind-farm markets develop elsewhere, more and more species could be affected and the effect on oceanic ecosystems may be profound albeit not necessarily in a negative way, especially in the context of ocean-scale changes initiated by global climate change.

In relation to wind farms, there are several rationales for monitoring marine mammals during the different phases of the wind-farm life cycle:

- **Pre-construction:** to establish a baseline of the species and populations present to later investigate the effect of construction and operational phases. These data can also assist in establishing the status of the species present, including whether they are also subject to other threats and therefore may be vulnerable. There are also a number of pre-construction activities that could affect the presence of marine mammals, such as an increased presence of vessels, sonar surveys to measure detailed bathymetry and sediment types, surveys for unexploded ordnance and, if located, detonation of such ordnance.

- **Construction:** to monitor the direct effect of multiple and concurrent construction activities including piling, installation of substation platform(s) and subsea cabling, and the effectiveness of mitigation measures. During this phase, monitoring is usually focused on noisy pile-driving activity, as this has the potential to harm all marine mammal species. In addition, scour protection, typically in the form of rock material, is generally installed to inhibit sediment loss from water movement around the base of wind-farm monopiles, thus also modifying the substrate.
- **Operation:** to establish whether any displaced marine mammals have returned to the area, which may include either the original animals or resettlement by new animals, and to investigate the long-term effects of the presence of wind farms on marine mammal habitat use. Effects may be positive as well as negative. For example, through the 'reef effect', wind farms change the abundance and distribution of many invertebrate and fish species (Dannheim *et al.* 2019; Gill & Wilhelmsson 2019), which can attract apex predators such as seals to wind farms (Russell *et al.* 2014; Nehls *et al.* 2019). Monitoring both during and after wind-farm construction also allows for predictions made in Environmental Impact Assessments to be tested. These data can then be used as the evidence base for future wind-farm developments, in addition to providing documentation of regulatory compliance.
- **Decommissioning:** to investigate the direct impacts of decommissioning activities on the population. This can also be considered an impact phase, in the same manner as construction, and may necessitate further baseline and impact studies.

Scope

This chapter provides an overview of the monitoring techniques that have been most commonly used to monitor marine mammal abundance, distribution and behaviour at wind-farm developments. These include line-transect surveys aboard vessels and aircraft, individual tracking via telemetry and acoustic monitoring. These are standard methods that have been applied for the past few decades to study marine mammals. The information on these methods is based on established reference books, peer-reviewed literature, publicly available reports, and the expertise and experiences of the authors. The intention of this chapter is to describe the key methods available to monitor marine mammals and to discuss the advantages and limitations of each method (Table 2.1). The review is not exhaustive, and it is hoped that the description provided of both key methods and emerging approaches will guide the reader to more detailed studies.

All of these techniques can be applied to the different research questions that emerge throughout wind-farm life cycles (see *Introduction* above). As also highlighted in the Introduction, the majority of wind-farm impact assessments have been conducted in European waters. As there are emerging developments outside Europe, a case study is included from Taiwan, where wind-farm development is in its initial phase.

It cannot be emphasised enough that there must be a clearly defined research question at the heart of any study focusing on the impact of wind-farm development on marine mammals. The type of monitoring methodology employed will also be influenced by the species under study and the time-frame available in which to conduct the study. Research questions may include the identification of species present, the distribution, in time and space, of these species and how far reaching the potential impacts of individual wind-farm development might be. Assessing cumulative impacts of wind-farm developments is also becoming increasingly important. There are existing techniques that can address such

Table 2.1 Methods for use in wind-farm studies on marine mammals

Method	Benefits	Limitations
Line-transect surveys	• Absolute and/or relative abundance • Distribution • Long term • Entire range of population • Robust methods/analyses	• Expensive • Weather and daylight restricted (unless acoustic component used) • High variability (difficult to detect trends) • Provides population 'snapshots' over the short term • Responsive movement (boat-based only) • Range restricted • Weather constraints
Acoustic surveys	• Not weather dependent • Non-invasive • Can be automated • Large sample sizes • Long deployment time	• Limited detection range • Restricted to species that vocalise regularly
Telemetry	• Detailed spatial and temporal distribution • Remote data retrieval	• Capture and handing required • Device attachment (abrasion, hydrodynamic modification) • Expensive (satellite time) • Small sample size (limited to number of marine mammals captured) • Mainly used for pinnipeds

questions; however, emerging tools are likely to improve data-collection efficiency and, therefore, increase the ability to detect changes in marine mammal population abundance, distribution and behaviour.

Themes

Power analyses

An important consideration in any study is to determine how much data is sufficient to answer your research question (e.g. Taylor *et al*. 2007). Inadequate data collection is due, in part, to the intrinsic difficulties of any assessment of marine mammal populations, as described later in this chapter. It can be very challenging to survey an offshore area for widely ranging species which usually occur at low densities and that are only visible for short periods in good weather. Nevertheless, it is important to consider at an early stage the amount of data that will be needed and to design the research project accordingly.

One way to gain a better understanding of this problem is to conduct a power analysis (e.g. Gerrodette 1987). This analysis will provide information on what kind of results a study design will be likely to yield. Some *a priori* information will be needed to do this, such as expected sighting rates per effort, and pilot studies may be needed to obtain this. For the power analysis, one also needs to consider what kind of change the study aims to detect. Obviously, with all other factors being the same, larger changes in occurrence or behaviour will be easier to detect than smaller changes.

There is a direct relationship between increasing the power of a design and the costs. In general, to improve data quality and quantity an increase in effort is needed, for example through the coverage of a larger number of transect lines. As budgets are limited, this leads to a trade-off between costs and statistical precision. A specific challenge for wind-farm studies is that the study area is often small, and that the monitoring period, in particular during construction work, is relatively short.

A number of approaches have been proposed and applied with regard to detecting changes in occurrence or behaviour of marine mammals in wind farms, such as the before–after control-impact (BACI) design or the use of Bayesian modelling (e.g. Carstensen *et al.* 2006; Sveegaard *et al.* 2013; Thompson *et al.* 2013). The decision on which monitoring method and analyses to use, and how much effort should be spent, needs to be considered for each case.

Mapping distribution and estimating abundance

Investigating the distribution and abundance of what are usually relatively small populations in a vast oceanic habitat has multiple challenges. The environment itself may only be accessible during certain seasonal periods and sudden poor weather intervals can delay or disrupt the best planned surveys. In addition, the species of interest may be widely dispersed and the ability to detect animals may be comprised by inclement weather conditions, the competence of the observers, the proficiency of recording equipment and, of course, the behaviour of the individual animals themselves.

Multiple techniques can be successfully employed in a variety of different marine mammal habitats that provide an understanding of distribution and habitat use; however, some general principles should be applied to all wildlife surveys. First, the data collected should be typical or representative of the habitat or population that is being surveyed and, where necessary, should be independent, so as to reduce biases. In addition, all data must be collected with the utmost accuracy possible and in accordance with the requirements of the technique being used. Detailed reviews on the advantages and disadvantages of the different methods, also with respect to wind-farm projects, can be found in a number of publications, such as Hammond (2002), Evans & Hammond (2004), Evans (2008) and Verfuß *et al.* (2015).

Survey techniques and tools

One of the most widely used techniques for estimating animal density or abundance is 'distance sampling'. The analysis is so called as it relies on knowing the recorded distance from the animal or 'cue', such as a group of dolphins, to a survey line or survey point (Buckland *et al.* 1993; 2001; 2004; Thomas *et al.* 2010). Data on marine mammals for distance sampling are commonly collected from visual and acoustic line-transect or point-transect surveys (e.g. Hammond *et al.* 2002; Southwell 2005; Kaschner *et al.* 2012; Marques *et al.* 2013; Bortolotto *et al.* 2017).

One of the most important assumptions of line-transect sampling is that animal density in the area being surveyed is representative of the entire habitat in which the population of interest occurs. As it is rarely the case that populations are distributed evenly throughout any given area, this assumption needs to be met by utilising a robust survey design, which ensures that all portions of the study area have an equal probability of being surveyed. At least 10–20 replicate transect lines should be placed in a systematic but randomised manner to provide a basis for an adequate variance of the encounter rate and a reasonable number of degrees of freedom for constructing confidence intervals. In addition, transects should not be placed parallel to any anticipated density gradient of the population, such as depth, as this could lead to high variance in encounter rates between the replicate transects. The layouts of transect lines in marine surveys are typically parallel or zigzag (sawtooth), the latter being the most efficient with regard to survey time, including fuel costs. The software program 'Distance' (http://distancesampling.org) (Thomas *et al.* 2010) can be used to create a survey design with an equal coverage probability.

It is also important that a sufficient number of sightings is obtained to calculate a detection function and to achieve the desired level of precision. For reliable estimation of the detection function, the absolute minimum sample size should be 60–80 sightings of groups of animals. To assess how much effort is needed to obtain this number of sightings, existing information should be analysed or pilot studies should be conducted. Stratification of a study area into smaller sub-areas can be helpful, for example, to accommodate discrete areas of certain habitat types or areas of anticipated high or low density. It should be noted that once a survey has been conducted using a stratified design, the data cannot then be pooled and analyses must account for this stratification.

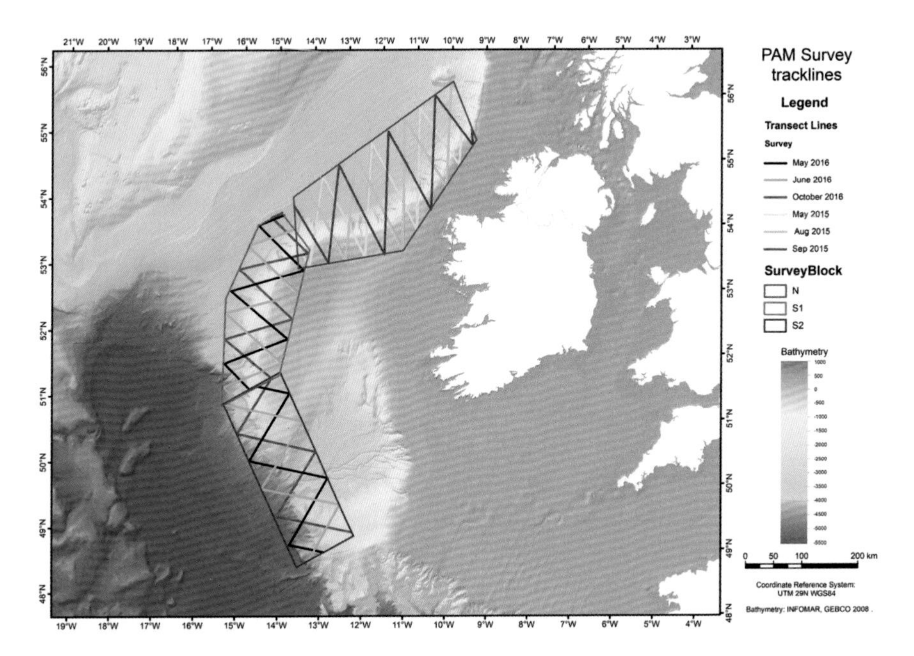

Figure 2.2 The survey blocks and designed transect lines for the ObSERVE Acoustic surveys 2015–2016. The northern region was considered as a single survey block, while the geometry of the shelf break in the southern region required stratification. (From Berrow *et al.* 2018; https://www.dccae.gov.ie/images/ObSERVE%20Acoustic%20Colour%20Coded%20Survey%20Lines.jpg)

The logistical constraints of any survey, such as the location of airports and ports for refuelling, closed areas for shipping or air traffic, and the endurance of the vessel or aircraft to remote survey sites, must also be considered and accommodated within the survey design. An example of an offshore survey design in Atlantic waters, using a stratified survey design, for a combined visual and acoustic survey is provided in Figure 2.2.

One challenge posed by cetaceans is that they spend a considerable amount of their lives underwater and are not always available to be observed. Furthermore, when cetaceans are at the surface, observations may be missed owing to poor weather conditions, making it difficult to detect their brief surfacing, or simply by the observer failing to detect an animal that is on the surface. If the survey aim is to obtain an absolute estimate of abundance, the chance of observing a sighting on the transect line, called $g(0)$, must be quantified. If $g(0)$ is not calculated, density will be underestimated. A number of different methods have been used to estimate $g(0)$ and these methods differ between ship and aerial surveys (Box 2.1).

Box 2.1 Correcting for animals that are not seen: the *g*(0) factor

During line-transect surveys, one of the assumptions is that all animals present on the transect line are detected; $g(0)=1$, where $g(x)$ is the probability of detecting an animal at perpendicular distance x. This assumption is rarely true for cetaceans, as a proportion of sightings will always be missed. There are two main reasons for this: (1) animals are present but are missed by the observer, known as *perception (or observer) bias*; or (2) animals are not detected as they cannot be seen as they are diving, known as *availability bias*.

It is possible to correct for the proportion of missed animals so that absolute abundance estimates can be calculated; for example, if $g(0)=0.30$, this means that 30% of all animals present on the transect line are sighted and the relative abundance estimate derived from the survey data can be corrected accordingly. The $g(0)$ factor should ideally be assessed per sighting condition, as a change in environmental factors, such as wind speed or turbidity, will have an impact. It is also specific for species (or taxa) and observers or observer teams.

In many instances, the perception bias alone can be reduced by using experienced observers, training, appropriate survey protocols (e.g. rotating observers to avoid fatigue), and suitable survey platform configuration and speed. The probability of an observer detecting an animal can be quantified by testing how many visible animals are missed. This has been done, for example, using a second independent observer team in a plane that covers the same field of view (e.g. Palka 2005; Thomson *et al.* 2012). The ratio of animals seen by one team, but missed by the other team, provides a correction factor for the number of undetected animals that was available to be seen.

The availability bias is sometimes quantified by investigating the behaviour of a species to determine how much time it spends at the surface (or close to the surface). For example, for Harbour Porpoise *Phocoena phocoena*, several studies have recorded breathing rates from observations (Barlow *et al.* 1988) or have used diving data obtained from tagged animals (Westgate *et al.* 1995; Teilmann *et al.* 2013) to obtain a value of availability. This value has then been applied to the collected survey data (Thomson *et al.* 2012). Caution must be exercised, however, when applying this correction factor to data sets collected elsewhere. Behaviour of animals, and thus the time they spend at the surface, is likely to vary between different areas, weather conditions or seasons. Therefore, a correction factor for the proportion of time an animal is available at the surface may not be applicable in general.

Perception and availability bias can also be estimated at the same time. One example of this is from Laake *et al.* (1997), who tracked porpoise sightings from land at the same time as an aerial survey was underway overhead. This provided data on the availability of the porpoises to the aerial observers as well as how many were missed by the observers. However, this approach requires the species or population of interest to be located close to a shore that has a suitably high vantage point, which is not applicable to most study areas. For Harbour Porpoise, the 'circle-back' or 'race-track' method was specifically developed for aerial surveys (Hiby & Lovell 1998; Hiby 1999). Originally, this method was applied using two aircraft flying in tandem. Currently, the survey aircraft circles back over a section of the transect line after an initial sighting and resamples it after a time-lag of about 3 minutes. This allows an estimate to be made of the re-sightability rate of an animal that is known to be present and visible during the first pass, but may not be visible to or may be missed by the observer on the second pass. The analysis also takes additional parameters into account, such as the swimming speed of the animals. In general, 50–100 race-tracks are needed to determine a reliable $g(0)$ for a particular survey. For boat-based surveys, independent observations can be made visually or acoustically, and are often referred to as dual- or double-platform techniques. For visual dual-platform surveys, the primary observation team searches with the naked eye close to the vessel, while a secondary 'tracker' team uses high-powered binoculars to search the transect line much farther in front of the boat (Figure 2.3). Once a detection is made, the tracker team continues to record the detected animal or group until the

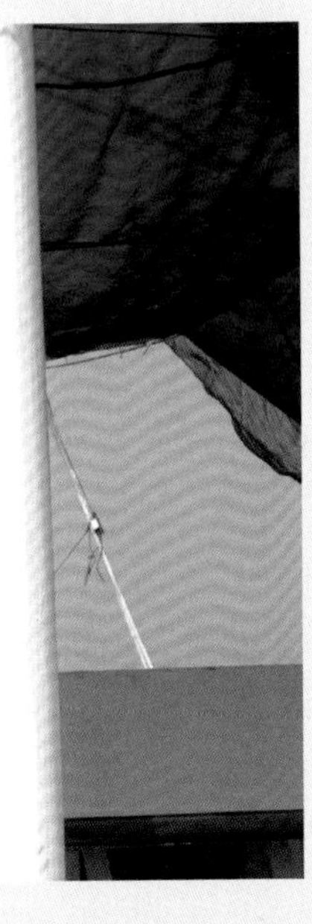

Figure 2.3 During double-platform surveys, high-powered binoculars are used by a secondary tracker team to determine the ability of the primary observer team to detect sightings and thus contribute to estimation of $g(0)$. (WWF Hong Kong)

sighting is seen by the primary team, or it disappears from view. In this way, the proportion of detections missed by the primary team can be calculated, including both perception and availability bias. It is also possible to use acoustic data as an independent detection method, provided that vocalisation behaviour is independent of surfacing behaviour. As yet, only a few studies have used dual acoustic–visual survey methods (Berrow *et al.* 2018) and more work is required to develop suitable analyses before this method can be used more widely.

Most of these methods need a minimum number of sightings to calculate a $g(0)$ value. In areas with a low density of the target species and in comparatively small survey areas this poses a major challenge, which often holds true for wind-farm developments. If it is not possible to obtain a $g(0)$ value, or a value of one of its components, it is important to state clearly that any estimates of abundance are minimum estimates.

Finally, a successful survey depends on the application of good field methods. Survey protocols for observers must be clearly defined. They need to include a description on how observers cover the search area and how measurements to determine the perpendicular distance of the sighting to the trackline are taken. Information on sightings typically details species identification, group size and composition, including the presence of calves or different sexes, when this can be determined, and behaviour. Equally important is the documentation of sighting conditions, such as sea state and visibility, throughout the survey. It is essential to provide sufficient training to observer teams before surveys, as well as monitoring observer performance during surveys, so that biases can be quantified, errors minimised and data collected consistently.

Vessel surveys

Distance sampling protocols are regularly applied during boat-based surveys to determine the abundance of cetaceans (e.g. Hammond *et al.* 2002; 2013; 2017; Gerrodette & Forcada 2005; Barlow & Forney 2007; Øien 2009; Pike *et al.* 2009). When the aim is to assess a population that ranges over a large area, such as the Harbour Porpoise *Phocoena phocoena* in the North Sea (see Figure 2.9), several vessels can be deployed to cover a large proportion of the study area simultaneously. In general, vessels may be of any size, noting that smaller vessels may not be able to accommodate a second platform and, therefore, values for $g(0)$ might not be determined. A good rule is for three observers to be active, one monitoring the trackline (0°) and two observing 90° on both port and starboard of the trackline; thus, the 180° area to the front of the vessel is continually scanned. If there are insufficient team numbers to allocate a dedicated trackline observer, two visual observers should overlap their observation zone by 10–20° across the trackline, thus providing additional visual coverage of this critical area. A data recorder notes all effort and environmental conditions throughout the survey and records all sightings. The data recorder should not simultaneously conduct visual observations, as sightings are likely to be missed. Thus, at a minimum, a team comprising two or three visual observers, one data recorder and one 'rest' position should conduct ship-based line-transect surveys. When a marine mammal (individual or group) is sighted, the time and position must be immediately recorded. An angle is measured to the individual (or centre of the group) and a distance from the observer to the group is estimated or measured. Angle and distance measurements can be taken using compass-reticle binoculars and distances can also be measured using range (range finding) sticks. In some cases, distance from observer to sighting can be estimated by eye, provided good distance estimation procedures are in place and are regularly practised. Since angle and distance values are later used to calculate the perpendicular distance from the trackline to the sighting, these estimates must be as accurate as possible. A common error is for observers to 'round up' angles to the nearest degree, which alters the detection function considerably, particularly when sightings are close to the trackline. Digital compass binoculars provide more accurate angle measurements and minimise rounding up errors. Although detection distances vary with species, observers should concentrate their observations close to the vessel. It is better to obtain consistent detections around the trackline than to be able to observe sightings at great distances; for example, observation zones should be 500–750 m radius from the observers' viewpoint (with the exception of double-platform protocols). Sightings that are made at great distances and those made behind the 180° area are noted but not included in distance analyses. The performance of individual observers should be assessed throughout the survey. Although various configurations of observers and data recorders can be deployed, depending on the vessel type, available space, species under study and resources available (Buckland *et al.* 1993; 2001; 2004), what is essential is consistency in survey protocols and measurement accuracy. The minimum data required are angle and distance to sighting, followed by species identification and group size.

One advantage of boat-based surveys is that additional environmental and ecological data relevant to marine mammal distribution can be collected, including water temperature, salinity, depth and measures of biological productivity, such as the abundance of phytoplankton (expressed as chlorophyll *a*) or even fish (often measured with hydroacoustics) (e.g. Scott *et al.* 2010; Certain *et al.* 2011). Often, marine mammal vessel surveys also provide a platform for other taxa, such as seabirds, to be observed. Conversely, in the UK for example, surveys for seabirds often accommodate marine mammal surveys (see Webb & Nehls, Chapter 3). A further advantage of using vessels is that passive acoustic monitoring (PAM) using towed hydrophone arrays can be conducted simultaneously with visual observations during the day, and as a sole means of detection both at night and when sea conditions are not conducive for visual observations (see *Passive acoustic monitoring surveys using towed arrays*, below).

Another important factor in survey design considerations is the speed at which surveys should be conducted. Ideally, survey speed must be twice that of the swim speed of the species likely to be encountered, so that bias due to animal speed might be avoided. Further biases may also derive from animals' responses to the boat itself, either by attraction or through avoidance. If an animal or a group of animals is attracted to a vessel to bow ride, a behaviour exhibited by many dolphin species (e.g. Short-beaked Common Dolphin *Delphinus delphis*), abundance can be considerably overestimated if not corrected for (Cañadas *et al.* 2004). Some species, such as Harbour Porpoise, may also avoid vessels and therefore abundance can be underestimated (Turnock & Quinn 1991; Hyrenbach 2001; Palka & Hammond 2001; Clarke *et al.* 2003). It is here that good survey design plays a critical role in minimising potential biases and by being able to detect and measure any biases.

Aerial surveys

Aerial surveys using distance sampling methods have been conducted from fixed-wing aircraft (e.g. Hammond *et al.* 2002), helicopters (Southwell 2005), microlights (Rowat *et al.*

2009; Jean *et al.* 2010) and blimps (Hain *et al.* 1999; Fürstenau Oliveira *et al.* 2017). Aerial surveys can also be conducted as manned or unmanned digital surveys (see *Digital aerial surveys*). A key consideration when planning aerial surveys is that fuel capacity is limited and, depending on the type of aircraft, surveys may be able to operate for only 7–8 hours. Working offshore in regions where weather conditions can change rapidly, this limitation poses safety concerns, again highlighting the importance of good survey design and planning.

For small cetaceans, good weather conditions are vitally important for successful data collection. For example, for the small (less than 2 m in length) and generally inconspicuous Harbour Porpoise, surveys are conducted only when sea state is Beaufort 3 or better. In some regions of the world, such oceanic conditions are rare. As aerial surveys are much faster than boat-based surveys, there is often greater flexibility in survey platform deployment, and aerial survey teams can wait for good weather windows and conduct surveys quickly before the weather changes. Aircraft configuration is also important as the area directly under the aerial platform, that is the transect line, should be visible to the observers. While some analytical methods deal with surveys that do not include viewing the transect line (Buckland *et al.* 2001), it is better to use high-winged planes that are equipped with bubble windows that allow an unrestricted view (Figure 2.4). For visual surveys, a common configuration is to situate an observer on each side of the aircraft, thus

Figure 2.4 A 'bubble-window' in an aircraft used for aerial surveys during the Small Cetaceans in European Atlantic waters and the North Sea (SCANS III) survey. The window allows the observer an unobstructed view of the transect line below the plane. (Nino Pierantonio)

enabling observation of the transect-line area under the plane. The observers can record information on environmental conditions, including sea state and glare, in addition to information on sightings or animals, such as species and group size. An inclinometer is used to measure the vertical angle to a sighting, which can then be converted to estimate the perpendicular distance from the sighting to the transect line. As aerial surveys are generally conducted at a speed of around 100 knots, large areas can be covered in a short time. A disadvantage of high-speed surveys is that there is limited time to detect marine mammals and observers must stay focused, so a dedicated data logger is also required to complete the survey team. At high speed, species identification can be challenging, but in some cases survey protocol can be adapted to allow for breaks in effort so that sightings can be photographed and verified (Figure 2.5) and observers allowed some moments of rest.

Figure 2.5 A group of White-beaked Dolphins *Lagenorhynchus albirostris* sighted during aerial line-transect surveys for the Small Cetaceans in European Atlantic waters and the North Sea (SCANS III) survey during the summer of 2016. (Hans Verdaat)

The optimal survey height during aerial surveys depends partly on the species of interest, as well as safety considerations. Observers will have a larger field of vision (and thus more time to scan the area) if the aircraft is at a higher altitude, but again, this can make species identification more challenging as the marine mammal is at considerable distance from the observer. For surveys focusing on small cetaceans, a height of 150–180 m is generally recommended (e.g. Hammond *et al.* 2017). This poses another issue in areas where wind farms are located, such as the North Sea, as flying at this level is generally not permitted because of safety concerns. As stated, when designing surveys, it is critical to assess what constraints there are throughout the entire cycle of the wind-farm development and which survey techniques are the most appropriate to use. In practice in the North Sea, the proportion of survey area missed owing to obstruction from existing wind farms is minor; however, as more wind farms are developed, this particular issue will play a greater role in survey design.

Digital aerial surveys

Digital aerial surveys have several advantages, the main one being that sightings can be replayed and reviewed multiple times, which increases observation objectivity and minimises some biases. Furthermore, for both manned and unmanned vehicles, only a flying crew is required as observer teams do not need to be on board, making flying the survey cheaper and logistically less challenging, with some health and safety risks also reduced. High-definition digital camera systems attached to aircraft can collect either moving or still imagery, and such systems have been used successfully to census birds (Buckland *et al.* 2012) and to estimate the relative abundance of Harbour Porpoise (Williamson *et al.* 2016). For wind-farm impact assessments, the German government requires that digital aerial surveys be conducted for birds and cetaceans (BSH 2013). As with aerial surveys with human observers, this method works best in good weather conditions. Analysis of the resultant data still requires considerable human input, which is a high-cost factor, but automated detection systems are on the horizon, which may streamline future analyses.

Unmanned aerial vehicles (UAVs) carrying camera systems have also been used to study several species of marine mammals (e.g. Koski *et al.* 2009; Linchant *et al.* 2015; Moreland *et al.* 2015; Pomeroy *et al.* 2015). The use of UAVs is limited for the purposes of wind-farm surveys as the flight range is short and good weather conditions are essential; for example, most UAVs cannot fly if it is too windy (Chabot & Bird 2015; Fiori *et al.* 2017). As both duration and endurance increase and costs reduce, the use of UAVs as a tool for marine mammal surveys is likely to increase, along with future applications for wind-farm surveys.

A number of challenges, such as the identification of cetacean species, the impact of environmental conditions on the sighting conditions and the calculation of correction factors for animals that are not visible, still need to be addressed for aerial digital imaging surveys. As this technique becomes more widespread however, it is likely that both data collection and analysis methods will be refined (Mackenzie *et al.* 2013). Importantly, as wind-farm coverage of areas such as the North Sea grows, aerial digital surveys will become increasingly useful, as these surveys can be flown at altitudes higher than the current 180 m requirement for visual aerial surveys, which is too low for flights over wind farms. Thus, the ability to undertake digital surveys at higher flight height enables coverage of both wind farms and their surrounds.

Satellite imagery surveys

Using images obtained from satellites to study marine mammals has been much discussed and reliable methodologies are beginning to emerge (Abileah 2002; Platonov *et al.* 2013; Cubaynes *et al.* 2019). One clear advantage is that large areas of extremely remote habitat can be surveyed, although as yet these areas are not likely to be developed as wind farms. As with aerial surveys, the observation platform, that is, the satellite, does not influence marine mammal behaviour or distribution. High-resolution images have been used successfully to count large whales, such as Southern Right Whale *Eubalaena australis*, Gray Whale *Eschrichtius robustus*, Fin Whale *Balaenoptera physalus* and Humpback Whale *Megaptera novaeangliae* (Fretwell *et al.* 2014; Cubaynes *et al.* 2019), as well as different pinniped species (LaRue *et al.* 2011; McMahon *et al.* 2014; Moxley *et al.* 2017). There is still much to be developed with regard to analytical protocols, such as correction factors for abundance estimates and, to be truly efficient, algorithms to automatically detect marine mammals. For cetaceans, this method has only been useful for larger whales; however, as technology improves, it is likely that satellite imagery will be of sufficient resolution to detect and identify even the smallest of species. This may be a useful method in the future for wind-farm monitoring, but has limited application at this time.

Passive acoustic monitoring surveys using towed arrays

Marine mammals, particularly toothed cetaceans (Odontoceti), are highly vocal and rely on the emission and reception of sound to communicate, forage and navigate. PAM techniques detect the presence of vocalising marine mammals and it is often possible to identify the different species present. From detailed analyses of vocalisations, some behavioural information can also be inferred. PAM systems deployed during boat-based surveys can provide spatial information on the abundance and distribution of the species that are acoustically detected (Figure 2.6). Static acoustic devices anchored to the seabed, or suspended in the water column, provide temporal information useful for monitoring species occurrence and trends in habitat use (see *Passive acoustic monitoring surveys using*

Figure 2.6 A hydrophone array (red line behind the vessel) deployed during combined acoustic visual surveys. (AER Corps, Ireland)

static devices, below). Using acoustic arrays with multiple hydrophones typically fixed on to long cables (400 m), it is possible to collect data that allow relative abundance estimation through distance sampling techniques (Buckland & York 2017). Since the 1970s, such systems have been used extensively for marine mammal research (e.g. Watlington 1979; Gordon *et al.* 1999; Gillespie & Chappell 2002) and more recently have become an integral part of cetacean monitoring in relation to wind farms. Differences in the time of arrival of cetacean vocalisations at stereo pairs of hydrophones can be used to calculate bearings to the vocalising animal. Target motion analysis can then be used to provide locational information and to estimate the distance to the detection from the transect line along which the vessel is travelling. These range data allow the calculation of detection functions that provide the basis of line-transect analyses (Thomas *et al.* 2010). Some species' vocalisations are particularly suitable for absolute abundance estimation, such as those from Sperm Whale *Physeter macrocephalus*; however, for some species, such as delphinids and pilot whales *Globicephala* spp., it is only possible to estimate indices of abundance for the purposes of mapping and habitat modelling (Aguilar de Soto *et al.* 2008; Gordon *et al.* 1999; Embling 2008). A combination of acoustic and visual boat-based surveys can be especially useful in filling in data gaps on deep-diving species, such as sperm and beaked whales (e.g. Berrow *et al.* 2018), although these are not currently target species for wind-farm developments. This may change in the future as wind farms expand in new areas and deeper waters, particularly with the advent of floating wind farms (Jameson *et al.* 2019).

Passive acoustic monitoring surveys using static devices

There are many acoustic devices that can be deployed in the long term at fixed locations (Sousa-Lima *et al.* 2013). These devices collect data continuously, which can be used to discern trends in habitat use and record behavioural patterns. For some species, such as deep-diving odontocetes, static devices have a higher detection rate than visual surveys (e.g. McDonald & Moore 2002; Mellinger & Barlow 2003; Barlow & Taylor 2005). As static devices do not produce any sound unlike survey vessels, these devices are less likely to disturb marine mammals away from the survey area. However, other analytical problems may occur if the devices become 'objects of interest' that marine mammals investigate acoustically. As most static acoustic devices archive data and need to be retrieved to obtain these data, there are considerable logistical challenges in both deployment and retrieval. Several new systems, for example, the Coastal Acoustic Buoy (Turner *et al.* 2019), are promising as they can relay acoustic information from remote areas and in real time, thus allowing reliable and fast data transfer without having to retrieve the device.

One disadvantage of static acoustic devices is that they have limited spatial coverage compared to towed acoustic surveys. Therefore, most static acoustic device surveys require the deployment of multiple devices to obtain sufficient spatial coverage and to account for variance in detection between locations. Estimation of population density using PAM is developing rapidly, particularly with regard to static acoustic devices, and new analytical methods, including spatially explicit capture–recapture models, are playing an increasing role in this field (e.g. McDonald & Fox 1999; Tougaard *et al.* 2006; Van Parijs *et al.* 2009; Marques *et al.* 2013).

Static acoustic devices play a dual role in wind-farm monitoring as they can also be used to measure construction noise, particularly high-impulse piling sounds. These data can then be used to validate noise-impact models and predicted impacts on marine mammal hearing apparatus (NMFS 2018), as well as to assess the effectiveness of mitigation measures. Noise modelling is now an essential component of every wind-farm assessment that has the potential to affect marine mammals (see Thomsen & Verfuβ, Chapter 7) (Box 2.2).

Box 2.2 Wind-farm piling noise characterisation and the Indo-Pacific Humpback Dolphin *Sousa chinensis* in Taiwan

Commencing in 2019, Taiwan will develop extensive areas of eastern Taiwanese waters into offshore wind farms (OWFs). There is concern that this development will detrimentally impact the Eastern Taiwan Strait (ETS) Humpback Dolphin *Sousa chinensis*, a critically endangered population which numbers some 60–70 individuals. The population has a restricted inshore range and is already subject to multiple threats from habitat degradation, coastal development, shipping and competition with fisheries (Figure 2.7). Although there is a port located within the dolphin's habitat, the current ambient underwater noise levels are low (Guan *et al.* 2015); therefore, both the piling and increase in vessel traffic associated with wind-farm construction have the potential to dramatically increase underwater noise levels. Any increase will pose an additional threat to an already compromised dolphin population. In 2016, a noise field characterisation study was conducted during test pile driving for the OWFs, to assess the potential impact on marine mammals (Chen *et al.* 2017). The aims of this study were: (1) to characterise the underwater soundscape during demonstration piling; (2) to identify the dominant sound sources in the vicinity of the piling area; and, from these data (3) to assess the risk of auditory damage, auditory masking and disturbance to dolphins.

Both static passive acoustic monitoring (PAM) devices and boat-based (shipboard) PAM devices were deployed at various distances from the piling activity. Two piling events were monitored: (1) of 14 hours' duration with 4,757 strikes at various powers, and a total of 1.9 hours' total piling; and (2) of 5 hours' duration with 3,778 strikes and 1.4 hours' total piling, again at various hammer powers. Using several independent

Figure 2.7 The Indo-Pacific Humpback Dolphin *Sousa chinensis*, known locally as the Chinese White Dolphin, in the vicinity of wind farms in the South China Sea. (Lindsay J. Porter, University of St Andrews)

means of calculation and direct spectral plotting, both sound exposure level (SEL) and peak sound pressure level (SPL_{pk}) were determined at various distances from the piling source. Few studies have examined the auditory physiology of *Sousa* spp.; however, several acoustic characterisations and audiograms have been obtained from free-ranging and stranded humpback dolphins outside Taiwan (Li *et al.* 2012; Wang *et al.* 2015). The recordings and observations available indicate that *Sousa* have a typical odontocete hearing range of 20 kHz to >100 kHz, with echolocation signals with peak energy above 100 kHz, but with significant energy down to 20 kHz. This is consistent with the classification of the species as a 'mid-frequency cetacean' for the purposes of assessing the effects of anthropogenic sound on marine mammal hearing (Finneran 2016). Using these published criteria, the study concluded that:

- Exposure to pile-driving activities at 750 m was unlikely to cause a permanent threshold shift (PTS) to humpback dolphins' hearing.
- Accumulation of acoustic energy at 750 m could, however, induce a temporary threshold shift (TTS) if dolphins remained in the area for some hours.
- Broadband increases in interpulse sound levels extended to 3 km from the piling source and would be likely to cause acoustic masking.

The ETS humpback dolphin population has a restricted nearshore range, and therefore is at greater risk than other wider ranging species that may be more flexible and thus better able to abandon noisy habitats. As such, there is a strong possibility that ETS dolphins may suffer auditory damage and will be disturbed during wind-farm construction.

The results of this study provided information on potential impact zones, which assisted in the design of baseline and impact monitoring surveys. The information also led to the establishment of strict mitigation protocols during wind-farm construction. At the time of writing, full-scale wind-farm development had not yet commenced; however, impact monitoring surveys comprising both static PAM and boat-based line-transect surveys are planned, with the joint aim of documenting impacts on the humpback dolphin population and the effectiveness of mitigation protocols.

A number of static acoustic devices have been used extensively in wind-farm marine mammal monitoring surveys, such as the C-POD and its predecessor, the T-POD (Figure 2.8). These have been used in multiple studies of Harbour Porpoise in north-western European waters, such as the study at Egmond aan Zee (OWEZ) in the Netherlands (Figure 2.9) (Scheidat *et al.* 2011; Lindeboom *et al.* 2011). The T-POD was developed in the 1990s to investigate Harbour Porpoise bycatch in set net fisheries (Tregenza 1998) and was designed to be robust, yet lightweight, and easy to deploy and retrieve by fishers. The T-POD has continually evolved, and today, the C-POD and Deep C-POD are commonly used in the monitoring of Harbour Porpoise and other toothed whales (e.g. Brandt *et al.* 2009; van Polanen Petel *et al.* 2012). Ideally, C-PODs are deployed in relative close proximity to each other within the wind-farm area, say at 5 km intervals, and then at greater distances (perhaps to 10 km) from the wind farm. For example, at the Gemini wind farm deployed montoring devices at 5 km, 7.5 km and 10 km (Figure 2.10). It is important

Figure 2.8 C-PODs deployed on large buoys during a monitoring project for Groningen Seaport. The German wind farm off the island of Borkum can be seen in the background. (Steve Geelhoed)

Figure 2.9 Harbour Porpoise *Phocoena phocoena* swimming near a turbine in the Dutch offshore wind farm Egmond aan Zee (OWEZ). (Erwin Winter)

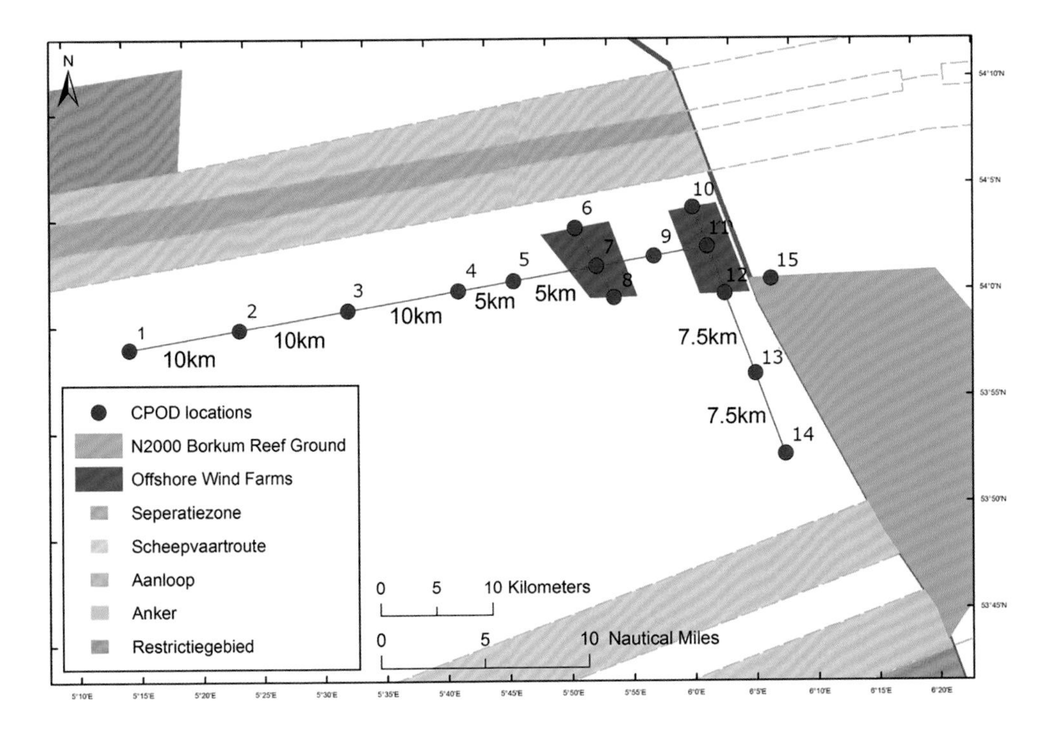

Figure 2.10 Example of C-POD placement at the Gemini wind farm off the Dutch coast (after Geelhoed *et al.* 2018). This study was designed to investigate changes in the distance of occurrence of species such as Harbour Porpoise *Phocoena phocoena* according to wind-farm construction and operation.

that the coverage of C-PODs extends to many tens of kilometres from the wind farm as the displacement distance for species such as Harbour Porpoise may be in excess of 20 km during construction (e.g. Tougaard *et al.* 2009; Dähne *et al.* 2013).

Harbour Porpoise are particularly easy to monitor acoustically as they produce stereotypical echolocation signals lasting for 15–50 μs, with a focused energy between 120 and 150 kHz. The inter-click interval and the amplitude of the clicks gradually change, which allows these vocalisations to be easily distinguished from other high-frequency sounds that occur at sea, such as boat sonar, which is of a similar frequency, but has a very consistent inter-click interval. C-POD analytical software allows a number of parameters to be extracted from the data collected (Figure 2.11). These can be used to describe the acoustic behaviour of Harbour Porpoise, such as porpoise-positive minutes (PPM). Acoustic data can also be expressed as encounters, which are defined as recordings separated by at least 10 minutes of silence. Basically, encounter duration is an expression of how long porpoise continuously click around the C-POD (number of minutes between two silent periods). A closer analysis of click trains can also provide information on foraging behaviour (e.g. Nuuttila 2013; Schaffeld *et al.* 2016; Weel *et al.* 2018; Berges *et al.* 2019). A number of studies have deployed C-PODs at wind-farm sites (Carstensen *et al.* 2006; Scheidat *et al.* 2011).

Anchoring devices such as C-PODs in the long term can be challenging. In some areas, such as the OWEZ wind farm, the most suitable deployment method was to suspend the C-POD from large buoys to reduce the possibility of equipment loss due to adverse weather conditions or trawling activities (Figure 2.8). In other areas, C-PODs are deployed with remote acoustic releases, or two anchors connected by a submerged float-line that can be hooked and hauled up (Anonymous 2016). In areas of sufficiently shallow water, divers can be used to retrieve acoustic devices (Le Double 2018).

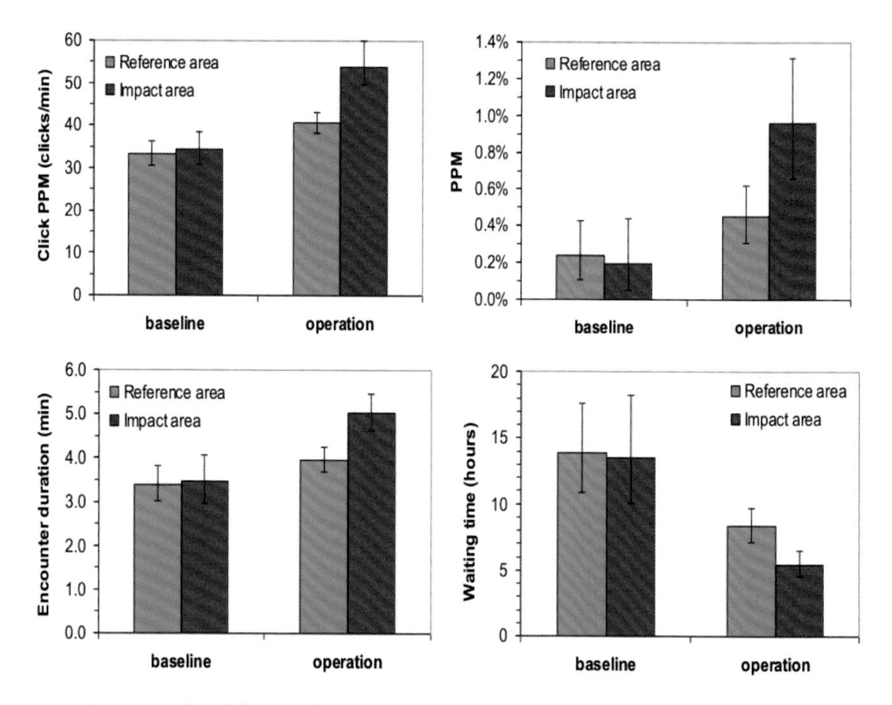

Figure 2.11 An example of four different C-POD parameters used to describe the acoustic activity (per day) of Harbour Porpoise *Phocoena phocoena* in a wind-farm area using a before–after control-impact (BACI) approach. PPM: porpoise-positive minutes. (From Scheidat *et al.* 2011)

Telemetry and tagging

The quantity and quality of data that telemetry studies facilitate are extremely detailed (Bailey *et al.* 2014; Aarts *et al.* 2016a) and provide the basis for in-depth investigations of the impact of human activities, including wind-farm development, on marine mammals (Aarts *et al.* 2016b; Wisniewska *et al.* 2016, 2018). There are multiple types of telemetry systems and electronic tags that may be attached to marine mammals, but as with all studies, the underpinning biological question and the characteristics of the species that is to be tagged will determine the specific method or technique used; in this case, the configuration of the telemetry system. Most tags actively record the location of the animal to which they are attached; however, with a variety of other sensors now available, other behavioural or physiological data from the tagged individual and/or the surrounding environment (e.g. time, depth, orientation, temperature, salinity, noise and vocalisations) can also be recorded (Madsen *et al.* 2006; Hussey *et al.* 2015; Read 2018). Information can be archived on tags for later retrieval and download, which may be difficult to achieve, or can be transmitted directly via cellular networks, wireless or satellite links. There are myriad options for telemetry studies and at the core of this field is the flexibility of modern tag design (McConnell *et al.* 2010).

While tag sizes have reduced considerably in recent years, as processors and storage components have been miniaturised, the issue of tag impact remains, both through attachment procedures and the wearing of the tag. In most cases, tag attachment necessitates capturing the individual. For pinnipeds, this can be accomplished during haul-out periods, when individuals can be more easily accessed on ice floes, sandbars or beaches. For small cetaceans, however, telemetry device attachment involves considerably more effort as individuals generally must be caught at sea and capture and attachment procedures can be highly stressful for the individuals concerned (Norman *et al.* 2004; Eskesen *et al.* 2009). Typically, long-term tag attachment to small cetaceans involves stitching and pinning tags on to the skin, often through the dorsal fin (Geertsen *et al.* 2004). For larger whales, where capture is not an option, tags can be lodged into the blubber layer using penetrative spars (Double *et al.* 2014). There are also several telemetry systems that have been successfully attached via suction cups, which is considerably less stressful to the individual, although the tag's attachment duration is generally shorter than with other attachment methods (Tyack *et al.* 2006; Wisniewska *et al.* 2016, 2018). It is also important to consider the effect that any attachment will have on an individual's hydrodynamic profile. The size and configuration of the tag, the attachment method and the attachment site can all result in significant drag, thus influencing both how well the tag remains on the individual and the normal behaviour of the animal (Tudorache *et al.* 2014). Bulky or poorly located tags may cause injury (Irvine *et al.* 1982; Chilvers *et al.* 2001) and may induce higher energetic costs to the individual during normal behaviours. As such, there is potential to cause long-term effects on the physiology and health status of a tagged individual (Scott & Chivers 2009; Balmer *et al.* 2010; Walker *et al.* 2012; van der Hoop *et al.* 2014; Gendron *et al.* 2015). Considerations of the impact on animals have led to the development of single-pin tags and attachment positions that are less likely to be energetically costly while still providing sufficient data (Balmer *et al.* 2014).

Despite the challenges of attaching telemetry devices, telemetry studies have been used to provide critical information on marine mammal behaviour and distribution in relation to wind farms during both construction and operational phases (e.g. Tougaard *et al.* 2003; Lindeboom *et al.* 2011). In particular, the long-term data sets now available for both Grey Seal *Halichoerus grypus* and Harbour Seal *Phoca vitulina* have provided researchers with a detailed overview of these species' behaviour at sea. Russell *et al.* (2014) used data from Global Positioning System (GPS) tags to show that Harbour Seals travel to and spend time around wind farms in the Netherlands. Studies on how marine mammals use operating wind farms are limited; however, it appears in this case, that the monopiles provide foraging opportunities for the seals. Tagging data have also shown that Harbour Seals avoided a wind-farm site during the construction phase (Russell *et al.* 2016), and the observed tracks of Harbour Seals during pile driving were used to predict sound exposure levels for this species (Hastie *et al.* 2015). Although telemetry studies can be extremely informative, they have a number of limitations when it comes to wind-farm studies. It is difficult to obtain enough samples, meaning that the number of individuals tagged may cover only a small proportion of the actual population, and individual seal behaviour can be highly variable (Sharples *et al.* 2012). This makes it challenging to obtain sufficient data to make generalised conclusions. Also, marine mammal populations rarely occur in any number in close vicinity to wind-farm development sites, so individuals are less likely to use a specific site on a regular basis and tagged animals may not even occur close to a wind farm during the construction phase. What has been very relevant to wind-farm development, however, is that the information obtained from tagged marine mammals can form the basis of larger scale models on habitat selection (e.g. Aarts *et al.* 2008; 2016a). These models both assist in our understanding of how animals occur naturally and can also predict how anthropogenic activities may impact them.

Haul-out counts for seals

Abundance and habitat use of different seal species can be estimated by regularly monitoring their haul-out habitats. Counts of colonies can be made directly from land

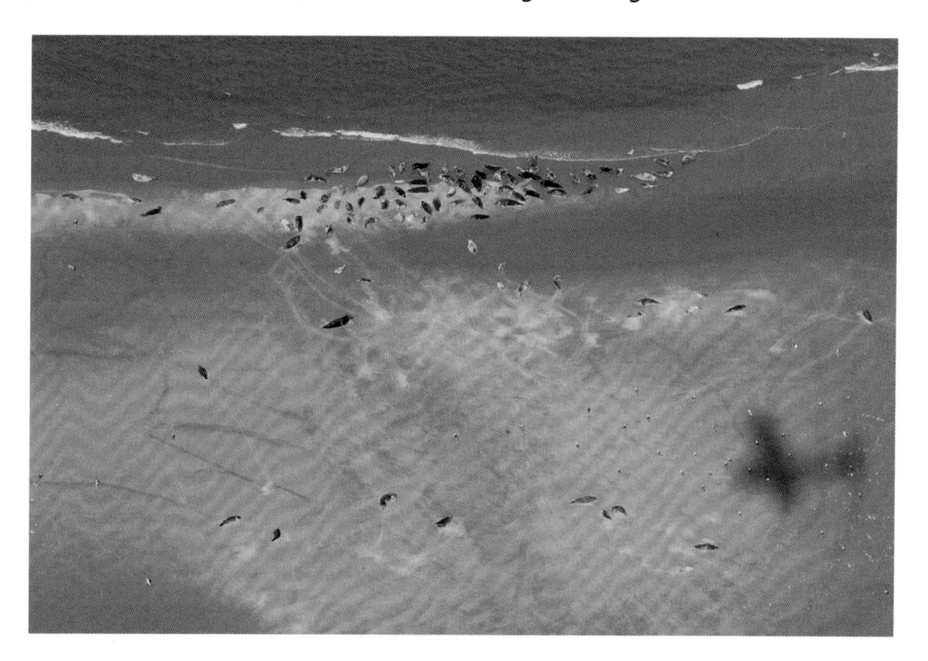

Figure 2.12 Aerial surveys recording a series of digital still images were used to determine the effects of the construction and operation of Scroby Sands wind farm on seals using the haul-out on the main Scroby Sands sandbank, around 2 km from the wind farm (Skeate *et al.* 2012). (Air Images Ltd)

or sea (e.g. Burn *et al.* 2006); however, it is more common to use aerial photographs of haul-out sites (Figure 2.12). Counts are made during specific times of the seals' life cycle, such as moulting or weaning, ideally when the highest and most stable numbers of seals are present (Thompson *et al.* 2005). Data from pup counts, for example, can be scaled up to estimate total population size using various statistical models (e.g. SCOS 2016).

Long-term studies have provided in-depth data on population sizes (e.g. Reijnders *et al.* 1997; Teilmann *et al.* 2010) that provide baselines that incorporate natural variation. However, it is challenging to link changes in a local seal population in terms of abundance, habitat use, reproductive success or population growth directly to developments of wind farms, or indeed any other anthropogenic activity. Nonetheless, long-term estimates of regional population size and vital parameters, in combination with data from telemetry studies, have been used to model the expected impact of wind-farm construction on Harbour Seals (Thompson *et al.* 2013). In areas where haul-out locations are close to wind-farm developments, direct effects on seal colonies have also been recorded. For example, the reduction in Harbour Seal numbers at Scroby Sands was most likely due to construction noise (Skeate *et al.* 2012) (Figure 2.12). During the pile-driving activities in the Nysted OWF in the Baltic Sea, a reduction in seals resting on land was also observed (Teilmann *et al.* 2006).

Cumulative impacts

One of the most important, and most difficult, questions in the study of marine mammals and wind-farm development is how to measure the impact of multiple sites. In addition, it is vital to consider other anthropogenic activities that may also affect marine mammal populations with ranges that overlap wind farms. Bycatch, compromised health status from contamination and stress, and other sources of underwater noise all contribute to the vulnerability of a population. It is not yet clear how additional impacts may exacerbate or contribute to populations that may already be under pressure. There are several large-scale wind-farm developments underway in Asia which are located within the habitat of the Indo-Pacific Humpback Dolphin *Sousa chinensis*, a species already threatened by a myriad of other anthropogenic activities (Wilson *et al.* 2008; Jefferson & Smith 2016) (Box 2.2). As yet, the effects of the introduction of wind farms as a novel source of disturbance to these at-risk populations are unknown.

Several modelling approaches have been developed to try to gain a better understanding of the issue of cumulative impacts. Nehls *et al.* (2019) describe two models that have been applied in the North Sea, where wind-farm development is extensive and ongoing. These are the interim Population Consequences of Disturbance (iPCoD) framework (King *et al.* 2015; Booth *et al.* 2017) and the DEPONS model (Nabe-Nielsen *et al.* 2018). Nabe-Nielsen & Harwood (2016) compare these modelling methodologies in detail.

Concluding remarks

Marine mammal monitoring methods applied in a wind farm vary depending on what is being assessed. During the construction phase, studies may focus on the immediate impact of noise on the avoidance behaviour of animals. For this, frequent or continuous monitoring before, during and after piling events is needed. This is especially important if any mitigation measures are taken and the effectiveness of these needs to be tested.

Once construction has ended, the focus should be on how the habitat use may have changed from a baseline, ideally including research on some of those parameters that could have an impact on the distribution of marine mammals (e.g. their prey). It is important to have a baseline, ideally over several years, to obtain an understanding of natural variations in occurrence. Currently most studies lack a good understanding of the distribution and abundance of marine mammals before the wind-farm construction.

With the number of wind farms being constructed, a standardised method for baseline survey data collection would be useful. The collection of the same type of data in different areas, and providing these data for synoptic meta-analyses, would allow a much better interpretation of the results than from one wind farm alone.

Wind farms are thought to have a direct impact, for example as a barrier to migrating animals. They could also serve as a type of refuge, with less boat traffic and shipping occurring inside the wind-farm area, as well as providing hard substrate that can serve as a type of artificial reef (Dannheim *et al.* 2019; Gill & Wilhelmsson 2019). How marine mammals adapt to this change in their ecosystem is poorly understood as the monitoring of operating wind farms is generally not a priority once construction is completed. This is especially important when it comes to decommissioning, the final phase of a wind-farm development, which, if undertaken, will involve noisy destruction activity (Topham & MacMillan 2017).

The expected lifetime of wind farms is generally a minimum of 25 years, but may be shorter, especially for the earliest 'old generation' sites (Topham & MacMillan 2017). Assuming that marine mammal populations have adapted to their altered habitat in the presence of wind farms, there could be impact if they are taken away. However, current evidence suggests that wholesale removal of subsurface structures will become unlikely, given its difficulty and the fact that there is increasing evidence of the value of wind farms to biodiversity as a result of the 'renewables to reefs' process. This may mean that there is a case for some sites becoming Marine Protected Areas in their own right (Perrow 2019).

Acknowledgements

The authors sincerely appreciate the editing skills and advice of Martin Perrow and thank him for his patience and for keeping this chapter on track. Many thanks also to Steve Geelhoed for his most useful comments on sections of this text.

References

Aarts, G., MacKenzie, M., McConnell, B., Fedak, M. & Matthiopoulos, J. (2008) Estimating space-use and habitat preference from wildlife telemetry data. *Ecography* 31: 140–160.

Aarts, G., Cremer, J., Kirkwood, R., van der Wal, J.T., Matthiopoulos, J. & Brasseur, S. (2016a) Spatial distribution and habitat preference of harbour seals (*Phoca vitulina*) in the Dutch North Sea. Report No. C118/16. Wageningen: IMARES, Wageningen University and Research Centre. Retrieved 1 April 2019 from http://dx.doi.org/10.18174/400306

Aarts, G., von Benda-Beckmann, A.M., Lucke, K., Sertlek, H.Ö., van Bemmelen, R., Geelhoed, S.C.V., Brasseur, S., Scheidat, M., Lam, F.-P.A. & Slabbekoorn, H. (2016b) Harbour porpoise movement strategy affects cumulative number of animals acoustically exposed to underwater explosions. *Marine Ecology Progress Series* 557: 261–275.

Abileah, R. (2002) Marine mammal census using space satellite imagery. *US Navy Journal of Underwater Acoustics* 52(3). Retrieved 7 December 2018 from http://jomegak.com/Publications/2002%20JUA%20-%20MARINE%20MAMMAL%20CENSUS%20USING%20SPACE%20SATELLITE%20IMAGERY.pdf

Aguilar de Soto, N., Johnson, M. P., & Madsen, P. T. (2008). Challenges in Population, Habitat Preference and Anthropogenic impact Assessment of Deep Diving Cetaceans. Selection Criteria For Marine Protected Areas For Cetaceans, 83. Retrieved 1 April 2019 from https://www.ascobans.org/sites/default/files/publication/MPA_Workshop2007_final.pdf#page=85

Anonymous (2016) LIFE+ SAMBAH project. Final report covering the project activities from 01/01/2010 to 30/09/2015. Reporting date 29/02/2016. Retrieved 1 April 2019 from http://www.sambah.org/SAMBAH-Final-Report-FINAL-for-website-April-2017.pdf

Bailey, H., Brookes, K.L. & Thompson, P.M. (2014) Assessing environmental impacts of offshore wind farms: lessons learned and recommendations for the future. *Aquatic Biosystems* 10: 1–13.

Balmer, B.C., Schwacke, L.H. & Wells, R.S. (2010) Linking dive behavior to satellite-linked tag condition for a bottlenose dolphin (*Tursiops truncatus*) along Florida's northern Gulf of Mexico coast. *Aquatic Mammals* 36: 1–8.

Balmer, B.C., Wells, R.S., Howle, L.E., Barleycorn, A.A., McLellan, W.A., Ann Pabst, D. & Zolman, E.S. (2014) Advances in cetacean telemetry: A review of single-pin transmitter attachment techniques on small cetaceans and development of a new satellite-linked transmitter design. *Marine Mammal Science* 30: 656–673.

Barlow, J. & Forney, K.A. (2007) Abundance and population density of cetaceans in the California current ecosystem. *Fishery Bulletin* 105: 509–526.

Barlow, J. & Taylor, B.T. (2005) Estimates of sperm whale abundance in the northeastern temperate Pacific from a combined acoustic and visual survey. *Marine Mammal Science* 21: 429–445.

Barlow, J., Oliver, C.W., Jackson, T.D. & Taylor, B.L. (1988) Harbor porpoise, *Phocoena phocoena*, abundance estimation for California, Oregon, and Washington: II. Aerial surveys. *Fishery Bulletin* 86: 433–444.

Berges, B.J.P., Geelhoed, S.C.V., Scheidat, M. & Tougaard, J. 2019. Quantifying harbour porpoise foraging behaviour in CPOD data identification, automatic detection and potential application. Report C039/19, Wageningen Marine Research (in preparation).

Berrow, S.D., O'Brien, J., Meade, R., Delarue, J., Kowarski, K., Martin, B., Moloney, J., Wall, D., Gillespie, D., Leaper, R., Gordon, J., Lee, A. & Porter, L. (2018) Acoustic surveys of cetaceans in the Irish Atlantic margin in 2015–2016: occurrence, distribution and abundance. Dublin: Department of Communications, Climate

Action & Environment and the National Parks and Wildlife Service (NPWS), Department of Culture, Heritage and the Gaeltacht. Retrieved 7 December 2018 from https://www.dccae.gov.ie/en-ie/natural-resources/topics/Oil-Gas-Exploration-Production/observe-programme/acoustic-survey/Pages/default.aspx

Booth, C., Harwood, J., Plunkett, R., Mendes, S. & Walker, R. (2017) Using the Interim PCoD framework to assess the potential impacts of offshore wind developments in eastern English waters on harbour porpoises in the North Sea. Natural England Joint Publication JP024. Retrieved 7 December 2018 from http://publications.naturalengland.org.uk/publication/4813967957950464

Bortolotto, G.A., Danilewicz, D., Hammond, P.S., Thomas, L. & Zerbini, A.N. (2017) Whale distribution in a breeding area: spatial models of habitat use and abundance of western South Atlantic humpback whales. *Marine Ecology Progress Series* 585: 213–227.

Brandt, M.J., Diederichs, A. & Nehls, G. (2009) Investigations into the effects of pile driving at the offshore wind farm Horns Rev II and the FINO III research platform. Report by BioConsult SH to DONG Energy. Retrieved 1 April 2019 from https://tethys.pnnl.gov/publications/investigations-effects-pile-driving-offshore-wind-farm-horns-rev-ii-and-fino-iii

BSH (2013) Untersuchung der Auswirkungen von Offshore-Windenergieanlagen auf die Meeresumwelt (StUK4), Hamburg und Rostock, Bundesamt für Seeschifffahrt und Hydrographie (BSH). Retrieved 7 December 2018 from http://www.energiewende-naturvertraeglich.de/fileadmin/Dateien/Dokumente/themen/Windenergie_Offshore/Artenschutz/StUK4_2013.pdf

Buckland, S. & York, A. (2017) Abundance estimation. In Würsig, B., Thewissen, J.G.M. & Kovacs, K.M. (eds) *Encyclopedia of Marine Mammals*. New York: Academic Press. pp. 1–6.

Buckland, S.T., Breiwick, J.M., Cattanach, K.L. & Laake, J.L. (1993) Estimated population size of the California gray whale. *Marine Mammal Science* 9: 235–249.

Buckland, S.T., Anderson, D.R., Burnham, K.P., Laake, J.L., Borchers, D.L. & Thomas, L. (2001) *Introduction to Distance Sampling: Estimating abundance of biological populations*. New York: Oxford University Press.

Buckland, S.T., Anderson, D.R., Burnham, K.P., Laake, J.L., Borchers, D.L. & Thomas, L. (2004) *Advanced Distance Sampling: Estimating abundance of biological populations*. New York: Oxford University Press.

Buckland, S.T., Burt, M.L., Rexstad, E.A., Mellor, M., Williams, A.E. & Woodward, R. (2012) Aerial surveys of seabirds: the advent of digital methods. *Journal of Applied Ecology* 49: 960–967.

Burn, D.M., Webber, M.A. & Udevitz, M.S. (2006) Application of airborne thermal imagery to surveys of Pacific walrus. *Wildlife Society Bulletin* 34: 51–58.

Cañadas, A., Desportes, G. & Borchers, D. (2004) The estimation of the detection function and g(0) for shortbeaked common dolphins (*Delphinus delphis*) using double platform data collected during the NASS-95 Faroese survey. *Journal of Cetacean Research and Management* 6: 191–198.

Carstensen, J., Henriksen, O.D. & Teilmann, J. (2006) Impacts on harbour porpoises from offshore wind farm construction: acoustic monitoring of echolocation activity using porpoise detectors (TPODs). *Marine Ecology Progress Series* 321: 295–308.

Certain, G., Masse, J., van Canneyt, O., Petitgas, P., Doremus, G., Santos, M.B. & Ridoux, V. (2011) Investigating the coupling between small pelagic fish and marine top predators using data collected from ecosystem-based surveys. *Marine Ecology Progress Series* 422: 23–39.

Chabot, D. & Bird, D.M. (2015) Wildlife research and management methods in the 21st century: where do unmanned aircraft fit in? *Journal of Unmanned Vehicle Systems* 3: 137–155.

Chen, C.F., Wang, W.J., Wu, C.H., Hu, W.C., Chen, N.C., Hwang, W.S., Chou, L.S., Guan, S., Lin, S.F. & Lin, D. (2017) Noise impact assessment on Indo-Pacific humpback dolphin in the habitat of the East Taiwan Strait during the first two pile driving activities of demonstration offshore wind farm. *Journal of the Acoustical Society of America* 141: 3922.

Chilvers, B.L., Corkeron, P.J., Blanshard, W.H., Long, T.R. & Martin, A.R. (2001) A new VHF tag and attachment technique for small cetaceans. *Aquatic Mammals* 27: 11–15.

Clarke, E.D., Spear, L.B., McCraken, M.L., Marques, F.F.C., Borchers, D.L., Buckland, S.T. & Ainley, D.G. (2003) Validating the use of generalized additive models and at-sea surveys to estimate size and temporal trends of seabird populations. *Journal of Applied Ecology* 40: 278–292.

Cubaynes, H.C., Fretwell, P.T., Bamford, C., Gerrish, L. & Jackson, J.A. (2019) Whales from space: four mysticete species described using new VHR satellite imagery. *Marine Mammal Science* 35: 466–491.

Dähne, M., Gilles, A., Lucke, K., Peschko, V., Adler, S., Krügel, K., Sundermeyer, J. & Siebert, U. (2013) Effects of pile-driving on harbour porpoises (*Phocoena phocoena*) at the first offshore wind farm in Germany. *Environmental Research Letters* 8: 025002.

Dannheim, J., Degraer, S., Elliot, M., Smyth, K. & Wilson, J.C. (2019) Seabed communities. In Perrow, M.R. (ed.) *Wildlife and Wind Farms, Conflicts and Solutions. Volume 3. Offshore: Potential effects*. Exeter: Pelagic Publishing. pp. 64–85.

Double, M.C., Andrews-Goff, V., Jenner, K.C.S., Jenner, M.N., Laverick, S.M., Branch, T.A. & Gales, N.J. (2014) Migratory movements of pygmy blue whales (*Balaenoptera musculus brevicauda*) between Australia and Indonesia as revealed by satellite telemetry. *PLoS ONE 9*(4): e93578. doi: 10.1371/journal.pone.0093578.

Embling, C.B. (2008) Predictive models of cetacean distributions off the west coast of Scotland (Doctoral dissertation, University of St Andrews).

Eskesen, I.G., Teilmann, J., Geertsen, B.M., Desportes, G., Riget, F., Dietz, R., Larsen, F. & Siebert, U. (2009) Stress level in wild harbour porpoises (*Phocoena phocoena*) during satellite tagging measured by respiration, heart rate and cortisol. *Journal of the Marine Biological Association of the United Kingdom* 89: 885–892.

Evans, P.G. (2008). Offshore wind farms and marine mammals: impacts & methodologies for assessing impacts. ECS Special publication series, (49). Retrieved 1 April 2019 from http://citeseerx.ist.psu.edu/viewdoc/download?doi=10.1.1.232.302&rep=rep1&type=pdf

Evans, P.G. & Hammond, P.S. (2004) Monitoring cetaceans in European waters. *Mammal Review* 34: 131–156.

Finneran, J.J. (2016) Auditory weighting functions and TTS/PTS exposure functions for marine mammals exposed to underwater noise. Technical Report No. 3026. San Diego, CA: Space and Naval Warfare Systems Center Pacific. Retrieved 1 April 2019 from https://apps.dtic.mil/dtic/tr/fulltext/u2/1026445.pdf

Fiori, L., Doshi, A., Martinez, E., Orams, M.B. & Bollard-Breen, B. (2017) The use of unmanned aerial systems in marine mammal research. *Remote Sensing* 9(6): 543.

Fretwell, P.T., Staniland, I.J. & Forcada. J. (2014) Whales from space: counting southern right whales by satellite. *PLoS ONE* 9(2): e88655. doi: 10.1371/journal.pone.0088655.

Fürstenau Oliveira, J.S., Georgiadis, G., Campello, S., Brandão, R.A. & Ciuti, S. (2017) Improving river dolphin monitoring using aerial surveys. *Ecosphere* 8(8): e01912. doi: 10.1002/ecs2.1912.

Geelhoed, S.C.V., Friedrich, E., Joost, M., Machiels, M.A.M. & Stöber, N. (2018) Gemini T-c: aerial surveys and passive acoustic monitoring of harbour porpoises. WMR Report No. C020/17. Wageningen: IMARES, Wageningen University and Research Centre. Retrieved 1 April 2019 from https://doi.org/10.18174/410635

Geertsen, B.M., Teilmann, J., Kastelein, R.A., Vlemmix, H.N.J. & Miller, L.A. (2004) Behaviour and physiological effects of transmitter attachments on a captive harbour porpoise (*Phocoena phocoena*). *Journal of Cetacean Research and Management* 6: 139–146.

Gendron, D., Serrano, I. M., de la Cruz, A. U., Calambokidis, J. & Mate, B. (2015). Long-term individual sighting history database: an effective tool to monitor satellite tag effects on cetaceans. *Endangered Species Research* 26(3): 235–241.

Gerrodette, T. (1987) A power analysis for detecting trends. *Ecology* 68: 1364–1372.

Gerrodette, T. & Forcada, J. (2005) Non-recovery of two spotted and spinner dolphin populations in the eastern tropical Pacific Ocean. *Marine Ecology Progress Series* 291: 1–21.

Gill, A.B. & Wilhelmsson, D. (2019) Fish. In Perrow, M.R. (ed.) *Wildlife and Wind Farms, Conflicts and Solutions. Volume 3. Offshore: Potential effects*. Exeter: Pelagic Publishing. pp. 86–111.

Gillespie, D. & Chappell, O. (2002) An automatic system for detecting and classifying the vocalisations of harbour porpoises. *Bioacoustics* 13: 37–61.

Gordon, J., Berrow, S.D., Rogan, E. & Fennelly, S. (1999). Acoustic and visual survey of cetaceans off the Mullet Peninsula, Co. Mayo. *Irish Naturalists' Journal* 26: 251–259.

Guan, S., Lin, T.H., Chou, L.S., Vignola, J., Judge, J. & Turo, D. (2015). Dynamics of soundscape in a shallow water marine environment: a study of the habitat of the Indo-Pacific humpback dolphin. *Journal of the Acoustical Society of America* 137: 2939–2949.

Hain, J.H.W., Ellis, S.L., Kenney, R.D. & Slay, C.K. (1999) Sightability of right whales in coastal waters of the southeastern United States with implications for the aerial monitoring program. In Garner, G.W., Amstrup, S.C., Laake, J.L.,

Manly, B.F.J., McDonald, L.L. & Robertson, D.G. (eds) *Marine Mammal Survey and Assessment Methods*. Rotterdam: Balkema. pp. 191–207.

Hammond, P.S. (2002) The assessment of marine mammal population size and status. In Evans, P.G.H. & Raga, J.A. (eds) *Marine Mammals: Biology and conservation*. London: Kluwer. pp. 269–291.

Hammond, P.S., Berggren, P., Benke, H., Borchers, D.L., Collet, A., Heide-Jorgensen, M.P., Heimlich, S., Hiby, A.R., Leopold, M.F. & Oien, N. (2002) Abundance of harbour porpoise and other cetaceans in the North Sea and adjacent waters. *Journal of Applied Ecology* 39: 361–376.

Hammond, P.S., Macleod, K., Berggren, P., Borchers, D.L., Burt, M.L., Cañadas, A., Desportes, G., Donovan, G.P., Gilles, A., Gillespie, D., Gordon, J., Hedley, S., Hiby, L., Kuklik, I., Leaper, R., Lehnert, K., Leopold, M., Lovell, P., Øien, N., Paxton, C., Ridoux, V., Rogan, E., Samarra, F., Scheidat, M., Sequeira, M., Siebert, U., Skov, H., Swift, R., Tasker, M.L., Teilmann, J., van Canneyt, O. & Vázquez, J.A. (2013) Distribution and abundance of harbour porpoise and other cetaceans in European Atlantic shelf waters: implications for conservation and management. *Biological Conservation* 164: 107–122.

Hammond, P.S., Lacey, C., Gilles, A., Viquerat, S., Boerjesson, P., Herr, H., Macleod, K., Ridoux, V., Santos, M., Scheidat, M. & Teilmann, J. (2017) Estimates of cetacean abundance in European Atlantic waters in summer 2016 from the SCANS-III aerial and shipboard surveys. Wageningen: IMARES, Wageningen University and Research Centre. Retrieved 7 December 2018 from https://library.wur.nl/WebQuery/wurpubs/fulltext/414756

Hastie, G.D., Russell, D.J.F., McConnell, B., Moss, S., Thompson, D. & Janik, V.M. (2015) Sound exposure in harbour seals during the installation of an offshore wind farm: predictions of auditory damage. *Journal of Applied Ecology* 52: 631–640.

Hiby, L. (1999) The objective identification of duplicate sightings in aerial survey for porpoise. In Garner, G.W., Amstrup, S.C., Laake, J.L., Manly, B.F.J., McDonald, L.L. & Robertson, D.G. (eds) *Marine Mammal Survey and Assessment Methods*. Rotterdam: Balkema. pp. 179–189.

Hiby, L. & Lovell, P. (1998) Using aircraft in tandem formation to estimate abundance of harbour porpoise. *Biometrics* 54: 1280–1289.

Hussey, N.E., Kessel, S.T., Aarestrup, K., Cooke, S.J., Cowley, P.D., Fisk, A.T., Harcourt, R.G., Holland, K.N., Iverson, S.J., Kocik, J.F., Mills, J.E. & Whoriskey, F.G. (2015) Aquatic animal telemetry: a panoramic window into the underwater world. *Science* 348: 1255642.

Hyrenbach, K.D. (2001) Albatross response to survey vessels: implications for studies of the distribution, abundance, and prey consumption of seabird populations. *Marine Ecology Progress Series* 212: 283–295.

Irvine, A.B., Wells, R.S. & Scott, M.D. (1982) An evaluation of techniques for tagging small odontocete cetaceans. *Fishery Bulletin* 80: 135–143.

Jameson, H., Reeve, E., Laubek, B. & Sittel, H. (2019) The nature of offshore wind farms. In Perrow, M.R. (ed.) *Wildlife and Wind Farms, Conflicts and Solutions. Volume 3. Offshore: Potential effects*. Exeter: Pelagic Publishing. pp. 1–29.

Jean, C., Ciccione, S., Ballorain, K., Georges, J-Y. & Bourjea, J. (2010) Ultralight aircraft surveys reveal marine turtle population increases along the west coast of Reunion Island. *Oryx* 44: 223–229.

Jefferson, T.A. & Smith, B.D. (2016) Re-assessment of the conservation status of the Indo-Pacific humpback dolphin (*Sousa chinensis*) using the IUCN Red List criteria. *Advances in Marine Biology* 73: 1–26.

Kaschner, K., Quick, N.J., Jewell, R., Williams, R. & Harris, C.M. (2012) Global coverage of cetacean line-transect surveys: status quo, data gaps and future challenges. *PLoS ONE* 7(9): e44075. doi: 10.1371/journal.pone.0044075.

King, S.L., Schick, R.S., Donovan, C., Booth, C.G., Burgman, M., Thomas, L. & Harwood, J. (2015) An interim framework for assessing the population consequences of disturbance. *Methods in Ecology and Evolution* 6: 1150–1158.

Koski, W.R., Allen, T., Ireland, D., Buck, G., Smith, P.R., Macrander, A.M., Halick, M.A., Rushing, C., Sliwa, D.J. & McDonald, T.L. (2009) Evaluation of an unmanned airborne system for monitoring marine mammals. *Aquatic Mammals* 35: 348–358.

Laake, J.L., Calambokidis, J., Osmek, S.D. & Rugh, D.J. (1997) Probability of detecting harbor porpoise from aerial surveys: estimating g(0). *Journal of Wildlife Management* 61: 63–75.

LaRue, M.A., Rotella, J.J., Siniff, D.B., Garrott, R.A., Stauffer, G.E., Porter, C.C. & Morinet, P.J. (2011) Satellite imagery can be used to detect variation in abundance of Weddell seals (*Leptonychotes weddellii*) in Erebus Bay, Antarctica. *Polar Biology* 34: 1727–1737.

Le Double, S. (2018) Diurnal pattern and seasonality in the acoustic behavior of Indo-Pacific finless porpoise (*Neophocaena phocaenoides)* in Hong Kong Special Administrative Region waters. MSc thesis, University of Amsterdam.

Li, S., Wang, D., Wang, K., Taylor, E.A., Cros, E., Shi, W., Wang, Z., Fang, L., Chen, Y. & Kong, F. (2012) Evoked-potential audiogram of an Indo-Pacific humpback dolphin (*Sousa chinensis*). *Journal of Experimental Biology* 215: 3055–3063.

Linchant, J., Lisein, J., Semeki, J., Lejeune, P. & Vermeulen, C. (2015) Are unmanned aircraft systems (UASs) the future of wildlife monitoring? A review of accomplishments and challenges. *Mammal Review* 45: 239–252.

Lindeboom, H.J., Kouwenhoven, H.J., Bergman, M.J.N., Bouma, S., Brasseur, S.M.J.M., Daan, R., Fijn, R.C., de Haan, D., Dirksen, S., van Hal, R., Lambers, R.H.R., ter Hofstede, R., Krijgsveld, K.L., Leopold, M. & Scheidat, M. (2011) Short-term ecological effects of an offshore wind farm in the Dutch coastal zone; a compilation. *Environmental Research Letters* 6(3): 035101.

Mackenzie, M.L., Scott-Hayward, L.A., Oedekoven, C.S., Skov, H., Humphreys, E. & Rexstad, E. (2013) Statistical Modelling of Seabird and Cetacean data: Guidance Document. Report SB9 (CR/2012/05), Centre for Research into Ecological and Environmental Modelling, University of St Andrews, St Andrews. Retrieved 1 April 2019 from https://www.researchgate.net/profile/Lindesay_Scott-Hayward/publication/276272284_Statistical_Modelling_of_Seabird_and_Cetacean_Data_Guidance_Document/links/5554636908ae6943a86f4f4b.pdf

Madsen, P.T., Johnson, M., Miller, P.J.O., Aguilar de Soto, N., Lynch, J. & Tyack, P. (2006) Quantitative measures of air-gun pulses recorded on sperm whales (*Physeter macrocephalus*) using acoustic tags during controlled exposure experiments. *Journal of the Acoustical Society of America* 120: 2366–2379.

Marques, T.A., Thomas, L., Martin, S.W., Mellinger, D.K., Ward, J.A., Moretti, D.J., Harris, D. & Tyack, P.L. (2013) Estimating animal population density using passive acoustics. *Biological Reviews* 88: 287–309.

Massachusetts Clean Energy Center (2018) 2018 Massachusetts offshore wind workforce assessment. Retrieved 1 April 2019 from http://files.masscec.com/2018%20MassCEC%20Workforce%20Study.pdf

McConnell, B.J., Fedak, M.A., Hooker, S.K. & Patterson, T. (2010) Telemetry. In Boyd, I.L., Bowen, W.D. & Iverson, S.J. (eds) *Marine Mammal Ecology and Conservation: A handbook of techniques.* New York: Oxford University Press. pp. 222–241.

McDonald, M.A. & Fox, C.G. (1999) Passive acoustic methods applied to fin whale population density estimation. *Journal of the Acoustical Society of America* 105: 2643.

McDonald, M.A. & Moore, S.E. (2002) Calls recorded from North Pacific right whales (*Eubalaena japonica*) in the eastern Bering Sea. *Journal of Cetacean Research and Management* 4: 261–266.

McMahon, C.R., Howe, H., van den Hoff, J., Alderman, R., Brolsma, H. & Hindell, M.A. (2014) Satellites, the all-seeing eyes in the sky: counting elephant seals from space. *PLoS ONE* 9(3): e92613. doi: 10.1371/journal.pone.0092613.

Mellinger, D.K., and J. Barlow. 2003. Future directions for marine mammal acoustic surveys: Stock assessment and habitat use. Report of a workshop held in La Jolla, CA, November 20–22, 2002. Technical contribution No. 2557, NOAA Pacific Marine Environmental Laboratory, Seattle, WA. Retrieved 1 April 2019 from https://repository.library.noaa.gov/view/noaa/11030/noaa_11030_DS1.pdf

Moreland, E.E., Cameron, M.F., Angliss, R.P. & Boveng, P.L. (2015) Evaluation of a ship-based unoccupied aircraft system (UAS) for surveys of spotted and ribbon seals in the Bering Sea pack ice. *Journal of Unmanned Vehicle Systems* 3: 114–122.

Moxley, J.H., Bogomolni, A., Hammill, M.O., Moore, K.M.T., Polito, M.J., Sette, L., Sharp, W.B., Waring, G.T., Gilbert, J.R., Halpin, P.N. & Johnston, D.W. (2017) Google haul out: earth observation imagery and digital aerial surveys in coastal wildlife management and abundance estimation. *BioScience* 67: 760–768.

Nabe-Nielsen, J. & Harwood, J. (2016) Comparison of the iPCoD and DEPONS models for modelling population consequences of noise on harbour porpoises. Scientific Report from DCE – Danish Centre for Environment and Energy, Aarhus University. No. 186. Retrieved 1 April 2019 from http://dce2.au.dk/pub/SR186.pdf

Nabe-Nielsen, J., van Beest, F.M., Grimm, V., Sibly, R.M., Teilmann, J. & Thompson, P.M. (2018) Predicting the impacts of anthropogenic disturbances on marine populations. *Conservation Letters* 11: e12563.

National Marine Fisheries Service (USA) (NMFS) (2018) 2018 Revision to: Technical guidance for assessing the effects of anthropogenic sound on marine mammal hearing (version 2.0): underwater thresholds for onset of permanent and temporary threshold shifts. NOAA Technical Memorandum NMFS-OPR-59. Silver Spring, MD: National Oceanic and Atmospheric Administration, US Department of Commerce. Retrieved 1 April 2019 from https://www.fisheries.noaa.gov/national/marine-mammal-protection/marine-mammal-acoustic-technical-guidance

Nehls, G., Harwood, A.J.P & Perrow, M.R. (2019) Marine mammals. In Perrow, M.R. (ed.) *Wildlife and Wind Farms, Conflicts and Solutions. Volume 3. Offshore: Potential effects*. Exeter: Pelagic Publishing. pp. 112–141.

Norman, S.A., Hobbs, R.C., Foster, J., Schroeder, J.P. & Townsend, F.I. (2004) A review of animal and human health concerns during capture–release, handling and tagging of odontocetes. *Journal of Cetacean Research and Management* 6: 53–62.

Nuuttila, H. (2013) Identifying foraging behaviour of wild bottlenose dolphins (*Tursiops truncatus*) and harbour porpoises (*Phocoena phocoena*) with static acoustic dataloggers. *Aquatic Mammals* 39: 147–161.

Øien, N. (2009) Distribution and abundance of large whales in Norwegian and adjacent waters based on ship surveys 1995–2001. *NAMMCO Scientific Publications* 7: 31–47.

Palka, D. (2005) Aerial surveys in the northeast Atlantic: estimation of g(0). In Thomsen, F., Ugarte, F. & Evans, P.G.H. (eds) Estimation of g(0) in line-transect surveys of cetaceans. *European Cetacean Society Newsletter* (Special Issue) 44: 12–17.

Palka, D.L. & Hammond, P.S. (2001) Accounting for responsive movement in line transect estimates of abundance. *Canadian Journal of Fisheries and Aquatic Sciences* 58: 777–787.

Perrow, M.R. (2019) A synthesis of effects and impacts. In Perrow, M.R. (ed.) *Wildlife and Wind Farms, Conflicts and Solutions. Volume 3. Offshore: Potential effects*. Exeter: Pelagic Publishing. pp. 235–277.

Pike, D.G., Gunnlaugsson, Th., Vikingsson, G.A., Desportes, G. & Bloch, D. (2009) Estimates of the abundance of minke whales (*Balaenoptera acutorostrata*) from Faroese and Icelandic NASS shipboard surveys. *NAMMCO Scientific Publications* 7: 81–93.

Platonov, N.G., Mordvintsev, I.N. & Rozhnov, V.V. (2013) The possibility of using high resolution satellite imagery for detection of marine mammals. *Biology Bulletin* 40: 197–205.

Pomeroy, P., O'Connor, L. & Davies, P. (2015) Assessing use of and reaction to unmanned aerial systems in gray and harbor seals during breeding and molt in the UK. *Journal of Unmanned Vehicle Systems* 3: 102–113.

Read, A.J. (2018) Biotelemetry. In Würsig, B., Thewissen, J.G.M. & Kovacs, K.M. (eds) *Encyclopedia of Marine Mammals*, 3rd edn. New York: Academic Press. pp. 103–106.

Reijnders, P.J.H., Ries, E.H., Tougaard, J., Norgaard, N., Heidemann, G., Schwarz, J.A., Vareschi, E. & Traut, I.M. (1997) Population development of harbour seals *Phoca vitulina* in the Wadden Sea after the 1988 virus epizootic. *Journal of Sea Research* 38: 161–168.

Rowat, D., Gore, M., Meekan, M.G., Lawler, I.R. & Bradshaw, C.J.A. (2009) Aerial survey as a tool to estimate whale shark abundance trends *Journal of Experimental Marine Biology and Ecology* 368: 1–8.

Russell, D.J.F., Brasseur, S.M.J.M., Thompson, D., Hastie, G.D., Janik, V.M., Aarts, G., McClintock, B.T., Matthiopoulos, J., Moss, S.E.W. & McConnell, B. (2014) Marine mammals trace anthropogenic structures at sea. *Current Biology* 24: R638–R639.

Russell, D.J., Hastie, G.D., Thompson, D., Janik, V.M., Hammond, P.S., Scott-Hayward, L.A., Matthiopoulos, J., Jones, E.L., McConnell, B.J. & Votier, S. (2016) Avoidance of wind farms by harbour seals is limited to pile driving activities. *Journal of Applied Ecology* 53: 1642–1652.

Schaffeld, T., Bräger, S., Gallus, A., Dähne, M., Krügel, K., Herrmann, A., Jabbusch, M., Ruf, T., Verfuß, U.K., Benke, H. & Koblitz, J.C. (2016) Diel and seasonal patterns in acoustic presence and foraging behaviour of free-ranging harbour porpoises. *Marine Ecology Progress Series* 547: 257–272.

Scheidat, M., Tougaard, J., Brasseur, S., Carstensen, J., van Polanen Petel, T., Teilmann, J. & Reijnders, P. (2011) Harbour porpoise (*Phocoena phocoena*) and wind farms: a case study in the Dutch North Sea. *Environmental Research Letters* 6: 025102.

Scott, B.E., Sharples, J., Ross, O.N., Wang, J., Pierce, G.J. & Camphuysen, C.J. (2010) Sub-surface hotspots in shallow seas: fine scale limited locations of top predator foraging habitat indicated by tidal mixing and sub-surface chlorophyll. *Marine Ecology Progress Series* 408: 207–226.

Scott, M.D. & Chivers, S.J. (2009) Movements and diving behavior of pelagic spotted dolphins. *Marine Mammal Science* 25: 137–160.

Sharples, R.J., Moss, S.E., Patterson, T.A. & Hammond, P.S. (2012) Spatial variation in foraging behaviour of a marine top predator

(*Phoca vitulina*) determined by a large-scale satellite tagging program. *PLoS ONE*, 7(5): e37216.

Skeate, E.R., Perrow, M.R. & Gilroy, J.J. (2012) Likely effects of construction of Scroby Sands offshore wind farm on a mixed population of harbour *Phoca vitulina* and grey *Halichoerus grypus* seals. *Marine Pollution Bulletin* 64: 872–881.

Slavik, K., Lemmen, C., Zhang, W., Kerimoglu, O., Klingbeil, K. & Wirtz, K.W. (2018) The large-scale impact of offshore wind farm structures on pelagic primary productivity in the southern North Sea. *Hydrobiologia* 1–19, doi: 10.1007/s10750-018-3653-5.

Sousa-Lima, R.S., Norris, T.F., Oswald, J.N. & Fernandes, D.P. (2013) A review and inventory of fixed autonomous recorders for passive acoustic monitoring of marine mammals. *Aquatic Mammals* 39: 23–53.

Southwell, C. (2005) Response of seals and penguins to helicopter surveys over the pack ice off East Antarctica. *Antarctic Science* 17: 328–334.

Special Committee on Seals (SCOS) (2016) Scientific advice on matters related to the management of seal populations, 2016. Sea Mammal Research Unit, University of St Andrews. Retrieved 7 December 2018 from http://www.smru.st-andrews.ac.uk/files/2017/04/SCOS-2016.pdf

Sveegaard, S., Teilmann, J. & Galatius, A. (2013) Abundance survey of harbour porpoises in Kattegat, Beld Seas and the Western Baltic, July 2012. Note from DCE – Danish Centre for Environment and Energy. 12 June 2013. Retrieved 7 December 2018 from http://dce.au.dk/fileadmin/dce.au.dk/Udgivelser/Abundance_survey_of_harbour_porpoises_2012_20130612.pdf

Taylor, B.L., Martinez, M., Gerrodette, T., Barlow, J. & Hrovat, Y.N. (2007) Lessons from monitoring trends in abundance of marine mammals. *Marine Mammal Science* 23: 157–175.

Teilmann, J., Tougaard, J., Carstensen, J., Dietz, R. & Tougaard, S. (2006) Summary on seal monitoring 1999–2005 around Nysted and Horns Rev Offshore Wind Farms. Report to Energi E2 A/S and Vattenfall A/S. Report No. 2389313244. National Environmental Research Institute. Retrieved 1 April 2019 from https://corporate.vattenfall.dk/globalassets/danmark/om_os/horns_rev/summary-on-harbour-porpoise-m.pdf

Teilmann, J., Riget, F. & Härkönen, T. (2010) Optimising survey design for Scandinavian harbour seals: population trend as an ecological quality element. *ICES Journal of Marine Science* 67: 952–958.

Teilmann, J., Christiansen, C.T., Kjellerup, S., Dietz, R. & Nachman, G. (2013) Geographic, seasonal and diurnal surface behavior of harbor porpoises. *Marine Mammal Science* 29: E60–E76.

Thomas, L., Buckland, S.T., Rexstad, E.A., Laake, J.L., Strindberg, S., Hedley, S.L., Bishop, J.R.B., Marques, T.A. & Burnham, K.P. (2010) Distance software: design and analysis of distance sampling surveys for estimating population size. *Journal of Applied Ecology* 47: 5–14.

Thomson, J.A., Cooper, A.B., Burkholder, D.A., Heithaus, M.R. & Dill, L.M. (2012) Heterogeneous patterns of availability for detection during visual surveys: spatiotemporal variation in sea turtle dive-surfacing behaviour on a feeding ground. *Methods in Ecology and Evolution* 3: 378–387.

Thompson, D., Lonergan, M. & Duck, C.D. (2005) Population dynamics of harbour seals (*Phoca vitulina*) in England: growth and catastrophic declines. *Journal of Applied Ecology* 42: 638–648.

Thompson, P.M., Hastie, G.D., Nedwell, J., Barham, R., Brookes, K.L., Cordes, L.S., Bailey, H. & McLean, N. (2013) Framework for assessing impacts of pile-driving noise from offshore wind farm construction on a harbour seal population. *Environmental Impact Assessment Review* 43: 73–85.

Topham, E. & McMillan, D. (2017) Sustainable decommissioning of an offshore wind farm. *Renewable Energy* 102: 470–480.

Tougaard, J., Tougaard, S., Jensen, T., Ebbesen, I. & Teilmann, J. (2003) Satellite tracking of harbour seals on Horns Reef: use of the Horns Reef wind farm area and the North Sea. Report No. NEI-DK-4700. Technical Report to Techwise A/S. Biological Papers from the Fisheries and Maritime Museum, Esbjerg. No. 3. Retrieved 7 December 2018 from https://inis.iaea.org/collection/NCLCollectionStore/_Public/37/113/37113071.pdf

Tougaard, J., Poulsen, L.R., Amundin, M., Larsen, F., Rye, J. & Teilmann, J. (2006) Detection function of T-PODs and estimation of porpoise densities. In *Proceedings of the Workshop Static Acoustic Monitoring of Cetaceans*, 20th Annual Meeting of the European Cetacean Society, Gdnyia, Poland, April 2006. ECS Newsletter Special Issue No. 46: 7–14.

Tougaard, J., Carstensen, J., Teilmann, J., Skov, H. & Rasmussen, P. (2009) Pile driving zone of responsiveness extends beyond 20 km for harbor porpoises (*Phocoena phocoena* (L.)). *Journal of the Acoustical Society of America* 126: 11–14.

Tregenza, N. (1998) Chapter 8.3. Site acoustic monitoring for cetaceans – a self-contained sonar click detector. In Tasker, M.L. & Weir, C. (eds) *Proceedings of the Seismic and Marine Mammals Workshop*, London, 23–25 June 1998. pp. 1–5. Retrieved 10 October 2018 from http://www.smru.st-and.ac.uk/seismic/seismicintro.htm

Tudorache, C., Burgerhout, E., Brittijn, S. & van den Thillart, G. (2014) The effect of drag and attachment site of external tags on swimming eels: experimental quantification and evaluation tool. *PLoS ONE* 9(11): e112280. doi: 10.1371/journal.pone.0112280.

Turner, J., Wood, J., Porter, L. & Lee, A. (2019) The coastal acoustic buoy: a new mitigation tool. In *Proceedings of Western Pacific Commission for Acoustics (WESPAC) 2018*, New Delhi, India, 11–15 November 2018. Retrieved 10 December 2018 from http://www.wespac2018.org.in/index.php

Turnock, B.J. & Quinn, T.J. II (1991) The effect of responsive movement on abundance estimation using line transect sampling. *Biometrics* 47: 701–715.

Tyack, P.L., Johnson, M., Soto, N.A., Sturlese, A. & Madsen, P.T. (2006) Extreme diving of beaked whales. *Journal of Experimental Biology* 209: 4238–4253.

van der Hoop, J.M., Fahlman, A., Hurst, T., Rocho-Levine, J., Shorter, K.A. & Moore, M.J. (2014) Bottlenose dolphins modify behavior to reduce metabolic effect of tag attachment. *Journal of Experimental Biology*. 217: 4229–4236.

van Parijs, S.M., Clark, C.W., Sousa-Lima, R.S., Parks, S.E., Rankin, S., Risch, D. & Van Opzeeland, I.C. (2009) Management and research applications of real-time and archival passive acoustic sensors over varying temporal and spatial scales. *Marine Ecology Progress Series* 395: 21–36.

van Polanen Petel, T., Geelhoed, S.C.V. & Meesters, H.W.G. (2012) Harbour porpoise occurrence in relation to the Prinses Amaliawindpark. Report No. C177/10. Wageningen: IMARES, Wageningen University and Research Centre. Retrieved 1 April 2019 from https://library.wur.nl/WebQuery/wurpubs/fulltext/245231

Verfuβ, U., Sparling, C., Arnot, C., Judd, A. & Coyle, M. (2015) Review of offshore wind Farm impact monitoring and mitigation with regard to marine mammals. *Advances in Experimental Medicine and Biology* 875: 1175–1182.

Walker, K., Trites, A., Haulena, M. & Weary, D. (2012) A review of the effects of different marking and tagging techniques on marine mammals. *Wildlife Research* 39: 15–30.

Watlington, F. (1979) *How to Build & Use Low Cost Hydrophones*. Blue Ridge Summit, PA: TAB Books.

Wang, Z.T., Nachtigall, P.E., Akamatsu, T., Wang, K.X., Wu, Y.P., Liu, J.C., Duan, G.Q., Cao, H.J. & Wang, D. (2015) Passive acoustic monitoring the diel, lunar, seasonal and tidal patterns in the biosonar activity of the Indo-Pacific humpback dolphins (*Sousa chinensis*) in the Pearl River estuary, China. *PLoS ONE 10*(11): e0141807. doi: 10.1371/journal.pone.0141807.

Weel, S.M.H., Geelhoed, S.C.V., Tulp, I. & Scheidat, M. (2018) Feeding behaviour of harbour porpoises (*Phocoena phocoena*) in the Ems estuary. *Lutra* 61: 137–152.

Westgate, A.J., Read, A.J. & Gaskin, D.E. (1995) Diving behaviour of harbour porpoises, *Phocoena phocoena*. *Canadian Journal of Fisheries and Aquatic Sciences* 52: 1064–1073.

Williamson, L.D., Brookes, K.L., Scott, B.E., Graham, I.M., Bradbury, G., Hammond, P.S. & Thompson, P.M. (2016) Echolocation detections and digital video surveys provide reliable estimates of the relative density of harbour porpoises. *Methods in Ecology and Evolution* 7: 762–769.

Wilson, B., Porter, L., Gordon, J., Hammond, P. S., Hodgins, N., Wei, L. & Wu, Y. P. (2008) A decade of management plans, conservation initiatives and protective legislations for Chinese white dolphin (*Sousa chinensis*): an assessment of progress and recommendations for future management strategies in the Pearl River Estuary, China. Retrieved 19 June 2019 from https://www.researchgate.net/publication/278037321_A_Decade_of_Management_Plans_Conservation_Initiatives_and_Protective_Legislation_for_Chinese_White_Dolphin_Sousa_chinensis_An_Assessment_of_Progress_and_Recommendations_for_Future_Management_Strategie

Wisniewska, D.M., Johnson, M., Teilmann, J., Rojano-Doñate, L., Shearer, J., Sveegaard, S., Miller, L.A., Siebert, U. & Madsen, P.T. (2016) Ultra-high foraging rates of Harbor Porpoises make them vulnerable to anthropogenic disturbance. *Current Biology* 26: 1441–1446. https://doi.org/10.1016/j.cub.2016.03.069

Wisniewska, D.M., Johnson, M., Teilmann. J., Siebert, U., Galatius, A., Dietz, R. & Madsen, P.T. (2018) High rates of vessel noise disrupt foraging in wild harbour porpoises (Phocoena phocoena). *Proceedings of the Royal Society B* 285: 20172314. http://dx.doi.org/10.1098/rspb.2017.2314

CHAPTER 3

Surveying seabirds

ANDY WEBB and GEORG NEHLS

Summary

Seabird surveys have a long history and have been adapted for use in relation to wind farms. At first, boat-based and aerial observer surveys were used, but digital aerial survey techniques have been developed that offer enhancements in terms of speed of survey and, in some cases, cost, accuracy and safety. With specific stand-alone examples presented in boxes, this chapter describes how these traditional and new methods may be used to provide abundance estimates for characterisation and impact assessment, and to monitor wind-farm effects and impacts as accurately and precisely as possible. Notwithstanding that it is important to tailor the methods to the specific circumstances and the questions being asked, this chapter concludes with a series of guiding principles on how to conduct surveys and how data should be analysed, as follows: (1) the size of the survey area must extend beyond the distance at which seabirds may respond to the wind farm, (2) aerial surveys are best suited to large study areas and species sensitive to vessels, (3) boat-based surveys are best suited to provide detailed information on seabird behaviour and flight height, (4) the proportion of transect area should sample at least 10% of the study area, (5) surveys should be carried out once or twice per month as an absolute minimum and generally be maintained throughout the annual cycle, (6) more than 2 years of surveys will be required to describe inter-annual variation, (7) where more than one site is present, projects should combine monitoring efforts over a large scale to understand cumulative effects, (8) power analysis should be used to estimate the ability of a given survey design to detect change, and (9) the use of sophisticated modelling techniques is likely to be needed to describe changes in spatiotemporal abundance patterns of seabirds.

Introduction

The census of seabirds at sea has been challenging since the first documented attempt in the North Atlantic in the 1920s (Jespersen 1924). Many seabird species are widely dispersed over a large sea area while others may be aggregated, sometimes close to the coast or in very shallow water (e.g. Skov *et al.* 1995; Stone *et al.* 1995). Seabirds are generally highly mobile and respond to changes in the location of their prey with consequent impacts on their spatial and temporal distribution patterns at sea. Short-term high-density foraging aggregations of one or more species often form at patches of abundant prey (Camphuysen & Webb 1999; Fauchald *et al.* 2000). Researchers charged with providing an unbiased estimate of the abundance of seabirds in a specific area to assess their sensitivity to specific threats, such as oil pollution, fishing or offshore renewable energy developments, thus face a somewhat daunting task.

A number of projects began during the 1970s to attempt to map the offshore distributions of seabirds, such as those by Gould (1974) in the eastern Pacific Ocean, Briggs *et al.* (1978) in the Southern California Bight, Brown *et al.* (1975) in the western Atlantic Ocean and the Natural Environment Research Council (NERC) (1977) in the eastern Atlantic. In the UK, the original drive to begin seabird surveys at sea came from the need for informed advice for the offshore oil and gas industry. The survey methods used for these projects advanced the standardisation of seabird abundance from the rudimentary 'number of birds per hour' to 'number of birds per square kilometre' as density. The advent of the Seabirds at Sea programme of the Nature Conservancy Council in 1979 in the North Sea resulted in a review of the potential survey methods for census of seabirds from ships and recommended a standardised approach to improve comparability between different and overlapping projects (Tasker *et al.* 1984). Most of these programmes used 'ships of opportunity', such as research ships engaged in other scientific pursuits. While these were an economical way to gather data, they often resulted in gaps in potentially critical sea areas. Many investigators resorted to the expense of dedicated ship charter to fill gaps, but others started to use faster moving and thus potentially cheaper aircraft for this purpose, employing strip-transect methods (e.g. Briggs *et al.* 1978; Gould *et al.* 1978).

The plea of Tasker *et al.* (1984) for standardisation of methods was mostly successful in northern European waters, with projects in the continental shelf waters of the Netherlands (Camphuysen & Leopold 1994), Germany (e.g. Garthe *et al.* 1995; Diederichs *et al.* 2002), Denmark (Skov, unpublished report 1992), Belgium (Offringa *et al.* 1995) and Norway (Barrett *et al.*2004) all adopting the basic foundations of the Tasker *et al.* (1984) method, comprising observations divided into 10 minute recording intervals, a 300 m wide strip transect and a 'snapshot' method for flying birds. Webb & Durinck (1992) described an addition to this standard in the form of a subdivision of the strip transect for birds sitting on the water to enable detection efficiency to be calculated. This provided compatibility with line-transect methods (Buckland *et al.* 2001), which some early investigators considered to be too cumbersome for use at sea (Wiens *et al.* 1978).

One of the outcomes of the standardisation of survey techniques was the formation of the European Seabirds at Sea Database Co-ordinating Group (ESAS) in 1991 (Reid & Camphuysen 1998), which focused initially on combining compatible data from different research groups in the North Sea, but has since expanded to much of the north-east Atlantic waters. The database has spawned many analyses, such as those by Camphuysen and Leopold (1994), Stone *et al.* (1995) and Skov *et al.* (2007). The database was not restricted to boat-based data, but also hosted visual aerial survey data, again collected using standardised methods. Initially, these standards, defined in Komdeur *et al.* (1992), were for strip transects, but owing to evidence that observers could not detect birds in the outer regions of these strips, surveyors adopted line-transect methods with distance estimation (Kahlert *et al.* 2000).

Beginning in the late 1990s, proposals to build offshore wind farms (OWFs) in European waters resulted in a need for dedicated surveys, often carried out by consultancies aboard vessels specifically chartered for the purpose, to meet the legislative requirements of

Environmental Impact Assessments (EIAs) to characterise the seabird communities associated with these developments. The surveys were generally undertaken at a much finer scale than normally considered for larger databases, such as ESAS. Nevertheless, there was still thought to be value in these surveys using ESAS-compatible methods, while incorporating additional approaches such as collecting data on flight heights, and more explicit standards to assist the offshore developers' assessment and their consultants. The Collaborative Offshore Wind Research Into the Environment (COWRIE) initiative, set up by the UK Crown Estate and supported by industry, commissioned Camphuysen *et al.* (2004) to provide a relatively detailed methodology for both boat (ship)-based and aerial surveys. The authors recommended an extension of the boat-based methods contained in Webb & Durinck (1992) and the visual aerial survey methods of Kahlert *et al.* (2000); and that EIAs should be based on the results from a combination of both types of survey. ESAS was thus retained as a focus for promoting common standards for data collection and initiated training programmes such as that offered by the Joint Nature Conservation Committee (JNCC 2017).

As wind farms started to be built in European offshore waters, mainly in the early 2000s onwards, attention turned to post-consent monitoring methods. In the UK, there were significant safety concerns and some methodological concerns about the prospect of aerial surveys being commissioned to fly between wind turbines at an altitude of 75 m above sea level (a.s.l.). This spawned a new approach to offshore seabird surveys for wind farms using digital video and stills cameras. These methods were accepted as a valid alternative to visual aerial survey methods in the UK (Thaxter & Burton 2009; Buckland *et al.* 2012) and Germany (BSH 2013). Initially, it was only possible to identify a relatively small proportion of birds to species level (Johnston *et al.* 2015), but new technology and techniques brought about a considerable improvement in image quality, resulting in identification rates of often more than 95% of birds to species level (e.g. Weiß *et al.* 2016; Webb *et al.* 2017; Zydelis *et al.* 2019). As a result of improved identification standards and the benefit of an auditable record of images, developers and regulators saw little need to commission boat-based surveys as part of either characterisation surveys or post-consent monitoring surveys. As a consequence, relatively few boat-based surveys are now undertaken in the UK, although such surveys are capable of delivering high-quality data in relation to determining the effects of wind farms on seabirds (and marine mammals) in the UK (Vallejo *et al.* 2017; Harwood *et al.* 2017; Perrow 2019) as well as in other North Sea countries such as Belgium (Vanermen *et al.* 2015a; 2015b).

Indeed, Vanermen & Steinen (2019) state that 'when aiming to monitor overall seabird responses to OWFs, there is not one method undeniably better than the other … In the end, the chosen method is likely to be a compromise between the study goals and the budget and logistics available to reach those goals.' These are important considerations as offshore wind, and thus the need to assess and monitor its effects upon seabirds, expands around the globe (e.g. Normandeau Associates 2013; Williams *et al.* 2015; Jameson *et al.* 2019).

Owing to the rapid expansion of offshore wind generation and still limited knowledge of seabird responses to these installations, the scope of seabird distribution survey in relation to OWFs often goes beyond classical impact assessments of defined projects, which are restricted to characterisation studies. Seabird surveys are also required to monitor the predictions of the impact assessment during construction and operation of OWFs, such as to measure displacement effects and reveal possible impacts on populations.

Scope

To facilitate the development of offshore wind energy with the lowest possible impacts on the marine environment, planning processes need to be informed by robust data on abundance and distribution of species that may be affected by this new marine industry. Legal frameworks require assessment of the impacts of plans and projects on the environment in general, and more specifically in regard to protected species and protected areas (e.g. Natura 2000: EU 1979; EU 1992). While the legal frameworks are general, the kind of information required needs to be defined in order to assess the impacts of OWFs. As a result, the first three themes of this chapter provide details of the survey requirements for Strategic Environmental Assessment (SEA), EIA and post-consent impact monitoring.

The remaining themes then detail the methods available to survey seabirds at sea by both boats and planes. While these methods to survey seabirds have been established for quite some time, the scope of surveying seabirds in relation to OWFs differs from former studies of seabird distribution. The purpose of seabird surveys in relation to OWFs is defined by the authorities to inform consenting processes involving a wide range of stakeholders, and the information needs perceived by these groups influence the design of baseline and monitoring studies with respect to the focal species, survey area, survey frequency, transect spacing and even choice of method. One aspect determining the choice may be whether surveys of seabirds and marine mammals should be combined, as seabird survey techniques usually cover marine mammals and most seabird observers are familiar with whales and dolphins, while dedicated marine mammal surveys, by definition, are restricted to this species group (see Scheidat & Porter, Chapter 2).

There has been a shift in the methods used for characterisation and monitoring of seabird distribution and abundance offshore since the first EIAs were carried out for OWFs, from visual boat and aerial surveys to digital aerial methods. In comparison to the visual methods, digital aerial survey standards have been superseded by technological developments since they were first set in 2009, so this chapter describes the most recent descriptions of the different methods employed for offshore digital aerial surveys. As such, this chapter aims to provide an update to the previous review of methods provided by Camphuysen *et al.* (2004).

Specific stand-alone examples of the different methods and analytical techniques available are also a key feature of the chapter. These are mostly generated by the personal experiences of the authors and their respective organisations, alongside additional invited contributions. The focus of the work presented is from European waters, especially the North Sea, where much of the offshore wind industry has been concentrated. However, the principles established are thought to be of relevance to the development of the industry in other seas. The chapter concludes with a series of guiding principles and current recommendations on how to conduct surveys and how data should best be analysed.

Themes

Surveys to inform strategic assessment

The first stage when planning OWF developments should be strategic investigations on a larger scale in order to identify areas of different sensitivity. The European Strategic Environmental Assessment Directive (Directive 2001/42/EC) provides the legal framework, but member states have applied this in different ways, reflecting their attitudes towards offshore wind and other factors (e.g. Albrecht 2007; DECC 2016). Particularly for mobile and often widely distributed seabirds, large-scale strategic surveys may provide highly relevant information to shape later EIA studies and to select or deselect areas for further planning. SEAs often do not include dedicated surveys but rely on available information from the literature. In this respect, the approach taken in the UK was a positive exception. During the early planning stages, large-scale aerial seabird surveys were conducted (Department of Business Energy and Regulatory Reform (BERR), unpublished report 2007) which, along with other sources of information, identified areas where conflicts might exist. This approach has been developed further using modelling techniques (Bradbury *et al.* 2014; Certain *et al.* 2015) (see Figure 3.1, representing Bradbury *et al.* 2014) and thus forms an important basis for developing future offshore wind energy capacity. The extent to which dedicated surveys are required depends on the information already available, the scope of intended OWF developments and on the specific situation of the marine waters

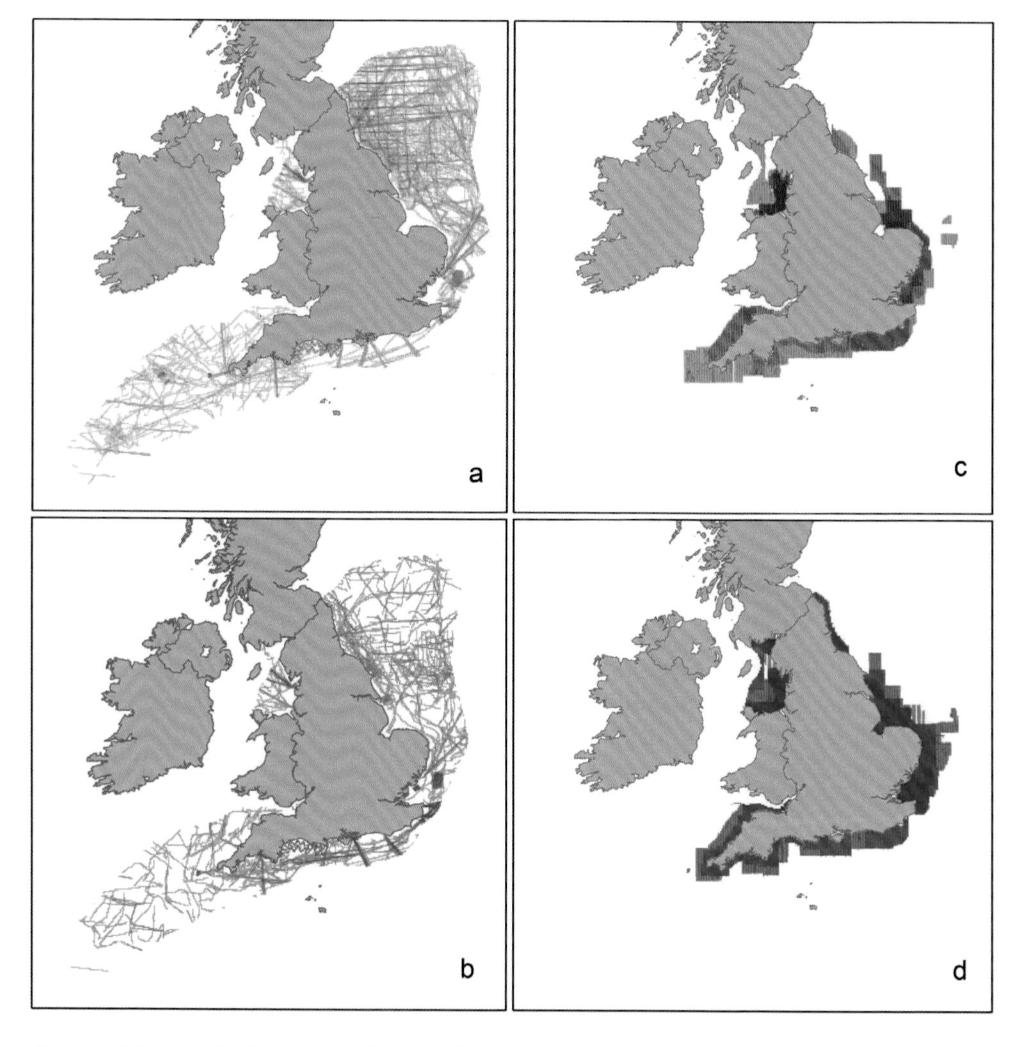

Figure 3.1 Extent of existing seabird survey data around England: (a) European Seabirds at Sea (ESAS) boat-based data in summer (April to September) 1979–2011; (b) ESAS boat-based data in winter 1979–2011; (c) WWT Consulting aerial survey data in summer 2001–2011; and (d) WWT Consulting aerial survey data in winter 2001–2011. (Bradbury *et al.* 2014)

where the development may take place. Large-scale strategic assessments also offer the most suitable base to account for cumulative impacts stemming from other activities, such as oil and gas installations, dredging or shipping, which are often difficult to include on the smaller spatial scale of project assessments.

Baseline studies for project-level Environmental Impact Assessments

EIAs rely on an evaluation of the presence of environmental receptors, such as seabirds, that may be exposed to the pressures of a given project such as the construction and operation of an OWF. This tends to require rather large survey areas, as seabird distribution, especially of pelagic feeding species, varies strongly in relation to a range of factors including hydrographic and meteorological factors. Camphuysen *et al.* (2014), for example, recommended that a study area of six times the area of the project site should be used for a characterisation study. An example of the scale of the surveys that were required for the large Round 3 Zones of development in the UK, each of which was to support several wind-farm sites, is provided in Box 3.1. Furthermore, as seabirds may respond to OWFs at substantial distances of even more than 10 km (e.g. Mendel *et al.* 2019), baseline studies must not focus on the wind farm itself but cover a large surrounding area. To cope with fluctuating seabird numbers within a defined area, a high frequency of surveys has also generally been requested by the statutory agencies. However, while careful planning should ensure that survey programmes are completed, this has not always been possible as a result of the limited availability of suitable weather conditions (Box 3.1). For German OWF studies, the Federal Maritime and Hydrographic Society of Germany (BSH) has drafted a Standard for Environmental Impact Assessments (StUK), which intends to balance the information needs to consent OWFs within ambitious targets for expanding renewable energy generation with the capabilities of different methods (BSH 2013). Although other countries have not fixed such detailed standards for their offshore industries, many (but by no means all) approaches are similar and methods usually follow international standards, as outlined in Camphuysen *et al.* (2004).

Box 3.1 Characterisation surveys of the Dogger Bank Zone

The Dogger Bank Zone was leased by the Crown Estate to the Forewind consortium for development in the UK sector of the North Sea in 2008 as the largest Round 3 zone (Forewind 2013). A potential 9 GW of electricity was proposed from the 8,660 km^2 sea area over the north-western part of the relatively shallow Dogger Bank, some 125 km from nearest land to the west in North Yorkshire, and near to the Flamborough Head Special Protection Area, internationally important for its nesting seabird assemblage. The zone was divided into four tranches of work (Tranches A–D), each containing two potential project sites. Of these, the greatest focus was placed on Tranche A, containing Creyke Beck A and Creyke Beck B, and on Tranche B, containing Teesside A and Teesside B.

A campaign of both boat-based and digital video aerial surveys was commissioned from Gardline Environmental Ltd and HiDef Aerial Surveying Ltd, respectively, to survey birds and marine mammals in the Round 3 zone and Tranches A and B. Initially for both survey platforms, 41 parallel, equally spaced transects 4 km apart, for a total transect length of about 2,430 km, were designed for survey on a monthly basis between February 2010 and June 2012 (Figure 3.2a). In the boat surveys, alternate transects, and especially those in the respective tranches, were given highest priority in the event of poor weather curtailing surveys before the end of each month. Survey coverage was achieved in all months, but all 41 transects could only be achieved in 5 months, and all of the priority transects in 20 out of 29 months. After the first 2 months, the survey design for digital aerial surveys was adapted to give 3 km transect spacing over the northern Dogger Bank Zone and either 1.4 km or 1.5 km spacing in Tranches A and B in the south. Between November 2013 and October 2014, HiDef adopted its higher specification GEN II camera system, again at 2 cm ground sample distance (GSD), but each transect was 500 m wide and transects were spaced at 6.75 km apart in the wider zone and 3.325 km apart in the north-western quarter or Tranche C (Figure 3.2b). Complete coverage of the study area was achieved in almost every month over several days with two or more planes collecting data concurrently.

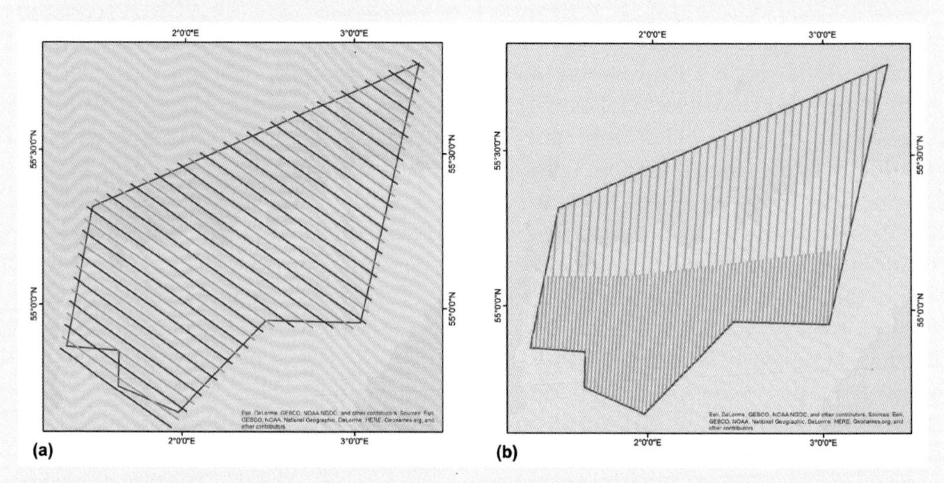

Figure 3.2 Location of transects used for surveys of the Dogger Round 3 Zone: (a) relative overall coverage intensity of boat-based transects that was achieved; (b) typical survey pattern followed by digital aerial surveys with intense coverage of Tranches A and B.

An environmental statement was submitted to the regulatory authority for the Creyke Beck projects in 2013. The analysis carried out for the ornithological technical report within this statement combined both boat-based and digital aerial survey data to assess the abundance and distribution of different seabirds present in the project areas (Forewind 2013). The analysis was based on data models in which it was assumed initially that the high detection rates from the digital aerial surveys were 100% efficient, but the low rate of identification to a species level from this survey platform in the early surveys (23%) required that the proportions of each species be taken from the boat-based surveys, where identifications were assumed to be 100% accurate. The novel method for analysis combined the boat-based spatially explicit abundance data with those of the digital aerial survey and spatially explicit environmental covariates using binomial generalised additive models (Johnston *et al.* 2015). The preference for digital aerial data for abundance estimation was made because of differences in calculated abundances of species groups, which were likely to have been caused by the greater detection efficiency of digital aerial surveys and the responsive behaviour of seabirds to the boat-based platform (Forewind 2013). The adoption of the HiDef GEN II technology in November 2013 meant that considerably improved identification rates to species level were possible using digital aerial surveys, and out of over 96,000 animals detected, some 94.9% of these were classified to species level, similar to the rate of the boat-based surveys (Johnston *et al.* 2015).

Post-consent impact monitoring

Because offshore wind farming is a new and rapid development, balancing this thriving industry with the demands of nature conservation relies essentially on detecting and assessing the impacts of the early projects to inform planning processes for the later ones. Monitoring of seabird distribution thus provides essential information if continued once the wind farm has been consented for development. OWFs attract some species, such as Great Cormorant *Phalacrocorax carbo* or gulls, especially Great Black-backed Gull *Larus marinus*, which perch on platforms and foundations (Leopold *et al.* 2013; Dierschke *et al.* 2016; Vanermen & Stienen 2019), but also displace other species through the presence of the turbines and the service vessels operating to and from the wind farms (e.g. Dierschke *et al.* 2016; Petersen *et al.* 2014; Webb *et al.* 2017; Vanermen & Stienen 2019).

Post-consent monitoring is practised in most countries, although with different intensity. The size of survey areas, frequency of surveys, and the monitoring during construction and operation are usually similar to those of the characterisation study. This should enable a proper comparison of baseline conditions with post-construction conditions and, if comparable reference areas are covered, a full before–after control-impact (BACI) analysis. However, as some studies have shown rather large displacement effects on species such as Red-throated Diver (Loon) *Gavia stellata* (Box 3.2), a thorough BACI approach is hampered by the difficulty in finding a reference area (or areas) that is similar in depth, seabed conditions, tidal flow patterns, prey density, distance to a colony, and so on, to the impact area (but see Vanermen *et al.* 2015a). An alternative to BACI is a before–after gradient (BAG) design in which the wind farm is located in the middle of a much larger survey area (Vanermen & Stienen 2019). In this approach, any differences between the pre- and post-construction period according to the effect of the wind farm are assumed to be a function of distance from the wind farm and that effects would be roughly the same in all directions (Vanermen & Stienen 2019). A significant before–after change (that may be positive as a result of attraction and negative as a result of displacement) that declines with distance from the wind farm provides compelling evidence that the wind farm is the cause of any change.

Box 3.2 Post-construction displacement studies at Lincs offshore wind farm using the spatially explicit MRSea modelling approach

Lincs Wind Farm is a 270 MW wind farm comprising 75 turbines situated approximately 8 km offshore due east of the coast of Lincolnshire, England. The development was constructed between March 2011 and March 2013 and is adjacent to the much smaller Lynn & Inner Dowsing (LID) wind farms, each with 27 wind turbines constructed between October 2006 and October 2008. Boat-based and visual aerial survey data, as well as radar studies, were used to inform the ornithological sections of the Environmental Impact Assessment of Lincs. However, because of restricted access to the airspace around the adjacent LID wind farms raising concerns over safety, higher flying (550 m) digital video aerial surveys were introduced and ultimately replaced visual aerial surveys for post-consent monitoring at Lincs in combination with LID.

In 2013, at the start of the post-construction monitoring programme, it was agreed to use HiDef's GEN II digital video aerial survey method and not use boat-based or visual aerial surveys, which were used extensively during baseline surveys, to investigate the post-construction distribution and abundance of key seabird species, including Common Scoter *Melanitta nigra*, Red-throated Diver *Gavia stellata*, Northern Fulmar *Fulmarus glacialis*, Northern Gannet *Morus bassanus*, Little Gull *Hydrocoloeus minutus*, Common Gull *Larus canus*, Lesser Black-backed Gull *Larus fuscus*, Common Tern *Sterna hirundo*, Common Guillemot *Uria aalge* and Razorbill *Alca torda*, with a focus on detecting and measuring potential displacement effects. Given the relatively small spatial extent of boat-based surveys around the wind farms, any changes in the distribution of seabirds were compared to the baseline visual aerial survey data only using spatially explicit modelling methods contained in the MRSea application

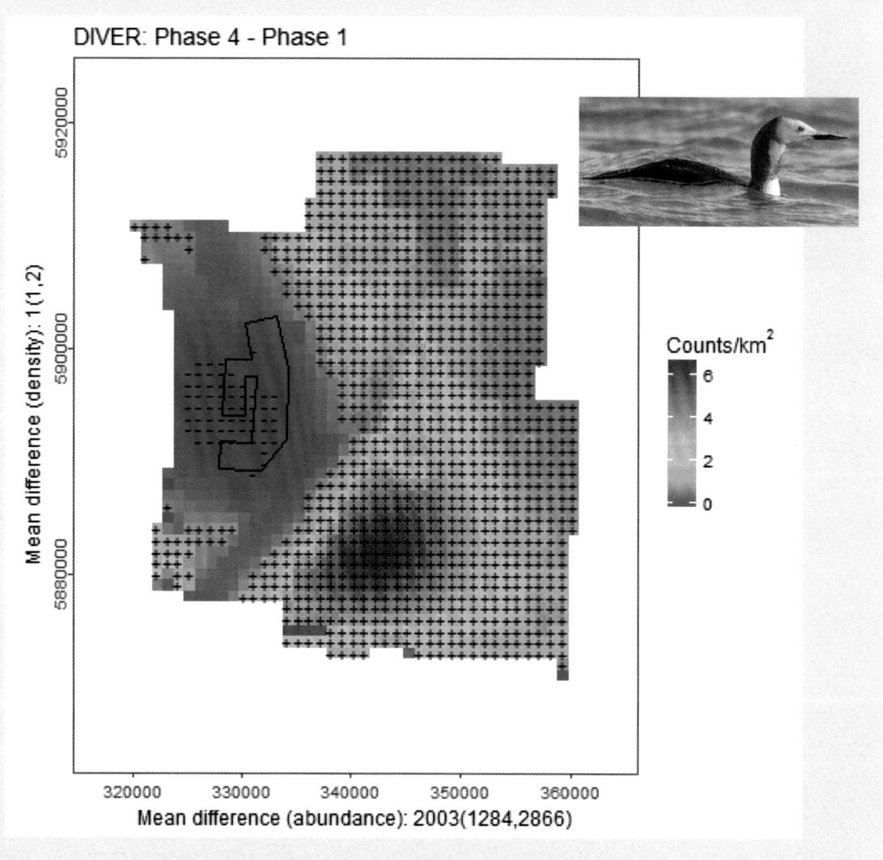

Figure 3.3 Changes in relative abundance of Red-throated Diver *Gavia stellata* (inset) between baseline and second post-construction year in relation to the Lincs and Lynn and Inner Dowsing combined wind-farm areas, with + signs showing significant increases relative to baseline and – signs showing significant decreases. (Inset: Martin Perrow)

(Scott-Hayward *et al.* 2013). It was important to account for some large changes in overall abundance of many species throughout the wider study area, but particularly of Red-throated Diver, in the wider Greater Wash, which otherwise would have swamped changes in relative distribution patterns within the region. Some of these changes were likely to be attributable to the use of improved digital aerial survey methods for the post-construction surveys, but also to real population changes in the Greater Wash with a marked increase between the baseline pre-construction phase and the construction phase which persisted into the operational phase of the project.

For most species and species groups, including gulls, terns and auks, no displacement effects were detected. In contrast, significant displacement of Red-throated Diver (Figure 3.3) and Northern Gannet was found. Effects were apparent by their geographical location and by comparison of the average abundance at different distances from the wind farms as a percentage of the average abundance in the Greater Wash. It was possible to detect these effects because of an appropriate number and geographical extent of post-construction surveys, each covering approximately 12.2% of the 1,282 km^2 Greater Wash study area, and the use of novel spatially explicit data modelling methods (Webb *et al.* 2017).

In the absence of 'before' data (for whatever reason), it is also possible to analyse the gradient of changing seabird densities in relation to distance to the wind farm to infer a wind-farm impact (Welcker & Nehls 2016), although any reference to baseline conditions will clearly help with the interpretation of the findings (Webb *et al.* 2017). When analysing large-scale effects extending 10 km or farther, such as observed at Lincs wind farm in the UK (Box 3.2), it is likely that environmental gradients such as water depth or salinity in estuarine environments will interact with wind-farm effects, and it becomes important to model seabird distribution in relation to these factors (Heinänen *et al.* 2017).

It is important to note that even with good survey design, detecting effects on the distribution of seabirds is not an easy task as a result of their generally patchy nature leading to high variance around mean values and a high proportion of zero values leading to overdispersion of data. As a result, increasingly sophisticated modelling techniques are now being used to tease out the effects of wind farms relative to other explanatory variables (Box 3.2). Vanermen & Steinen (2019) provide some further insight into suitable data processing and modelling.

Boat-based survey methods

The ESAS method for surveying seabirds from boats consists of a scan by two surveyors in a 90° arc from the bow to either the port or starboard side of the ship, whichever has the better viewing conditions for detecting sitting and flying birds. However, some survey teams operate a strip transect on both sides of the ship over 180°, especially in areas of low seabird density, because low encounter rates of birds at sea can result in low precision of the abundance estimates derived from the surveys. Moreover, a similar issue can occur where birds occur in infrequent groups and surveying a larger area may help to reduce variation around the mean.

The scan is carried out primarily by the naked eye, but with the additional support of binoculars on occasion, especially where seaduck and divers (loons) (Gaviidae) are known to be present. Those birds that occur within a 300 m wide strip parallel to the transect line within that arc are recorded as being 'in transect' using different methods according to whether they are recorded as flying or sitting on the sea (Figure 3.4). All sitting birds are recorded as being in transect, whereas those flying birds that occur within the limits of a defined zone in the 300 m strip at the time of a 'snapshot count' are recorded as being in transect. A snapshot count 'fixing the bird in time and space' is designed to compensate for the slow speed of the survey platform relative to the bird and allow calculation of density, as otherwise simply recording all birds seen is only a relative measure of flux, which may be correlated with density, but is not a measure of absolute density.

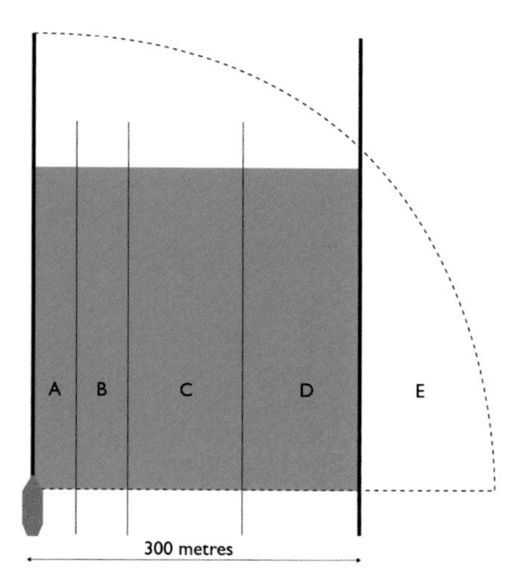

Figure 3.4 The strip transect, scan and snapshot zones used in the European Seabirds at Sea (ESAS) method. The transect zone is depicted by all areas within 300 m of the transect line and subdivided into smaller zones. The scan zone is depicted by the dotted line and the snapshot zone by the shaded rectangle.

The limits of the snapshot count area vary according to the ship's speed and are defined by the distance or time in the forward axis and by the 300 m strip width in the perpendicular axis. If the forward distance is defined by time, then usually the surveyor will calculate the average distance travelled by the ship every minute, normally between 250 and 350 m ahead of the ship, and conduct a snapshot at 1 minute intervals within that defined zone. If the forward distance is defined by distance, usually 300 m, the surveyor will calculate how long it takes on average to travel that distance and conduct the snapshot counts in a 300×300 m block at that calculated frequency. While all birds observed to occur in the 90° arc are recorded by the surveyor, only those that are in transect are used to calculate the density of birds.

The original methodology of Tasker *et al.* (1984) recommended that birds were classed as sitting if they made contact with the sea at any time when in view, regardless of how brief this contact was. However, many foraging behaviours, such as plunging, dipping, pattering, hydroplaning and skimming (Camphuysen & Garthe 2004), result in the bird making contact with the sea, but take place 'on the wing' as a bird is moving relative to the ship. Recording birds that were essentially in flight as 'sitting' would overestimate birds on the surface and underestimate birds in flight recorded by the snapshot method. The standards issued by the JNCC for ESAS trainers from 2005 onwards at least partly compensate for this problem by stating that birds must be assessed for flying or sitting behaviour at the time of first sighting. However, there may still be differences in the relative proportions of feeding to non-feeding birds recorded by the snapshot method.

To allow for the detection efficiency of the surveyor, birds sitting on the water are assigned to distance bands (or 'bins'), which are subdivisions of the 300 m strip transect according to their distance from the transect line when first sighted. These are defined as: Band A (0–50 m), Band B (50–100 m), Band C (100–200 m), Band D (200–300 m) and Band E (>300 m and not in transect). Transverse distance to the distance bands is measured at sea using methods described by Heinemann (1981). Using these distance bands to account for detection efficiency using distance analysis (Buckland *et al.* 2004; 2012; Thomas *et al.* 2009) (Box 3.3) essentially changes the method from a strip transect to a line transect in that the nearest distance band equates to the transect line. Line-transect theory requires that specific assumptions regarding the detection, lack of movement and accurate measurement of distance of animals are met (Buckland *et al.* 2001) (see Box 3.3). These assumptions can be met in part for sitting birds by good training and execution of the ESAS method. It is also possible to test the assumption of whether all sitting seabirds are being detected on the transect line in ESAS surveys by the use of double-platform methods (Borchers *et al.* 2002). Although this was recommended by Camphuysen *et al.* (2004), it is doubtful whether this has ever been performed for boat-based surveys of birds in relation to OWFs. Nonetheless, to maximise the detections, particularly in the closest distance bands to the transect line, which are fundamental to analysis of detection efficiency and assumed to be 100%, the maximum wind speed recommended for boat-based surveys is Beaufort force 4 (30 kph) (Camphuysen *et al.* 2004; BSH 2013). For the most part, this standard appears to have been followed by practitioners surveying around OWFs, although there are examples of surveys in higher wind speeds.

Box 3.3 Distance sampling for boat-based surveys

Andrew J. P. Harwood

Boat-based transect and snapshot methods can be used to derive simple density estimates by simply dividing the counts of birds seen in each by the survey effort applied, and combining the values. An average density and measure of variance can be estimated from densities calculated for individual transects (or smaller spatial units). However, decreases in detectability of animals with distance can result in negatively biased population estimates (Skov *et al.* 1995; Ronconi & Burger 2009). This is especially likely for relatively small species that are mainly recorded on the water, such as auks. Detection may also be influenced by many other factors, such as sea state, glare and surveyor ability. Distance analysis, usually undertaken in specially designed Distance software (Thomas *et al.* 2009) or the 'Distance' package (Miller 2017) in R (R Core Team 2018), can be used to analyse variations in the detectability of birds and correct density estimates accordingly. Buckland *et al.* (2001) define the central concept of distance analysis as the modelling of the detection function, $g(x)$, which is the probability of detecting an object (a bird or group of birds), given that it is at distance x from a transect line or point (in the case of snapshots) (see Box 2.1 in Scheidat & Porter, Chapter 2).

Distance correction analysis of bird survey data makes several important assumptions about the nature of the underlying data: (1) objects on the line (or point) are detected with 100% certainty; that is, all animals are detected; (2) objects are detected at their initial location; that is, animals do not move in response to the survey platform (flushing, swimming or diving) prior to being observed; and (3) measurements are exact; that is, the distance to the animal is measured accurately (Buckland *et al.* 2001). Thus, it is important that any birds that may have been attracted or associated with the survey vessel or any other vessel or structure are excluded from any analysis.

Distance corrections are largely only applied to birds on the water in line-transect surveys. In general, more than 60 observations are needed to generate a reasonable model (Thomas *et al.* 2009). Data from multiple surveys can be pooled for analyses if there is no reason to believe that detectability would vary as a function of the survey. Where there are changes between surveys, if they can be parameterised they may be included as explanatory variables in the models. Observation group size, or cluster size, is also accounted for within analyses either by using a mean value or size-biased regression, or using it as an explanatory variable in the model itself. Sea state, being a key variable that may affect the detectability of birds, is also often included in analyses where there are sufficient observations in different sea states, although data can be pooled if necessary.

Models are fitted using various key functions (uniform, half-normal, hazard rate or negative exponential), with or without adjustment terms (e.g. cosine, simple polynomial or hermite polynomial). Model selection is based on evaluation of the shape of the detection functions, Akaike information criterion values and chi-squared test results for grouped data and coefficients of variation. Density and population estimates and associated confidence intervals can then be derived for the desired survey/study area component. For further details of the application of Distance analysis, see Buckland *et al.* (2001; 2004).

Figure 3.5 illustrates the difference between fitted detection functions for Northern Gannet *Morus bassanus* (n=258 observations) and Common Guillemot *Uria aalge* (n=1,654 observations) on the water, using boat-based line-transect survey data from the survey of one wind-farm site in the UK North Sea. There is no drop-off in detectability (overall probability of detection of 1) for Northern Gannet, a large-bodied (94 cm in length and 3 kg; BTO 2018) and conspicuous species, while the much smaller Common Guillemot (40 cm in length and 690 g; BTO 2018) had an overall probability of detection of 0.48, and applying the correction would more than double the density estimate.

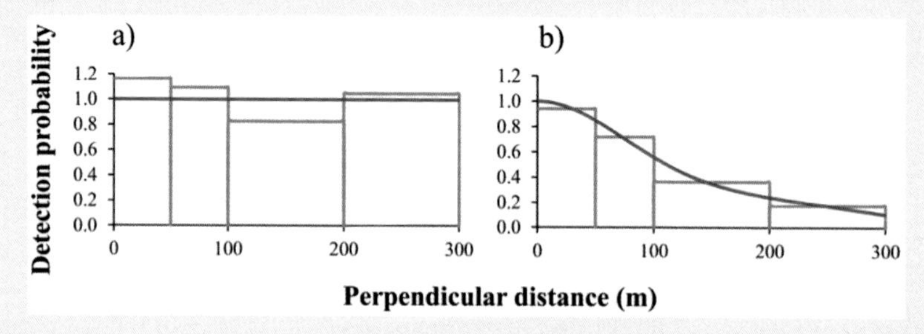

Figure 3.5 Examples of fitted detection functions from distance analysis for: (a) Northern Gannet *Morus bassanus* and (b) Common Guillemot *Uria aalge* on the water derived from boat-based line-transect surveys at one wind farm in the UK North Sea.

Flying birds tend to be excluded from distance analyses owing to their movements and potential for attraction to the survey vessel. Furthermore, appropriate distance data are not routinely collected during ESAS snapshot counts, as birds are not assigned to the radial distance bands required for point-sample analyses (see Box 3.5). Instead, it is assumed that all flying birds within a given box are seen and that there is no reduction in detectability with distance, even for very inconspicuous species, although Barbraud & Thiebot (2009) show that this is unlikely to be the case for many species and therefore densities may be underestimated.

However, distance corrections using point transect analyses can be performed where radial snapshots and distance bands have been adopted (see Spear *et al.* 2004) (Box 3.5). For point-transect analyses using a radial snapshot approach that surveys a semicircle, rather than a full 360° scan, a sampling fraction of 0.5 needs to be applied. In point-transect analysis the areas surveyed in each radial band will increase with distance from the vessel. Thus, while the detectability of the birds may decrease with distance, there are proportionally more birds available to detect. Conversely, the nearest band should contain the smallest numbers of birds, but they should all be detected. This means that a plot of probability density should approximate to a normal distribution, with most birds detected at an intermediate distance from the vessel. If there is positive movement bias to the vessel, that is, if birds are attracted (as is often suggested for large *Larus* gulls), then the distribution of probability density would be skewed to the left. Alternatively, if birds are repelled, then the probability density would be skewed to the right. Either attraction or repulsion will result in biased density estimates.

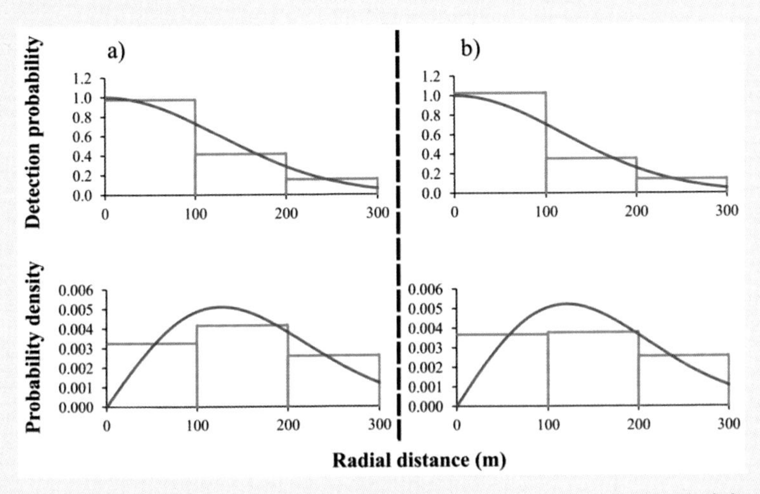

Figure 3.6 Examples of fitted detection functions (top) and the associated probability density functions (bottom) from point transect distance analysis for: (a) European Storm Petrel *Hydrobates pelagicus* and (b) Common Guillemot *Uria aalge* derived from boat-based radial snapshot surveys in the Celtic Sea, UK.

Trial analyses of birds in flight in radial snapshots, based on survey data from the Celtic Sea, UK, identified clear decreases in detectability with distance for some seabird species such as the small-sized (27 g and 38 cm wingspan; BTO 2018) European Storm Petrel *Hydrobates pelagicus* (*n*=77 observations) as well as the medium-sized (690 g and 67 cm wingspan; BTO 2018) Common Guillemot *Uria aalge* (*n*=313 observations) (Figure 3.6). The probability density functions for these species appear to be reasonable as they approximate to a normal distribution and reflect the limited availability of birds close to the vessel and reduced detectability with distance (Figure 3.6). Observations at the same site supported the view that neither European Storm Petrel nor Common Guillemot were attracted or repelled by the vessel. Overall probabilities of detection were similar for both species, at 0.33 and 0.31, respectively. Clearly, correcting density estimates based on these analyses could lead to much larger population estimates within the surveyed areas, with possible implications for impact assessments.

Line-transect methods do not fit well for flying birds recorded in snapshots by the ESAS method, partly because of the difficulty of estimating the distance to flying birds. For example, the Heinemann (1981) method cannot easily be applied to flying birds at variable distances from the water surface and is too slow for use in moderate or higher bird density. The use of rangefinders at sea shows promise however (Borkenhagen *et al.* 2018; Harwood *et al.* 2018), but will also be slow and difficult to use at moderate or high bird density. Moreover, it is difficult to account for responsive movement of seabirds to the survey platform, which may take place at several kilometres beyond the visual range of humans (Skov & Durinck 2001; Hyrenbach *et al.* 2007). Nevertheless, flying birds that show no obvious response to the vessel and that are assigned to radial distance bands over 180° during snapshots (and not perpendicular distance bands), essentially as 'point' samples, may allow correction for detection efficiency (Box 3.4).

Box 3.4 A comparison between box and radial snapshots to estimate densities of flying birds in boat-based surveys

Richard J. Berridge, Martin R. Perrow & Andrew J. P. Harwood

To standardise boat-based surveys for use in Environmental Impact Assessment (EIA) of wind farms, Camphuysen *et al.* (2004) recommended the basic European Seabirds at Sea (ESAS) method (Tasker *et al.* 1984; Webb & Durinck 1992). To negate the bias introduced by the movement of flying birds relative to the survey vessel, instantaneous 'snapshots' along the line transect are conducted to allow density estimates of flying birds to be calculated. The method assumes that all birds are visible for the full transect width, with snapshot counts conducted in continuous abutting 300×300 m boxes (Figure 3.7).

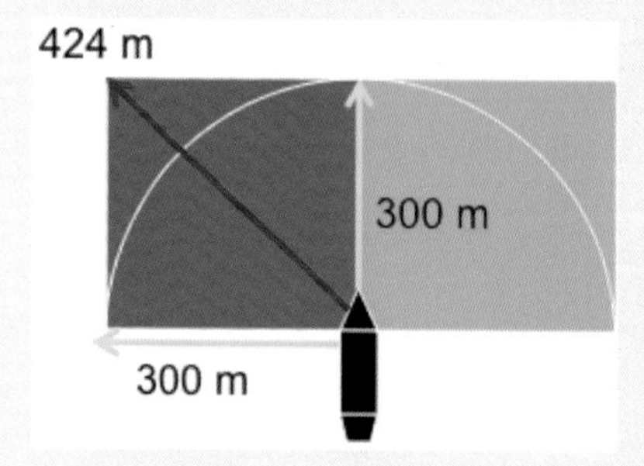

Figure 3.7 Diagrammatic representation of the area surveyed by a 300 m radial snapshot over 90° on one side of the vessel (blue) or over 180° on both sides of the vessel (blue and green combined) demarcated by a white line, compared to the equivalent box snapshot on one or both sides of the vessel. The maximum distance across the box is shown. In this study, each snapshot type sampled both sides of the vessel.

As collision risk is almost invariably a critical part of wind-farm EIA, it is crucial that the density of flying birds is accurately estimated. However, while the box method allows relative measures of flying bird abundance to be estimated, which may be perfectly adequate for the purposes for which it was designed, there are conceptual issues with its use to provide accurate density measures. To begin with, the maximum distance across a 300×300 m box at 45° from the surveyor is actually 424 m (Figure 3.7), with 20% of the area to be sampled in the box beyond 300 m, the distance set for transect width perpendicular to the vessel. This means that the surveyors are required to continually adjust the distance at which they should be scanning in order to place birds within an imaginary box. Even if this could be achieved, surveyors are therefore required to detect all flying birds irrespective of size and colour, position relative to the observer (perpendicular, flying away) and flight behaviour (high against the sky, low against the sea surface or a combination of the two) to a distance of 424 m. In the case of Barbraud & Thiebot (2009), the detection function by eye was virtually 1 for a variety of species at a distance of 100 m, but ranged from 0.87 for large (albatross-sized) species down to 0.69 for small species (small petrels) at 300 m. In other words, in this case, 100% detection of any bird group was not achieved up to 300 m. It is also important to note that Barbraud & Thiebot (2009) surveyed from a relatively high eye-height of 17.5 m, much greater than the 5–8 m that is typical of seabird surveys at offshore wind farms, and probabilities of detection from smaller platforms may be lower; although this may be offset by generally better conditions in wind-farm studies in the North Sea for example, compared to the southern Indian Ocean. Potentially, all birds could be seen beyond 300 m using binoculars, but surveyors would still be required to effectively place birds at different distances.

A conceptually easier alternative to the box is simply to use a radial (arc) snapshot with a fixed distance such as 300 m or less for small-bodied species, as has been adopted in a number of studies (e.g. Spear *et al.* 2004; Parsons *et al.* 2015). Indeed, the use of radial snapshots was subsequently approved by regulators and their advisory bodies in England, Wales and Scotland for use in seven different wind farms, including three of the original nine Round 3 zones in the UK (e.g. Centrica Energy 2007; Dudgeon Offshore Wind Limited 2010; Seagreen Wind Energy 2012).

However, to the authors' knowledge, there had been no comparative test of the counts and densities supplied by the two different techniques until their preliminary investigation in the Firth of Forth off the east coast of Scotland in October 2017. In the one-day survey aboard a 17 m Severn class ex-lifeboat with an eye-height of 6.1 m, a transect length of 148 km was surveyed using 283 snapshots at fixed 500 m intervals, adopting both box and radial snapshot methods simultaneously. Two surveyors accustomed to the box method (one exclusively so) each surveyed one side of the boat, which were combined to produce counts and according density estimates for a 600×300 m box with an area of 0.18 km^2. A single surveyor accustomed only to the radial method sampled both sides of the boat, that is over 180°, using a radial distance of 300 m, giving a snapshot area of 0.141 km^2.

Only pairs of snapshots in which at least one flying bird was recorded by either technique were taken forward for analysis. The recordings included n=120 for all birds combined, n=63 for Black-legged Kittiwake *Rissa tridactyla*, n=35 for Razorbill *Alca torda*, n=21 for Common Guillemot *Uria aalge* and n=13 for Common Gull *Larus canus*. Too few records of European Storm Petrel *Hydrobates pelagicus*, Northern Fulmar *Fulmarus glacialis*, Northern Gannet *Morus bassanus*, Great Black-backed Gull *Larus marinus*, European Herring Gull *Larus argentatus*, Little Gull *Hydrocoloeus minutus*, Little Auk *Alle alle* and Atlantic Puffin *Fratercula arctica* were obtained to enable species-specific comparison, but these were all included in the 'all birds' comparison.

Shapiro–Wilk tests showed that both count and density data were not normally distributed and thus non-parametric methods were used. Wilcoxon signed-rank tests for paired data with continuity correction in R software (R Core Team 2018) revealed no significant differences in counts between the two techniques (Table 3.1) despite the difference in the area being surveyed. A closer investigation of the data revealed that there were only three occasions on which additional single birds were recorded in a box relative to a radial snapshot, strongly suggesting that surveyors did not generally see birds at distances greater than 300 m. As a result, significantly higher

Table 3.1 Results of Wilcoxon signed rank tests for paired data derived from box and radial snapshots for each species and all birds combined.

Species	Wingspan (m)	Body length (m)	*n*	Count	Density
Black-legged Kittiwake	1.08	0.39	63	V=30.5, P=0.527	V=54, P<0.001
Common Gull	1.20	0.41	13	V=0, P=NA	V=0, P<0.001
Common Guillemot	0.67	0.40	21	V=9.5, P=0.916	V=18, P<0.001
Razorbill	0.66	0.38	35	V=63, P=0.517	V=0, P<0.001
All birds			120	V=86.5, P=0.190	V=653, P<0.001

Data on wingspan and length from the British Trust for Ornithology (2018). The V-statistic is derived from the pairwise difference between observations made using the two methods. A P-value could not be calculated (P=N/A) during the analysis for counts of Common Gull.

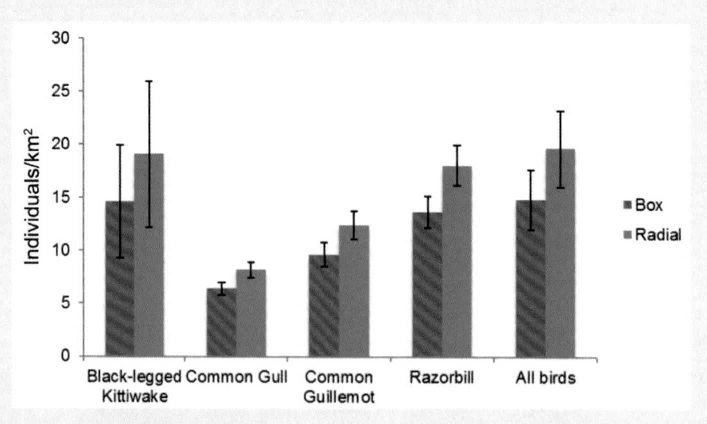

Figure 3.8 Mean density (±1 SE) derived using the box and radial snapshot methods for the variety of medium-sized seabirds sampled, as well as 'all birds' comprising a range of large to small species.

densities of all species (and group) considered were recorded in radial compared to box snapshots (Table 3.1, Figure 3.8), simply because of the greater area used in the latter relative to the similar number of birds seen.

It was therefore concluded that the use of the box snapshot method and the use of a larger sample area that is not effectively sampled in density calculations, underestimated the density of mostly medium-sized flying birds (see Table 3.1) in this case. It is important to note, however, that it was not possible to make a comparison for both small and large species such as Northern Gannet or large *Larus* gulls, and further tests need to be undertaken under different conditions (e.g. vessels and eye-heights) to generate a larger data set for a more detailed comparison.

Seagreen Wind Energy Limited funded the study. We are also grateful to Andrew Chick and Rab Shand, the additional surveyors other than the authors, and Jim Keenan and Danny Brand aboard the Eileen May.

Even if distance correction is not, or cannot be, performed on flying birds, it has been suggested that a radial snapshot is likely to provide more accurate measures of density than the conventional rectangular snapshot zone originally designed within the framework of a strip-transect method to achieve 100% longitudinal coverage of the strip transect for flying birds. This is because the many seabirds, especially small or indistinct ones, may not detected at the farthest limits of the strip (Barbraud & Thiebot 2009) and, even if they could be, surveyors are required to account for the distance of the bird relative to its angle from the observer in order for it to be accurately placed within the transect strip (Box 3.4). To account, at least, for the detectability of small indistinct seabirds such as European Storm Petrels *Hydrobates pelagicus*, surveyors may resort to using binoculars (van der Meer & Leopold 1995).

Some surveyors have used conventional line-transect methods in which the range and bearing are calculated for each individual bird sighting, for surveys around prospective wind-farm sites in the 'mid-Atlantic' region of the eastern USA (Connelly *et al.* 2015). Here, the authors used the vector analysis method of Spear *et al.* (1992) to account for flux of flying birds. However, experience suggests that 'swamping' of observers may occur, even at moderate seabird density, and Connelly *et al.* (2015) reported that they had to reduce their focus to a 300 m strip transect when swamping occurred. Line-transect methods may, however, still be achievable when seabirds are at low density, such as described by Barbraud & Thiebaut (2009) and Black *et al.* (2015).

One important metric for assessing the potential impact of OWFs on seabirds other than their density at sea is the height at which they fly. Camphuysen *et al.* (2004) recommended that estimates of bird flight heights be assigned to the following bands (after Lensink *et al.* 2002): 0–2 m, 2–10 m, 10–25 m, 25–50 m, 50–100 m, 100–200 m and >200 m a.s.l. These height bands were selected on the assumption that the typical minimum height of an offshore turbine would be 25 m, but designs differ between projects, and even change within a project, meaning that calculation of the collision risk for seabirds is problematic. Although there was likely to be a degree of false accuracy, many projects adopted a more pragmatic approach of assigning bird heights to 5 m bands, and in some cases 1 m bands. Given the difficulty of estimating flight height and potential observer bias aboard survey platforms of variable height, the approach of using generic flight-height distributions of different seabird species generated from large sample sizes during a range of boat-based surveys (Johnston *et al.* 2014) has often been adopted.

Nevertheless, Thaxter *et al.* (2016) criticised boat-based estimation of flight heights for being biased low, suggesting that this was attributable to many flying seabirds having responded to the survey vessel and being recorded shortly after taking off from the sea, reduced detection rates at greater altitudes above the observation platform (similar to perpendicular reductions in detection rates of sitting birds), and responsive movement towards the ship in a vertical axis. A comparison between observer flight heights and those generated by a laser rangefinder provides some assurance of the accuracy of observers (Harwood *et al.* 2018). For example, the estimates of flight height of a range of species by experienced surveyors matched reasonably well with rangefinders in the same 5 m height band (58% agreement), with agreement increasing (to 92%) if the adjacent height bands were also considered. Agreement did, however, vary between different seabird groups and there was a tendency for observers to underestimate the flight height of some species. Similarly, Cleasby *et al.* (2015), using altimeters on breeding adult Northern Gannets *Morus bassanus*, highlighted a disparity with Johnston *et al.* (2014) using data sets generated from boat-based surveys. However, the Cleasby *et al.* (2015) study may also be criticised for its low sample size and the methods of validation for the altimeters. Overall, the use of rangefinding equipment during surveys (see also Borkenhagen *et al.* 2018) to train observers, assess their accuracy and consistency, and calibrate the flight-height distributions that they deliver is to be recommended (Harwood *et al.* 2018).

Camphuysen *et al.* (2004) made a clear recommendation about the size, speed and type of ship that was acceptable for use in ornithological surveys at OWFs, with a preference for vessels >20 m in length and the need for surveyor eye-height >5 m from the sea surface, which is particularly important for detection of birds on the sea surface (Figure 3.9). This recommendation arose because many small and often unstable fishing boats were chartered for some of the first ornithological surveys at wind farms, potentially compromising data quality. Following this recommendation, the size and stability of the boats chartered improved for the large part.

The relatively slow speed of boats means that it is not possible to cover the large territory that it is possible to survey from aircraft in a short period of time. Conversely, the longer time spent at sea during boat-based surveys can detect potential diurnal (e.g. Schwemmer & Garthe 2005) and tidal (e.g. Embling *et al.* 2012) patterns in the presence of birds that cannot be revealed by the rapid snapshot character of aerial surveys. Of course, it is possible to adapt aerial surveys for this purpose by increasing the frequency of aerial surveys by undertaking multiple flights per day or over a few days, although this may prove to be prohibitively expensive (Vanermen & Stienen 2019). Migration of both seabirds and landbirds may also be better detected by boat-based surveys of longer duration, especially if these are conducted in more challenging weather conditions. The sample sizes of flying birds also tend to be higher for boat-based surveys on account of the flux of birds relative to the speed of the platform.

Perhaps most importantly, with high-quality and well-trained surveyors on board it is possible to obtain considerably more data on the behaviours of birds from a boat-based platform than is possible during aerial surveys. Behavioural data can be used to infer the purpose of the birds' use of a proposed development site, such as its importance as a foraging area. Vessels may also carry other instrumentation in order to collect contemporaneous habitat data in the form of a number of environmental variables that may explain seabird distribution, including water depth, temperature, salinity, abundance of phytoplankton (expressed as chlorophyll a), measures of tidal mixing (Scott *et al.* 2010; Embling *et al.* 2012) and even fish as measured by hydroacoustic sampling (Certain *et al.* 2011; Krägefsky 2014). This is a step towards ecosystem-based research, which is now seen as key to unravelling the true extent of what are now seen to be generally hitherto undescribed, but significant physical and indirect impacts of large-scale wind farms upon ecosystem function and thus upon seabirds (Perrow 2019).

Figure 3.9 Boat-based survey in progress in the North Sea. The survey platform is a 40 m former fishing vessel. Note the observers in the box on the A-frame. (Graem Pegram)

Visual aerial survey methods

The line-transect methods employed for visual aerial surveys for OWFs were first introduced for characterisation of seabird communities around the first proposed wind farms in Danish and Swedish waters (e.g. Kahlert *et al.* 2000). These methods were then adopted by other aerial survey teams in Germany (Diederichs *et al.* 2002), the Netherlands (Poot *et al.* 2011) and the UK (Department of Business, Energy and Regulatory Reform [BERR], unpublished report 2007), and reinforced by Camphuysen *et al.* (2004). Where visual aerial surveys continue, at the time of writing the same methods are still employed.

In essence, the methods of Kahlert *et al.* (2000) require that a twin-engined, fixed-wing, high-winged aircraft (Figure 3.10) be flown at a speed of 185 kph 80 m above the sea. Two observers in the aircraft, with one looking at each side, record birds continuously along the transect, dividing observations into three distance bands delimited by angles of 60–25°, 25–10° and 10–4° from the horizon, which at the recommended survey altitude correspond to 44–163 m, 164–432 m and 433–1,000 m from the transect line. Where the aircraft was fitted with bubble windows (Figure 3.10) surveyors attempted to count within an additional band, 0–43 m from the transect line. In the UK, the middle band was subdivided to give bands of 164–262 m and 263–432 m (Department of Business, Energy and Regulatory Reform [BERR], unpublished report 2007) to give more degrees of freedom to modelled detection functions used for estimating detection rates of birds at greater distances from the transect line (Dean *et al.* 2004). This approach was not adopted for surveys in continental European waters because of the higher species diversity potentially leading to observer swamping (I.K. Petersen, personal communication, 2007).

The maximum wind speed recommended for visual aerial surveys of birds is Beaufort 3 or a maximum wind speed of 18.5 kph (Camphuysen *et al.* 2004; BSH 2013). The main reason for this is the increased probability of not detecting all birds on the transect line at higher wind speeds. However, this standard was stretched to a maximum wind speed of 15 knots (27.8 kph) in surveys conducted around the UK (Department of Business Energy and Regulatory Reform (BERR), unpublished report 2007), potentially resulting in missed detections on the transect line.

The same assumptions relating to line-transect theory for boat-based surveys (see *Boat-based survey methods* above, and Box 3.3) also apply to visual aerial surveys. It is likely that the speed of survey prevents positive responsive behaviour by seabirds, but it remains open to question whether negative responsive behaviour occurs in visual surveys for some species. Comparisons between visual and digital aerial survey methods both in

Figure 3.10 Twin-engined Partenavia P.68 equipped with bubble-windows used for visual observer aerial surveys. (Ansgar Diederichs)

the UK (Buckland *et al.* 2012) and Germany (Box 3.5) reported much higher abundance of Common Scoter *Melanitta nigra* for digital aerial methods than for equivalent visual aerial methods. What is not clear is whether the differences reported are attributable to lower detection rates, survey error or responsive behaviour. The assumption most likely to be violated is that all birds on the transect line are detected (perception bias). Detection of birds by visual methods may be reduced if weather conditions are poor, such as higher wind speeds or if carried out with strong sun glare obscuring the transect line on one side or even both sides of the aircraft. It would be very valuable to employ double-platform methods to determine detection probability on the transect line, as recommended by Camphuysen *et al.* (2004). Unfortunately, no such surveys are known to have taken place around OWFs.

Box 3.5 Comparing aerial digital video and observer surveys for Common Scoter *Melanitta nigra* in the German Bight

Because a large number of aerial observer surveys have already been conducted to inform offshore wind-farm planning processes, it is of great interest to compare the results with new digital techniques. A direct comparison using fully matching double surveys is not possible because, owing to different demands on flight height, observers and cameras cannot be placed in the same plane. Weiß *et al.* (2016) made the first comparison of an observer survey, followed about 30 minutes later by a digital video survey using the HiDef system, in the German Bight, north of the island of Helgoland. Sighting rates for all species were two to four times higher for the digital video technique and differences were especially striking for Common Scoter *Melanitta nigra* (Figure 3.11). On an effective transect length of 324 km, 2,203 scoters were counted by observers, whereas 15,341 scoters were later detected on the video footage. Both methods provided largely similar distributions of scoter in the survey area, but calculated densities were about 50% higher for the digital video method (Figure 3.12). The large difference between methods is probably caused by the disturbance of Common Scoter by the low-flying observer plane, which usually

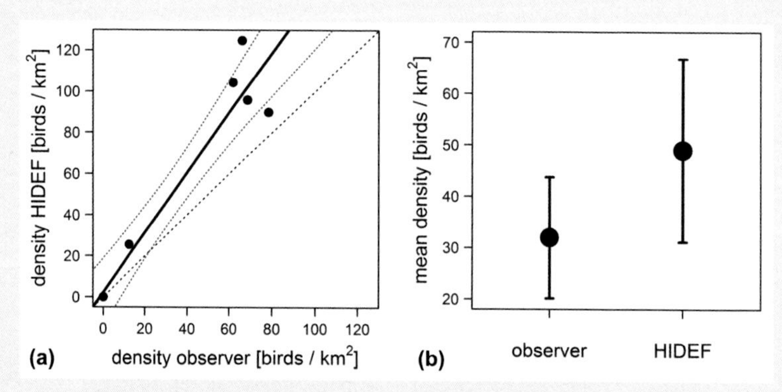

Figure 3.11 Relationships between the density of Common Scoter *Melanitta nigra* in six synchronous aerial observer and digital video surveys carried out on 2 December 2013 (*N* = 17,544) (a) point estimates of density with regression line and 95% confidence limits compared to parity between two methods; and (b) mean density of scoters recorded using the two methods. (After Weiß *et al.* 2016)

flushes most Common Scoters on approach. The footage of the digital videos, which were acquired from a height of about 550 m (1,800 feet), did not indicate any response to the plane.

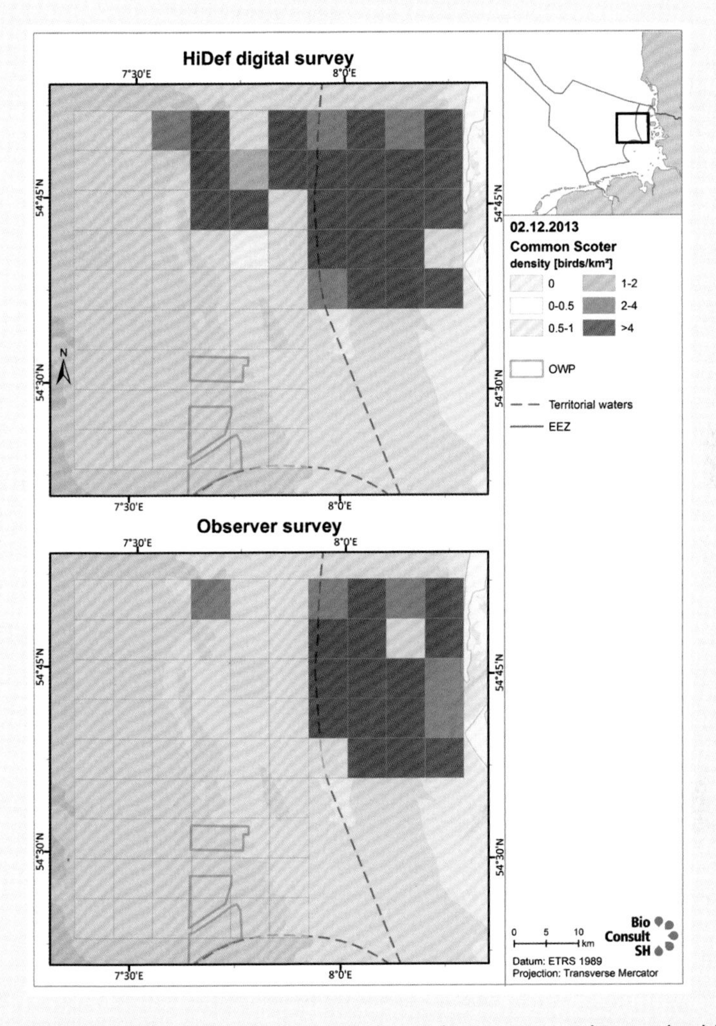

Figure 3.12 Density distribution of Common Scoter *Melanitta nigra* within 5×5 km blocks in an area of the German Bight derived from comparable digital video (above) and aerial observer (below) surveys.

This comparison indicates substantial differences between observer surveys and digital video surveys. Differences between survey techniques may result from reduced disturbance from digital surveys operating at flight heights above 450 m (1,500 feet), but also from different detection rates, which are likely to differ between species.

A further known bias from visual aerial surveys is one of availability bias, whereby not all animals present in the survey are available for detection (Marsh & Sinclair 1989). Typically, this occurs when diving seabirds are foraging. Availability bias can be corrected using the following formula (Barlow *et al.* 1988):

$$Pr = \frac{(s+t)}{(s+d)}$$

where *Pr* is the probability of being visible, *s* is the average time spent below the surface per dive cycle, *t* is the window of time that the bird is within view and *d* is the average time spent at the surface per dive cycle. In practice, the length of time that the bird is within view is difficult to calculate in an aerial survey, because this is a function of the distance from the transect line. In practice, availability bias is better treated with perception bias to determine detection rates using double-platform methods (e.g. Borchers *et al.* 2002; Hiby & Lovell 1998).

The main advantage of using aerial surveys is that a large area may be covered in a relatively short space of time. This means that it is possible to obtain a snapshot of seabird distribution over a wide area around the potential development area, which is important in characterisation surveys to provide context to the distribution patterns observed at the development site (see *Baseline studies for project-level Environmental Impact Assessments*, above). Similarly, during post-consent monitoring surveys, it may be difficult to detect changes in seabird distribution patterns around the wind farm unless a sufficiently large buffer area is surveyed (see *Post-consent impact monitoring*, above). For the most sensitive species, such as Red-throated Diver, this has been measured at 9 km from the wind farm (Petersen *et al.* 2014; Webb *et al.* 2017) and at 12 km by Mendel *et al.* (2019).

The speed of aerial surveys also means that it is possible to take advantage of short weather windows that boat-based surveys may find difficult to exploit, particularly if the site is far from port (Box 3.1), which could result in missed surveys. The other key advantage of the visual aerial survey is the cost, as it may be the cheapest survey method. However, this should be offset against a number of disadvantages, of which safety concerns are foremost. Historically, low-level aerial survey has been beset by accidents, often resulting in fatalities (Hodgson *et al.* 2013). This is without the additional potential hazard of flying aircraft between wind turbines at the same altitude as rotors. Other concerns relate to the accuracy of this method, given the large volume of data that must be processed by observers in very little time, which leads to concerns about the accuracy of abundance estimates. The identification rates of birds to species level are also typically much lower than in boat-based surveys (e.g. in the UK, 0.3% of auks and 28% of gulls were identified to species level; Department of Business, Energy and Regulatory Reform [BERR], unpublished report 2007), partly because of the time constraints during surveys. The last potential concern relates to survey design when flying between turbines, as transect line positioning and orientation is necessarily constrained by turbine location within a wind farm, meaning that, if bird distribution within the wind farm is influenced by the turbine location then it is unlikely that the transect positioning will be independent of the distribution of the birds.

Digital aerial survey methods

Digital aerial survey methods have evolved rapidly since commercial surveys began in offshore waters in 2009. The initial standards for these surveys in the UK were set out in Thaxter & Burton (2009) and updated in Thaxter *et al.* (2016), and by BSH (2013) for surveys in German waters with reference to Groom *et al.* (2013) and Buckland *et al.* (2012). These standards have been superseded by technological developments, so this section describes the most recent descriptions of the different methods used for digital aerial surveys.

There are two approaches to digital aerial survey, using either bespoke digital video camera technology or digital stills camera technology. The digital stills camera technologies are largely based on systems used for aerial photogrammetry, and all successful digital aerial surveys are operated by commercial companies in the UK, Germany, Denmark and the USA. Both approaches can be employed for a strip-transect sampling or plot sampling regime. In transect-based surveys, birds will be present in one to two frames in digital stills camera systems or seven to ten frames in video camera systems, depending on the frame-rate capability of the camera and data-storage systems.

The sensors used in cameras are always very high end, based on charge-coupled devices (CCDs) or complementary metal-oxide semiconductors (CMOS), and have arrays of pixels significantly greater than normally seen in domestic camera systems. The image quality is dependent on a number of factors, but key to this is the image resolution on the ground, or the ground sample distance (GSD), which is the dimensions of the image on the ground for each pixel. The distance to the object or ground, the focal length of the camera lenses and the physical size of the camera sensor pixels determine the GSD. Most digital aerial surveys are now operated at an image resolution of 2 cm GSD, and this is generally sufficient to identify most birds to species level. The number of pixels in the sensor and the image resolution determine the size of the image, which, in effect, determines the transect strip width, or the dimensions of the plot.

Trained operatives undertake post-survey review of digital images under laboratory conditions. All digital aerial survey operators review their imagery to search for objects that are likely to be birds or other marine mega-vertebrates, then pass these objects for identification to the lowest order possible, preferably species, and apply a confidence to that identification. In both phases, there is a quality-assurance (QA) process, usually consisting of a double-blind review or identification of either 10% or 20% of the material, and requiring at least 90% agreement between the first and second reviews. Unlike visual-based survey methods, the detection of objects is uniform across the transect or plot, avoiding the need for calculating a detection function to correct for perception bias. Perception bias is, in effect, enumerated by the QA process in the first search phase.

All digital aerial survey providers are able to use transect-based and quadrat (or plot) methods for sampling bird densities at sea, although most tend to use transect-based sampling methods. Both sampling approaches require that individual samples (transects or plots) are equally spaced. For plots, this spacing is required in two dimensions to avoid errors that may arise from pseudoreplication (Buckland *et al.* 2001). When using design-based analysis methods (Burt *et al.* 2010), plot-based sampling works well for very abundant species, but less well when there is a low encounter rate of fewer than 60 individuals (Buckland *et al.* 2001). Transect-based sampling can cause lower precision in abundance estimates at small sites when there is a small number (fewer than 13) of transects or samples (Buckland *et al.* 2001). Plot-based and transect-based sampling methods are insensitive to the number of samples when using model-based analytical methods, simply because subsections of transects (or segments) are treated as samples.

Precision in abundance estimates is affected by the amount of survey coverage and this can be varied in any survey method by reducing the distance between sample locations, as long as there is no responsive movement by birds to the survey platform. In most surveys, a minimum target coverage of 10% of the offshore study site is required in the UK and in Germany (BSH 2013). This should be seen as the minimum coverage.

Digital video

All digital video surveys use what is known as the 'HiDef method', first developed by HiDef Aerial Surveying Ltd in the UK, and later also operated by BioConsult SH in Germany. The surveys can be flown in different twin-engined aircraft types and do not require the high-winged varieties of visual aerial surveys (Figure 3.10). Typical aircraft are the Diamond DA42, Vulcanair P68 and Piper PA-23 (Aztec), flown at a height of approximately 550 m (1,800 feet) a.s.l. Flying above 500 m ensures that there is no risk of flushing those species known to be easily disturbed by aircraft noise (Thaxter *et al.* 2016; Weiß *et al.* 2016). Position data for the aircraft are captured from a Global Positioning System (GPS) receiver with differential GPS enabled to give 1 m accuracy, and recording updates in location at 1 second intervals for matching later to bird and marine mammal observations. Surveys are flown in conditions of moderate to high cloud (at least above 550 m), with no precipitation and clear visibility. While the maximum potential operational wind speed for surveying is Beaufort force 6 (14 m/s), in practice, slower maximum wind speeds of Beaufort force 4 (8 m/s) are preferred.

The survey aircraft is equipped with four HiDef Gen II cameras with sensors set to a resolution of 2 cm GSD. Each camera samples a strip of 125 m width, separated from the next camera by approximately 25 m, thus providing a combined sampled width of 500 m within a 575 m overall strip. However, in Germany the full potential strip width of these cameras has been used to give a total strip width of 544 m. This comb pattern with gaps between the cameras is employed to ensure that no individual birds are double counted between cameras and to target a wider area for detection of birds than the limits of the camera strip width, thus increasing the encounter rate for rarer species that occur in dense flocks (Thaxter & Burton 2009). The cameras are angled at 30° from vertical, and are mounted on a rotating plinth. The angle is useful for providing more than just a plan view of birds, thus giving more potential features to support identification. The plinth is rotated between transects to angle the cameras away from any sun glare, which can be a significant impairment to detection in the resulting images (Thaxter *et al.* 2016). The cameras record images at a rate of seven frames per second, which means that birds will be visible in about seven frames, depending on a range of factors. An example of the images produced is shown in Figure 3.13.

All review and identification is done manually as, in the experience of HiDef, all automated systems tested to date do not perform sufficiently well against human

Figure 3.13 Still grab from a digital video aerial survey showing a group of Northern Gannet *Morus bassanus* in flight.

counterparts in terms of efficiency and accuracy. All video footage is reviewed and all objects of interest are marked for subsequent analysis. Some 20% of video reels (a reel is the video material from a single camera along a single transect) are subjected to a random blind audit by an experienced operator, and if there is less than 90% agreement between review and audit, then all of that reviewer's material for the day is reassessed.

All objects detected during the review phase are assessed by expert ornithologists and marine mammologists who classify these to the lowest taxonomic order possible, assigning these to both species group and, where possible, to species. Each identification is given a confidence rating of possible, probable or definite, equivalent to low, medium or high confidence. Additional details are recorded at this stage, where possible, of age, sex, behaviour and direction of travel. As with the review stage, 20% of all objects are subjected to a random blind audit by the most experienced identification team members and more than 10% disagreement will result in reidentification of the species that caused the disagreement by a third member who adjudicates on any disagreements. A regular assessment (every 500 frames), of the sea conditions is made from the video footage by members of the identification team, recording the sea state according to the Douglas sea scale (Met Office 2016), the amount of sun glare, the water turbidity and the air clarity. The last three are assessed on a four-point scale and all are referenced to a set of standard images to remove subjectivity of the assessment.

Identification rates differ between species groups. As a mean of 50 surveys conducted in the German North Sea and the Baltic Sea, about 90% of all detected birds were determined to species level. Identification rates are obviously high for large birds, such as Northern Gannet (Figure 3.13) and Northern Fulmar *Fulmarus glacialis*, and lower for smaller and less conspicuous birds, such as grebes and waders (Table 3.2). Overall identification rates will thus differ between areas in relation to the species composition. The low identification rate for tern species results from the fact that Arctic Tern *Sterna paradisaea* and Common Tern *Sterna hirundo* could not be separated, while the discrimination of Sandwich Tern

Table 3.2 Proportion of individuals within taxonomic groups determined to species level from 50 surveys in the German part of the North Sea and the Baltic Sea (Weiß *et al.* 2016)

Taxonomic group	Number of individuals	Proportion (%) determined to species level
All species	351,240	90.0
Divers	10,857	88.5
Grebes	1,310	64.0
Procellariformes	101	100.0
Gannets	875	100.0
Ducks	227,355	99.4
Raptors	71	94.4
Waders	25,669	58.7
Skuas	30	66.7
Gulls	57,816	85.8
Terns	5,435	18.4
Auks	13,673	80.7
Marine mammals	7,837	88.1
Seals	2,106	71.9

Thalasseus sandvicensis from the smaller tern species is relatively straightforward. It needs to be noted that the identification rate of divers and auks is high, which is an important improvement over visual aerial surveys for which separating Common Guillemot *Uria aalge* and Razorbill *Alca torda*, the predominant auk species in the north-west European surveys, is rarely possible. In terms of identification, digital video surveys offer marked advantages against observer surveys for three reasons: (1) more time can be spent on identifying the bird on the screen compared to observer flights, 2) the bird can be viewed from different angles as an individual is usually present in at least five frames, and 3) a tool permits precise measurement of the bird's size, which is an important criterion for similar species such as European Herring Gull *Larus argentatus* and Common Gull *Larus canus*.

Zydelis *et al.* (2019) compared results from contemporaneous digital video aerial and visual aerial surveys in the German North Sea and found that, apart from grebes, digital video methods were able to identify more seabird taxonomic groups to species than visual aerial methods. They also found that density estimates were higher for most taxonomic groups using digital video compared to visual aerial methods, providing further support for using digital video methods when performing bird and marine mammal surveys at OWFs.

A comparison of the size of a flying bird in the image with known or reference sizes of each species has been used to estimate the flight height of a bird, at least in broad terms of whether it falls in the collision risk zone of turbines (Thaxter *et al.* 2016). While this method provides empirical measurements for use in collision risk modelling, it risks being biased if flying birds appear shortened in any way, such as if looking downward when searching for food, or their body is angled upward or downward. This method is also fairly imprecise on account of normal variation in the reference size of each species and because measurement error has the greatest effect farthest away from the aircraft, critically in the zone where birds are recored at or below collision risk height, where greatest precision is needed.

Digital stills

There are three providers of digital stills aerial survey data for seabirds and marine mammals: IfAÖ GmbH, using a bespoke 'DAISI' system in Germany; APEM Ltd, using off-the-shelf camera systems in the UK, the USA and Germany; and Aarhus University, also using off-the-shelf camera systems in the UK and Denmark.

The DAISI system employed by IfAÖ consists of two cameras with lenses mounted to point downwards from the instrument hatch of a Vulcanair P68, pointing at the starboard and port sides of the aircraft either side of the transect line so that a total image width of 407 m is obtained when the aircraft is flown at 423 m a.s.l. with 2 cm GSD. If the aircraft is flown at a ground speed of 180 kph, and images are recorded at 1–2 second intervals, 50% overlap is obtained between images along the length of the transect. The cameras are mounted on a gyroscopic platform and the cameras have forward-motion compensation to reduce motion blur (Coppack *et al.* 2015).

APEM has employed a single medium-format camera system with a standard lens that is pointed directly downwards from the camera hatch in a Vulcanair P68 or Britten-Norman Islander Mk II. The total image width for this system is approximately 400 m if flown from an altitude of 400 m, achieving an image resolution of 3 cm GSD (Mendel *et al.* 2018). Typically, these surveys are flown using a strip-transect design or as a plot-based survey design. The camera system uses forward-motion compensation to reduce image blur (Busch *et al.* 2014).

Aarhus University uses a system based on one developed originally by the Natural Environmental Research Institute (NERI) in Denmark. Their system uses a Vexcel Ultracam system which, when flown at an altitude of 470 m and image resolution of 3 cm GSD, obtains an image dimension of 520×340 m (Groom *et al.* 2013). Like the DAISI system, an image is recorded along a strip transect such that at least 50% overlap between consecutive images is obtained (Groom *et al.* 2013).

All three systems use automated methods for detecting objects. The system used by DAISI and Blom is the same Trimble Envision Object-based Image Analysis platform developed at Aarhus University to locate animals in the images using a process of segmentation and then labelling objects in the form of polygons. A single rule set is used for identifying these images that is knowledge based and adaptive but does not require training typical of most artificial intelligence systems. Some 10% or occasionally 20% of all images are selected for manual QA, requiring at least 90% agreement between the automated and manual detection systems. A manual system is used for identification of objects to the lowest taxon order possible (Groom *et al.* 2013). The system employed by APEM uses a manual system of screening for candidate animals; 10% of blank images and 100% of all images are subjected to independent QA, requiring that 90% agreement is obtained. Bespoke software is used that automates recognition of species, measures the length and wingspan of objects as well as the direction of travel, and estimates the flight height of birds based on their size and compared to known sizes for a given species (Johnston & Cook 2016). This software supports manual identification of species, and 10% of these objects undergo external QA, again requiring 90% agreement between both parties (Busch *et al.* 2014).

Sun glare or glint is more of an issue for digital stills than for video aerial surveys, because all operators use cameras in plan view, and thus they cannot be directed away from the sun, as is the case for angled cameras. Severe sun glare can cause significant problems with object detection and reduce the effective coverage of the survey. The DAISI and Aarhus University systems both use a complex technique developed at NERI in which any glare-affected portions of images are cropped from overlapping sections, thus leaving most of the strip transect intact and free from significant glare (Groom *et al.* 2013). APEM avoids sun glare by not surveying during the middle part of the day and selecting, when possible, overcast days for surveying (Busch *et al.* 2014).

All stills digital aerial survey methods are operated in relatively light winds, up to Beaufort Force 4 (up to 8 m/s) (Busch *et al.* 2014).

Concluding remarks

By the end of 2017, 4,149 turbines had been installed at 92 wind farms in the territorial waters of 11 European countries, mostly in the North and Baltic Seas, as the global focus of offshore wind (WindEurope 2018). Further expansion is envisaged by other European countries and in many others around the world, including China, Taiwan, Japan and the USA (Jameson *et al.* 2019). After two decades of offshore wind farming, developments of 20–80 turbines have been constructed along the coasts in European waters. While planning of wind farm results in clustering of their locations in some countries, in others wind farms have a more scattered location. As displacement effects in some seabird species reach considerable distances (>10 km) beyond the wind farms (Vanermen & Stienen 2019; Mendel *et al.* 2019), scattering small wind farms maximises border effects and thus total impacts on seabirds, and leads to a growing demand to account for cumulative effects.

To some degree, survey techniques have been standardised to allow direct comparison and even joint analysis of seabird surveys between research groups and countries. However, the scope of seabird surveys in relation to offshore wind should extend beyond projects to enable monitoring of seabird numbers and populations at regional and transboundary scales. There are risks that project-oriented monitoring schemes may miss cumulative impacts even on regional scales if they are not synchronised in time and space, and countries intending to develop OWFs are encouraged either to coordinate the monitoring obligations of the industry or to launch survey programmes filling knowledge gaps. Survey programmes investigating impacts from OWFs should be combined with national obligations to monitor protected species and maintain the integrity of protected sites such as Natura 2000 areas designated under the European Habitats Directive (EU 1992), as well as monitoring of the effects of other industries; in spite of the ongoing risks and impacts on seabirds associated with offshore oil exploration (Haney *et al.* 2014), there is a significant shortage of contemporary data on seabird distribution around many offshore oil facilities (Webb *et al.* unpublished report 2014).

Based on the information provided in this chapter, the following characteristics of seabird surveys in relation to OWFs are recommended.

- In both characterisation and impact monitoring studies, the size of survey areas must extend beyond the area where seabirds may respond to the wind farm. In this respect, the size of survey areas will differ between projects and depend on the target species. Many studies of changes in the distribution of Red-throated Diver following wind-farm construction suggest that effects can be found at up to 12 km from the wind turbines (Petersen *et al.* 2014; Welcker & Nehls 2016; Webb *et al.* 2017; Mendel *et al.* 2019). In contrast, 2 km appears to be the limit of Northern Gannet displacement at least at one wind farm (Webb *et al.* 2017). Clearly, it is important that sufficient data are collected in the region beyond the displacement distance for the key species in the project. In German OWF projects, aerial survey sites are required to extend at least 20 km beyond wind-farm boundaries, covering at least 2,000 km^2. This appears to be sufficient to cover the full gradient of seabird response to the wind farm, even for sensitive species, but is already at the limit of covering full response ranges of marine mammals to unmitigated pile driving noise (e.g. Brandt *et al.* 2011). Camphuysen *et al.* (2004) recommended that the study area should be six times the area of the development in the UK, but this recommendation has rarely been followed. Smaller study areas may be chosen in areas where surveys would extend beyond the suitable habitat of the species of interest, where these have restricted habitat choices.
- In relation to large study areas and where the focus is on the abundance and potential displacement of seabirds, aerial surveys using a fast-moving platform are likely to be the preferred method. In contrast, the relatively small area that can be covered by a boat-based survey in one day makes this method less attractive. Further, as digital aerial surveys flying above 500 m do not cause disturbance and achieve a very high identification rate, they are especially useful in covering the area affected by multiple neighbouring wind farms.
- Boat-based surveys are, however, best suited to provide detailed information on seabird behaviour and flight height. Boats provide an opportunity to measure seabird flight height with some degree of accuracy when supported by laser rangefinders (Borkenhagen *et al.* 2018; Harwood *et al.* 2018). There are also a number of environmental parameters that can be measured during boat surveys that may influence seabird distribution at a project site, such as tidal variation, sea salinity and temperature, chlorophyll *a* and, of course, fish abundance and distribution.
- The proportion of area covered by either boat-based or aerial transects may differ between projects in relation to size of the project area as well as abundance and distribution pattern of species of interest. It is recommended that this be at least

10% of the study area (valid effort accounting for glare) (BSH 2013). Modelling techniques, such as the MRSea suite of tools (Scott-Hayward *et al.* 2013), may allow robust predictions of species distribution and densities at lower area coverage, but in order to measure avoidance ranges accurately, sample size inevitably becomes the limiting factor at larger distances from the OWF. Thus, higher coverage may be preferred during construction and operation of the wind farms. Power analysis is an important statistical technique for estimating the power of a given survey design to detect change, if a representative set of survey data is available of known precision and, ideally, error structure (Mackenzie *et al.* 2017) which can be used to design impact studies at OWFs.

- Surveys are usually carried out once or twice per month during the period that the species of interest occur. Most EIA studies require that this is maintained through a whole annual cycle as different species may occur at different times of the year. More frequent surveys may be required during the breeding season.
- Weather may restrict the number of surveys that can be completed in 1 year and, given the large natural inter-annual fluctuations in seabird numbers and distribution, baseline studies are often conducted over a minimum of 2 years, but there is no international standard on this and consenting authorities in different countries decide differently on their information needs. It is questionable whether 2 years of survey data are sufficient to characterise inter-annual variation in highly mobile species such as seabirds, and more survey years are recommended.
- In areas where more than one wind farm is being built, it may not be possible to monitor the effects on seabirds of each wind farm separately. In such circumstances, it is strongly recommended that monitoring projects be combined to reduce the overall amount of surveying and (in the case of boat-based surveys) of disturbance to seabirds, rather than implemented over different areas, such as has been the case in Germany, encouraged by BSH. Here, almost all seabird and marine mammal monitoring is carried out in clusters; in some cases, more than ten projects have agreed on a joint monitoring approach, covering all projects in coordinated boat-based and digital aerial surveys. Such an approach is recommended to be adopted in other areas as it is understood to be the best way to assess cumulative impacts from different projects. When covering large survey areas, surveys should extend across the entire survey area in a single day, using more than one platform if needed, as changes in seabird distribution between days may bias survey results when extended over longer periods (Thaxter & Burton 2009). The survey design can achieve finer scale resolution than is possible in one day by repeating surveys, but offsetting transects to fill in between the transects for the first day.

Acknowledgements

We are grateful to a number of people who contributed to this chapter, in particular Felix Weiß, Alex Schubert and Jorg Welcker of BioConsult SH, and to Andrew Harwood, Richard Berridge and Martin Perrow of ECON Ecological Consultancy Ltd for their additional text. Many people have contributed to the evolution of different methods for census surveys of birds at sea and are referenced as much as possible in the text. Much of the intellectual input on survey methods at sea has evolved from long discussions with Martin Perrow, many staff at the Joint Nature Conservation Committee, staff at the Centre for Ecological and Environmental Modelling (CREEM) at the University of St Andrews, but especially Professor Steve Buckland, Kees Camphuysen of the Netherlands Institute for Sea Research and Ib Krag Petersen of the

University of Århus. Digital video aerial survey methods have evolved rapidly over time, but initially benefited greatly from work by David Baillie of Wildcat Films, Matt Mellor of Createc Ltd and Rhys Hexter of RH Business Services Ltd, and latterly with Felix Weiß and many others at BioConsult SH. The ornithological studies at Dogger Bank (Box 3.1) were supported by Alastair Mackay and Gareth Lewis formerly of Forewind and permission to include details from this study were kindly provided by Scottish and Southern Energy, Equinor and Innogy. Many people at Gardline Limited, WWT Consulting and HiDef Aerial Surveying Ltd, too many to mention by name, contributed their considerable time and effort to this study. The studies at Lincs (Box 3.2) were conceived and supported by Jen Snowball, Kit Hawkins and Ben Coulston, all of Centrica Renewable Energy Limited and permission to reproduce these results was granted by Ørsted; special thanks to Madeleine Hodge for her input to the text. MRSea modelling was carried out by Monique Mackenzie and others at DMP Statistical Solutions Ltd.

References

Albrecht J. (2007) Guidelines for SEA in Marine Spatial Planning for the German Exclusive Economic Zone (EEZ) — with Special Consideration of Tiering Procedure for SEA and EIA. In: Schmidt M., Glasson J., Emmelin L., Helbron H. (eds) *Standards and Thresholds for Impact Assessment. Environmental Protection in the European Union*, vol 3. Springer, Berlin, Heidelberg

Barbraud, C. & Thiebot, J.-B. (2009) On the importance of estimating detection probabilities from at-sea surveys of flying seabirds. *Journal of Avian Biology* 40: 584–590.

Barlow, J., Oliver, C.W., Jackson, T.D. & Taylor, B.L. (1988) Harbor porpoise, *Phocoena phocoena*, abundance estimation for California, Oregon, and Washington: II. Aerial surveys. *Fishery Bulletin* 86: 433–444.

Barrett, R.T., Anker-Nilssen, T., Erikstad, K.E. Lorentsen, S.-H. & Strøm, H. 2004. Initiating SEAPOP in the Lofoten and Barents Sea area? Report from the OLF study in 2004. NINA Minirapport 86, 11 pp. Retrieved 29 March 2019 from http://www.seapop.no/opencms/export/sites/SEAPOP/no/filer/pdf/nina-minirapport-86.pdf

Black, J., Dean, B.J., Webb, A., Lewis, M., Okill, D. & Reid, J.B. (2015) Identification of important marine areas in the UK for red-throated divers (*Gavia stellata*) during the breeding season. JNCC Report No 541. Peterborough: Joint Nature Conservation Committee. Retrieved 14 March 2018 from http://jncc.defra.gov.uk/page-6971

Borchers, D.L., Buckland, S.T. & Zucchini, W. (2002) *Estimating Animal Abundance: Closed Populations*. Berlin: Springer.

Borkenhagen, K., Corman, A.-M. & Garthe, S. (2018) Estimating flight heights of seabirds using optical rangefinders and GPS data loggers: a methodological comparison. *Marine Biology* 165: 17. doi: 10.1007/s00227-017-3273-z.

Bradbury, G., Trinder, M., Furness, R., Banks, A.N., Caldow, R.W.G. & Hume, D. (2014) Mapping seabird sensitivity to offshore wind farms. *PLoS ONE* 9(9): e106366. doi: 10.1371/journal.pone.0106366.

Brandt, M.J., Diederichs, A., Betke, K. & Nehls, G. (2011) Responses of harbour porpoises to pile driving at the Horns Rev II offshore wind farm in the Danish North Sea. *Marine Ecology Progress Series* 421: 205–216.

Briggs, K.T., Chu, E.W., Lewis, D.B., Tyler, W.B., Pitman, R.L. & Hunt, G.L., Jr (1978) *Distribution, numbers and seasonal status of seabirds of the Southern California Bight. Book 1, Part 3, Volume 3. Investigators' Reports, Summary of Marine Mammal and Seabird Surveys of the Southern California Bight Area, 1975–1978*. Santa Cruz, CA: University of California.

British Trust for Ornithology (BTO) (2018) BirdFacts. Retrieved 2 November 2018 from https://www.bto.org/about-birds/birdfacts

Brown, R.G.B., Nettleship, D.N., Germain, P., Tull, C.E. & Davis, T. (1975) *Atlas of Eastern Canadian Seabirds*. Ottawa: Canadian Wildlife Service.

BSH (2013) *Standard for Environmental Impact Assessment*. Hamburg: Federal Maritime and Hydrographic Society of Germany (BSH). Retrieved 27 June 2014 from http://www.bsh.de/en/Products/Books/Standard/7003eng.pdf

Buckland, S.T., Anderson, D.R., Burnham, K.P., Laake, J.L., Borchers, D.L. & Thomas, L. (2001) *Introduction to Distance Sampling: Estimating abundance of biological populations*. Oxford: Oxford University Press.

Buckland, S.T., Anderson, D.R., Burnham, K.P., Laake, J.L., Borchers, D.L. & Thomas, L. (2004) *Advanced Distance Sampling: Estimating abundance of biological populations*. Oxford: Oxford University Press.

Buckland, S.T., Burt, L.M., Rexstad, E.A., Mellor, M., Williams, A.E. & Woodward, R. (2012) Aerial surveys of seabirds: the advent of digital methods. *Journal of Applied Ecology* 49: 960–967.

Burt, M.L, Rexstad, E. & Buckland, S.T. (2010) Comparison of design- and model-based estimates of seabird abundance derived from visual, digital still transects and digital video aerial surveys in Carmarthen Bay. COWRIE Ltd. Retrieved 27 March 2019 from https://tethys.pnnl.gov/sites/default/files/publications/Burt,%20Rexstad,%20and%20Buckland%202010.pdf

Busch, M., Clough, S. & Rehfisch, M. (2014) Digital aerial surveys in Germany: the APEM experience. Presentation at BSH Workshop on the use of standardised digital survey methods for environmental impact assessment studies in German offshore wind farms, Hamburg, 15 October 2014. Retrieved 5 January 2015 from http://www.bsh.de/de/Meeresnutzung/Wirtschaft/Windparks/Windparks/Workshops/Digital_Survey_Methods/APEM_experiences_after_one_year_of_digital_aerial_surveys_in_Germany.pdf

Camphuysen, C.J. & Garthe, S. (2004) Recording foraging seabirds at sea standardised recording and coding of foraging behaviour and multi-species foraging associations. *Atlantic Seabirds* 6: 1–32.

Camphuysen, C.J. & Leopold, M.F. (1994) *Atlas of Seabirds in the Southern North Sea*. IBN Research Report 94/6, NIOZ-Rapport 1994-8. Texel: Institute for Forestry and Nature Research, Netherlands Institute for Sea Research and Dutch Seabird Group.

Camphuysen, C.J. & Webb, A. (1999) Multi-species feeding associations in North Sea seabirds: jointly exploiting a patchy environment. *Ardea* 87: 177–198.

Camphuysen, C.J., Fox, A.D., Leopold, M.F. & Petersen, I.K. (2004) Towards standardised seabirds at sea census techniques in connection with environmental impact assessments for offshore wind farms in the UK: a comparison of ship and aerial sampling methods for marine birds, and their applicability to offshore wind farm assessments. NIOZ Report No. BAM-02-2002 to COWRIE. Texel: Royal Netherlands Institute for Sea Research (NIOZ). Retrieved 28 April 2004 from http://jncc.defra.gov.uk/PDF/Camphuysenetal2004_COWRIEmethods.PDF

Centrica Energy (2007) *Environmental Statement for Lincs Offshore Wind Farm: Volume 1. Offshore Works*. Uxbridge: Centrica (in association with RES & AMEC).

Certain, G., Masse, J., van Canneyt, O., Petitgas, P., Doremus, G., Santos, M.B. & Ridoux, V. (2011) Investigating the coupling between small pelagic fish and marine top predators using data collected from ecosystem-based surveys. *Marine Ecology Progress Series* 422: 23–39.

Certain, G., Jørgensen, L.L., Christel, I., Planque, B. & Bretagnolle, V. (2015) Mapping the vulnerability of animal community to pressure in marine systems: disentangling pressure types and integrating their impact from the individual to the community level. *ICES Journal of Marine Science* 72: 1470–1482.

Cleasby, I.R., Wakefield, E.D., Bearhop, S., Bodey, T.W., Votier, S.C. & Hamer, K.C. (2015) Three-dimensional tracking of a wide-ranging marine predator: flight heights and vulnerability to offshore wind farms. *Journal of Applied Ecology* 52: 1474–1482.

Connelly, E.E., Stenhouse, I.J., Williams, K.A. & Veit, R.R. (2015) Boat survey protocol for mid-Atlantic baseline studies. In Williams, K.A., Connelly, E.E., Johnson, S.M. & Stenhouse, I.J. (eds) Wildlife densities and habitat use across temporal and spatial scales on the mid-Atlantic outer continental shelf. Final Report to the Department of Energy EERE Wind & Water Power Technologies Office. Award No. DE-EE0005362. Report BRI 2015-11. Portland, ME: Biodiversity Research Institute. Retrieved 15 January 2016 from http://www.briloon.org/uploads/BRI_Documents/Wildlife_and_Renewable_Energy/MABS%20Project%20Chapter%207%20-%20Connelly%20et%20al%202015.pdf

Coppack, T., Weidauer, A. & Kemper, G. (2015) Erfassung von Seevogel- und Meeressäugerbeständen mittels georeferenzierter Digitalfotografie. *AGIT – Journal für Angewandte Geoinformatik* 1-2015. doi: 10.14627/537557050.

Dean, B.J., Webb, A., McSorley, C.A. & Reid, J.B. (2004) Surveillance of wintering seaduck, divers and grebes in UK inshore areas: aerial surveys 2003/03. JNCC Report No. 345. Peterborough: Joint Nature Conservation Committee. Retrieved 1 August 2014 from http://jncc.defra.gov.uk/page-3626

Department of Energy and Climate Change (DECC) (2016) UK offshore energy strategic environmental assessment 3 post-consultation report July 2016. London: DECC. Retrieved 14 February 2017 from https://www.gov.uk/government/uploads/system/uploads/attachment_data/file/536672/OESEA3_Post_Consultation_Report.pdf

Diederichs, A., Nehls, G. & Petersen, I.K. (2002) Flugzeugzählungen zur großflächigen Erfassung von Seevögeln und marinen Säugern als Grundlage für Umweltverträglichkeitsstudien im Offshorebereich. *Seevögel* 23: 38–46.

Dierschke, V., Furness, R.W. & Garthe, S. (2016) Seabirds and offshore wind farms in European waters: avoidance and attraction. *Biological Conservation* 202: 59–68.

Dudgeon Offshore Wind Limited (2010) Update to the ornithological assessment of the Dudgeon Offshore Wind Farm with 2009 survey data. Technical Report. Wellesbourne: Dudgeon Offshore Wind Ltd, c/o Warwick Energy Ltd. Retrieved 15 July 2017 from http://www.marinedataexchange.co.uk/ItemDetails.aspx?id=871

Embling, C.B., Illian, J., Armstrong, E., van der Kooij, J., Sharples, J., Camphuysen, C.J. & Scott, B.E. (2012) Investigating fine-scale spatio-temporal predator–prey patterns in dynamic marine ecosystems: a functional data analysis approach. *Journal of Applied Ecology* 49: 481–492.

EU (1979) Council Directive 79/409/EEC of 2 April 1979 on the conservation of wild birds Official Journal L103, 25/04/1979 0001-0018 (The 'Birds Directive').

EU (1992) Council Directive 92/43/EEC of 21 May 1992 on the conservation of natural habitats and of wild fauna and flora. Official Journal L206, 22/07/1992 0007-0050 (The 'Habitats Directive').

Fauchald, P., Erikstad, K.E. & Skarsfjord, H. (2000) Scale-dependent predator–prey interactions: the hierarchical spatial distribution of seabirds and prey. *Ecology* 81: 773–783.

Forewind (2013) Dogger Bank Creyke Beck Environmental Statement Chapter 11. Appendix A: BTO Ornithology Technical Report, sub-Appendix 5. Retrieved 13 December 2015 from http://www.forewind.co.uk/uploads/files/Creyke_Beck/Application_Documents/6.11.1_Chapter_11_Appendix_A_Creyke_Beck_A_and_B_Ornithology_Technical_Report_-_Application_Submission_F-OFC-CH-011.pdf

Garthe, S., Alicki, K., Hüppop, O. & Sprotte, B. (1995) Die Verbreitung und Häufigkeit ausgewählter See- und Küstenvogelarten während der Brutzeit in der südöstlichen Nordsee. *Journal of Ornithology* 136: 253–266.

Gould, P.J. (1974) Introduction. In Kind, W.B. (ed.) *Pelagic Studies of Seabirds in the Central and Eastern Pacific Ocean*. Smithsonian Contributions to Zoology No. 158. Washington, DC: Smithsonian Institution Press. pp. 1–5.

Gould, P.J., Harrison, C.S. & Forsell, D.J. (1978) Distribution and abundance of marine birds – south and east Kodiak Island waters. In Annual Report of the Research Unit N. 337. Annual reports of the principal investigators for the year ending March 1978, Volume 2. Boulder, CO: NOAA. pp. 614–710.

Groom, G., Stjernholm, M., Nielsen, R.D., Fleetwood, A. & Petersen, I.K. (2013) Remote sensing image data and automated analysis to describe marine bird distributions and abundances. *Ecological Informatics* 14: 2–8.

Haney, J.C., Geiger, H.J. & Short, J.W. (2014) Bird mortality from the Deepwater Horizon oil spill. I. Exposure probability in the offshore Gulf of Mexico. *Marine Ecology Progress Series* 513: 225–237.

Harwood, A.J.P., Perrow, M.R., Berridge, R., Tomlinson, M.L. & Skeate, E.R. (2017) Unforeseen responses of a breeding seabird to the construction of an offshore wind farm. In Köppel, J. (ed.) *Wind Energy and Wildlife Interactions. Presentations from the CWW2015 conference*. Cham: Springer. pp. 19–41.

Harwood, A.J.P., Perrow, M.R. & Berridge, R.J. (2018) Use of an optical rangefinder to assess the reliability of seabird flight heights from boat-based surveyors: implications for collision risk at offshore wind farms. *Journal of Field Ornithology* 89: 372–383.

Heinänen, S., Zydelis, R., Dorsch, M., Nehls, G. & Skov, H. (2017) High-resolution sea duck distribution modeling: relating aerial and ship survey data to food resources, anthropogenic pressures, and topographic variables. *The Condor: Ornithological Applications* 119: 175–190.

Heinemann, D. (1981) A rangefinder for pelagic bird censusing. *Journal of Wildlife Management* 45: 489–493.

Hiby, L. & Lovell, P. (1998) Using aircraft in tandem formation to estimate abundance of harbor porpoise. *Biometrics* 54: 1280–1289.

Hodgson, A., Kelly, N. & Peel, D. (2013) Unmanned aerial vehicles (UAVs) for surveying marine fauna: a Dugong case study. *PLoS ONE* 8(11): e79556. doi: 10.1371/journal.pone.0079556.

Hyrenbach, D., Henry, M.F., Morgan, K.H., Welch, D.W. & Sydemann, W.J. (2007) Optimizing the width of strip transects for seabird surveys from vessels of opportunity. *Marine Ornithology* 35: 29–38.

Jameson, H., Reeve, E., Laubek, B. & Sittel, H. (2019) The nature of offshore wind farms. In Perrow,

M.R. (ed.) *Wildlife and Wind Farms, Conflicts and Solutions. Volume 3. Offshore: Potential effects.* Exeter: Pelagic Publishing. pp. 1–29.

Jesperson, P. (1924) The frequency of birds over the high Atlantic Ocean. *Nature* 114: 281–283.

Johnston, A., Cook, A.S., Wright, L.J., Humphreys, E.M. & Burton, N.H. (2014) Modelling flight heights of marine birds to more accurately assess collision risk with offshore wind turbines. *Journal of Applied Ecology* 51: 31–41.

Johnston, A., Thaxter, C.B., Austin, G.E., Cook, A.S.C.P., Humphreys, E.M., Still, D.A., Mackay, A., Irvine, R., Webb, A. & Burton, N.H.K. (2015) Modelling the abundance and distribution of marine birds accounting for uncertain species identification. *Journal of Applied Ecology* 52: 150–160.

Johnston, A. & Cook, A.S.P.C. (2016). How high do birds fly? Development of methods and analysis of digital aerial data of seabird flight heights. British Trust for Ornithology Research Report Number 676. Retrieved 10 April 2018 from https://www.bto.org/file/337905/download?token=bAbLpuAG

Joint Nature Conservation Committee (JNCC) (2017) Training in marine bird surveys from boats. Peterborough: JNCC. Retrieved 30 April 2018 from http://jncc.defra.gov.uk/page-4568.

Kahlert, J., Desholm, M., Clausager, I. & Petersen, I.K. (2000) Environmental impact assessment of an offshore wind park at Rødsand. Technical report on birds. Rønde: NERI. Retrieved 27 January 2002 from http://www.folkecenter.eu/FC_old/www.folkecenter.dk/mediafiles/folkecenter/pdf/Final_results_of_bird_studies_at_the_offshore_wind_farms_at_Nysted_and_Horns_Rev_Denmark.pdf

Komdeur, J., Bertelsen, J. & Cracknell, G. (1992) *Manual for Aeroplane and Ship Surveys of Waterfowl and Seabirds.* IWRB Special Publication No. 19. Slimbridge: International Wildfowl Research Bureau.

Krägefsky, S. (2014) Effects of the Alpha Ventus offshore test site on pelagic fish. In BSH & BMU (eds) *Ecological Research at the Offshore Windfarm* Alpha Ventus*: Challenges, results and perspectives*. Federal Maritime and Hydrographic Society of Germany (BSH) and Federal Ministry for the Environment, Nature Conservation and Nuclear Safety (BMU). Weisbaden: Springer Spektrum. pp. 83–94.

Lensink, R., van Gasteren, H., Hustings, F., Buurma, L., van Duin, G., Linnartz, L., Vogelzang, F. & Witkamp, C. (2002) *Vogeltrek over Nederland 1976–1993*. Haarlem: Schuyt & Co.

Leopold, M.F., van Bemmelen, R.S.A. & Zuur, A.F. (2013) Responses of local birds to the offshore wind farms PAWP and OWEZ off the Dutch mainland coast. Report C151/12. Wageningen: IMARES. Retrieved 10 July 2017 from http://edepot.wur.nl/279573

Mackenzie, M.L., Scott-Hayward, L.A.S., Paxton, C.G. & Burt, M.L. (2017) Quantifying the power to detect change: methodological development and implementation using the R package MRSeaPower. Retrieved 25 May 2018 from https://www.creem.st-andrews.ac.uk/software/

Marsh, H. & Sinclair, D.F. (1989) Correcting for visibility bias in strip transect aerial surveys of aquatic fauna. *Journal of Wildlife Management* 53: 1017–1024.

Mendel, B., Peschko, V., Kubetzki, U., Weiel, S. & Garthe, S. (2018) Untersuchungen zu möglichen Auswirkungen der Offshore-Windparks im Windcluster nördlich von Helgoland auf Seevögel und Meeressäuger (HELBIRD). University of Kiel. Retrieved 12 December 2018 from http://www.ftz.uni-kiel.de/de/forschungsabteilungen/ecolab-oekologie-mariner-tiere/abgeschlossene-projekte/helbird

Mendel, B., Schwemmer, P., Peschko, V., Muller, S., Schwemmer, H., Mercker, M. & Garthe, S. (2019) Operational offshore wind farms and associated ship traffic cause profound changes in distribution patterns of Loons (Gavia spp.). *Journal of Environmental Management* 231: 429-438.

Met Office (2016) Marine forecasts glossary. Retrieved 20 July 2017 from http://www.metoffice.gov.uk/guide/weather/marine/glossary

Miller, D.L. (2017) Distance: distance sampling detection function and abundance estimation. R package version 0.9.7. Retrieved 6 November 2018 from https://CRAN.R-project.org/package=Distance

Natural Environment Research Council (NERC) (1977) The report of a working group on ecological research on seabirds. NERC Publication Series C, No. 18. London: Natural Environment Research Council.

Normandeau Associates (2013) High-resolution aerial imaging surveys of marine birds, mammals, and turtles on the US Atlantic outer continental shelf – utility assessment, methodology recommendations, and implementation tools. Report prepared under BOEM Contract No. M10PC00099. Herndon, VA: US Department of the Interior, Bureau of Ocean Energy Management,. Retrieved 9 March 2014 from https://www.boem.gov/ESPIS/5/5272.pdf

Offringa, H., Seys, J., Bossche, W. & van den Meire, P. (1995) *Seabirds on the Channel Doormat*. Report IN 95.12. Hasselt: Instituut voor Natuurbehoud.

Parsons, M., Lawson, J., Lewis, M., Lawrence, R. & Kuepfer, A. (2015) Quantifying foraging areas of little tern around its breeding colony SPA during chick-rearing. JNCC Report No. 548. Peterborough: Joint Nature Conservation Committee. Retrieved 3 February 2017 from http://jncc.defra.gov.uk/page-6976

Percival, S. (2014) Kentish Flat offshore wind farm: diver surveys 2011–12 and 2012–13. Durham: Ecology Consulting. Retrieved 25 January 2016 from https://corporate.vattenfall.co.uk/globalassets/uk/projects/redthroated-diver-2014.pdf

Perrow, M.R. (2019) A synthesis of effects and impacts. In Perrow, M.R. (ed.) *Wildlife and Wind Farms, Conflicts and Solutions. Volume 3. Offshore: Potential effects*. Exeter: Pelagic Publishing. pp. 235–277.

Petersen, I.K., Nielsen, R.D. & Mackenzie, M.L (2014) Post-construction evaluation of bird abundances and distribution in the Horns Rev 2 offshore wind farm area, 2011 and 2012. Report commissioned by DONG Energy Ltd. Arhus University, Danish Centre for Environment and Energy. Retrieved 29 March 2019 from http://birdlife.se/wp-content/uploads/2019/01/Bird-abundances-and-distributions_Evaluation_Horns_Rev_2.pdf

Poot, M.J.M., Fijn, R.C., Jonkvorst, R.J., Heunks, C., Collier, M.P., de Jong, J. & van Horssen, P.W. (2011) Aerial surveys of seabirds in the Dutch North Sea May 2010–April 2011: seabird distribution in relation to future offshore wind farms. Report No. 10-235 to IMARES. Culemborg: Bureau Waardenburg. Retrieved 21 June 2016 from https://www.researchgate.net/profile/Martin_Poot2/publication/306038249_Aerial_surveys_of_seabirds_in_the_Dutch_North_Sea_May_2010_-_April_2011_-_Seabird_distribution_in_relation_to_future_offshore_wind_farms/links/57ac33d408ae42ba52b0df4b/Aerial-surveys-of-seabirds-in-the-Dutch-North-Sea-May-2010-April-2011-Seabird-distribution-in-relation-to-future-offshore-wind-farms.pdf

R Core Team (2018) R: a language and environment for statistical computing. Vienna: R Foundation for Statistical Computing. Retrieved 6 November 2018 from https://www.R-project.org/.

Reid, J.B. & Camphuysen, C.J. (1998) The European Seabirds at Sea database. *Biological Conservation* 102: 291.

Ronconi, R.A. & Burger, A.E. (2009) Estimating seabird densities from vessel transects: distance sampling and implications for strip transects. *Aquatic Biology* 4: 297–309.

Schwemmer, P. & Garthe, S. (2005) At-sea distribution and behaviour of a surface-feeding seabird, the lesser black-backed gull *Larus fuscus*, and its association with different prey. *Marine Ecology Progress Series* 285: 245–258.

Scott, B.E., Sharples, J., Ross, O.N., Wang, J., Pierce, G.J. & Camphuysen, C.J. (2010) Sub-surface hotspots in shallow seas: fine-scale limited locations of top predator foraging habitat indicated by tidal mixing and sub-surface chlorophyll. *Marine Ecology Progress Series* 408: 207–226.

Scott-Hayward, L.A.S., Oedekoven, C.S., Mackenzie, M.L., Walker, C.G. & Rexstad, E. (2013) User guide for the MRSea Package: statistical modelling of bird and cetacean distributions in offshore renewables development areas. University of St Andrews contract for Marine Scotland. SB9 (CR/2012/05). Retrieved 27 December 2017 from https://tethys.pnnl.gov/sites/default/files/publications/Scott-Hayward-et-al-2013-b.pdf

Seagreen Wind Energy (2012) ES volume I. Chapter 10. Ornithology. Seagreen Wind Energy Ltd. Document No. A4MR-SEAG-Z-DOC100-SPR-060. Retrieved 5 November 2018 http://marine.gov.scot/datafiles/lot/SG_FoF_alpha-bravo/SG_Phase1_Offshore_Project_Consent_Application_Document%20(September%202012)/006%20ES/Volume%20I_Main%20Text/A4MRSEAG-Z-DOC100-SPR-060_ES_10.pdf

Skov, H. & Durinck, J. (2001) Seabird attraction to fishing vessels is a local process. *Marine Ecology Progress Series* 214: 289–298.

Skov, H., Durinck, J. Leopold, M.F. & Tasker, M.L. (1995) *Important Bird Areas for Seabirds in the North Sea*. Cambridge: BirdLife International.

Skov, H., Durinck, J., Leopold, M. & Tasker, M.L. (2007) A quantitative method for evaluating the importance of marine areas for conservation of birds. *Biological Conservation* 136: 362–371.

Spear, L.B., Nur, N. & Ainley, D.G. (1992) Estimating absolute densities of flying seabirds using analyses of relative movement. *Auk* 109:385–389.

Spear, L.B., Ainley, D.G., Hardesty, B.D., Howell, S.N.G. & Webb, S.W. (2004) Reducing biases affecting at-sea surveys of seabirds: use of multiple observer teams. *Marine Ornithology* 32: 147–157.

Stone, C J., Webb, A., Barton, C., Ratcliffe, N., Reed, T.C., Tasker, M.L., Camphuysen, C.J. & Pienkowski, M.W. (1995) *An Atlas of Seabird Distribution in Northwest European Waters*. Peterborough: Joint Nature Conservation Committee.

Tasker, M.L., Jones, P.H., Dixon, T.J. & Blake, B.F. (1984) Counting seabirds at sea from ships: a review of methods employed and a suggestion for a standardized approach. *Auk* 101: 567–577.

Thaxter, C.B. & Burton, N.H.K. (2009) High definition imagery for surveying seabirds and marine mammals: a review of recent trials and development of protocols. British Trust for Ornithology Report Commissioned by COWRIE Ltd. Thetford: British Trust for Ornithology. Retrieved 14 October 2011 from https://tethys.pnnl.gov/sites/default/files/publications/Thaxter-Burton-2009.pdf

Thaxter, C.B., Ross-Smith, V.H. & Cook, A.S.C.P. (2016) How high do birds fly? A review of current datasets and an appraisal of current methodologies for collecting flight height data: literature review. BTO Research Report No. 666. Retrieved 18 March 2016 from https://www.bto.org/research-data-services/publications/research-reports/2016/how-high-do-birds-fly-review-current

Thomas, L., Buckland, S.T., Rexstad, E.A., Laake, J.L., Strindberg, S., Hedley, S.L., Bishop, J.R.B, Marques, T.A. & Burnham, K.P. (2009) Distance software: design and analysis of distance sampling surveys for estimating population size. *Journal of Applied Ecology* 47: 5–14.

Vallejo, G.C., Grellier, K., Nelson, E.J., McGregor, R.M., Canning, S.J., Caryl, F.M. & McLean, N. (2017) Responses of two marine top predators to an offshore wind farm. *Ecology and Evolution* 7: 8698–8708.

van der Meer, J. & Leopold, M.F. (1995) Assessing the population size of the European storm petrel (*Hydrobates pelagicus*) using spatial autocorrelation between counts from segments of criss-cross ship transects. *ICES Journal of Marine Science* 52: 809–818.

Vanermen, N. & Stienen, E.W.M. (2019) Seabirds: displacement. In Perrow, M.R. (ed.) *Wildlife and Wind Farms, Conflicts and Solutions. Volume 3. Offshore: Potential effects.* Exeter: Pelagic Publishing. pp. 174–205.

Vanermen, N., Onkelinx, T., Courtens, W., van de Walle, M., Verstraete, H. & Stienen, E.W.M. (2015a) Seabird avoidance and attraction at an offshore wind farm in the Belgian part of the North Sea. *Hydrobiologia* 756: 51–61.

Vanermen, N., Onkelinx, T., Verschelde, P., Courtens, W., van de Walle, M., Verstraete, H. & Stienen, E.W.M. (2015b) Assessing seabird displacement at offshore wind farms: power ranges of a monitoring and data handling protocol. *Hydrobiologia* 756: 155–167.

Webb, A. & Durinck, J. (1992) Counting birds from ship. In Komdeur, J., Bertelsen, J. & Cracknell, G. (eds) *Manual for Aeroplane and Ship Surveys of Waterfowl and Seabirds.* IWRB Special Publication No. 19. Slimbridge: International Wildfowl Research Bureau. pp. 24–37.

Webb, A., Irwin, C., Mackenzie, M., Scott-Hayward, L., Caneco, B. & Donovan, C. (2017) Lincs Wind Farm: third annual post-construction aerial ornithological monitoring report. Report by HiDef Aerial Surveying Ltd to Centrica Renewable Energy Ltd. CREL Report No. LN-E-EV-013-0006-400013-007. Retrieved 2 February 2017 from http://www.marinedataexchange.co.uk/ItemDetails.aspx?id=7022

Weiß, F., Buettger, H., Baer, J., Welcker, J. & Nehls, G. (2016) Erfassung von Seevögeln und Meeressäugetieren mit dem HiDef Kamerasystem aus der Luft. *Seevögel* 37(2): 14–21.

Welcker, J. & Nehls, G. (2016) Displacement of seabirds by an offshore wind farm in the North Sea. *Marine Ecology Progress Series* 554: 173–182.

Wiens, J.A., Heinemann, D. & Hoffman, W. (1978) Community structure, distribution and interrelationships of marine birds in the Gulf of Alaska. Final reports of the principal investigators, Volume 3. Boulder, CO: NOAA.

Williams, K.A., Connelly, E.E., Johnson, S.M. & Stenhouse, I.J. (eds) (2015) Wildlife densities and habitat use across temporal and spatial scales on the mid-Atlantic outer continental shelf (2012–2014): final report to the Department of Energy Office of EERE Wind & Water Power Technologies Office. Award No. DE-EE0005362. Report BRI 2015-11. Portland, ME: Biodiversity Research Institute. Retrieved 28 November 2015 from http://www.briloon.org/uploads/BRI_Documents/Wildlife_and_Renewable_Energy/MABS%20Project%202015.pdf

WindEurope (2018) Offshore wind in Europe: key trends and statistics 2017. Brussels: WindEurope. Retrieved 27 July 2018 from https://windeurope.org/wp-content/uploads/files/about-wind/statistics/WindEurope-Annual-Offshore-Statistics-2017.pdf

Zydelis, R., Dorsch, M., Heinänen, S., Nehls, G. & Weiss, F. (2019) Comparison of digital video surveys with visual aerial surveys for bird monitoring at sea. *Journal of Ornithology* 160: 567–580. https://doi.org/10.1007/s10336-018-1622-4.

CHAPTER 4

Telemetry and tracking of birds

CHRIS B. THAXTER and MARTIN R. PERROW

Summary

Tracking and telemetry are routinely used in ecological studies of birds. The attachment of telemetry devices (or 'tags') to individuals, can, at its simplest level, record the spatial movement of a species of interest, which has numerous applications in the assessment of interactions of marine birds and migrant waterbirds with offshore wind farms. The initial application of telemetry in Europe has expanded to include other countries where developments are being proposed. Telemetry is useful in: (1) scoping species and sites that could be affected by developments throughout the year at different scales, (2) identifying the exposure of species to particular effects of wind farms prior to construction, and (3) quantifying impacts and monitoring behaviour once the wind farm has been constructed. To scope species and protected sites, breeding season foraging range reviews and direct tracking information have been used to identify potential or realised connectivity between protected sites and wind farms, to quantify intensity of habitat use and potential overlap with wind farms, and to assess the scale at which wind farms may have bearing on species' populations. Telemetry has also been used to quantify the potential effects of displacement or avoidance, collision risk and barrier effects associated with constructed wind farms, thereby helping to validate assessment predictions. Fine-scale movements around individual wind turbines are now being revealed. Information from positional telemetry can be supplemented with additional sensors, such as those that record diving patterns or time–activity, which can help to identify behaviour and refine understanding of habitat use. The use of telemetry will undoubtedly increase in the future, facilitated by continuing technological developments that will allow new species to be tracked over longer periods, and advancements in analytical techniques within an increasingly diverse monitoring toolkit, thus providing new solutions to existing problems.

Introduction

The marine environment is coming under increasing pressure from human activities such as shipping, oil and gas, and offshore renewable energy developments (Halpern *et al.* 2008; Korpinen & Andersen 2016). In relation to wind farms, the birds present in the marine environment (Paleczny *et al.* 2015), including both seabirds and terrestrial or coastal birds migrating or moving over the surface of the sea (see the definition of 'marine birds' in *Scope*, below), are at risk of a range of impacts, including through direct collision (King 2019), barrier effects and displacement (Vanermen & Stienen 2019), with seabirds also subject to potential indirect and ecosystem effects, particularly through their prey (Perrow 2019). The position of a bird through time can yield valuable information on proximity or overlap with anthropogenic factors, including wind farms (Hart & Hyrenbach 2009; Waggit & Scott 2014; Cleasby *et al.* 2015; Thaxter *et al.* 2018a). While traditional survey methods using boats or aircraft (see Webb & Nehls, Chapter 3) remain crucial in determining baseline distribution and abundance in the area of interest for Environmental Impact Assessment (EIA) of wind farms, such surveys cannot generally 'track' birds, which requires a sequence of positional data to be gathered ('tracking'). This is the basic remit of 'telemetry', as derived from the Greek words *tele*, meaning far, and *metros*, meaning measurement. Although the resulting definition of 'remote measurement of data' can therefore cover a range of survey platforms, such as aerial surveys using high-definition imagery (Webb & Nehls, Chapter 3) as well as radar studies (e.g. Masden *et al.* 2010; Molis *et al.*, Chapter 6), the term 'telemetry' is generally associated with the more specific use of animal-borne telemetry. In this, the individual itself provides information that is stored or relayed back to the observer, generally via some form of 'tag', here also referred to as a tracking or telemetry device (Box 4.1), thus allowing the observation of animals from their own perspective as opposed to direct observations from a certain distance (Ropert-Coudert & Wilson 2005).

Box 4.1 A brief review of the development of telemetry studies on marine birds and migratory waterbirds

Pioneering telemetry studies, for instance using archival depth recorders and capillary tubes, were conducted as early as the 1940s (e.g. Scholander 1940; Eliassen 1960). Since the 1980s, passive integrated responder (PIT) tags have been widely used to study animal movements (Gibbons & Andrews 2004), although as these use communication over very short distances between the tag and receiver, the animal either has to be captured or must pass close to the receiver. For birds in relation to marine environment, such devices have been limited to studying aspects such as nest-site attendance (e.g. Le Maho *et al.* 1993). Tracking studies that measured the location of animals were also first undertaken in the 1980s (Ropert-Coudert & Wilson 2005) using radio-tracking and ground-based receivers (Amlaner & McDonald 1980; Kenward 1987; Wanless *et al.* 1990). Radio-tracking has limitations of distance due to signal retention and line-of-sight issues, and radio signals cannot penetrate seawater (Wilson & Vandenabele 2012). Paradoxically, this limitation has proved useful in observing the behaviour of diving species, as the gap between signals may be used to indicate diving duration (Wanless *et al.* 1993). Furthermore, radio-telemetry is often an attractive, or in some cases the only, method available for smaller species, owing

Figure 4.1 Examples of different telemetry devices used to track species at sea: (a) Lesser Black-backed Gull *Larus fuscus* with a GPS tag (Gary Clewley); (b) Great Skua *Stercorarius skua* with a GPS tag (Tim Stenton); (c) Whooper Swan *Cygnus cygnus* with a GPS/PTT transmitter (WWT); (d) Atlantic Puffin *Fratercula arctica* with a geolocator (Eleanor Wood); (e) Little Tern *Sternula albifrons* with a radio-transmitter (Martin Perrow); (f) Barnacle Goose *Branta leucopsis* with a GPS/PTT transmitter (WWT); (g) Razorbill *Alca torda* with a time–depth recorder (Chris Thaxter); and (h) Black-legged Kittiwake *Rissa tradactyla* with a GPS tag (Kane Brides).

to the availability of very small tags. Individual birds that have been tagged with radio-transmitters have also been followed offshore from boats or aircraft (Heath & Randall 1989; Perrow *et al.* 2006; Adams & Takekawa 2008) (Figure 4.1). Coordinated radio-telemetry networks have been developed more recently, providing a powerful tool to understand broader migration patterns (Taylor *et al.* 2017).

The onset of satellite relay systems permitted the gathering of locational data on animals anywhere on the planet (e.g. Jouventin & Weimerskirch 1990; Hart & Hyrenbach 2009). Platform terminal transmitters (PTTs) record location through the Doppler effect, whereby a shift in wavelength radio transmissions occurs as the satellite passes overhead, allowing position to be determined (Maxwell 1971). Data can be transmitted back to the user through the Argos system, which has been widely used on pelagic seabirds and waterbirds (Phillips *et al.* 2008; Griffin *et al.* 2011) (see Figure 4.1). Tracking devices that use dead-reckoning methods with heading and speed can also provide similar locational data (Wilson *et al.* 1991; Thaxter *et al.* 2010). In the early 1990s, the geolocation sensor (GLS) method was developed using light sensors that enable latitude and longitude to be derived using daylength and local timing of midday (Hill 1994). Geolocation is an archival technology requiring recovery of the tag by the user to obtain the data. Several important geolocation studies have taken place since the 1990s revealing the migration behaviour of many marine birds (Guilford *et al.* 2009; Harris *et al.* 2010) (Figure 4.1).

Geolocation suffers in accuracy of position (*ca.* 200 km), as issues of constant daylight or darkness around the poles and equal daylength at the equinoxes, respectively, prevent longitude and latitude being determined (e.g. Phillips *et al.* 2004). PTTs are advantageous in that they relay information directly to the observer, allowing completely remote capability. However, PTTs using the Doppler effect to determine position suffer from problems of accuracy, typically up to a few hundred metres (Weimerskirch *et al.* 1992), down to 150 m at best (Bridge *et al.* 2011). Further refinements in locational precision have been made through archival Global Positioning System (GPS) technology, which can have locational accuracy of up to a few metres (Bridge *et al.* 2011). Early GPS tags were expensive, but now a wider range of more affordable devices has allowed many individuals of study populations to be tracked over shorter periods. For example, using low-cost and lightweight archival GPS tags (see Figure 4.1), the collaborative Future of the Atlantic Marine Environment (FAME) project, at the time of writing, has tracked more than 2,000 individuals of 12 species from 40 marine bird colonies since 2010 (RSPB 2018). Archival telemetry devices require recapture of the bird to recover the data. Combined approaches, however, have brought together the benefits of more accurate GPS position and the transmission advantages of satellite delivery systems such as Argos into one device (e.g. Microwave Telemetry 2018). GPS tags have also been developed that can record data very frequently, such as every few seconds, and relay that information back to the user remotely while birds are within range of a receiver (Bouten *et al.* 2013; UvA-BiTS 2018).

Further steps forward in remote capabilities have also been made through the use of Global System for Mobile Communications, previously Groupe Spécial Mobile, otherwise known as GSM mobile phone cellular networks using GPS/GSM tags (Griffin *et al.* 2016; Scragg *et al.* 2016; Spiegel *et al.* 2017). These tags relay information directly to the user and, although dependent on network coverage, are useful in situations where there may be uncertainty over whether a bird will return within the vicinity of a local fixed short-range receiver, or where access to locations to download data remotely from individual birds is not feasible. Further initiatives are continually being developed. For example, the Icarus (International Cooperation for Animal Research Using Space) project (Icarus 2018) aims to monitor the migratory movements of small animals and involves deployment of a satellite system on the International Space Station. National programmes are also being developed to facilitate monitoring of aquatic and marine animal movements; for example, the US Animal Telemetry Network seeks to track the movements of species over different spatial and temporal scales for species across multiple sites, facilitating data sharing and interdisciplinary research (Block *et al.* 2016). A cited use of this approach is to ascertain potential interactions of species with wind farms. The Motus Wildlife Tracking System is another example of a network that uses radio-telemetry arrays to investigate the study of small animals, such as passerines fitted with tags of <0.21 g (Taylor *et al.* 2017). Indeed, the weight of all tags is continually reducing, with geolocators at <1 g and GPS devices <3 g now available, permitting ever more marine bird species to be considered for tagging (e.g. Bridge *et al.* 2011).

Telemetry sensors can also reveal many other aspects of species biology and behaviour. Early studies using depth gauges based on capillary tubes have been superseded by hi-tech pressure loggers that record time and depth for species diving underwater (Daunt *et al.* 2003) (Figure 4.1). Simply by using only the GPS fixes where diving data were also recorded by time–depth recorders (TDRs), Harris *et al.* (2012) were better able to identify key foraging areas of Atlantic Puffins from the Isle of May (see Figure 4.3 below). Diving activity may also reveal valuable information on predator–prey interactions, such as likely food obtained, through diving activity signatures (Elliott *et al.* 2008). Activity loggers have been used to reveal different behaviours while birds are at sea (Benvenuti *et al.* 2001). Accelerometery has been widely used to study diving and flight behaviour and refine our understanding of activity patterns, using measurements of gravity and speed in multiple axes of movement (Watanuki *et al.* 2003; Weimerskirch *et al.* 2005; Shamoun-Baranes *et al.* 2017). Other examples of telemetry include altimeters to measure flight height (Weimerskirch *et al.* 2005; Cleasby *et al.* 2015), heart-rate loggers to measure metabolic rates and energy expenditure (e.g. Bevan *et al.* 1997; Ponganis 2007), prey ingestion loggers, including beak sensors (Wilson *et al.* 2007) and internal temperature loggers (Wilson *et al.* 1992) to understand prey consumption, wet/dry loggers or saltwater switches to indicate movements in and out of water (Pinaud & Weimerskirch 2005) and tags that measure environmental features (Boehlert *et al.* 2001). Finally, camera-logger technology is also now permitting a direct 'bird's-eye view' of habitat use and diet (e.g. Watanuki *et al.* 2008; Votier *et al.* 2013).

As well as answering the fundamental question of how far birds range or travel (Wakefield *et al.* 2009), data obtained from telemetry can identify key habitat used by marine birds offshore that is essential to determine whether birds are using areas of sea that may also be developed for offshore renewable energy, and if particular habitats within those areas are used more than others. Understanding the impacts of an offshore wind farm (OWF) development on a particular population is a crucial aspect to quantify before any wind farm is consented (Bailey *et al.* 2014; Horswill *et al.* 2017). In Europe, important populations of birds, including breeding marine birds, are protected as features of Special Protection Areas (SPAs), designated under the European Union's (EU's) Birds Directive (2009/147/EC), which together with Special Areas of Conservation (SACs), designated under the EU's Habitats Directive (92/43/EEC), form the Europe-wide Natura 2000 network. If birds within the population are to be affected, with implications for the integrity of the site, this may trigger a Habitats Regulations Assessment (HRA).

Studies using telemetry are increasingly being conducted to determine location in space and time and quantify potential bird–wind-farm interactions (e.g. Perrow *et al.* 2006; Langston *et al.* 2013; Wade *et al.* 2014; Thaxter *et al.* 2015), and are providing further information using other sensors, greatly enhancing our knowledge of how animals are using their environment. For example, acceleration or activity loggers can distinguish behaviours and refine our understanding of habitat use (Weimerskirch *et al.* 2005; Bouten *et al.* 2013; Shamoun-Baranes *et al.* 2016), and time-depth recorders monitor diving behaviour and together with positional information can help to reveal important foraging sites in relation to OWFs (Harris *et al.* 2012; T. Cook *et al.* 2012). Altimeters and Global Positioning System (GPS)-derived flight height information are increasingly being used to demonstrate flight heights of marine birds to determine whether this brings them into potential conflict with turbine rotors (e.g. Cleasby *et al.* 2015; Ross-Smith *et al.* 2016; Thaxter *et al.* 2018a). In turn, estimating the risk of collision remains a key component of wind-farm EIA and assessment of risks for protected species such as within HRAs for birds from SPAs within Europe (see Cook & Masden, Chapter 5).

The considerable and rapid advances in telemetry systems used to track marine birds, especially since the 1990s, has been the subject of a number of reviews (Ropert-Coudert & Wilson 2005; Hooker *et al.* 2007; Burger & Shaffer 2008; Hart & Hyrenbach 2009; Bridge *et al.* 2011; Brown *et al.* 2013). Development continues apace, particularly as a result of miniaturisation of tags allowing smaller, more efficient devices to be used on a wider range of species, and a wider range of sensors within them to study different aspects of behaviour (Box 4.1, Figure 4.1). Box 4.1 provides a chronological overview of the advancement of different technologies and retrieval systems and the evolution of additional telemetry sensors that have been used to date on marine birds. These developments go hand-in-hand with improvements in the ways that tags are attached to marine birds, and the critical aspects of bird welfare considerations that are further outlined in Box 4.2. Telemetry methods can be grouped by the different tracking technologies used and the data-retrieval system to acquire the data, such as satellite relay, fixed or mobile receivers or archival storage (see Box 4.1). When referring to individual types of tags, 'satellite tags', 'satellite tracking' and 'satellite telemetry' are sometimes used as a shorthand to refer to devices that collect information using the Doppler effect (Maxwell 1971) or just through general transmission of information back to the user via the Argos satellite tracking system (Box 4.1). Therefore, the term platform terminal transmitter (PTT) is used when referring to tags using the Doppler effect and transmission of data via satellite, noting that PTTs can be combined with other systems such as GPS, which are then referred to as GPS/PTT, for example.

Box 4.2 Considering the effects that tags and attachment methods may have on marine birds

Attachment methods for tags on marine birds depend on the goals of the research, and thus its duration, and the species' ecology and anatomy. For short periods during the breeding season, tags may be attached using a temporary fixture to feathers, for example by tesa tape or glue (Figure 4.1), which are lost when the bird moults (e.g. Hamer *et al.* 2007). For longer term attachments beyond the duration of the breeding season, other methods are needed. Some tags may require harnessing (e.g. Klaassen *et al.* 2012; Thaxter *et al.* 2015) and for larger species, some devices can be fitted using neck collars (e.g. Griffin *et al.* 2016). Surgical implantation of tags such as satellite platform terminal transmitter (PTTs) may be undertaken to study movement into the longer term (White *et al.* 2013; Speigel *et al.* 2017) and may be useful for diving species such as Red-throated Diver (Loon) *Gavia stellata* (Žydelis *et al.* 2018), as well as other tags such as those used for studying heart rate and energy expenditure (e.g. Bevan *et al.* 1997). Telemetry devices that are small and light enough, such as geolocators, may be attached on a leg-ring (Guilford *et al.* 2009; Harris *et al.* 2010).

In all cases, it is important to recognise that the attachment method and the type of tag used could potentially impact animals and alter their behaviour and survival. A general rule is that together, tags and attachments should not weigh more than 3% of an animal's body mass (Vandenabele *et al.* 2012). However, the situation is complicated by the numerous tag types and designs available (weight and shape of tags), the placement position on the animal, the type of attachment method, duration of attachment (e.g. short or long term), species-specific flight costs, and different interactions that species may have with their environment (Vandeabele *et al.* 2012; Thaxter *et al.* 2016). Inappropriate positioning of tags, for example, may affect balancing, travelling and foraging efficiency and energetic expenditure (Vandenabele *et al.* 2014). The tags themselves may also increase the wing-loading (wing area to body mass ratio) of a bird, which may compromise flight ability or movement underwater (Vandenabele *et al.* 2012; Thaxter *et al.* 2016).

The most serious effects of tags and their attachments are through direct physical injury, a reduction in body condition, and compromised feeding ability leading to reduced nesting success and lower adult survival (Thaxter *et al.* 2016). However, a wide variation of species responses has been recorded, and consequently, there have been numerous reviews on this subject (e.g. Barron *et al.* 2010; Vandenabele *et al.* 2011). The picture is complicated further by the range of metrics often used to investigate the effects of tags and their attachments, which also vary between studies. Many short-term studies during breeding, for example using tags attached through tape to feathers, have reported no effects on metrics such as foraging activity, breeding-site attendance or prey delivery to chicks (e.g. Benvenuti *et al.* 2001; Hamer *et al.* 2007), although effects on such metrics have been reported in other studies (e.g. Massey *et al.* 1988; Hamel *et al.* 2004; Harris *et al.* 2012). The picture is also mixed for longer term studies. For example, a GPS tag attached using a wing harnesses to Great Skuas *Stercorarius skua* in Scotland (Figure 4.1) had no short-term effects on breeding productivity during the year of marking, but reduced return rates in comparison to control birds the following year. By contrast, for Lesser Black-backed Gulls *Larus fuscus* tagged using the same tag type and attachment (Figure 4.1), survival and breeding success was comparable to control birds (Thaxter *et al.* 2016). For geolocators attached to leg-rings (Figure 4.1), Weiser *et al.* (2016) found that return rates and hatching success were lowered in only the three smallest species of 16 shorebirds studied, with results also dependent on the attachment method to the leg-ring and the overall weight of the tag (e.g. Weiser *et al.* 2016). More subtle responses of species to tag and attachment methods could also go unnoticed by studying only coarser measures such as breeding productivity and survival. Detailed investigations into behaviour, such as flight duration (Chivers *et al.* 2016), or ecophysiologcal measures, such as corticosterone levels and leucocyte blood-cell counts (Elliott *et al.* 2012; Ludynia *et al.* 2012), may also be needed.

Following a meta-analysis of over 200 tagging studies, Bodey *et al.* (2018) provide recommendations on the essential information that should be gathered to facilitate the assessment of impacts of the attachment of tags to birds in the future. This accepts that telemetry studies must always seek to minimise potential impacts of the tag and their attachment method on the species studied for both the welfare of the bird and the scientific veracity of data gathered. Advances are also being made in tag designs and attachment methods. For instance, the use of weak-link harnesses (e.g. Scragg *et al.* 2016) allows the tag and harness to detach from the bird after a period of time, thus helping to minimise potential long-term impacts. Tagging may only be undertaken under licence following independent assessment of the project, such as by the Special Methods Technical Panel of the British Trust for Ornithology Ringing Committee panel in the UK. Such assessment is especially important when new species are studied or new methods of attachment are being suggested.

Scope

This chapter aims to review and detail how tracking and telemetry technology is being used, and also how it could be used, in many aspects of wind-farm studies in relation to birds, with a focus on marine birds and some groups of migratory birds. Marine birds are defined as those that regularly use the marine environment, including species that are only found in such habitats, such as 'true' seabirds in the families Procellariidae, Hydrobatidae, Sulidae, Phalacrocoracidae, Laridae and Alcidae in a European context, but also including waterbirds, such as seaduck (Anatidae), divers (loons) (Gaviidae) and grebes (Podicipedidae). The migratory birds considered include waterfowl such as swans and geese (Anatidae) and wading birds (of several families within the order Charadriiformes) that may interact with the marine environment at particular times of the year, especially by migrating over the sea surface. Although terrestrial species including raptors (Accipitridae, Pandionidae and Falconidae) and a wide range of passerines also have the potential to interact with OWFs on migration, they are not specifically covered as part of this chapter. The reader is referred to Molis *et al.* (Chapter 6) for further consideration of monitoring collision of terrestrial migratory birds at OWFs.

Relevant information was sourced by extensive searches of research archives such as Web of Science and Google Scholar, using combinations of: (1) general key words, such as 'telemetry', 'tracking', 'bird-borne', 'bio-logging', 'logger', 'instrument', 'tag' and 'device'; (2) specific groups of telemetry devices, such as 'GPS', 'Doppler', 'PTT', 'geolocator' and 'GSM' (e.g. Bridge *et al.* 2011); (3) data-retrieval capability of systems: 'satellite', 'archival', 'data storage', 'download', 'transmitter' and 'receiver' (e.g. Bridge *et al.* 2011); (4) terms such as: 'offshore renewable energy', 'development', 'installation', 'wind farm', 'turbine', 'construction', 'operation' and 'impact assessment'; (5) terms for sensitivity and exposure to, or regarding specific effects, such as 'connectivity', 'interaction', 'sensitivity', 'exposure', 'vulnerability', 'overlap', 'flight height', 'barrier effect', 'collision risk', 'displacement' and 'habitat loss'; (6) specific species groupings, including 'seabird', 'marine bird', 'wader' and 'waterbird'; and (7) phases of the annual cycle, including 'migration', 'breeding' and 'non-breeding'.

Although telemetry and tracking studies on relevant bird species have been undertaken all over the world, most of the work focused on interactions between bird and wind farms has been conducted in Europe, simply because this has been the epicentre of the global industry (Jameson *et al.* 2019). As such, most work cited in this chapter is European in origin. However, several other countries such as China and the USA now have some existing sites, with many more proposed (e.g. Da *et al.* 2011; Winiarski *et al.* 2014). At the time of writing, numerous tracking projects have recently taken place or are ongoing in the USA (e.g. Spiegel *et al.* 2017; UMass Amherst 2018; USGS 2018) to assess potential risks of proposed wind farms, and these are included where relevant.

The information gathered illustrated that tracking technology may be used in all stages of different stages of the planning process and development stages. During scoping, for example, telemetry may be used to determine spatial and temporal exposure of sensitive birds to potential effects, which then may assist in the prediction of effects in EIA or HRA concerning a site of European importance for particular bird populations. Once a wind farm has been consented and constructed, studying any changes in the movements and behaviour of target species especially in comparison to the baseline pre-construction scenario, can help to show whether birds have been displaced or attracted or broadly unaffected, and may help to elucidate the mechanisms involved. The rapid advance in the use of telemetry to determine fine-scale movements offers the opportunity to learn how birds respond to the presence of turbines offshore, although this is still perhaps in its infancy compared to the use of radar (e.g. Plonczkier & Simms 2012) and multi-sensor systems (see Molis *et al.*, Chapter 6).

The theme headings provided below are designed to be applicable to one or more of broad stages of wind-farm development, from scoping to assessment, pre-construction, construction and operation.

Themes

Establishing foraging ranges and connectivity in the breeding season

During the breeding season, most bird species are central-place foragers (Orians & Pearson 1979), meaning that they are constrained to repeatedly return to a central place, that is, the nest site. The mystery of how far birds travel to find food within a 'foraging range' and the habitat they use away from the colony during breeding has long fascinated ornithologists. Early studies on foraging range of marine birds focused on indirect methods, such as visually following individuals, observing distances where birds were noticed carrying food (e.g. Bradstreet & Brown 1985) or recording the time elapsed between successive nest-site visits, to estimate potential straight-line distance travelled using a constant mean flight speed (e.g. Corkhill 1973). The development of telemetry methods (Box 4.1), however, has enabled more accurate estimations of foraging range to be estimated for populations of marine birds.

Tracking individuals of a particular species from a breeding colony can directly determine linkages between a wind farm and the colony, broadly termed 'connectivity', which can then be used to partition potential effects to these populations within the impact assessment process. This has proved to be of particular interest in determining potential impacts upon protected sites, such as SPAs in Europe within a stringent HRA process (see *Introduction*, above). An important first step in assessment has often been to use a generic foraging range derived from a number of studies (e.g. Thaxter *et al.* 2012) to provide a species-specific representative estimate (e.g. RWE 2011; ERM 2012; Natural Power 2012; SMartWind 2013; Forewind 2014). The use of generic foraging range data occurs mainly in two stages: (1) a scoping phase, whereby protected sites and species potentially impacted by the effects associated with a development are identified, followed by (2) more detailed consideration of pathways of potential effects on protected sites within the foraging range during the breeding season for particular species (Figure 4.2). The latter provides a means of apportioning effects, such as collision risk or displacement, to particular sites to determine the numbers of individuals that may be affected (Figure 4.2). Generic foraging range information is ideally superseded by direct information from telemetry where connectivity between a protected site and a development has been demonstrated. This is because birds generally favour particular areas within their range and rarely display an even fan-shaped distribution from colonies delivered by the use of a single range metric (Perrow *et al.* 2015). Representative values can be used to determine an area of potential habitat use that can then be investigated further by more intensive study (Thaxter *et al.* 2012). Generic species foraging ranges obtained from reviews, however, are inevitably constrained by data availability at the time, and are assumed to be maximal for a given species.

Telemetry studies provide a bank of information that is useful to determine connectivity with proposed, consented and operational wind-farm developments. For example, the

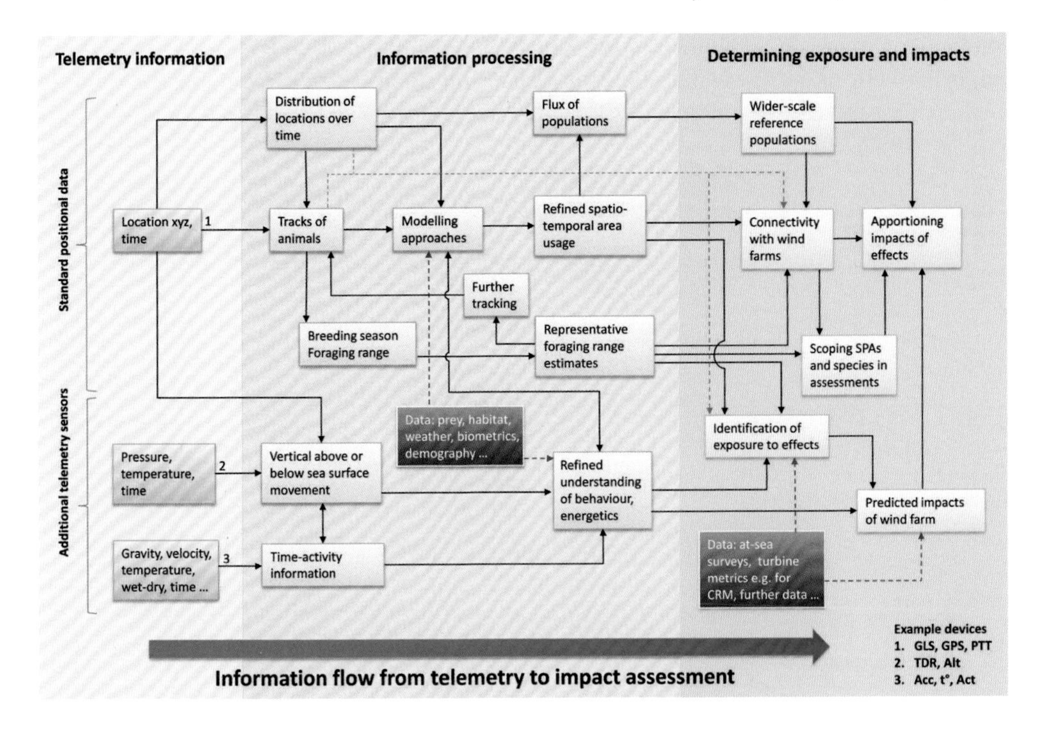

Figure 4.2 Example of information flow from initial tracking telemetry data obtained from a set of individual animals (orange box) to eventual use of this information (blue box) and then to determine spatial and temporal exposure to effects and impacts (green box). A range of telemetry devices may be used (see also Box 4.1); examples are given for: geolocation [global local sensor (GLS)], Global Positioning System (GPS), platform terminal transmitter (PTT), time–depth recorder (TDR), altimeter (Alt), Accelerometer (Acc), temperature sensor (t°) and activity logger (Act); ... denotes any further information that could be included; SPA = Special Protection Area and CRM = collision risk model. Grey dashed lines highlight potential uses of data, ahead of further processing or modelling, for example investigating connectivity with a wind-farm development using initial tracks of animals; the red highlighted boxes and red dashed lines indicate points at which external data sets may be introduced.

studies of Northern Gannet *Morus bassanus* tracked from Bass Rock and other colonies (e.g. Hamer *et al.* 2009; Votier *et al.* 2017) have enabled the spatial area usage of this wide-ranging species to be mapped. In addition, wider scale area usage during breeding of many different colonies has now been assessed, which highlights variations in foraging range between colonies and potential overlap with specific wind farms (Wakefield *et al.* 2013). A simple overlay of tracks of marine birds and wind-farm sites or potential development zones is a straightforward Geographic Information System (GIS) exercise, with subsequent modelling used to generate density contours (Figure 4.3; see Box 4.3 and Box 4.4). A number of such studies have been used in EIAs in the UK, such as the tracking of Lesser Black-backed Gulls *Larus fuscus* from Orford Ness (Box 4.3) used to assess interactions with the East Anglia OWF development zone (ERM 2012) and for a proposed extension to the Galloper wind farm in south-east England (RWE 2011). Similarly, information on Northern Gannet and Black-legged Kittiwake *Rissa tridactyla* tracked during the breeding season from Bempton Cliffs (Langston *et al.* 2013) was included to assess the potential for barrier effects for the Hornsea OWF development zone (SMartWind 2013), and telemetry data from several species, including Northern Fulmar *Fulmarus glacialis*, Black-legged

Kittiwake, Common Guillemot *Uria aalge* and Razorbill *Alca torda*, were used to assess (and apportion) effects for the Moray Firth and Neart na Gaoithe OWF development zones in eastern Scotland (Natural Power 2012; Mainstream Renewable Power 2012).

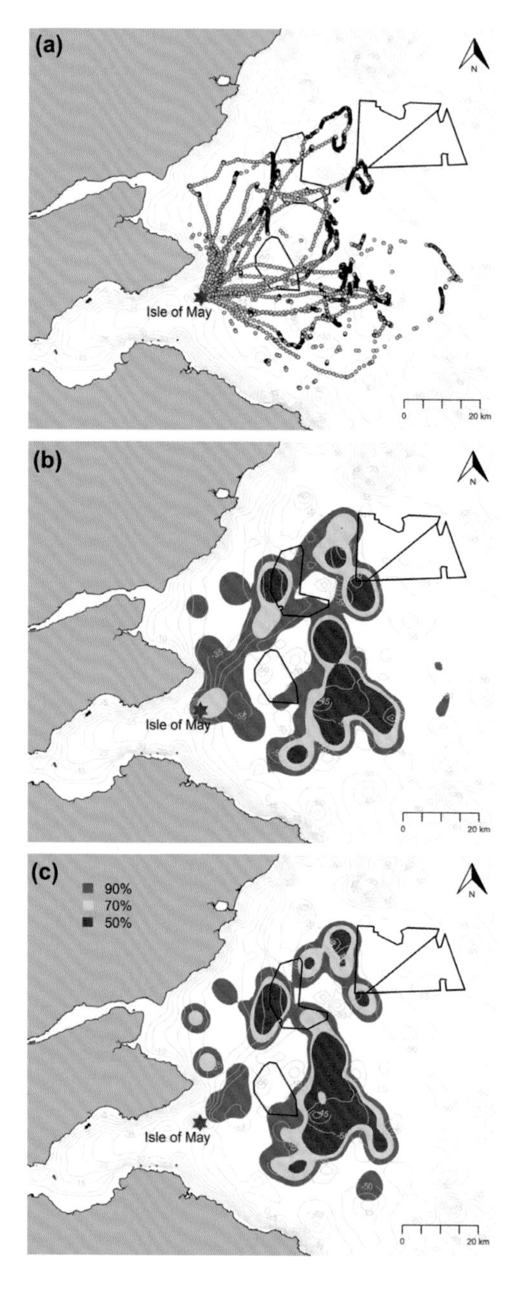

Figure 4.3 Example tracks and kernel density distributions for Atlantic Puffin *Fratercula arctica* from GPS tags (i-got-U GT-120) attached to birds at the Isle of May, south-east Scotland, overlain with areas of proposed offshore wind-farm developments (© Crown Estate 2012 updated). Maps show (a) all fixes of Atlantic Puffins, (b) density contours based on all locations, and (c) density contours based on diving locations. In (b) and (c), 50% (red), 70% (yellow) and 90% (blue) contours are shown. (Adapted from Figure 2 in Harris *et al.* 2012, reproduced with permission of the British Trust for Ornithology)

Box 4.3 Tracking Lesser Black-backed Gulls from breeding grounds at Orford Ness, UK, to understand their interactions with wind farms over the annual cycle

Chris B. Thaxter & Viola H. Ross-Smith

Since 2010, the British Trust for Ornithology (BTO), in collaboration with the University of Amsterdam, has tracked the movements of Lesser Black-backed Gulls *Larus fuscus graellsii* in relation to proposed, consented and operational offshore wind farm (OWF) areas. This study, carried out on behalf of the Department of Energy and Climate Change (DECC), superseded by the Department for Business, Enterprise and Industrial Strategy (BEIS), had the following main aims: (1) to investigate connectivity between protected seabird populations and OWFs throughout the year, (2) to determine the variability in exposure to effects of OWFs, and (3) to examine flight heights to inform collision risk.

The study was conducted at Orford Ness, part of the Alde-Ore Estuary Special Protection Area on the Suffolk coast, at a declining colony of 550–640 actual occupied territories. A total of 25 birds (11 in 2010 and 14 in 2011) were captured at the nest during incubation using a walk-in cage trap, and a GPS tag (Bouten *et al.* 2013) weighing 19 g was attached using a wing harness (see Thaxter *et al.* 2015 for more information) (Figure 4.4). Data from the GPS tags of 24 birds were downloaded remotely to a base station and processed; one tag did not produce any data.

Figure 4.4 Photographs showing attachment of GPS tags to Lesser Black-backed Gulls *Larus fuscus* before (left; Chris Thaxter) and after release (right; Dave Crawshaw).

During the breeding season, most birds (up to 80% in a given year) showed connectivity with the proposed East Anglia OWF zone, located approximately 40 km offshore from Orford Ness, while they were associated with their breeding colony (*ca.* March to August). Some individuals also showed connectivity with smaller consented and operational sites in the Thames estuary. However, the proportion of time that birds spent in all proposed, consented and operational OWF areas was quite small, ranging from 5% in 2010 to <1% in 2012. Using kernel density estimation, spatial overlaps of the 95% contour with OWF areas were highest in 2010 (up to 6%) and lowest in 2012 (<1%) (Figure 4.5) (Thaxter *et al.* 2014; 2015). General additive mixed-effects models were then used to analyse time budgets and spatial overlaps

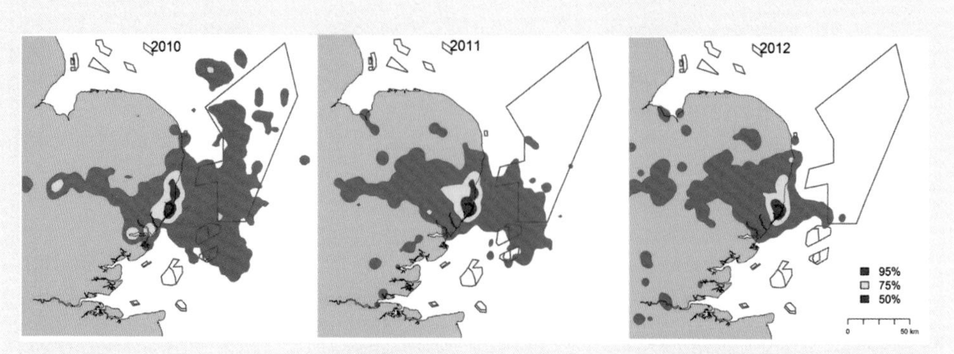

Figure 4.5 Overlap of 24 GPS-tagged Lesser Black-backed Gulls with offshore wind farms during the breeding season (*ca.* March to August) across three years, 2010–2012. Kernel density estimate contours are shown, with 95% in blue representing total area use, 75% in yellow, and 50% in red representing core area use. (Reproduced with permission of Department of Energy and Climate Change, University of Amsterdam and British Trust for Ornithology, from Thaxter *et al.* 2014, Crown copyright). See also Thaxter *et al.* (2015).

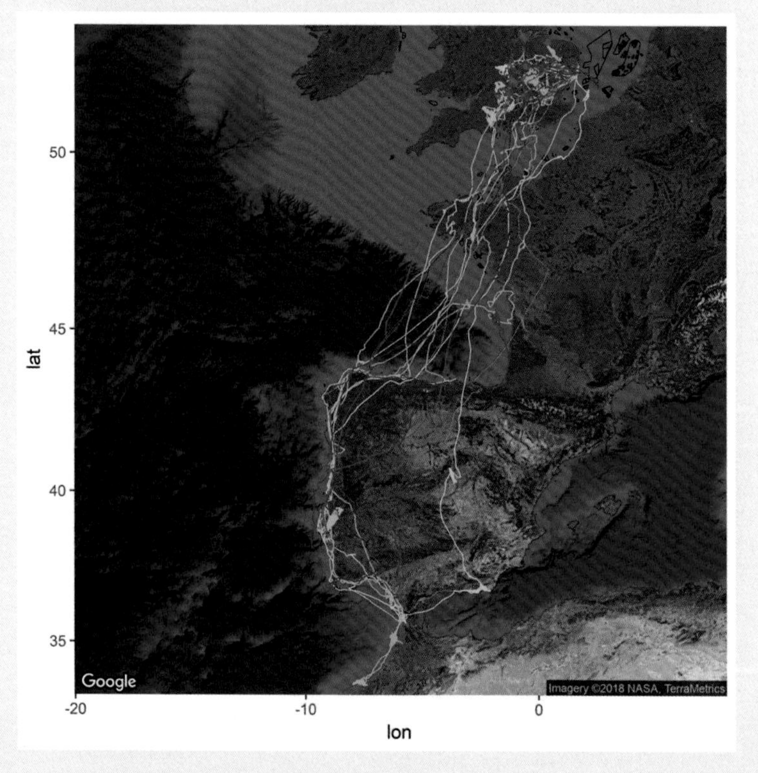

Figure 4.6 Non-breeding movements (migration and overwintering) of 15 GPS-tagged Lesser Black-backed Gulls that provided data during the 2011/12 non-breeding period (*ca.* September to February) from the Orford Ness breeding colony (part of the Alde-Ore Estuary Special Protection Area); all birds were tagged during the 2010 and 2011 breeding seasons (see also Figure 4.5). (Redrawn from Thaxter *et al.* 2018b with permission from the Department of Energy and Climate Change, superseded by the Department for Business, Enterprise and Industrial Strategy, University of Amsterdam and British Trust for Ornithology, Crown copyright; wind-farm areas © Crown Estate 2012, and redrawn from 4cOffshore 2013)

with OWFs in 5 day time slices across the breeding period, which revealed significant seasonal patterns, with overlaps restricted to the early chick-rearing period. A mean peak overlap of 36% in time budgets was observed during the 28 June–2 July time-section in 2010 (Thaxter *et al.* 2015). Compared to females, males also spent twice as much time in wind-farm areas. Consequently, both inter-annual and within-breeding season exposure was highly variable.

Of the birds initially tagged, a total of 19 birds provided further data during non-breeding periods in subsequent years. Six birds tagged in 2010 provided data for the 2010/11 non-breeding period. Thereafter, 15 birds provided data in 2011/12 and 11 in 2012/13, including birds tagged in the 2010 or 2011 breeding seasons. These non-breeding movements were then mapped in relation to OWFs (see Figure 4.6 for 2011/12), and revealed that birds from Orford Ness showed connectivity with proposed wind farms in the English Channel and off the coast of France, as they migrated mostly to and from Iberia and into North Africa (Figure 4.6).

Collision risk is particularly influenced by avoidance rate, flight height and flight speed of birds (see *Informing collision risk modelling*, below, and Cook & Masden, Chapter 5). The flight height of breeding Lesser Black-backed Gulls was examined using altitude recorded by the GPS tags. A Bayesian state–space model based on speed and location, with >4 km/h classed as flight (after Shamoun-Baranes *et al.* 2011), showed that gulls flew higher over land than at sea, and lower after dark than during the day. In more detail, mean flight altitudes were around 10 m at sea and during the day, and 50% of observations classed as 'marine' were within 12.8 m of sea level, in contrast to 50% of observations within 5.6 m after dark (Ross-Smith *et al.* 2016). A total of 31% of daytime flights at sea overlapped with the turbine rotor-swept risk area of 30–258 m, which concurred with previous studies using boat-based estimates (Johnston *et al.* 2014). Overall, the study was highly informative in gauging exposure to collision risk.

Connectivity of birds to source colonies during the non-breeding period

While it is easier to imagine a wind farm near to a breeding colony having clear connectivity and a potential definable impact on the colony, the impacts of developments are not likely to be restricted to breeding birds. During the breeding season, populations will also comprise juvenile and immature birds and other non-breeding individuals. Outside the breeding season, all individuals including adults and non-adults have no central-place foraging restriction and may disperse over wide areas, where they may interact with OWFs. Understanding the wider-scale movements of non-breeding birds when on migration and in overwintering areas in the context of connectivity with source colonies of interest, is also of importance from both the individual project level and the wider cumulative perspective.

Movements of adult birds during the non-breeding season have traditionally been less well studied than adults during the breeding season. This is due to the numerous technical challenges such studies involve. Obvious issues include the safe attachment of a tag to provide information over a longer period when individuals may be away from a breeding location for many years (see also Box 4.2), and whether a species can be studied with current technology, given a finite battery life, the associated size/weight restrictions of tags and how the data may be retrieved. These challenges are now being resolved for adult

Box 4.4 Insights into the variety of analytical approaches

There are many ways in which telemetry data can be analysed to reveal potential interactions with offshore wind farms. Often an initial step is to simply overlay tracks with an area of interest, to determine 'connectivity' with locations of interest (see *Establishing foraging ranges and connectivity in the breeding season* and *Connectivity of birds to source colonies during the non-breeding period*). Kernel-type approaches including Brownian bridges, or gridded quantification of intensity of use such as time spent (e.g. Soanes *et al.* 2013; Warwick-Evans *et al.* 2016; Thaxter *et al.* 2017) can then be used to further determine the core and total areas used within the home range, alongside time budgets for a given protected population (e.g. Harris *et al.* 2012; Langston *et al.* 2013; Wade *et al.* 2014; Thaxter *et al.* 2015) (Figure 4.2) (see Box 4.3).

Additional sensors on tags can provide deeper understanding of behaviour and potential use of locations in space and time (Harris *et al.* 2012; T. Cook *et al.* 2012; Brown *et al.* 2013; Waggitt & Scott 2014; Shamoun-Baranes *et al.* 2017) (see Box 4.1). However, even in the absence of such sensors and the information they provide, movement models, such as hidden Markov methods based on speed and turning angles of tracks (e.g. McClintock & Michelot 2018), first passage time analyses (Pinaud & Weimerskirch 2007) and Bayesian state–space models (e.g. Patterson *et al.* 2008), can be used to categorise different behaviours. Environmental factors that may affect distributions and behaviours such as bathymetry, sea surface temperature, and productivity and prey distribution, can then be modelled with telemetry and other marine bird data sets to reveal the reasons underpinning space use (e.g. Skov *et al.* 2008; Wakefield *et al.* 2009; Louzao *et al.* 2009), which may, in turn, inform marine spatial planning (Loring *et al.* 2014).

Theoretical models can also be used to predict likely responses of marine birds to offshore structures. For instance, Chimienti *et al.* (2014) used models to understand the implications of hypothetical disturbance scenarios from underwater renewable devices (such as wave and tidal devices) for the foraging efficiency of diving marine birds. Tracking data can also be incorporated within individual-based models (IBMs) to predict potential consequences of environmental change on specific populations. A key example of this is the study by Warwick-Evans *et al.* (2017), who used telemetry data of Northern Gannet *Morus bassanus* at Alderney in the Channel Islands, UK, to parameterise a spatially explicit IBM to investigate potential impacts of wind farms on body mass and vital rates for the protected population, under scenarios of avoidance behaviour and collision. Here, telemetry data were separated into flight resting and foraging, with information also included for adult and chick components of daily activity budgets, prey and energetics (Warwick-Evans *et al.* 2017). In this case, wind farms were predicted to have little impact on the Alderney Northern Gannet population. Nevertheless, this approach shows how telemetry data can be used in an applied context at the individual bird level to assess population-level consequences.

birds of some species outside the breeding season, and important non-breeding areas and migration routes have been identified using geolocation (e.g. Guilford *et al.* 2009; Harris *et al.* 2010), GPS/PTT devices (Klaassen *et al.* 2012; Spiegel *et al.* 2017) and GPS (Steinen *et al.* 2016; Shamoun-Baranes *et al.* 2017; Thaxter *et al.* 2018b) (Figure 4.6 in Box 4.3). The collation of such data through portals such as the Seabird Tracking Database (Birdlife International

2018) constitutes a powerful conservation tool. Longer term studies are now permitting longitudinal investigation into the repeatability of migration routes (Guilford *et al.* 2009).

Such long-term studies are now being used directly to assess overlaps with wind farms. Some species of waterfowl have been extensively tracked using GPS/PTT tags between breeding and wintering grounds (Griffin *et al.* 2011; 2016) (Figure 4.7 in Box 4.5). Even small wading birds such as the Piping Plover *Charadrius melodus* that could interact with the Block Island Wind Farm in the USA are now being studied throughout the year (UMass Amherst 2018). Combining geographic position and flight height information from GPS telemetry (see also *Informing collision risk modelling*, below) offers a meaningful way of determining potential vulnerability of species to wind farms through their annual cycle. For example, Thaxter *et al.* (2018b), in studying movements of Lesser Black-backed Gulls through the year for three breeding colonies, quantified the distance that individual birds travelled at altitudes between the minimum and maximum rotor-swept zone of wind turbines, as an index of sensitivity to collision risk; this layer was then combined with the location of wind turbines (exposure) to estimate potential spatiotemporal vulnerability. Such mapping exercises could inform the siting of wind farms to minimise potential impact, and areas where mitigation may be explored further at locations where vulnerability may be greatest.

Box 4.5 Tracking swans and geese in relation to wind-farm development along migration routes

Eileen Rees, Larry Griffin & Baz Hughes

The Wildfowl & Wetlands Trust (WWT) has used tracking data recorded since 2006 to provide information on swan and goose migration routes in relation to offshore (and onshore) wind-farm sites. The studies, initially on behalf of COWRIE Ltd and more recently for the Department for Business, Energy and Industrial Strategy (BEIS) (formerly the Department of Energy and Climate Change), are being used to inform the Offshore Energy Strategic Environmental Assessment (SEA) programme being undertaken by BEIS, by highlighting key areas where the proposed offshore wind farms (OWFs) may pose a risk, and the potential for cumulative effects alongside existing wind farms located on the flyway.

In winter 2008/09, forty 70 g solar GPS/PTT-100 Argos transmitters were fitted with harnesses to Whooper Swans *Cygnus cygnus* (see Figure 4.1c in Box 4.1) at three Special Protection Areas (SPAs) of international importance for the species: 20 at Martin Mere SPA in north-west England; 15 at Welney, Ouse Washes SPA in south-east England; and five at Caerlaverock, Upper Solway Flats and Marshes SPA in south-west Scotland. Ten transmitters were also fitted to swans at breeding sites in Iceland during summer 2009 to provide additional data on their autumn migration, and five were fitted to swans at Martin Mere in 2009/10 to determine consistency in migration routes. The transmitters provided hourly location data between 18:00 and 11:00 h in spring (8 March to 15 May), and every 2 hours between 06:00 and 20:00 h in autumn (15 October to 29 November). Whooper Swans were thought to be at particular risk of collision because of their large size and low manoeuvrability, but many other species are similarly at risk. Therefore, existing tracking data held by WWT for Svalbard Barnacle Geese *Branta leucopsis*, Greenland White-fronted

Geese *Anser albifrons flavirostris* and East Canadian Light-bellied Brent Geese *Branta bernicla hrota* were also examined. Data quality and sample sizes were most useful for Barnacle Goose as a result of location data recorded for 27 Svalbard Barnacle Geese fitted with 30 g or 45 g solar GPS/PTT-100 Argos transmitters (see Figure 4.1f in Box 4.1) from April 2006 onwards at the main wintering grounds on the Solway Firth. Location data were investigated to see whether a bird crossed the footprints of potential wind farms, defined as where the extrapolated track between two consecutive GPS fixes (minimum 1–2 hour interval) intersected the wind farm.

During the 2008/09 spring migration, a high proportion (75%) of swans migrating from north-west England crossed at least one OWF footprint, compared with 7.1% tracked along the east coast from south-east England (Griffin *et al.* 2010). All wind farms traversed were those either consented or operational wind farms (Round 1 or Round 2) off the Cumbrian coast or in proposed Scottish territorial waters sites. No Whooper Swans were recorded flying across Round 3 wind-farm footprints (Figure 4.7). The autumn 2009 and spring 2010 data indicated some variability in flight lines, with the autumn migration being more westerly. Swans from north-west England also took a more westerly migration route in spring 2010 than in spring 2009.

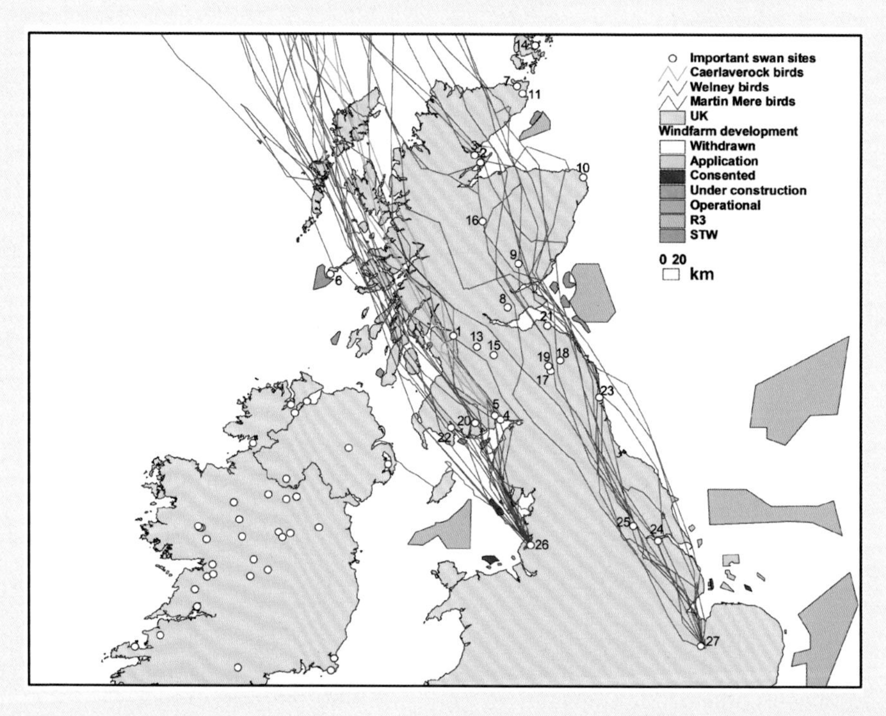

Figure 4.7 Whooper Swan *Cygnus cygnus* migration routes in relation to offshore wind-farm sites in spring 2009, showing movements to and from overwintering sites at Caerlaverock, Welney and Martin Mere.

Barnacle Geese migrated across seven prospective UK OWF sites (Figure 4.8), of which five were located in the Firth of Forth, and a further 15 proposed sites along the coast of Norway. Birds interacted most with the proposed Firth of Forth Round 3 zone (13 of 26 tracks), with tracks mainly crossing the mid-southern part of the footprint. Of spring tracks that missed the zone, 87% were also to the south, indicating that the

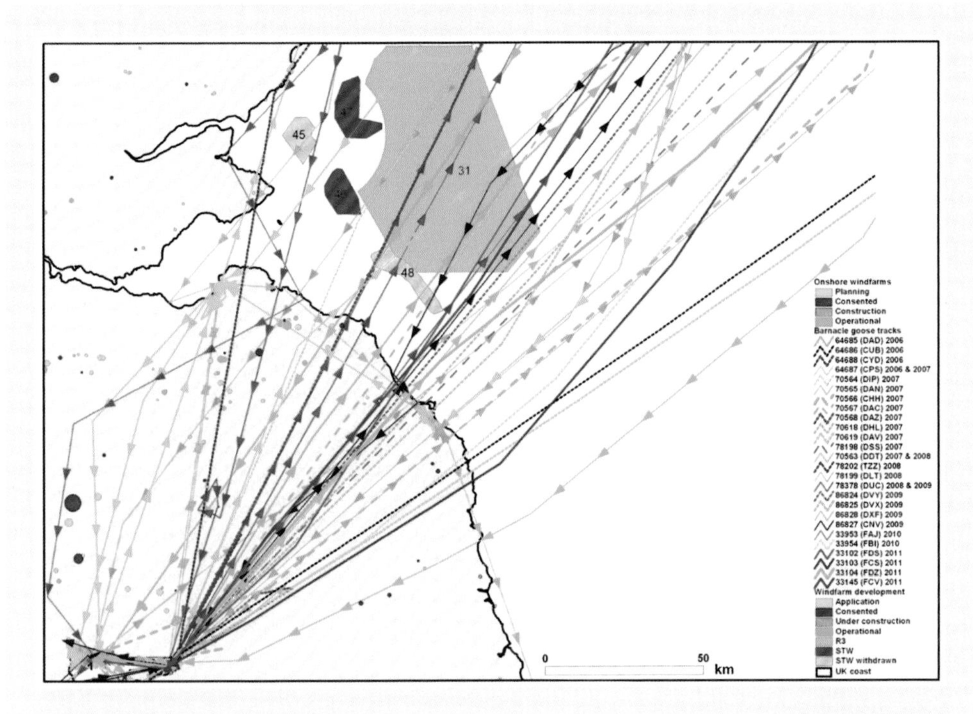

Figure 4.8 Barnacle Goose *Branta leucopsis* migration movements across the Firth of Forth and interactions with proposed offshore wind-farm areas; example tracks and direction of travel are shown for nine birds, including spring and autumn migration route distinctions for birds with more than one migration period studied.

southern part of this zone is most sensitive for the geese (Griffin *et al.* 2011). It is of note that the only sites yet proposed in this zone are in the north, where no migrating Barnacle Geese were observed on surveys across all seasons (Seagreen Wind Energy 2012), suggesting negligible risk of interaction.

Considering terrestrial wind farms, 80% of the Whooper Swans in 2009 and 2010 crossed at least one proposed or operational site. For individuals tracked from Martin Mere and Welney, 39% and 22%, respectively, crossed at least three wind-farm sites/footprints (offshore/onshore) during a single migration. Nearly all 21 Svalbard Barnacle Geese from the Solway Firth to Svalbard passed within 2 km of at least one proposed/operational wind farm (onshore and offshore). Four geese (19%) crossed wind farms on at least six occasions during a return migration. An estimated 50% and 60% of birds crossed wind farms (existing or planned) in the UK and Norway, respectively. These data all illustrate the need to consider both terrestrial and OWFs in an international context when considering cumulative effects.

Further work on Bewick's Swans *Cygnus bewickii* has involved tracking a total of 22 birds (8 in winter 2013/14 and 14 in 2014/15) with 50 g GPS/GSM tags, attached using neck-collars, to determine potential impacts of OWFs between south-east England and the Netherlands in relation to the movements of swans migrating to and from UK protected sites, including the Ouse Washes SPA, which holds up to 33% of the north-west European population in mid-winter. Birds crossed a total of 52 wind-farm footprints, including 11 operational sites and many European North Sea wind-farm zones, with the greatest interaction seen for the consented East Anglia

ONE zone in UK waters. Several birds migrated to the Schleswig-Holstein area in Germany, also bringing them into overlap with sites in the German Bight. Flight altitude data were also collected for five birds, revealing that 93% of offshore flights were <150 m, with most (mean and median) less than 50 m, potentially below the vertical rotor-swept zone of rotors. Given the number of wind farm areas crossed within and outside the UK, this telemetry study highlights the need for careful cumulative assessment of potential impacts on migrating species (see Griffin *et al.* 2016 for more information).

Studies also need to be extended to juvenile and immature individuals simply because, as shown through some telemetry studies, the movements of these groups may be quite different from those of adults. For example, the migration routes of Scopoli's Shearwater *Calonectris diomedea* differ among adults, immatures and juveniles (Péron & Grémillet 2013). In addition, Votier *et al.* (2017) showed the foraging site fidelity of Northern Gannets fitted with GPS/Global System for Mobile Communications (GSM) tags varied across different age classes, including immature birds. Although the challenges of tracking birds in these periods can be substantial, the opportunity nonetheless exists to investigate wind-farm interactions across different ontogenetic phases.

Spatial and temporal variation in movements

Appreciating the complexity of movements of marine birds is important when evaluating the potential for specific wind-farm effects, such as collision (see *Informing collision risk modelling*, below), to have impacts on given populations. For example, any variability in the flyways of migratory waterbirds may lead to considerable differences in likely predicted interactions with wind farms (Griffin *et al.* 2016) (Box 4.5). Studies on seabirds such as Northern Gannets (Hamer *et al.* 2007; Langston *et al.* 2013), Lesser Black-backed Gulls (Thaxter *et al.* 2015) (Box 4.3) and European Shags *Phalacrocorax aristotelis* (Bogdanova *et al.* 2014) have highlighted the variability of movements of these species from key colonies between years, and thus the implications of potential interactions with OWF areas. In the case of Lesser Black-backed Gulls, which have the potential to use both marine and terrestrial environments, the pattern of use of wind farms by birds from Orford Ness was found to vary considerably, both within the breeding season and between years (Box 4.3) (Thaxter *et al.* 2015). Great Skua *Stercorarius skua* at Foula and Hoy in northern Scotland also showed variation in the use of proposed and consented OWFs, which was greater after breeding had finished and birds made much wider movements (Wade *et al.* 2014).

In general, the use of the area surrounding one breeding colony of a particular species may be very different from the use around another colony, linked to density-dependent competition for resources (Wakefield *et al.* 2013) as well as differences in marine systems and abundance and distribution of particular prey. Changing patterns of the use of different marine areas for a single colony may also be driven by variation in prey resources between years or across seasons (e.g. Hamer *et al.* 2007; Warwick-Evans *et al.* 2016). For example, using geolocators to determine wintering areas of Atlantic Puffin *Fratercula arctica* originating from the Isle of May, in Scotland, Harris *et al.* (2010) suggested that a reduction in food availability may have led to changes in migration behaviour and increased mortality, with carry-over effects of poor individual condition impacting breeding season productivity. This is aside from the potential effects of displacement from the proposed wind-farm developments within the foraging range of breeding birds. Understanding the impact of prey resources on population-level processes would better allow assessment of the potential impacts of the effects associated with OWFs.

Given the complexities and variations in the movements of a wide spectrum of marine birds, the extent of wind-farm interactions is unlikely to be a constant that can be ascertained through monitoring in a single given year, or through a short-scale study within a year. An appreciation of annual variation in distribution of species from at-sea surveys is inherent within impact assessments (Camphuysen *et al.* 2004). Similar congruency should be applied to telemetry methods to allow suitable appraisal of the interactions between marine birds and wind farms.

Defining the spatial scale of affected populations

Impact assessment often requires consideration of the consequences of effects at a number of spatial scales, ranging from protected sites, through suites of protected sites, to larger scales such as national and biogeographic populations (see Figure 4.2). More recent approaches have sought to identify a biologically defined minimum population scale (BDMPS) to serve as a reference population to judge specific effects of developments (e.g. Forewind 2014). Tracking is particularly useful to identify which populations may be affected by a development and over what geographic scale. For example, the relative use of key areas such as the North Sea by birds from different populations is not well known, with even less known about the influx and outflux of particular populations to and from these key areas.

A good example where telemetry has tackled this issue is the study by Frederiksen *et al.* (2012) that used geolocators to examine the non-breeding season movements of 236 Black-legged Kittiwakes from 19 different colonies. The study identified the non-breeding areas, in particular large marine ecosystems (LMEs) thereby allowing the size and composition of non-breeding populations in each area to be estimated through proportional area use (Frederiksen *et al.* 2012). It was estimated that 40% of the North Sea LME Black-legged Kittiwake population present in December 2009 originated from the Barents Sea LME, whereas 18% of the breeding North Sea population remained in the North Sea over the winter (Frederiksen *et al.* 2012). Such information allows an assessment of the connectivity between birds wintering in particular areas and different source breeding populations (or colonies), and thus the assignment of potential impacts to non-breeding birds from particular populations (or colonies) (see Figure 4.2). This example also highlights the potential value of geolocation, even though it is not as precise as GPS (Box 4.1). Future tracking studies will no doubt be important in helping to identify use of specific areas and thus to refine definitions of BDMPS in relation to wind-farm development.

Displacement and barrier effects

Displacement of birds from areas that they would naturally occupy is perceived as a key effect of wind farms (Vanermen & Stienen 2019). Displacement can take several forms, including as a barrier to bird movements over an area, often assessed separately within EIAs. Tracking data sets that inform the potential connectivity of birds from protected sites with proposed developments are often used for assessing the potential impacts of barrier effects (see also *Informing collision risk modelling*, below), such as in the case of Whooper Swans *Cygnus cygnus* (Griffin *et al.* 2011) (Box 4.5) in relation to the Moray Firth OWF zone (Natural Power 2012). Given the marked and highly informative response of birds in the few studies undertaken, further tracking studies of marine birds and waterbirds during both the breeding and non-breeding season would be highly welcomed to add to the radar studies that have measured the increase in flight distance as a result of barrier effects (Masden *et al.* 2009).

The energetic consequences of the increase in flight distance as a result of barrier effects have generally been considered to be minimal (Masden *et al.* 2009; 2010), although they may come into play at the cumulative level (Masden *et al.* 2009), with impacts on individual fitness perhaps being expressed on the population as a whole. By contrast, McDonald *et al.* (2012) developed an approach incorporating both barrier and displacement effects from proposed wind farms in the Forth of Tay (south-east Scotland) on the time–energy budgets of Common Guillemots breeding on the Isle of May. This method combined information on foraging behaviour and time spent in flight from telemetry data with information on prey distribution. Assuming a random prey distribution, displacement was predicted to increase flight and foraging costs by 36% and 18%, respectively, hence potentially negatively impacting time–energy budgets with consequences for breeding success and/or adult survival.

Informing collision risk modelling

Collision risk is a key concern for some species (Furness *et al.* 2013; King 2019), which is assessed using collision risk models (CRMs). CRMs typically draw upon species-specific representative values for a number of parameters that are often taken from reviews of information of variable quality (Band 2012; A. Cook *et al.* 2012; 2014; 2018; Masden & Cook 2016; Cook & Masden, Chapter 5). CRMs, however, can be sensitive to the choice of input parameters (Masden & Cook 2016). Telemetry offers the key advantage of providing detailed site-specific information, albeit often from a small number of individuals, for some key model parameters, that may then facilitate more robust estimation of collision risk. These include: (1) flight height, which will inform whether and how often a species flies at the altitude of rotating turbine blades, (2) flight speed, which influences the chances of collision when passing through the rotor-swept area, (3) time–activity budgets, which provide potential time spent in flight and time spent in flight at night, and (4) overall avoidance rate and behaviour (Thaxter *et al.* 2018a), especially if studies cover pre- and post-construction periods (Harwood *et al.* 2017). The combination of these parameters within CRMs can be used to estimate the probability of collision and potential numbers of species from protected sites that may be affected by wind-farm developments.

The flight height of birds is a key factor in determining their collision risk with turbines offshore (Furness *et al.* 2013; Cook *et al.* 2014; 2018). Information on flight heights has historically been derived from boat-based surveys (Johnston *et al.* 2014; Webb & Nehls, Chapter 3) or from vertical radar studies (see Molis *et al.*, Chapter 6), but is increasingly being recorded through aerial digital survey methods (see Webb & Nehls, Chapter 3). Bird-borne telemetry, however, provides the potential for estimating flight altitude across all height ranges and weather conditions, for example using altimeters measuring differences in barometric pressure, and through GPS, which record geographic position in three dimensions. Altimeters have been used to estimate flight heights of a range of marine bird and waterbird species, including Whooper Swans (Pennycuick *et al.* 1999), American White Pelicans *Pelecanus erythrorhynchos* (Shannon *et al.* 2002), Red-footed Boobies *Sula sula* (Weimerskirch *et al.* 2005) and Northern Gannets (Cleasby *et al.* 2015).

Altimeters have also been used to assess wind-farm interactions, such as by Cleasby *et al.* (2015), who used dual attachment of separate GPS tags and altimeters to determine three-dimensional movements of Northern Gannets from Bass Rock in Scotland. In that study, flight heights were higher during periods of foraging than commuting, and spatially explicit flight heights, when combined with estimates of density of birds in a CRM, revealed that conventional flight height distributions from species-specific representations may underestimate potential collisions. The altimeters used by Cleasby *et al.* (2015) weighed 18 g; however, further miniaturisation is enabling sensors to be incorporated directly into GPS tags, which would then allow flight heights of lighter species to be investigated using altimetry.

Altitude has also been recorded directly from GPS and GPS/PTT tags, for example for Frigatebirds *Fregata* spp. (de Monte *et al.* 2012), Lesser black-backed Gulls (Klaassen *et al.* 2012; Corman & Garthe 2014) and Bar-headed Geese *Anser indicus* (Bishop *et al.* 2015). GPS-derived altitude can be used to estimate potential interaction of species with the vertical rotor-swept zone of turbines (e.g. Corman & Garthe 2014; Griffin *et al.* 2016) (Box 4.5), but may also suffer from vertical measurement error of around 15–20 m (e.g. Corman & Garthe 2014). While this error seems problematic, it is dependent on a number of factors, including the sampling rate of tags (see Thaxter *et al.* 2018a for further discussion). Faster sampling protocols recording more fixes over a shorter period can increase the precision of estimates (Bouten *et al.* 2013; Thaxter *et al.* 2018a). This has facilitated the use of GPS altitude to study three-dimensional activity of Lesser Black-backed Gulls at operational OWFs in north-west England at very high resolution (e.g. at 10 second rates; Thaxter *et al.* 2018a).

Advances in modelling methods are also permitting investigation of flight altitude using GPS. For instance, Ross-Smith *et al.* (2016) used a state–space Bayesian modelling framework to account for some known sources of error, such as position of satellites in the sky, which allowed the production of flight height distributions and assessment of potential wind-farm interactions. Differences in flight heights between day and night and between land, coast and at-sea locations were revealed (Ross-Smith *et al.* 2016) (Box 4.3). Similar results were obtained from another Lesser Black-backed Gull colony in Germany, where birds also flew lower over the sea than land and flew lower at night (interquartile range –2 to 8 m) than during the day (2 to 36 m) (Corman & Garthe 2014).

The Band (2012) CRM assumes a constant 'flux' of birds and therefore the speed of birds flying through the wind farm and rotor-swept zone is assumed to be constant. In reality, birds are unlikely always to conform to this assumption, with variations in behaviour being almost inevitable, such as tortuosity of track influencing speed (Masden & Cook 2016). Thus, instead of a static component for speed of birds, CRMs could be better informed by more precise species- and location-specific information on speed (similar to bespoke estimates on flight altitude, above) (A.Cook *et al.* 2012). Trajectory speed is recorded automatically through telemetry when date–time and position information are combined. Additional vector-based instantaneous speeds, however, can also be calculated by some devices (e.g. Klaassen *et al.* 2012; Bouten *et al.* 2013).

The percentage of time in flight of different species has been reviewed in relation to OWFs (Garthe & Hüppop 2004; Furness *et al.* 2013) because the higher the proportion of time that a species spends in flight, the greater will be its likely risk of collision (Band 2012; A.Cook *et al.* 2012). The extent of nocturnal flight activity also increases the degree of exposure to collision risk, although information on nocturnal activity is lacking for many species. It has been suggested that collision risk may be considerably increased in low light visibility (Garthe & Hüppop 2004; Furness *et al.* 2013). Telemetry can provide data on flight activity and time budgets of marine birds throughout the day and night (Thaxter *et al.* 2010). The time spent in flight can be approximated for a given species simply by considering the minimum speeds thought to possible allowing sustained powered flight (e.g. Shamoun-Baranes *et al.* 2011). Movement modelling approaches, or using information from multi-axial accelerometer sensors (Shamoun-Baranes *et al.* 2016) (Box 4.4; see review by Brown *et al.* 2013), can build up a more detailed spatial picture of movement and behaviour. Such information can be summarised across individual birds for specific wind farms, which can, in turn, feed into collision risk modelling (Cleasby *et al.* 2015).

Box 4.6 outlines how information on the relative time spent at sea, relative use of a wind farm, and time spent at risk height generated during an radio-telemetry study of Little Tern *Sternula albifrons* (Perrow *et al.* 2006) was incorporated into an alternative version of the Band model (Band *et al.* 2007), originally derived for use in onshore studies for birds making less predictable movements; that is, without constant flux. This model, based on time spent in the risk window to generate a transit rate for use within the model, was presented as being appropriate for hunting raptors, which can share similar flight characteristics with foraging seabirds. In the case of Little Terns, the study proved useful in scoping the likely number of birds colliding with turbines at Scroby Sands.

Box 4.6 Estimating collision risk of Little Tern *Sternula albifrons* at Scroby Sands wind farm

Martin R. Perrow, Andrew J.P. Harwood & Eleanor R. Skeate

The Scroby Sands wind farm, comprised of 30 2 kV Vestas V80 turbines, was installed between October 2003 and August 2004 approximately 2–3 km east of North Denes, part of the only SPA designated solely for Little Tern *Sternula albifrons* in the UK. At that time, the colony was the UK's largest, averaging 188 pairs from 1986 to 2006 representing more than 10% of the UK population. A 5-year study was initiated from 2002 to 2006 inclusive to understand the interactions between breeding Little Terns and the wind farm. After 2 years of an intensive boat-based survey programme with six to eight surveys per season, there had been no records of Little Terns within the prospective wind farm and there was concern that a low intensity or intermittent use of the wind farm once it was built would go undetected, leading to the conclusion that there was no possible risk of collision with turbines, when this might not have been the case. As a result, the paradigm of 'monitoring the bird using its habitat' rather than 'monitoring the habitat and describing its use by the bird' was adopted and radio-tagging of Little Terns was attempted for the first time. To the authors' knowledge, this was the first application of tracking in the study of the interactions between birds and OWFs.

After capture at the nest in 2004–2006 inclusive, a total of 22 birds (*n*=4, 9 and 9, respectively) were fitted with 1 g radio-tags (Biotrack Ag376/379) comprising about 2% body weight, which were glued to the back feathers and designed to be shed naturally after a short time (Figure 4.9). Tags were operational for an average of 12 days. Tracking was undertaken from a 6.3 m rigid-hulled inflatable boat capable of high speed (>30 knots). Details of tagging and tracking procedures are provided in Perrow *et al.* (2006).

No use of the wind farm was recorded in 2004, with use increasing annually thereafter (Figure 4.10), mirroring the increase in the use of offshore areas farther from the colony in later years. Pile-driving during construction was thought to have affected the adult population of locally spawning Atlantic Herring *Clupea harengus* that, in turn, reduced the abundance of young-of-the-year, the most important prey item for breeding Little Terns at North Denes (Perrow *et al.* 2011a). But whereas prey supply was low in all years, the colony was small in 2004 (peak of 17 nests), of average size in

Figure 4.9 Radio-tracked Little Tern *Sternula albifrons* at sea after a dive. (Martin Perrow)

2005 (peak of 196 nests) and very large in 2006 (369 nests at peak), potentially increasing the range of individual foraging trips. Furthermore, while heavy predation of chicks by Common Kestrel *Falco tinnunculus* was noted in 2005, when an estimated 455 chicks were taken and only 11 fledged, supplementary feeding of the main Common Kestrel pair was thought to reduce chick predation to low levels in 2006, when an estimated 673 fledged. High demand for prey by Little Terns against a backdrop of low availability seemed to increase the use of the wind farm, where the development of a series of subsidiary sandbars formed as deposition 'tails' downdrift of the turbine bases provided suitable foraging habitat (Perrow 2019).

Importantly, tracking data also provided a means of estimating collision risk in 2005 and 2006 when some use of the wind farm was recorded. The methods of Band *et al.* (2007) for less predictable bird movements recorded as timed observations were used to determine the transit rate needed as an output of stage 1 of the modelling. Stage 2 was then completed using the spreadsheet provided by Scottish Natural Heritage (2014). The data sources and treatments and outputs of each step required are outlined in Figure 4.11. An equivalent collision risk was also calculated for fledged juveniles, assuming that these accompanied their adults on foraging trips over a 30 day occupancy period (16 July–15 August). The avoidance rate of Little Terns is not known, but a generic

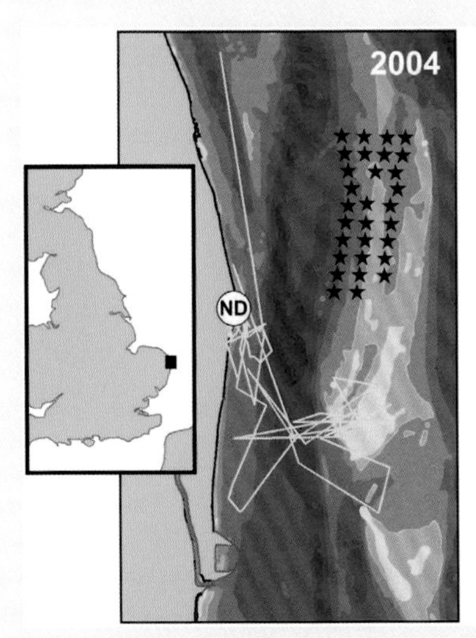

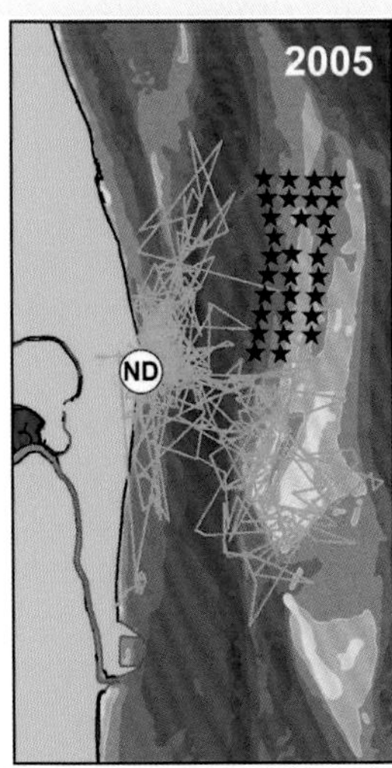

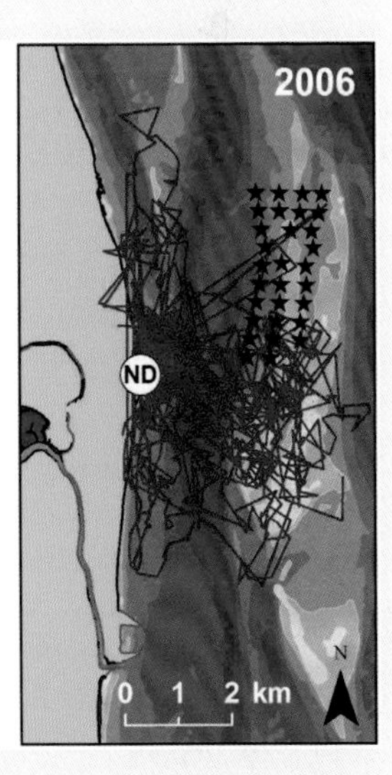

Figure 4.10 Foraging tracks (n=128 individual bouts) of tagged Little Tern from the North Denes colony (ND) in 2004 (n=10), 2005 (n=46) and 2006 (n=72), showing increasing use of the wind farm.

precautionary value of 98% predicted the loss of 6 adults and 0 juveniles in 2005 and 25 adults and 6 juveniles in 2006, for an annual average of 10 adults and 2 juveniles over the three years. As a result of the many uncertainties and approximations of the modelling process in particular, it is important to use such a prediction only as a rough guide to indicate the scope of collision risk. The actual extent of losses would require verification. Nevertheless, the predicted risk to Little Terns from Scroby Sands wind farm is in keeping with the observed loss of 2–10 birds per year (2001–2005 inclusive) from the Zeebrugge breeding colony of 11–150 pairs as a result of collision with turbines along a breakwater between the colony and the sea (Everaert & Stienen 2007).

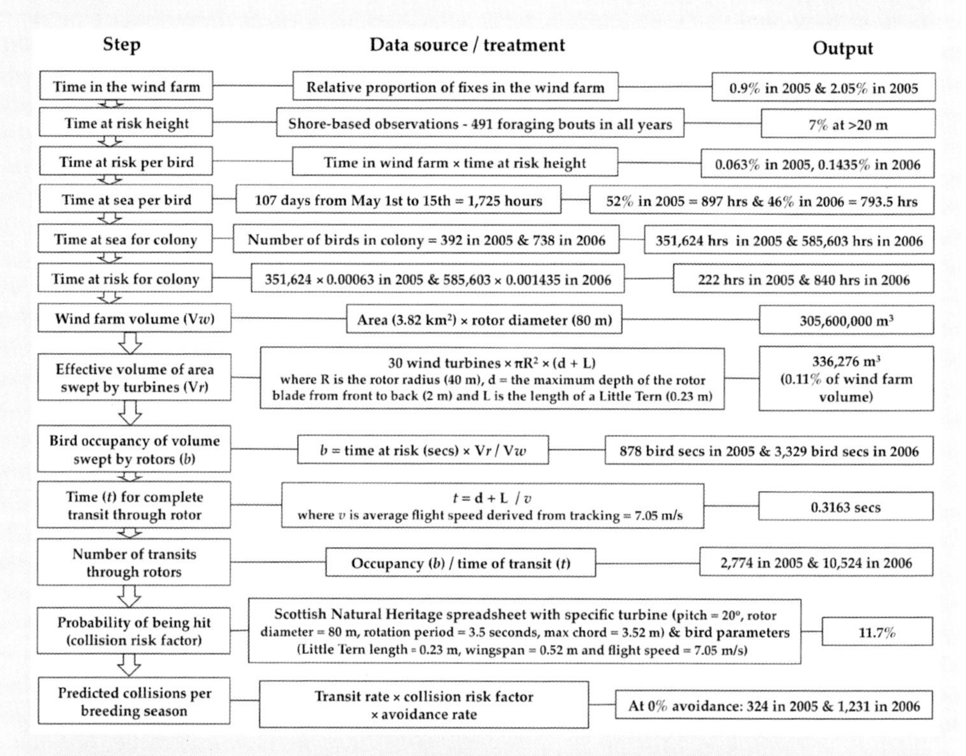

Figure 4.11 Details of the collision modelling process following Band *et al.* (2007) for Little Tern tracked from North Denes in 2005 and 2006, when birds entered the Scroby Sands wind farm. In 2004, no recorded use meant no collisions were predicted.

Finally, it is important to note that in 2013 Natural England (unpublished report) commissioned further boat-based surveys to estimate density of Little Terns within and immediately around the wind farm. Application of the standard Band (2012) collision risk model on all data estimated 3 collisions per annum at 98% avoidance; although no Little Terns were actually recorded in the wind farm itself. Moreover, a more typical avoidance rate now applied to seabirds (98.9–99.8% from Cook *et al.* 2018 and 99.6-99.9% from Skov *et al.* 2018; as derived from a variety of gulls and Northern Gannet *Morus bassanus* in both cases) suggests that the collision rate of Little Terns would likely be <1 bird per annum. It was therefore concluded that collision is of negligible threat to the local Little Tern population, some of which now breed on Scroby Sands themselves (see Perrow 2019).

Three-dimensional avoidance behaviour and collision risk

As outlined by Molis *et al.* in Chapter 6, a variety of methods have been used to assess avoidance behaviour and collision risk from direct observation, and in combination with radar (Petersen *et al.* 2006; Krijgsveld *et al.* 2011; Plonczkier & Simms 2012); culminating in advanced combinations of radar and video and thermal cameras in digital communication, used in conjunction with laser rangefinders (Skov *et al.* 2018). However, relatively few studies at OWFs have yet to investigate directly the avoidance behaviour and potential collision risk of marine birds using bird-borne telemetry. The radio-telemetry of Little Terns provided some insight into movements of tracked individuals in relation to the wind farm (Box 4.6), but the birds did not routinely enter the site and spend much time there. In contrast, the Lesser Black-backed Gulls tracked by Camphuysen (2011) frequently entered the operational Egmond aan Zee and Prinses Amalia wind farms, which enabled a spatial 'intensity risk map' to be produced derived from flight heights and locational positions. However, it is the extension of the work detailed in Box 4.3 on Lesser Black-backed Gulls

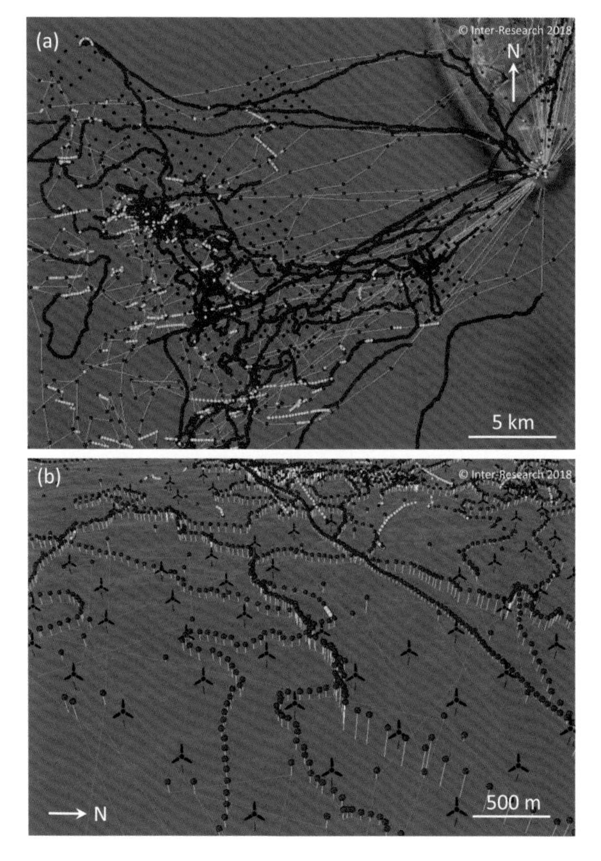

Figure 4.12 Example foraging tracks of one tagged Lesser Black-backed Gull *Larus fuscus* from South Walney, UK, part of the Morecombe Bay and Duddon Estuary Special Protection Area, showing use of existing wind farms during the 2014 breeding season. Two scales are shown: (a) localised two-dimensional movements in relation to operational wind turbines and; (b) fine-scale three-dimensional movements within operational wind farms in relation to locations of a subset of turbines. Locations are shown as likely flight (red; travel speed >4 km/h) or likely resting, bathing or swimming (yellow; travel speed ≤4 km/h); orange lines link consecutive GPS points, The pink star indicates the location of the breeding colony. (Reproduced with permission from Thaxter *et al.* 2018a; images © Inter-Research 2018; Wind-farm turbine locations © Crown Estate; map backgrounds © GoogleEarth)

to other colonies in the UK, including South Walney, a part of the Morecombe Bay and Duddon Estuary SPA for the species that is close (<20 km) to five operational wind farms, that has enabled the development of three-dimensional tracking for potential assessment of avoidance behaviour (Thaxter *et al.* 2018a). Combining information on flight altitude and geographic position within and outside wind farms, Thaxter *et al.* (2018a) examined the movements of birds at different scales, including macro-avoidance (wind-farm scale) and meso-avoidance (avoidance of rows of turbines once in a wind farm), with potential to consider micro-avoidance (last-minute near-scale response) (see May 2015 for definitions). Results from a subset of birds showed no apparent macro-scale avoidance of wind farms (Figure 4.12), but once within the wind farm there was more than 50% overlap with the vertical rotor-swept zone, which is higher than in previous studies (Box 4.3). However, importantly, very few GPS locations were recorded within the three-dimensional rotor-swept 'volume' around turbines, with this being significantly different from a random distribution and indicating an apparent meso-avoidance response (Thaxter *et al.* 2018a). Overall differences in the flight heights of birds between colonies could also be related to variation in foraging and commuting patterns of birds and distance of the wind farm to shore, as well as altered meteorological conditions due to the wind farm itself (Broström *et al.* 2019) influencing flight behaviour, or the provision of additional foraging opportunities, such as reef habitats at the base of turbines (Dannheim *et al.* 2019).

The level of fine-scale information gained for Lesser Black-backed Gulls above is similar to that obtained by Harwood *et al.* (2017), who employed visual tracking (see also Perrow *et al.* 2011b) from a rigid-hulled inflatable boat to follow breeding Sandwich Terns *Thalasseus sandvicensis* before, during and after construction of Sheringham Shoal wind farm on the east coast of the UK. During construction, terns demonstrated a macro-avoidance response, with the proportion of birds crossing the site reducing by 30%. Within the site, birds also avoided areas of construction activity. Both responses were unexpected considering there was no risk of collision from what were non-operational standing structures and that Sandwich Tern is typically classed as being insensitive to human activity (Garthe & Hüppop 2004). Results of the continuation of tracking during the operational phase are awaited, with the anticipation that the rate of collision would be very low in accordance with the avoidance behaviour observed. Visual tracking, whereby birds are followed as they are encountered, which readily provides hundreds if not thousands of independent samples, overcomes the frequent criticisms of telemetry studies that the number of individuals with tracks on which analysis is based may be small, and data are repeat measures of the same individuals. An opportunity to undertake both visual tracking and telemetry simultaneously on a target species would be particularly valuable.

Finally, it should be noted that the behaviour of species may vary between locations, while time and weather patterns may also influence the flight behaviour of marine birds (e.g. Kahlert *et al.* 2012). Intrinsic variation may also exist in patterns of habitat use and environmental interactions between individual birds, age classes and sexes. The picture of three-dimensional habitat use is therefore likely to be complex and specific to individual locations and developments.

Indirect responses through changes to habitat and prey

Marine birds are part of a wider ecosystem, being at or near the top of their trophic system. As summarised by Perrow (2019), OWFs have the potential to alter the physical environment, thus causing potential changes to prey resources that seabirds rely on. Such processes are linked particularly to displacement effects (see *Displacement and barrier effects*, above), but have only occasionally been given separate consideration within impact assessments (but see Forewind 2014). The mechanisms by which species are affected through ecosystem or habitat changes are often poorly known (Perrow 2019). As such, there are few studies for OWFs that have been conducted at a sufficient level of detail (but see Perrow *et al.* 2011a). Wind farms could, for example, alter the available substrate through the installation of foundations, particularly monopole turbines and scour protection, which could, in turn, create reef habitats for colonising invertebrates and fish, potentially attracting marine birds to a new food supply (Dannheim *et al.* 2019; Gill & Wilhelmsson 2019). Alternatively, the presence of wind turbines could alter habitat use and spawning and migration of prey species (Gill & Wilhelmsson 2019), thus altering the use of offshore areas by marine birds and mammals (Nehls *et al.* 2019). Birds may also use offshore structures in novel ways, such as perching and resting (e.g. Krijgsveld *et al.* 2011; Vanermen & Stienen 2019), which is a form of habitat alteration that could offer advantages such as extending foraging ranges and is already suspected to have occurred for Great Cormorant *Phalacrocorax carbo* (Vanermen & Stienen 2019).

Telemetry, alongside other data sets, offers the potential to investigate such fine-scale predator–prey interactions and relationships of marine birds with their environment. At the simplest level, telemetry can identify whether birds may be using wind farms and offshore structures, which is important to understand, particularly for birds originating from nearby protected breeding locations. However, with the advent of advanced modelling techniques (see Box 4.4) it is now possible to consider many aspects of a marine bird's life in holistic modelling approaches, such as diet, foraging behaviour, and demography and vital rates (McDonald *et al.* 2012; Warwick-Evans *et al.* 2017). As also noted in Box 4.4, telemetry methods can be used together with other data sets to build up a picture of trophic linkages, throughout all phases of wind-farm development. Understanding the fundamental (baseline) reasons for marine species utilising offshore areas as well as the effects that wind farms may have is therefore crucial. Developing further process-based modelling approaches, identifying mechanistic links for different species, may also allow the effects of wind farms on prey and, in turn, on marine birds to be better predicted in the future.

Concluding remarks

The first application of telemetry upon a marine bird to judge potential effects of an OWF appears to be the study by Perrow *et al.* (2006) on Little Terns at Scroby Sands. Since then, there has been a rapid increase in the scope and diversity of telemetry and tracking studies of marine birds to address a number of key issues surrounding the interactions between marine birds and OWFs. These run right through the phases of wind-farm impact assessment, from initial screening and scoping of species and sites through to post-consent monitoring and continued assessment for constructed wind farms.

At the outset, the assessment of species foraging ranges and connectivity using telemetry during the breeding season is an important first step in identifying potential conflict between protected populations of marine birds and proposed developments (Thaxter *et al.* 2012). Spatial locations of birds around the colony can then be used to identify key areas of habitat use. However, telemetry can provide further insight into the ecology of a species, including behaviour (Weimerskirch *et al.* 2005), physiology (Bevan *et al.* 1997), diet (Watanuki *et al.* 2008), environmental relationships (Skov *et al.* 2008) and predator–prey interactions (Elliott *et al.* 2008). This eclectic mix and the ability to combine different devices and sensors (such as GPS and diving activity) are enabling habitat relationships to be modelled in greater detail, allowing potential population consequences of wind farms to be better appraised. A refined spatial understanding can inform assessment of displacement and barrier effects (McDonald *et al.* 2012), while understanding of flight height, speed and diurnal activity from telemetry can inform predictions of collision risk (Cleasby *et al.* 2015; Warwick-Evans *et al.* 2016). Tracking of the migratory and non-breeding passage movements of marine birds crossing potential development areas can similarly be used to understand potential barrier and potential collision risk effects (Griffin *et al.* 2016).

Once a wind farm has been constructed, telemetry can also contribute substantially to the understanding of flight heights, travel speeds and likely avoidance behaviour of different species, which can be valuable in refining CRMs (Thaxter *et al.* 2018a). Further assessment of these key factors influencing collision for different species throughout the year is an important research gap to fill, and could lead to much needed site-specific assessments for different species throughout the year. With the increasing longevity of tracking devices, such assessments are now possible for some species through the annual cycle, which will give a relative perspective of vulnerability between different life-history phases, such as breeding and non-breeding periods (Thaxter *et al.* 2018b). Such approaches would be useful to inform both spatial planning ahead of development through sensitivity mapping, and also mitigation for wind farms already constructed in areas of potentially high vulnerability. Moreover, telemetry used in this way can determine the network of wind farms, and other marine activities, that a population may come into contact with, which is useful when considering potential cumulative impacts of developments on an international scale on protected populations. Further study of movements of different demographic components of populations (such as adults, immatures and juveniles) and of multiple populations would also refine predictions of the potential demographic impacts of developments.

More work is still needed for many species, so the job is not yet done. Key migration corridors of marine birds need to be identified to avoid reliance on precautionary information for wind-farm interactions. A greater appreciation of the likely variation inherent in movements through time is also required to build up a baseline understanding, and the consequences of displacement are far from being fully understood for many species. More assessments of sample sizes are also required. Individual (or agent)-based telemetry has traditionally suffered from the repeated sampling of a small number of birds and the question of how representative this is of an entire colony (Soanes *et al.* 2013). Only recently have specific studies been conducted to begin to address these issues for marine birds (Soanes *et al.* 2013; Bogdanova *et al.* 2014; Thaxter *et al.* 2017). However, increasingly complex data sets are being collected for long periods within a given season (such as breeding) over multiple years, adding further complexity. The sex, age, number of chicks and breeding site have also been found to have an influence on behaviour, highlighting the importance of developing a sampling regimen to reflect variability in movements and ensure that the tagged component is representative of the population studied (Soanes *et al.* 2014).

Telemetry can also be used in conjunction with other technologies and survey platforms (Louzao *et al.* 2009). For example, visual tracking of birds to provide GPS positional data has been used alongside hand-held laser rangefinders to provide flight height information (Perrow *et al.* 2017). Telemetry is also congruent to radar studies (e.g. Krijgsveld *et al.* 2011; Plonczkier & Simms 2012) to assess influx and outflux of birds across the year as they cross wind farms. Further, for diving species such as auks, divers, grebes and seaduck that spend a proportion of their time underwater, the abundance figures derived from aerial and boat-based surveys used in impact assessments may potentially be underestimated (see Webb & Nehls, Chapter 3). One possibility is to use tracking and telemetry data in combination, to assess the proportion of time spent underwater for a species and then apply a correction to the original number estimated. Different survey platforms are therefore very complementary. A joint approach combining different methods could therefore be a powerful one and could potentially improve sample sizes.

Under the current trajectory of telemetry and tracking developments, such ambition is possible. If the past is a precursor to the future, new families of devices with ever-improving technologies, including smaller, more streamlined designs, lighter in weight with more efficient battery life, more sensors and greater functionality, can be expected. Smarter ways of attaching tags, such as the use of different harness materials or weak-link harnesses are being explored, crucial also for minimising the potential impacts that tags and attachments may have on birds (Box 4.2). Not only would this allow more species, including lighter ones, to be tracked, but a greater volume of information would be produced, giving an enhanced understanding of ecology and interactions between marine birds and their environment. It is likely that tracking and telemetry will continue to provide ever more useful solutions to resolve conflicts between marine bird populations and OWFs.

Acknowledgements

The case studies on Lesser Black-backed Gulls (Box 4.3) and migratory waterbirds (Box 4.5) presented in this chapter were funded by the Department of Energy and Climate Change (DECC), superseded by the Department for Business, Enterprise and Industrial Strategy (BEIS), with thanks in particular to John Hartley (Hartley Anderson), Emma Cole, Mandy King, Sophie Thomas and James Burt (DECC) for their support. Also in relation to Lesser Black-backed Gull research, thanks to the National Trust, Royal Society for the Protection of Birds (RSPB) and Natural England for permissions, and to all who have helped with discussions and fieldwork, in particular to Willem Bouten and Judy Shamoun-Baranes (University of Amsterdam, UvA-BiTS) and Mark Rehfisch (formerly BTO) for support. The UvA-BiTS tracking studies are facilitated by infrastructures for e-Science, developed with support of the NLeSC (http://www.esciencecenter.com/) and LifeWatch, carried out on the Dutch national e-infrastructure with support of SURF Foundation. The study on Little Terns at Scroby Sands (Box 4.6) was funded and supported by E.ON Renewables Development Ltd, with additional data for 2013 commissioned by Natural England. Andrew Green, formerly of ECON Ecological Consultancy Ltd made an essential contribution to the modelling described. Thanks also to Philip Pearson and Daniel Hercock of the RSPB for kindly providing Little Tern breeding data gathered by RSPB staff and volunteers.

References

4cOffshore (2013) Global offshore wind farms database. Retrieved 3 March 2013 from www.4coffshore.com/offshorewind/

Adams, J. & Takekawa, J.Y. (2008) At-sea distribution of radio-marked ashy storm-petrels *Oceanodroma homochroa* captured on the California Channel Islands. *Marine Ornithology* 36: 9–17.

Amlaner, C.J. & McDonald, D.W. (1980) *A Handbook on Biotelemetry and Radio Tracking*. Oxford: Pergamon Press.

Bailey, H., Brookes, K.L. & Thompson, P.M. (2014) Assessing environmental impacts of offshore wind farms: lessons learned and recommendations for the future. *Aquatic Biosystems* 10: 8. doi: 10.1186/2046-9063-10-8.

Band, W. (2012) Using a collision risk model to assess bird collision risks for offshore windfarms. Strategic Ornithological Support Services. Project SOSS-02. Thetford: British Trust for Ornithology. Retrieved 11 December 2018 from https://www.bto.org/sites/default/files/u28/downloads/Projects/Final_Report_SOSS02_Band1ModelGuidance.pdf

Band, W., Madders, M. & Whitfield, D.P. (2007) Developing field and analytical methods to assess avian collision risk at wind farms. In de Lucas, M., Janss, G.F.E. & Ferrer, M. (eds) *Birds and Wind Farms: Risk assessment and mitigation*. Madrid: Quercus/Servicios Informativos Ambientales. pp. 259–275.

Barron, D.G., Brawn, J.D. & Weatherhead, P.J. (2010) Metaanalysis of transmitter effects on avian behaviour and ecology. *Methods in Ecology and Evolution* 1: 180–187.

Benvenuti, S., Dall'Antonia, L. & Lyngs, P. (2001) Foraging behaviour and time allocation of chick-rearing razorbills *Alca torda* at Græsholmen, central Baltic Sea. *Ibis* 143: 402–412.

Bevan, R.M., Boyd, I.L., Butler, P.J., Reid, K., Woakes, A.J. & Croxall, J.P. (1997) Heart rates and abdominal temperatures of free-ranging South Georgian shags, *Phalacrocorax georgianus*. *Journal of Experimental Biology* 200: 661–675.

Birdlife International (2018) Seabird Tracking Database: tracking ocean wanderers. Cambridge: Birdlife International. Retrieved 27 February 2018 from http://www.seabirdtracking.org/

Bishop, C.M., Spivey, R.J., Hawkes, L.A., Batbayar, N., Chua, B., Frappell, P.B., Milsom, W.K., Natsagdorj, T., Newman, S.H., Scott, G.R., Takekawa, J.Y., Wikelski, M. & Butler, P.J. (2015) The roller coaster flight strategy of bar-headed geese conserves energy during Himalayan migrations. *Science* 347: 250–254.

Block, B.A., Holbrook, C.M., Simmons, S.E., Holland, K.N., Ault, J.S., Costa, D.P., Mate, B.R., Seitz, A.C., Arendt, M.D., Payne, J.C., Mahmoudi, B., Moore, P., Price, J.M., Levenson, J., Wilson, D. & Kochevar, R.E. (2016) Toward a national animal telemetry network for aquatic observations in the United States. *Animal Biotelemetry* 4: 6. doi: 10.1186/s40317-015-0092-1.

Bodey, T.W., Cleasby, I.R., Bell, F., Parr, N., Schultz, A., Votier, S.C. & Bearhop, S. (2018) A phylogenetically-controlled meta-analysis of biologging device effects on birds: deleterious effects and a call for more standardized reporting of study data. *Methods in Ecology and Evolution* 9: 946–955.

Boehlert, G.W., Costa, D.P., Crocker, D.E., Green, P., O'Brien, T., Levitus, S. & Le Boeuf, B.J. (2001) Autonomous pinniped environmental samplers: using instrumented animals as oceanographic data collectors. *Journal of Atmospheric and Oceanic Technology* 18: 1882–1893.

Bogdanova, M.I., Wanless, S., Harris, M.P., Lindström, J., Butler, A., Newell, M.A., Sato, K., Watanuki, Y., Parsons, M. & Daunt, F. (2014) Among-year and within-population variation in foraging distribution of European shags *Phalacrocorax aristotelis* over two decades: implications for marine spatial planning. *Biological Conservation* 170: 292–299.

Bouten, W., Baaij, E.W., Shamoun-Baranes, J. & Camphuysen, K.C.J. (2013) A flexible GPS tracking system for studying bird behaviour at multiple scales. *Journal of Ornithology* 154: 571–580.

Bradstreet, M.S.W. & Brown, R.G.B. (1985) Feeding ecology of the Atlantic Alcidae. In Nettleship, D.N. & Birkhead, T.R. (eds) *The Atlantic Alcidae*. London: Academic Press. pp. 264–318.

Bridge, E.S., Thorup, K., Bowlin, M.S., Chilson, P.B., Diehl, R.H., Fléron, R.W., Hartl, P., Kays, R., Kelly, J.F., Robinson, W.D. & Wikelski, M. (2011) Technology on the move: recent and forthcoming innovations for tracking migratory birds. *BioScience* 61: 689–698.

Broström, G., Ludewig, E., Schneehorst, A. & Pohlmann, T. (2019) Atmosphere and ocean dynamics. In Perrow, M.R. (ed.) *Wildlife and Wind Farms, Conflicts and Solutions. Volume 3. Offshore: Potential effects*. Exeter: Pelagic Publishing. pp. 47–63.

Brown, D.D., Kays, R., Wikelski, M., Wilson, R. & Klimley, A.P. (2013) Observing the unwatchable through acceleration logging of animal behaviour. *Animal Biotelemetry* 1: 20. doi: 10.1186/2050-3385-1-20.

Burger, A.E. & Shaffer, S.A. (2008) Application of tracking and data-logging technology in research and conservation of seabirds. *Auk* 125: 253–264.

Camphuysen, C.J. (2011) Lesser black-backed gulls nesting at Texel: foraging distribution, diet, survival, recruitment and breeding biology of birds carrying advanced GPS loggers. NIOZ Report 2011-05. Texel: Royal Netherlands Institute for Sea Research. Retrieved 11 December 2018 from http://imis.nioz.nl/imis.php?module=ref&refid=253221

Camphuysen, C.J., Fox, A.D., Leopold, M.F. & Petersen, I.K. (2004) Towards standardised seabirds at sea census techniques in connection with environmental impact assessments for offshore wind farms in the UK: a comparison of ship and aerial sampling methods for marine birds, and their applicability to offshore wind farm assessments. NIOZ Report No. BAM-02-2002 to COWRIE. Texel: Royal Netherlands Institute for Sea Research (NIOZ). Retrieved from http://jncc.defra.gov.uk/PDF/Camphuysenetal2004_COWRIEmethods.PDF

Chimienti, M., Bartoń, K.A., Scott, B.E. & Travis, J.M.J. (2014) Modelling foraging movements of diving predators: a theoretical study exploring the effect of heterogeneous landscapes on foraging efficiency. *PeerJ* 2: e544. doi: 10.7717/peerj.544.

Chivers, L.S., Hatch, S.A. & Elliott, K.H. (2016) Accelerometry reveals an impact of short-term tagging on seabird activity budgets. *Condor* 118: 159–168.

Cleasby, I.R., Wakefield, E.D., Bearhop, S., Bodey, T.W., Votier, S.C. & Hamer, K.C. (2015) Three dimensional tracking of a wide-ranging marine predator: flight heights and vulnerability to offshore wind farms. *Journal of Applied Ecology* 52: 1474–1482.

Cook, A.S.C.P., Wright, L.J. & Burton, N.H.K. (2012) A review of flight heights and avoidance rates in relation to offshore wind farms. Strategic Ornithological Support Services. Project SOSS-02. BTO Research Report 618. Thetford: British Trust for Ornithology. Retrieved 11 December 2018 from https://www.bto.org/research-data-services/publications/research-reports/2013/strategic-ornithological-support-services-

Cook, A.S.C.P., Humphreys, E.M., Masden, E.A., Band, W. & Burton, N.H.K. (2014) The avoidance rates of collision between birds and offshore turbines. BTO Research Report No. 656 on behalf of Marine Scotland. Thetford: British Trust for Ornithology. Retrieved 11 December 2018 from https://www.gov.scot/publications/scottish-marine-freshwater-science-volume-5-number-16-avoidance-rates/

Cook, A.S.C.P., Humphreys, E.M., Bennet, F., Masden, E.A. & Burton, N.H.K. (2018) Quantifying avian avoidance of offshore wind turbines: Current evidence and key knowledge gaps. *Marine Environmental Research* 140: 278–288.

Cook, T., Hamann, M., Pichegru, L., Bonadonna, F., Grémillet, D. & Ryan, P.G. (2012) GPS and time-depth loggers reveal underwater foraging plasticity in a flying diver, the Cape cormorant. *Marine Biology* 159: 373–387.

Corkhill, P. (1973) Food and feeding ecology of puffins. *Bird Study* 20: 207–220.

Corman, A.-M. & Garthe, S. (2014) What flight heights tell us about foraging and potential conflicts with wind farms: a case study in lesser black-backed gulls (*Larus fuscus*). *Journal of Ornithology* 155: 1037–1043.

Da, Z., Xiliang, Z., Jiankun, H. & Qimin, C. (2011) Offshore wind energy development in China: current status and future perspective. *Renewable and Sustainable Energy Reviews* 15: 4673–4684.

Dannheim, J., Degraer, S., Elliot, M., Smyth, K. & Wilson, J.C. (2019) Seabed communities. In Perrow, M.R. (ed.) *Wildlife and Wind Farms, Conflicts and Solutions. Volume 3. Offshore: Potential effects*. Exeter: Pelagic Publishing. pp. 64–85.

Daunt, F., Peters, G., Scott, B., Grémillet, D. & Wanless, S. (2003) Rapid-response recorders reveal interplay between marine physics and seabird behaviour. *Marine Ecology Progress Series* 255: 283–288.

de Monte, S., Cotté, C., d'Ovidio, F., Lévy, M., Le Corre, M. & Weimerskirch, H. (2012) Frigatebird behaviour at the ocean–atmosphere interface: integrating animal behaviour with multi-satellite data. *Journal of the Royal Society Interface* 9: 3351–3358.

Eliassen, E. (1960) *A Method for Measuring the Heart Rate and Stroke/Pulse Pressures in Birds in Normal Flight*. Årbok for Universitetet i Bergen/Matematisk-naturvitenskapelig Series 12. Bergen: Norwegian University Press.

Elliott, K.H., Davoren, G.K. & Gaston, A.J. (2008) Time allocation by a deep-diving bird reflects prey type and energy gain. *Animal Behaviour* 75: 1301–1310.

Elliott, K.H., McFarlane-Tranquilla, L., Burke, C.M., Hedd, A., Montevecchi, W.A. & Anderson, W.G. (2012) Year-long deployments of small geolocators increase corticosterone levels in murres. *Marine Ecology Progress Series* 466: 1–7.

Environmental Resource Management (ERM) (2012) East Anglia ONE Offshore Wind Farm. Environmental statement. Volume 2 Offshore. Chapter 12: Ornithology marine and coastal. East Anglia Offshore Wind Ltd, APEM Ltd. Retrieved 4 March 2018 from https://infrastructure.planninginspectorate.gov.uk/wp-content/ipc/uploads/projects/EN010025/EN010025-000478-7.3.7%20Volume%202%20Chapter%2012%20Ornithology%20Marine%20and%20Coastal.pdf

Everaert, J. & Stienen, E.W.M. (2007) Impact of wind turbines on birds in Zeebrugge (Belgium): significant effect on breeding tern colony due to collisions. *Biodiversity and Conservation* 16: 3345–3359.

Forewind (2014) Dogger Bank Teesside A & B. Environmental statement. Chapter 11: Marine and coastal ornithology. F-OFC-CH-011_Issue 4.1. Forewind Ltd. Retrieved 7 July 2018 from https://infrastructure.planninginspectorate.gov.uk/wp-content/ipc/uploads/projects/EN010051/EN010051-000225-6.11%20ES%20Chapter%20

11%20Marine%20and%20Coastal%20Ornithology.pdf

Frederiksen, M., Moe, B., Daunt, F., Phillips, R.A., Barrett, R.T., Bodganova, M.I., Boulinier, T., Chardine, J.W., Chastel, O., Chivers, L.S., Christensen-Dalsgaard, S., Clément-Chastel, C., Colhoun, K., Freeman, R., Gaston, A.J., Gonzalez-Solis, J., Goutte, A., Grémillet, D., Guilford, T., Jensen, G.H., Krasnov, Y., Lorentsen, S.-H., Mallory, M.L., Newell, M., Olsen, B., Shaw, D., Steen, H., Strøm, H., Systad, G.H., Thórarinsson, T. & Anker-Nilssen, T. (2012) Multicolony tracking reveals the winter distribution of a pelagic seabird on an ocean basin scale. *Diversity and Distributions* 18: 530–542.

Furness, R.W., Wade, H.M. & Masden, E.A. (2013) Assessing vulnerability of marine bird populations to offshore wind farms. *Journal of Environmental Management* 119: 56–66.

Garthe, S. & Hüppop, O. (2004) Scaling possible adverse effects of marine wind farms on seabirds: developing and applying a vulnerability index. *Journal of Applied Ecology* 41: 724–734.

Gibbons, J.W. & Andrews, K.M. (2004) PIT tagging: simple technology at its best. *BioScience* 54: 447–454.

Gill, A.B. & Wilhelmsson, D. (2019) Fish. In Perrow, M.R. (ed.) *Wildlife and Wind Farms, Conflicts and Solutions. Volume 3. Offshore: Potential effects.* Exeter: Pelagic Publishing. pp. 86–111.

Griffin, L., Rees, E. & Hughes, B. (2010) The migration of whooper swans in relation to offshore wind farms. WWT Final Report to COWRIE Ltd. Slimbridge: Wildfowl & Wetlands Trust. Retrieved 11 December 2018 from http://www.swansg.org/wp-content/uploads/2017/01/WWT-Final-Report-to-COWRIE-Ltd-Whooper-Swan-tracking.pdf

Griffin, L., Rees, E. & Hughes, B. (2011) Migration routes of whooper swans and geese in relation to wind farm footprints. WWT Final Report to the Department of Energy and Climate Change. Slimbridge: Wildfowl & Wetlands Trust. Retrieved 11 December 2018 from https://slidelegend.com/migration-routes-of-whooper-swans-and-geese-in-relation-to-govuk_5a263a7c1723dd19ed0e6ca1.html

Griffin, L., Rees, E. & Hughes, B. (2016) Satellite tracking Bewick's swan migration in relation to offshore and onshore wind farm sites. WWT Final Report to the Department of Energy and Climate Change. Slimbridge: Wildfowl & Wetlands Trust. Retrieved 11 December 2018 from https://assets.publishing.service.gov.uk/government/uploads/system/uploads/attachment_data/file/584607/WWT_2016_Bewicks_Swan_GPS_tracking_in_relation_to_offshore_and_onshore_wind_farms.pdf

Guilford, T., Meade, J., Willis, J., Phillips, R.A., Boyle, D., Roberts, S., Colle, M., Freeman, R. & Perrins, C.M. (2009) Migration and stopover in a small pelagic seabird, the Manx shearwater *Puffinus puffinus*: insights from machine learning. *Proceedings of the Royal Society B* 276: 1215. doi: 10.1098/rspb.2008.1577.

Halpern, B.S., Walbridge, S., Selkoe, K.A., Kappel, C.V., Micheli, F., D'Agrosa, C., Bruno, J.F., Casey, K.S., Ebert, C., Fox, H.E., Fujita, R., Heinemann, D., Lenihan, H.S., Madin, E.M.P., Perry, M.T., Selig, E.R., Spalding, M., Steneck, R. & Watson, R. (2008) A global map of human impact on marine ecosystems. *Science* 319: 948–952.

Hamel, N.J., Parrish, J.K., Conquest, L.L. & Burger, A.E. (2004) Effects of tagging on behavior, provisioning, and reproduction in the common murre (*Uria aalge*), a diving seabird. *Auk* 121: 1161–1171.

Hamer, K.C., Humphreys, E.M., Garthe, S., Hennicke, J., Peters, G., Grémillet, D., Phillips, R.A., Harris, M.P. & Wanless, S. (2007) Annual variation in diets, feeding locations and foraging behaviour of gannets in the North Sea: flexibility, consistency and constraint. *Marine Ecology Progress Series* 338: 295–305.

Hamer, K.C., Humphreys, E.M., Magalhães, M.C., Garthe, S., Hennicke, J., Peters, G., Grémillet, D., Skov, H. & Wanless, S. (2009) Fine-scale foraging behaviour of a medium-ranging marine predator. *Journal of Animal Ecology* 78: 880–889.

Harris, M.P., Daunt, F., Newell, M., Phillips, R.A. & Wanless, S. (2010) Wintering areas of adult Atlantic puffins *Fratercula arctica* from a North Sea colony as revealed by geolocation technology. *Marine Biology* 157: 827–836.

Harris, M.P., Bogdanova, M.I., Daunt, F. & Wanless, S. (2012) Using GPS technology to assess feeding areas of Atlantic puffins *Fratercula arctica*. *Ringing & Migration* 27: 43–49.

Hart, K.M. & Hyrenbach, K.D. (2009) Satellite telemetry of marine megainvertebrates: the coming of age of an experimental science. *Endangered Species Research* 10: 9–20.

Harwood, A.J.P., Perrow, M.R., Berridge, R., Tomlinson, M.L. & Skeate, E.R. (2017) Unforeseen responses of a breeding seabird to the construction of an offshore wind farm. In Köppel, J. (ed.) *Wind Energy and Wildlife Interactions. Presentations from the CWW2015 conference.* Cham: Springer. pp. 19–41.

Heath, R. & Randall, R. (1989) Foraging ranges and movements of jackass penguins (*Spheniscus demersus*) established through radio telemetry. *Journal of the Zoological Society of London* 217: 367–379.

Hill, R.D. (1994) Theory of geolocation by light levels. In Le Boeuf, B.J. & Laws, R.M. (eds) *Elephant Seals: Population, ecology, behavior and physiology*. Berkeley, CA: University of California Press. pp. 237–236.

Hooker, S.K., Biuw, M., McConnell, B.J., Miller, P.J.O. & Sparling, C.E. (2007) Bio-logging science: logging and relaying physical and biological data using animal-attached tags. *Deep-Sea Research II* 54: 177–182.

Horswill, C., O'Brien, S. & Robinson, R.A. (2017) Density-dependence and marine bird populations: are wind farm assessments precautionary? *Journal of Applied Ecology* 54: 1406–1414.

Icarus (2018) Icarus Initiative, International Cooperation for Animal Research Using Space: about Icarus. Retrieved 3 March 2018 from https://icarusinitiative.org/about-icarus

Jameson, H., Reeve, E., Laubek, B. & Sittel, H. (2019) The nature of offshore wind farms. In Perrow, M.R. (ed.) *Wildlife and Wind Farms, Conflicts and Solutions. Volume 3. Offshore: Potential effects*. Exeter: Pelagic Publishing. pp. 1–29.

Johnston, A., Cook, A.S.C.P., Wright, L.J., Humphreys, E.M. & Burton, N.H.K. (2014) Modelling flight heights of marine birds to more accurately assess collision risk with offshore wind turbines. *Journal of Applied Ecology* 51: 1126–1130.

Jouventin, P. & Weimerskirch, H. (1990) Satellite tracking of wandering albatrosses. *Nature* 343: 746–748.

Kahlert, J.A., Leito, A., Laubek, B., Luigujoe, L., Kuresoo, A., Aaen, K. & Luud, A (2012) Factors affecting the flight altitude of migrating waterbirds in western Estonia. *Ornis Fennica* 89: 241–253.

Kenward, R. (1987) *Wildlife Radio Tagging: Equipment, field techniques and data analysis*. London: Academic Press.

King, S. (2019) Seabirds: collision. In Perrow, M.R. (ed.) *Wildlife and Wind Farms, Conflicts and Solutions. Volume 3. Offshore: Potential effects*. Exeter: Pelagic Publishing. pp. 206–234.

Klaassen, R.H.G., Ens, B.J., Shamoun-Baranes, J., Exo, K.-M. & Bairlein, F. (2012) Migration strategy of a flight generalist, the lesser black-backed gull *Larus fuscus*. *Behavioral Ecology* 23: 58–68.

Korpinen, S. & Andersen, J.H. (2016) A global review of cumulative pressure and impact assessments in marine environments. *Frontiers in Marine Science* 3: 153. doi: 10.3389/fmars.2016.00153.

Krijgsveld, K.L., Fijn, R.C., Japink, M., van Horssen, P.W., Heunks, C., Collier, M.P., Poot, M.J.M., Beuker, D. & Dirksen, S. (2011) Effect studies offshore wind farm Egmond aan Zee. Final report on fluxes, flight altitudes and behaviour of flying birds. Bureau Waardenburg Report 10-219 NZW-Report R_231_T1_flu&flight. Culemborg: Bureau Waardenburg. Retrieved 29 March 2019 from https://tethys.pnnl.gov/sites/default/files/publications/Krijgsveld%20et%20al.%202011.pdf

Langston, R.H., Teuten, E. & Butler, A. (2013) Foraging ranges of northern gannets *Morus bassanus* in relation to proposed offshore wind farms in the UK: 2010–2012. RSPB Report to DECC. Sandy: Royal Society for the Protection of Birds. Retrieved 11 December 2018 from https://tethys.pnnl.gov/publications/foraging-ranges-northern-gannets-relation-proposed-offshore-wind-farms-uk-2010-2012

Le Maho, Y., Gendner, J.-P., Challet, E., Bost, C.-A., Gilles, J., Verdon, C., Plumere, C., Robin, J.-P. & Handrich, Y. (1993) Undisturbed breeding penguins as indicators of changes in marine resources. *Marine Ecology Progress Series* 95: 1–6.

Loring, P.H., Paton, P.W.C., Osenkowski, J.E., Gilliland, S.G., Savard, J.-P.L. & McWilliams, S.R. (2014) Habitat use and selection of black scoters in southern New England and siting of offshore wind energy facilities. *Journal of Wildlife Management* 78: 645–656.

Louzao, M., Bécares, J., Rodríguez, B., Hyrenbach, K.D., Ruiz, A. & Arcos, J.M. (2009) Combining vessel-based surveys and tracking data to identify key marine areas for seabirds. *Marine Ecology Progress Series* 391: 183–197.

Ludynia, K., Dehnhard, N., Poisbleau, M., Demongin, L., Masello, J.F. & Quillfeldt, P. (2012) Evaluating the impact of handling and logger attachment on foraging parameters and physiology in southern rockhopper penguins. *PLoS ONE* 7(11): e50429. doi: 10.1371/journal.pone.0050429.

Mainstream Renewable Power (2012) Mainstream Neart na Gaoithe Offshore Wind Farm. Ornithology Technical Report June 2012. Banchory: Natural Research Projects Ltd; Cork: Cork Ecology. Retrieved 7 December 2018 from http://

marine.gov.scot/datafiles/lot/nng/Environmental_statement/Appendices/Appendix%20 12.1%20-%20Ornithology%20Technical%20 Report.pdf

Masden, E.A. & Cook, A.S.C.P. (2016) Avian collision risk models for wind energy impact assessments. *Environmental Impact Assessment Review* 56: 43–49.

Masden, E.A., Haydon, D.T., Fox, A.D., Furness, R.W., Bullman, R. & Desholm, M. (2009) Barriers to movement: impacts of wind farms on migrating birds. *ICES Journal of Marine Science* 66: 746–753.

Masden, E.A., Haydon, D.T., Fox, A.D. & Furness, R.W. (2010) Barriers to movement: modelling energetic costs of avoiding marine wind farms amongst breeding seabirds. *Marine Pollution Bulletin* 60: 1085–1091.

Massey, B.W., Keane, K. & Boardman, C. (1988) Adverse effects of radio transmitters on the behaviour of nesting least terns. *Condor* 90: 945–947.

Maxwell, J.C. (1971) A Doppler satellite system design for animal tracking. In *National Telemetering Conference*, Washington DC, April 12–15. New York: Institute of Electrical and Electronics Engineers. pp. 269–270.

May, R. (2015) A unifying framework for the underlying mechanisms of avian avoidance of wind turbines. *Biological Conservation* 190: 179–187.

McClintock, B.T. & Michelot, T. (2018) momentuHMM: R package for generalized hidden Markov models of animal movement. *Methods in Ecology and Evolution* 9: 1518–1530.

McDonald, C., Searle, K., Wanless, S. & Daunt, F. (2012) Effects of displacement from marine renewable development on seabirds breeding at SPAs: a proof of concept model of common guillemots breeding on the Isle of May. Final Report to MSS. Edinburgh: Centre for Ecology & Hydrology. Retrieved 11 December 2018 from https://www.gov.scot/binaries/content/documents/govscot/publications/research-publication/2012/10/effects-displacement-marine-renewable-developments-seabirds-breeding-isle/documents/00404982-pdf/00404982-pdf/govscot%3Adocument

Microwave Telemetry (2018) Solar Argos/GPS PTTs. Columbia, MD: Microwave Telemetry Inc. Retrieved 11 December 2018 from https://www.microwavetelemetry.com/solar_argos_gps_ptts

Natural Power (2012) Technical Appendix 4.5 A – Ornithology baseline and impact assessment. Moray Offshore Renewables Ltd. Environmental statement. Telford, Stevenson, MacColl Offshore Wind Farms and transmission infrastructure. Edinburgh: Moray Offshore Renewables. Retrieved 11 December 2018 from http://marine.gov.scot/datafiles/lot/MORL/Environmental_statement/Volumes%208%20to%2011%20-%20Technical%20Appendices/Volume%20 10%20Part%203%20-%20Biological%20Environment%20Technical%20Appendices/Appendix%204.5%20A%20-%20Ornithology.pdf

Nehls, A., Harwood, A.J.P. & Perrow, M.R. (2019) Marine mammals. In Perrow, M.R. (ed.) *Wildlife and Wind Farms, Conflicts and Solutions. Volume 3. Offshore: Potential effects*. Exeter: Pelagic Publishing. pp. 112–141.

Orians, G.H. & Pearson, N.E. (1979) On the theory of central place foraging. In Horn, D.J., Stairs, G.R. & Mitchelle, R.G. (eds) *Analysis of ecological systems*. Columbus, OH: Ohio State University Press. pp. 155–177.

Paleczny, M., Hammill, E., Karpouzi, V. & Pauly, D. (2015) Population trend of the world's monitored seabirds 1950–2010. *PLoS ONE* 10(6): e0129342. doi: 10.1371/journal.pone.0129342.

Patterson, T.A., Thomas, L., Wilcox, C., Ovaskainen, O. & Matthiopoulos, J. (2008) State-space models of individual animal movement. *Trends in Ecology and Evolution* 23: 87–94.

Pennycuick, C.J. & Bradbury, T.A.M., Einarsson, O. & Owen, M. (1999) Response to weather and light conditions of migratory whooper swans *Cygnus cygnus* and flying height profiles, observed with the Argos satellite system. *Ibis* 141: 434–443.

Péron, C. & Grémillet, D. (2013) Tracking through life stages: adult, immature and juvenile autumn migration in a long-lived seabird. *PLoS ONE* 8(8): e72713. doi: 10.1371/journal.pone.0072713.

Perrow, M.R. (2019) A synthesis of effects and impacts. In Perrow, M.R. (ed.) *Wildlife and Wind Farms, Conflicts and Solutions. Volume 3. Offshore: Potential effects*. Exeter: Pelagic Publishing. pp. 235–277.

Perrow, M.R., Skeate, E.R., Lines, P., Brown, D. & Tomlinson, M.L. (2006) Radio telemetry as a tool for impact assessment of wind farms: the case of little terns *Sterna albifrons* at Scroby Sands, Norfolk, UK. *Ibis* 148: 57–75.

Perrow, M.R., Gilroy, J.J., Skeate, E.R. & Tomlinson, M.L. (2011a) Effects of the construction of Scroby Sands offshore wind farm on the prey base of little tern *Sternula albifrons* at its most important UK colony. *Marine Pollution Bulletin* 62: 1661–1670.

Perrow, M.R., Skeate, E.R. & Gilroy, J.J. (2011b) Novel use of visual tracking from a rigid-hulled inflatable boat (RIB) to determine foraging movements of breeding terns. *Journal of Field Ornithology* 82: 68–79.

Perrow, M.R., Harwood, A.J.P., Skeate, E.R., Praca, E. & Eglington, S.M. (2015) Use of multiple data sources and analytical approaches to derive a marine protected area for a breeding seabird. *Biological Conservation* 191: 729–738.

Perrow, M.R., Harwood, A.J.P., Berridge, R. & Skeate, E.R. (2017) The foraging ecology of sandwich terns in north Norfolk. *British Birds* 110: 257–277.

Petersen, I.K., Christensen, T.K., Kahlert, J., Desholm, M. & Fox, A.D. (2006) Final results of bird studies at the offshore wind farms at Nysted and Horns Rev, Denmark. National Environmental Research Institute, Denmark. Retrieved 11 December 2018 from https://tethys.pnnl.gov/publications/final-results-bird-studies-offshore-wind-farms-nysted-and-horns-rev-denmark

Phillips, R.A., Silk, J.R.D., Croxall, J.P., Afanasyev, V. & Briggs, D.R. (2004) Accuracy of geolocation estimates for flying seabirds. *Marine Ecology Progress Series* 266: 265–272.

Phillips, R.A., Croxall, J.P., Silk, J.R.D. & Briggs, D.R. (2008) Foraging ecology of albatrosses and petrels from South Georgia: two decades of insights from tracking technologies. *Aquatic Conservation: Marine and Freshwater Ecosystems* 17: S6–S21.

Pinaud, D. & Weimerskirch, H. (2005) Scale-dependent habitat use in a long-ranging central place predator. *Journal of Animal Ecology* 74: 852–863.

Pinaud, D. & Weimerskirch, H. (2007) At-sea distribution and scale-dependent foraging behaviour of petrels and albatrosses: a comparative study. *Journal of Animal Ecology* 76: 9–19.

Plonczkier, P. & Simms, I.C. (2012) Radar monitoring of migrating pink-footed geese: behavioural responses to offshore wind farm development. *Journal of Applied Ecology* 49: 1187–1194

Ponganis, P.J. (2007) Bio-logging of physiological parameters in higher marine vertebrates. *Deep-Sea Research II* 54: 183–192.

Ropert-Coudert, Y. & Wilson, R.P. (2005) Trends and perspectives in animal-attached remote sensing. *Frontiers in Ecology and the Environment* 3: 437–444.

Ross-Smith, V.H., Thaxter, C.B., Masden, E.A., Shamoun-Baranes, J., Burton, N.H.K., Wright, L., Rehfisch, M.M. & Johnston, A. (2016) Modelling flight heights of lesser black-backed gulls and great skuas from GPS: a Bayesian approach. *Journal of Applied Ecology* 53: 1676–1685.

Royal Society for the Protection of Birds (RSPB) (2018) Tracking seabirds to inform conservation of the marine environment. Sandy: RSPB. Retrieved 1 March 2018 from https://www.rspb.org.uk/our-work/conservation/projects/tracking-seabirds-to-inform-conservation-of-the-marine-environment/

RWE (2011) Galloper Wind Farm Project. Environmental Statement – Technical Appendices 2. Offshore Ornithology – Ornithological Technical Report – 11.A. Galloper Wind Farm Ltd. Retrieved 7 December 2018 from http://www.galloperwindfarm.com/assets/images/documents/GWF%20Environmental%20Statement/ES_Appendices_Technical_Appendix_2.pdf

Scholander, P.E. (1940) Experimental investigations on the respiratory function in diving mammals and birds. *Hvalrådets Skrifter* 22: 1–131.

Scottish Natural Heritage (2014) Wind farm impacts on birds. Inverness: Scottish Natural Heritage. Retrieved 5 January 2015 from http://www.snh.gov.uk/planning-and-development/renewable-energy/onshore-wind/bird-collision-risks-guidance/

Scragg, E.S., Thaxter, C.B., Clewley, G.D. & Burton, N.H.K (2016) Assessing behaviour of lesser black-backed gulls from the Ribble and Alt estuaries SPA using GPS tracking devices. BTO Research Report No. 689, on behalf of Natural England. Thetford: British Trust for Ornithology. Retrieved 11 December 2018 from https://www.bto.org/sites/default/files/publications/rr689.pdf

Seagreen Wind Energy Ltd (2012) Seagreen Firth of Forth offshore wind farm development offshore transmission assets project – Ornithological Technical Report. Seagreen Wind Energy Ltd, Niras Consulting Ltd. Retrieved 11 December 2018 from http://marine.gov.scot/datafiles/lot/SG_FoF_alpha-bravo/SG_Phase1_Offshore_Project_Consent_Application_Document%20(September%202012)/006%20ES/Volume%20III_Technical%20Appendices/Part%202_Technical%20Appendices/Appendix%20F2.pdf

Shamoun-Baranes, J., Bouten, W., Camphuysen, C.J. & Baaij, E. (2011) Riding the tide: intriguing observations of gulls resting at sea during breeding. *Ibis* 153: 411–415.

Shamoun-Baranes, J., Bouten, W., van Loon, E., Meijer, C. & Camphuysen, C.J. (2016) Flap or soar? How a flight generalist responds to its aerial environment. *Philosophical Transactions of the Royal Society of London B* 371(1704): 20150395.

Shamoun-Baranes, J., Burant, J.B., van Loon, E.E., Bouten, W. & Camhuysen, C.J. (2017) Short distance migrants travel as far as long distance migrants in lesser black-backed gulls *Larus fuscus*. *Journal of Avian Biology* 48: 49–57.

Shannon, H.D., Young, G.S., Yates, M.A., Fuller, M.R. & Seegar, W.S. (2002) American white pelican soaring flight times and altitudes relative to changes in thermal depth and intensity. *Condor* 104: 679–683.

Skov, H., Humphreys, E., Garthe, S., Geitner, K., Hamer, K., Hennicke, J., Pamer, H., Grémillet, D. & Wanless, S. (2008) Application of habitat suitability modelling to tracking data of marine animals as a means of analyzing their feeding habitats. *Ecological Modelling* 212: 504–512.

Skov, H., Heinänen, S., Norman, T., Ward, R., Méndez-Roldán, S. & Ellis, I. (2018) ORJIP bird collision and avoidance study. Final Report April 2018. London: Carbon Trust. Retrieved 20 June 2018 from https://www.carbontrust.com/media/675793/orjip-bird-collision-avoidance-study_april-2018.pdf

SMartWind (2013) Hornsea Project One ES: Volume 2, Offshore. Chapter 5: Ornithology. PINS Document Reference: 7.2.5 July 2013. London: SMart Wind. Retrieved 11 December 2018 from https://infrastructure.planninginspectorate.gov.uk/wp-content/ipc/uploads/projects/EN010033/EN010033-000518-7.2.5%20Ornithology.pdf

Soanes, L.M., Arnould, J.P.Y., Dodd, S.G., Sumner, M.D. & Green, J.A. (2013) How many seabirds do we need to track to define home-range area? *Journal of Applied Ecology* 50: 671–679.

Soanes, L.M., Arnould, J.P.Y., Dodd, S.G., Milligan, G. & Green, J.A. (2014) Factors affecting the foraging behaviour of the European shag: implications for seabird tracking studies. *Marine Biology* 161: 1335–1348.

Spiegel, C.S., Berlin, A.M., Gilbert, A.T., Gray, C.O., Montevecchi, W.A., Stenhouse, I.J., Ford, S.L., Olsen, G.H., Fiely, J.L., Savoy, L., Goodale, M.W. & Burke, C.M. (2017) Determining fine-scale use and movement patterns of diving bird species in federal waters of the mid-Atlantic United States using satellite telemetry. OCS Study BOEM 2017-069. Sterling, VA: US Department of the Interior, Bureau of Ocean Energy Management. Retrieved 11 December 2018 from https://www.boem.gov/espis/5/5635.pdf

Stienen, E.W.M., Desmet, P., Aelterman, B., Courtens, W., Feys, S., Vanermen, N., Verstraete, H., van de Walle, M., Deneudt, K., Hernandez, F., Houthoofdt, R., Vanhoorne, B., Bouten, W., Buijs, R.J., Kavelaars, M.M., Müller, W., Herman, D., Matheve, H., Sotillo, A. & Lens, L. (2016) GPS tracking data of lesser black-backed gulls and herring gulls breeding at the southern North Sea coast. *ZooKeys* 555: 115–124.

Taylor, P.D., Crewe, T.L., Mackenzie, S.A., Lepage, D., Aubry, Y., Crysler, Z., Finney, G., Francis, C.M., Guglielmo, C.G., Hamilton, D.J., Holberton, R.L., Loring, P.H., Mitchell, G.W., Norris, D., Paquet, J., Ronconi, R.A., Smetzer, J., Smith, P.A., Welch, L.J. & Woodworth, B.K. (2017) The Motus Wildlife Tracking System: a collaborative research network to enhance the understanding of wildlife movement. *Avian Conservation and Ecology* 12(1): 8. doi: 10.5751/ACE-00953-120108.

Thaxter, C.B., Wanless, S., Daunt, F., Harris, M.P., Benvenuti, S., Watanuki, Y., Grémillet, D. & Hamer, K.C. (2010) Influence of wing loading on the trade-off between pursuit-diving and flight in common guillemots and razorbills. *Journal of Experimental Biology* 213: 1018–1025.

Thaxter, C.B., Lascelles, B., Sugar, K., Cook, A.S.C.P., Roos, S., Bolton, M., Langston, R.H.W. & Burton, N.H.K. (2012) Seabird foraging ranges as a preliminary tool for identifying candidate Marine Protected Areas. *Biological Conservation* 156: 53–61.

Thaxter, C.B., Ross-Smith, V.H., Clark, N.A., Conway, G.J., Johnston, A., Wade, H.M., Masden, E.A., Bouten, W. & Burton, N.H.K. (2014) Measuring the interaction between marine features of Special Protection Areas with offshore wind farm development sites through telemetry. Final Report to DECC. BTO Research Report 649. Thetford: British Trust for Ornithology. Retrieved 11 December 2018 from https://assets.publishing.service.gov.uk/government/uploads/system/uploads/attachment_data/file/657524/BTO_Research_Report_649_-_Interactions_between_SPA_features_and_offshore_windfarms_final_report.pdf

Thaxter, C.B., Ross-Smith, V.H., Bouten, W., Rehfisch, M.M., Clark, N.A., Conway, G.J. & Burton, N.H.K. (2015) Seabird–wind farm interactions during the breeding season vary within and between years: a case study of lesser black-backed gull *Larus fuscus* in the UK. *Biological Conservation* 186: 347–358.

Thaxter, C.B., Ross-Smith, V.H., Clark, J.A., Clark, N.A., Conway, G.J., Masden, E.A., Wade, H.M., Leat, E.H.K., Gear, S.C., Marsh, M., Booth, C., Furness, R.W., Votier, S.C. & Burton, N.H.K. (2016) Contrasting effects of GPS device and harness attachment on adult survival of lesser black-backed gulls *Larus fuscus* and great skuas *Stercorarius skua*. *Ibis* 158: 279–290.

Thaxter, C.B., Clark, N.A., Ross-Smith, V.H., Conway, G.J., Bouten, W. & Burton, N.K.H. (2017) Sample size required to characterize area use of tracked seabirds. *Journal of Wildlife Management* 81: 1098–1109.

Thaxter, C.B., Ross-Smith, V.H., Bouten, W., Masden, E.A., Clark, N.A., Conway, G.J., Barber, L., Clewley, G.D. & Burton, N.H.K. (2018a) Dodging the blades: new insights into three dimensional space use of offshore wind farms by lesser black-backed gulls *Larus fuscus*. *Marine Ecology Progress Series* 587: 247–253.

Thaxter, C.B., Scragg, E.S., Clark, N.A., Clewley, G., Humphreys, E.M., Ross-Smith, V.H., Barber, L., Conway, G.J., Harris, S.J., Masden, E.A., Bouten, W. & Burton, N.H.K. (2018b) Measuring the interaction between lesser black-backed gulls and herring gulls from the Skomer, Skokholm and the seas off Pembrokeshire SPA and Morecambe Bay and Duddon Estuary SPA and offshore wind farm development sites. BTO Research Report No. 702 on behalf of BEIS. Thetford: British Trust for Ornithology. Retrieved 14 December 2018 from https://www.gov.uk/guidance/offshore-energy-strategic-environmental-assessment-sea-an-overview-of-the-sea-process#offshore-energy-sea-research-programme

University of Amsterdam Bird Tracking System (UvA-BiTS) (2018) University of Amsterdam Bird Tracking System. Retrieved 3 March 2018 from http://www.uva-bits.nl/

University of Massachusetts Amherst (UMass Amherst) (2018) Tracking endangered birds. Retrieved 27 February 2018 from http://videos.umass.edu/invideo/detail/video/4380925212001/tracking-endangered-birds?autoStart=true&q=tracking

US Geological Survey (USGS) (2018) Ecosystems: energy and wildlife: research. Reston, VA: US Geological Survey. Retrieved 27 February 2018 from https://www2.usgs.gov/ecosystems/energy_wildlife/seabirds.html

Vandenabeele, S.P., Wilson, R.P. & Grogan, A. (2011) Tags on seabirds: how seriously are instrument-induced behaviours considered? *Animal Welfare* 20: 559–571.

Vandenabeele, S.P., Shepard, E.L., Grogan, A. & Wilson, R.P. (2012) When three per cent may not be three per cent; device-equipped seabirds experience variable flight constraints. *Marine Biology* 159: 1–14.

Vandenabeele, S.P., Grundy, E., Friswell, M.I., Grogan, A., Votier, S.C. & Wilson, R.P. (2014) Excess baggage for birds: inappropriate placement of tags on gannets changes flight patterns. *PLoS ONE* 9(3): e92657. doi: 10.1371/journal.pone.0092657.

Vanermen, N. & Stienen, E.W.M. (2019) Seabirds: displacement. In Perrow, M.R. (ed.) *Wildlife and Wind Farms, Conflicts and Solutions. Volume 3. Offshore: Potential effects*. Exeter: Pelagic Publishing. pp. 174–205.

Votier, S.C., Bicknell, A., Cox, S.L., Scales, K.L. & Patrick, S.C. (2013) A bird's eye view of discard reforms: bird-borne cameras reveal seabird/fishery interactions. *PLoS ONE* 8(3): e57376. doi: 10.1371/journal.pone.0057376.

Votier, S.C., Fayet, A.L., Bearhop, S., Bodey, T.W., Clark, B.L., Grecian, J., Guilford, T., Hamer, K.C., Jeglinski, J.W.E., Morgan, G., Wakefield, E. & Patrick, S.C. (2017) Effects of age and reproductive status on individual foraging site fidelity in a long-lived marine predator. *Proceedings of the Royal Society B* 284: 20171068. doi: 10.1098/rspb.2017.1068

Wade, H.M., Masden, E.A., Jackson, A.C., Thaxter, C.B., Burton, N.H.K., Bouten, W. & Furness, R.W. (2014) Great skua (*Stercorarius skua*) movements at sea in relation to marine renewable energy developments. *Marine Environmental Research* 101: 69–80.

Waggit, J.J. & Scott, B.E. (2014) Using a spatial overlap approach to estimate the risk of collisions between deep diving seabirds and tidal stream turbines: a review of potential methods and approaches. *Marine Policy* 44: 90–97.

Wakefield, E.D., Phillips, R.A. & Matthiopoulos, J. (2009) Quantifying habitat use and preferences of pelagic seabirds using individual movement data: a review. *Marine Ecology Progress Series* 391: 165–182.

Wakefield, E.D., Bodey, T.W., Bearhop, S., Blackburn, J., Colhoun, K., Davies, R., Dwyer, R.G., Green, J.A., Grémillet, D., Jackson, A.L., Jessopp, M.J., Kane, A., Langston, R.H.W., Lescroel, A., Murray, S., Le Nuz, M., Patrick, S.C., Peron, C., Soanes, L.M., Wanless, S., Votier, S.C. & Hamer, K.C. (2013) Space partitioning without territoriality in gannets. *Science* 341: 68–70.

Wanless, S., Harris, M.P. & Morris, J.A. (1990) A comparison of feeding areas used by individual common murres (*Uria aalge*), razorbills (*Alca torda*) and an Atlantic puffin (*Fratercula arctica*) during the breeding season. *Colonial Waterbirds* 13: 16–24.

Wanless, S., Corfield, T., Harris, M.P., Buckland, S.T. & Morris, J.A. (1993) Diving behaviour of the shag *Phalacrocorax aristotelis* (Aves: Pelecaniformes) in relation to water depth and prey size. *Journal of the Zoological Society of London* 123: 11–25.

Warwick-Evans, V., Atkinson, P.W., Arnould, J.P.Y., Gauvain, R.D., Soanes, L., Robinson, L.A. & Green, J.A. (2016) Changes in behaviour drive inter-annual variability in the at-sea distribution of northern gannets. *Marine Biology* 163: 156.

Warwick-Evans, V., Atkinson, P.W., Walkington, I. & Green, J.A. (2017) Predicting the impacts of wind farms on seabirds: an individual-based model. *Journal of Applied Ecology* 55: 503–515.

Watanuki, Y., Niizuma, Y., Gabrielsen, G.W., Sato, K. & Naito, Y. (2003) Stroke and glide of wing propelled divers: deep diving seabirds adjust surge frequency to buoyancy change with depth. *Proceedings of the Royal Society B* 270: 483–488.

Watanuki, Y., Daunt, F., Takahashi, A., Newei, M., Wanless, S., Sato, K. & Miyazaki, N. (2008) Microhabitat use and prey capture of a bottom-feeding top predator, the European shag, shown by camera loggers. *Marine Ecology Progress Series* 356: 283–293.

Weimerskirch, H., Salamolard, M. & Jouventin, P. (1992) Satellite telemetry of foraging movements in the wandering albatross. In Priede, I.G. & Swift, S.M. (eds) *Wildlife Telemetry: Remote monitoring and tracking of animals*. Chichester: Ellis Horwood. pp. 185–198.

Weimerskirch, H., Le Corre, M., Ropert-Coudert, Y., Kato, A. & Marsac, F. (2005) The three-dimensional flight of red-footed boobies: adaptations to foraging in a tropical environment? *Proceedings of the Royal Society B* 272: 53–61.

Weiser, E.L., Lanctot, R.B., Brown, S.C., Alves, J.A., Battley, P.F., Bentzen, R., Bêty, J., Bishop, M.A., Boldenow, M., Bollache, L., Casler, B., Christie, M., Coleman, J.T., Conklin, J.R., English, W.B., Gates, H.R., Gilg, O., Giroux, M.-A., Gosbell, K., Hassell, C., Helmericks, J., Johnson, A., Katrínardóttir, B., Koivula, K., Kwon, E., Lamarre, J.-F., Lang, J., Lank, D.B., Lecomte, N., Liebezeit, J., Loverti, V., McKinnon, L., Minton, C., Mizrahi, D., Nol, E., Pakanen, V.-M., Perz, J., Porter, R., Rausch, J., Reneerkens, J., Rönkä, N., Saalfeld, S., Senner, N., Sittler, B., Smith, P.A., Sowl, K., Taylor, A., Ward, D.H., Yezerinac, S. & Sandercock, B.K. (2016) Effects of geolocators on hatching success, return rates, breeding movements, and change in body mass in 16 species of Arctic-breeding shorebirds. *Movement Ecology* 4: 12.

White, C.R., Cassey, P., Schimpf, N.G., Halsey, L.G., Green, J.A. & Portugal, S.J. (2013) Implantation reduces the negative effects of bio-logging devices on birds. *Journal of Experimental Biology* 216: 537–542.

Wilson, R.P. & Vandenabele, S.P. (2012) Technological innovation in archival tags used in seabird research. *Marine Ecology Progress Series* 451: 245–262.

Wilson, R.P., Wilson, M.P.T., Link, R., Mempel, H. & Adams, N.J. (1991) Determination of movements of African penguins *Spheniscus demersus* using a compass system: dead reckoning may be an alternative to telemetry. *Journal of Experimental Biology* 157: 557–564.

Wilson, R.P., Cooper, J. & Plotz, J. (1992) Can we determine when marine endotherms feed? A case study with seabirds. *Journal of Experimental Biology* 167: 267–275.

Wilson, R.P., Jackson, S. & Thor Straten, M. (2007) Rates of food consumption in free-living Magellanic penguins. *Marine Ornithology* 35: 109–111.

Winiarski, K.J., Miller, D.L., Paton, P.W.C. & McWilliams, S.R. (2014) A spatial conservation prioritization approach for protecting marine birds given proposed offshore wind energy development. *Biological Conservation* 169: 79–88.

Žydelis, R., Heinänen, S., Dorsch, M., Nehls, G., Kleinschmidt, B., Quillfeldt, P. & Morkūnas, J. (2018) High mobility of red-throated divers revealed by satellite telemetry, Danish Hydraulic Institute, Denmark. Retrieved 3 December 2018 from http://www.divertracking.com/wp-content/uploads/Presentation_Zydelis_mobility.pdf

CHAPTER 5

Modelling collision risk and predicting population-level consequences

AONGHAIS S. C. P. COOK and ELIZABETH A. MASDEN

Summary

Collision risk is seen as a key consenting risk in relation to offshore wind farms. As part of the Environmental Impact Assessment process, the potential collision risk to seabird populations must be estimated both at the level of the individual project and at a cumulative level across all projects. Having assessed collision risk in terms of the total number of birds likely to be affected, these numbers must then be put in the context of the populations concerned. Models are central to this process, from the collision risk models used to assess risk to the population models used to assess the consequences of an impact at a colony level. In this chapter, the process by which collision risk is assessed using a collision risk model is described, the results of which are then used in conjunction with a population model to assess the likely population-level consequences of any collisions. As with any models, both the collision risk models and the population models are simplifications of reality and, as a result, often subject to significant uncertainty. The types of model available to practitioners are outlined and the limitations associated with them are discussed. As the number of operational wind farms increases, assessing the cumulative effect of multiple projects becomes a more pressing concern. However, this remains a challenge as a result of the difficulty in defining the processes acting on the population(s) concerned and, the challenges associated with combining lethal and sub-lethal impacts from multiple sources. While it is clearly desirable that the models used more accurately reflect biological reality, it is important to ensure that rigorous evidence is available to support any of the assumptions that are made.

Introduction

The renewable energy sector has undergone rapid expansion as part of measures to combat anthropogenic climate change (Obama 2017). However, while providing clean energy, renewable energy schemes can still present a number of risks to biodiversity and the environment, resulting in the need for an appropriate regimen of environmental impact assessment (Gibson *et al.* 2017). Wind power and, in particular, offshore wind power is likely to play a key role in meeting targets for renewable energy growth in coming years (Toke 2011). Collisions between birds and turbine blades have long been identified as one of the key environmental impacts of wind farms (Drewitt & Langston 2006; de Lucas & Perrow 2017); and as a result of a number of high-profile incidents and studies (e.g. de Lucas *et al.* 2008; Smallwood & Thelander 2008; Stienen *et al.* 2008; Dahl *et al.* 2012; Katzner *et al.* 2017), they are one of the impacts that attracts greatest public attention. Despite this, there are relatively few studies with the specific aim of documenting bird collisions in the offshore environment (King 2019). Collecting data to inform collision estimates in the offshore environment is extremely challenging owing to technical limitations (Collier *et al.* 2011). The authors are aware of only four studies (Pettersson 2005; Desholm 2006; Newton & Little 2009; Skov *et al.* 2018) that have recorded collisions of birds in the offshore environment.

Uncertainty over likely collision rates has meant that predicting collision risk as part of pre-construction Environmental Impact Assessment (EIA) has become a key part of the planning process (Masden & Cook 2016). This typically involves using data collected during pre-construction surveys within a collision risk modelling framework. In the UK, the most widely used model is the Band collision risk model (CRM) (Band 2012), although other models (e.g. Desholm 2006; Masden 2015; Kleyheeg-Hartman *et al.* 2018; McGregor *et al.* 2018) have been developed for use in the offshore environment. Given the prominence of collision risk in the decision-making process for offshore wind farms (OWFs), the outputs from these models often form a key part of impact assessments and contributed to the decision to refuse planning consent for the Docking Shoal OWF (DECC 2012), and to the decision by the Royal Society for the Protection of Birds (RSPB) to launch a judicial review of the decision to grant planning consent to four wind farms in the Firth of Forth (Scottish Courts and Tribunals 2016). Such problems may arise as a result of unresolved disagreements between different stakeholders about the parameters and models to be used in the assessment process (Box 5.1).

Box 5.1 The stakeholder perspective on the use of collision risk modelling and population modelling in the consenting process for an offshore wind farm

Sue King

The questions below were derived by the chapter authors based on experiences with a similar exercise as part of Masden (2015). The responses reflect the opinions of Sue King, based on her experience and discussions with the representatives of the offshore wind industry.

- *How do you think the outputs from collision/population models should be applied in the consenting process?*
  Outputs should reflect the uncertainty inherent in both types of model as many of the parameters used are not well known or understood, or may vary according to different circumstances. Consent documents should provide a detailed explanation of how and why this uncertainty, represented by a range of possible outcomes, has been taken into account even where, perhaps by necessity, the final decision is based on a single value, for example a threshold of mortality for a species at a Special Protection Area.

- *What do you think are the key issues with how the model outputs are currently applied?*
  Model outputs tend to cause problems when there are unresolved disagreements between the parties involved about the appropriate model to be used and/or the way outputs are presented and interpreted. Standardised advice is helpful; for example, in the UK, the Statutory Nature Conservation Bodies' advice on collision risk modelling (SNCBs 2014).

- *In your opinion, what problems have these issues caused the consenting process?*
  Unresolved issues can cause significant time and economic costs to all participants both pre- and post-consent. They are exacerbated where positions become entrenched or, by contrast, methods evolve during the determination process. If specific thresholds are used, they may lead to developers committing to equally specific project modifications which are embedded as consent conditions.

Example: difficulties in agreeing parameters/changing parameters

Docking Shoal (UK) was refused consent in 2012 owing to its effect on the in-combination collision mortality of breeding Sandwich Tern *Thalasseus sandvicensis* at the North Norfolk Coast Special Protection Area (SPA). The determination process took 3.5 years at a cost to the developer (among others) of millions of pounds (Macalister 2012). This was, in large part, due to the difficulties of agreeing the in-combination collision mortality and consequent population effects on the colony at the same time as offshore collision risk modelling was evolving and project modifications were being undertaken to reduce collision effects.

Ironically, subsequent changes to collision risk modelling using generic flight heights as recommended by SNCB (2014) guidance would have reduced collision estimates and may have allowed the project to be consented. However, by 2017 the subsequent use of laser rangefinders to estimate the height of Sandwich Terns at sea (Perrow *et al.* 2017) were more in line with the original flight heights presented in the Environmental Statement, tending to confirm the basis of the original decision. This illustrates how the timing of planning applications while guidance and evidence evolves can lead to different project outcomes.

Example: unresolved differences

Firth of Forth. In October 2014, the Scottish Government consented four offshore wind projects in the Firth of Forth. These consents were challenged by the Royal Society for the Protection of Birds (RSPB) via judicial review. Although this is a review of process, much was made of whether stakeholders had been permitted sufficient consultation on the methods used for collision and population modelling. The final result, in favour of the Scottish Government, was reached by the UK Supreme Court in November 2017. During this three-year period, significant legal costs were incurred by all parties including the petitioner (RSPB) and one project risked the loss of its Contract for Difference. The fact that Scoping Opinions for these projects were received in 2010 and issues identified at an early stage illustrates the difficulties of rationalisation and agreement on process.

Example: project modifications

To reduce collision numbers for Northern Gannet *Morus bassanus*, the East Anglia THREE committed to raising the rotor-tip clearance of 70% of its turbines from 22 m to 24 m above mean high water springs, reducing effects on the species at the closest SPA to less than eight birds. This, together with a reduction in the number of turbines from 240 to 150 at its sister project East Anglia ONE, reduced Northern Gannet (and Black-legged Kittiwake *Rissa tridactyla*) collisions to a level at which Natural England could state that the effect of the project on the Flamborough and Filey Coast SPA 'while not *de minimis*, is so small as to not materially alter the significance or the likelihood of an adverse effect on the (SPA's) integrity'. This modification was embedded in the requirements of the project's Development Consent Order (DCO).

- *What do you see as the single biggest improvement that could be made to these models in order to support the consenting process?*
  The greatest improvement to these models would come from the use of empirical data, particularly for flight heights, flight speed, nocturnal activity and avoidance rate. The technical difficulties and costs of collecting such data are well known and, as a result, very few projects have consent conditions requiring such use. However, where this has been done, for example in radar studies of Pink-footed Goose *Anser brachyryhnchus* avoidance rate at Lynn and Inner Dowsing offshore wind farms, results have shown that previous avoidance rates were conservative and have provided the evidence to raise them. In 2018, the Offshore Renewables Joint Industry Project on bird collision avoidance reported (Skov *et al.* 2018) and, following further review commissioned by the SNCBs, allowed recalculation of avoidance rates for some species (a variety of large *Larus* gulls, Black-legged Kittiwake and Northern Gannet) based on empirical evidence (Bowgen & Cook 2018). In addition, stochastic collision risk models have been developed (Masden 2015; McGregor *et al.* 2018), allowing variation around each parameter to be incorporated.

As part of the planning process, predicted collision rates must be considered in the context of the population(s) concerned, and assessed at a cumulative scale as well as at the level of the individual project. A variety of approaches exist to enable this. In recent years, assessments for OWFs in Europe have used potential biological removal (PBR) (e.g. Skov *et al.* 2012; Ministry of Economic Affairs 2015; Busch & Garthe 2016), whereby predicted collision rates are assessed against estimates of the number of additional mortalities a population can sustain. Some projects have also used population viability analysis (PVA), whereby predicted collision rates are incorporated into population models and projected over the life of a development. These analyses can consider either the collisions associated with single projects or the cumulative consequences associated with multiple projects.

Uncertainty is inherent in the assessment process for OWFs (Masden *et al.* 2015). In relation to collision risk, it is drawn into the process through a range of parameters, including: (1) those used in the CRM itself, (2) demographic parameters used in the population model, (3) the demographic consequences of any predicted impacts, and (4) decisions about how to incorporate cumulative impacts. This has contributed to significant concerns regarding the assessment of impacts associated with OWFs (Green *et al.* 2016; O'Brien *et al.* 2017).

Scope

Based on the authors' experiences providing advice as part of the consenting process for OWFs, this chapter highlights the role of CRMs in decision making, drawing from recently published papers on the topic. How collision risk modelling has evolved as the wind industry has developed, both onshore and offshore, is discussed, highlighting differences between models that are used onshore and offshore and possible reasons for these differences. However, it is worth noting that 'all models are wrong, but some are useful' (Box *et al.* 2005). With this in mind, this chapter highlights the limitations of CRMs, many of which relate to uncertainties associated with the data that are used and a lack of model validation. The implications of these limitations for the planning process are considered, drawing in the views of stakeholders, where appropriate (Box 5.1).

For the purposes of the planning process, it is not sufficient to present a raw estimate of the number of birds that may collide with wind turbines; it is important to understand the consequences of these collisions at a population level. As with the initial models, putting collision estimates into a population context can be fraught with difficulty, with a variety of different approaches available and broad uncertainty surrounding many of the model parameters. Some of the approaches that have been used to assess the population-level consequences of collisions are described, which leads to discussion of their relative strengths and limitations.

The question of cumulative impacts remains. While a single wind farm on its own may have a limited impact on a population, the offshore wind sector is expanding rapidly, with 4,149 turbines installed in European waters to the end of 2017 (WindEurope 2018). The cumulative consequences of this expansion are far from certain, but are likely to outweigh the summed impacts associated with individual projects. As many seabirds are migratory, they may be exposed to different offshore (and onshore) wind farms at different points in their annual cycles (Busch & Garthe 2017). The majority of studies have taken place in north-western Europe, reflecting the dominance of this region in the deployment of offshore wind to date. However, tracking studies (e.g. Frederiksen *et al.* 2012; Jessopp *et al.* 2013) have suggested the potential for trans-Atlantic migrations to the east coast of the USA and Canada, where the offshore wind industry is still at a relatively early stage. Such results highlight the need for careful consideration of spatial scale in the assessment of cumulative impacts. Additional challenges relate to the temporal scales over which impacts may occur and understanding how these impacts may interact with one another (Masden *et al.* 2010). Consequently, the approaches used to assess cumulative impacts to date have largely been found to be inadequate (Willsteed *et al.* 2017). Some of the key data gaps relating to effective cumulative impact assessment are briefly described in this chapter.

Themes

Collision modelling

Collision risk models

Masden & Cook (2016) reviewed avian CRMs for wind-energy assessments and of the ten models that they included, only three (McAdam 2005; Desholm 2006; Band 2012) were designed with the offshore environment in mind. A key issue in relation to identifying the models that had been used in the assessment of collision risk in relation to wind energy was found to be a lack of documentation surrounding them. This leads to concern in relation to the transparency of any assessments, increasing uncertainty in the consenting process.

In general, the models are based on the probability of a turbine blade occupying the same space as a bird during the time it takes the bird to pass through the rotor-swept volume of the turbine. The probability of collision relies on information about the bird (wingspan, body length, flight speed, flight height, nocturnal flight activity) and the turbine (blade width, blade length, blade pitch, rotor speed, hub height, operational time) (Figure 5.1). The bird is assumed to be cruciform, although this may result in an underestimate of collision risk. The turbine blade is assumed to have a width (chord) and a pitch angle, but no thickness. The model only considers flights that are perpendicular to the turbine blades and that the effects of approaching the turbine at oblique angles will cancel each other out (Band 2012). It also considers only the moving rotor, excluding the stationary elements such as the tower, although other models such as Smales *et al.* (2013) include these.

Of the models identified, Band (2012) is the most commonly used, particularly in the UK, but also recommended for use elsewhere (e.g. Ministry of Economic Affairs 2015). The approach used by the Band model was originally devised for onshore wind farms and promoted as guidance by Scottish Natural Heritage (SNH) but was later modified (Band *et al.* 2007) and then updated for OWFs (Band 2012) (Figure 5.2). A key change between the onshore and offshore models was the incorporation of density data in the offshore model.

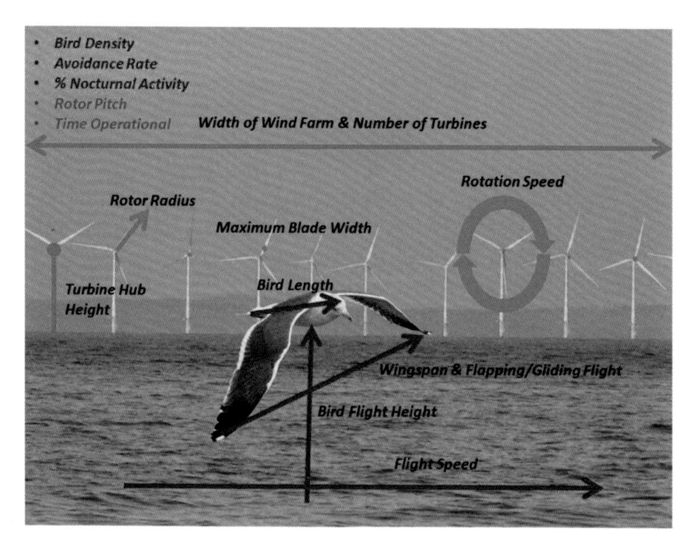

Figure 5.1 Range of parameters required to predict collision rates within a wind farm using a collision risk model. (Picture of Lesser Black-backed Gull *Larus fuscus* and wind farm: Tommy Holden and Moss Taylor, BTO)

This reflects differences in data collection in the onshore and offshore environments. In the onshore environment, data on bird abundance are typically collected using vantage-point surveys over a minimum of 72 hours, split between the breeding and non-breeding seasons (SNH 2014). In contrast, such surveys are not possible in the offshore environment and data are collected from moving platforms. Initially, this was achieved through the use of monthly boat surveys following a standard methodology (Camphuysen *et al.* 2004). However, there has been a move towards collecting data using digital aerial surveys, which can cover the wind farm and buffer areas over the space of a few hours, rather than the several days often required for boat surveys (see Webb & Nehls, Chapter 3).

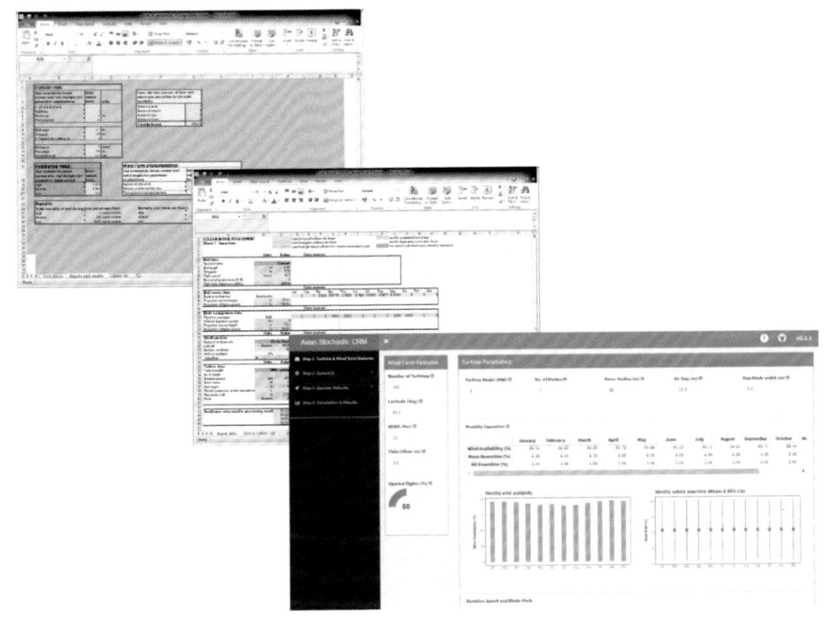

Figure 5.2 Evolution of the Band collision risk model. Initially, this was available for the onshore environment as an Excel spreadsheet (Band *et al.* 2007). The model was then refined as part of the Strategic Ornithological Support Services (SOSS) process to more accurately reflect data collection in the offshore environment (Band 2012). A stochastic version of the offshore model has since been developed (Masden 2015) and made available as a web application (McGregor *et al.* 2018).

Boat-based survey data collected as part of wind-farm pre-construction surveys also enabled the development of continuous flight height distributions (Johnston *et al.* 2014), an approach subsequently applied to data collected using Global Positioning System (GPS) tags (Ross-Smith *et al.* 2016), digital aerial survey (Johnston & Cook 2016) and light detection and ranging (lidar) (Cook *et al.* 2018b). These have been incorporated as an add-on to the offshore Band model (Band 2012), referred to as 'option 3' or the 'extended' Band model. This has two key advantages. First, it allows the proportion of birds at collision risk height to be inferred in situations where data are limited and/or unavailable. Secondly, it allows the model to account for the fact that birds towards the lower edge of a turbine's rotor-swept area are at a lower risk of collision, because the blades occupy a lower proportion of the total rotor-swept area than they do towards the midpoint of the rotor-swept area (Band 2012). The onshore Band CRM and both versions of the offshore version are deterministic. However, an update of the offshore version (Masden 2015; McGregor *et al.* 2018) has enabled the introduction of stochasticity into the modelled estimates of collision risk (Figure 5.2). This enables predicted collision rates to be expressed as an estimate with associated confidence intervals.

Onshore versus offshore

One difference between the onshore and offshore versions of the Band model is that the data available to parameterise or input into the model are collected using different methods. The Band model can essentially be broken down into two parts: (1) the probability of a single bird colliding with the turbine rotor during a single passage, and (2) a calculation of the number of birds flying through the rotor and thus the expected number of collisions, after accounting for any avoidance behaviour. When calculating the probability of a single bird colliding with the rotor, the models for onshore and offshore are similar except for the fact that the height of the rotor will change relative to the sea level over a tidal cycle for turbines fixed to the seabed (rather than floating), which is factored into the model. When considering the number of birds flying through the rotor, the data-collection methods differ.

For the original Band model, estimates of the number of birds flying through the rotor-swept area were dependent on whether the birds were making predictable or less predictable movements, for example migration versus foraging. For the predictable movements, the mean number of birds per hour of observation to fly through the risk window was determined. For less predictable movements, vantage-point surveys were used to estimate bird activity in an area. For offshore environments, the methods are more similar to the predictable movement methods for onshore and generally are calculated from at-sea surveys, and assume a constant flux of birds through the area derived from a measure of density and a known flight speed of the bird species concerned; that is, predictable movements. While updating the Band model for the offshore environment, other elements of the model were modified including the inclusion of flight height distribution data. There are uncertainties associated with all CRMs; however, those used in the offshore environment have the challenge that it is difficult to collect empirical mortality data, meaning that model validation is difficult.

Limitations and assumptions

Although CRMs are a useful tool to estimate likely collisions of birds with wind turbines, and provide information on the potential environmental impacts of wind-farm developments, they also have limitations. It is important that these are recognised to ensure that the data outputs are used appropriately. Potentially the greatest limitation of CRMs is that they generally assume much about bird behaviour. For example, the majority of models assume a linear relationship between bird abundance and collision risk, which may not be true for all situations (de Lucas *et al.* 2008; Ferrer *et al.* 2012) as there may be interactions between topography, species-specific behaviour, turbine layout and wind parameters (Barrios & Rodríguez 2004; Smallwood *et al.* 2009; Ferrer *et al.* 2012; Schaub 2012). In addition, many of the models include avoidance behaviour in the form of an avoidance rate, which assumes that a certain proportion of those birds on a collision path will take avoiding action before a collision occurs. Most models assume that avoidance behaviour is constant across all individuals within a species, which is unlikely. However, there are limited data on avoidance rates and estimating variability both between and within species is difficult (Cook *et al.* 2014; 2018a).

Models, such as Band (2012), assume a constant flux of birds through a wind farm. Such models assume that there are *X* birds within the wind farm at any given time, each of which takes *Y* seconds to fly through the rotor. From these data, it is possible to estimate the total number of birds passing through a wind farm in a day or year based on the speed of the birds, with a mean flight speed typically assumed. However, if the birds are not flying at the reported or constant speed, or not commuting through the wind farm but instead moving tortuously within the area (Patrick *et al.* 2014), it is possible to overestimate the number of birds flying through the wind farm in a given period. As a consequence, the total number of collisions will be elevated, and these inflated estimates of collision rates can have serious consequences and costs for a wind-farm developer. This is likely to be the case also for models that use data from vantage-point surveys; for example, the model developed by the US Fish and Wildlife Service (2013) uses data on the number of expected exposure events. It may be difficult to distinguish whether two observations of a bird are the same individual or different unless birds have distinct and unique marks. This distinction is important because each bird can only collide once, if it is assumed that collision equates to mortality, and if the number of birds using the area is overestimated, it is possible to overestimate the total number of collisions. Eichhorn *et al.* (2012) circumvent this limitation using an agent-based model to describe movements of individual birds through a landscape and applying a collision risk to each interaction of an individual with a wind turbine, although such a method can be computationally intensive.

Another limitation of CRMs is that they are frequently 'data hungry' in situations where data availability is often limited. For example, mechanistic models such as Band (2012) have many input parameters relating both to the birds, such as flight speed and morphometrics, and to the turbines, including rotor speed and blade width (Figure 5.1). In relation to birds, there is still much to be learned about behaviour, and therefore knowledge of aspects required within the models, such as flight speed, is limited. This can have significant consequences for the consenting processes as stakeholders take different positions on what represents realistic values to use in the models and, therefore, what reflect realistic, but precautionary, assessments of collision risk (Box 5.1). This is improving with the development of individual tracking and telemetry technologies such as miniaturised GPS tags, but often data are limited to the breeding season. In addition, many of the turbines suggested for offshore projects are still under development and therefore only a design 'envelope' can generally be provided, in which a range of values rather than specifics for any given parameter are presented and modelled. The approach developed by US Fish and Wildlife Service (2013) removes some of these data requirements by using a Bayesian framework, although it relies on the ability to collect data on actual collisions to validate the model, and thus limits the approach to onshore sites. As a result, a Bayesian method is unlikely to be suitable for offshore sites without development of methods to collate data on collisions; therefore, a mechanistic theoretical modelling approach will still be required, otherwise a constant collision risk is to be applied.

CRM is thus limited not only by a lack of data on model inputs, but also by a lack of opportunities for model validation. CRMs are rarely validated, but where they have been, predictions from EIA often show only a weak relationship with observed effects and predictor variables. Ferrer *et al.* (2012) found no relationship between variables predicting risk from EIAs and actual mortality, and only a weak relationship between mortality and species passage rates. Similarly, de Lucas *et al.* (2008) also found a weak relationship between abundance and recorded collisions for vultures, *Gyps* spp. In addition, Everaert (2014) found that large gulls were more likely to collide than smaller species. More specifically, opportunities for model validation are particularly limited for OWFs because bird mortality events are more difficult to document in the offshore environment as corpses do not remain in the area. In the terrestrial environment, the Bayesian framework employed by the US Fish and Wildlife Service (2013) model allows information to be updated. Although this is not validation, the input of empirical mortality data from corpse searches allows for the collision estimate to be refined. The use of camera systems offers a promising approach to detecting collisions (e.g. Skov *et al.* 2018); however, until the collection of mortality data from offshore developments becomes standard practice, Bayesian CRMs are unlikely to be a feasible option.

The use of model outputs

In general, the output from CRMs is a single estimate of the number of likely collisions. The uncertainty associated with these estimates is addressed to varying degrees within different CRMs. The majority of models applied onshore, such as Tucker (1996), Podolsky (2008), Holmstrom *et al.* (2011) and Smales *et al.* (2013), are deterministic and do not consider uncertainty within the model. The Band model is also deterministic but in the latest iteration of the model, the guidance provides a method to express the uncertainty associated with a collision estimate *post hoc* (Band 2012). By contrast, McAdam (2005) used a Monte Carlo model to consider joint distributions of wind speed and direction and distributions of flight height to produce collision risk estimates with associated measures of uncertainty. By using Bayesian methods, The US Fish and Wildlife Service (2013) model naturally allows for the consideration of uncertainty. Statutory and non-statutory stakeholders with a duty or interest in bird conservation have for some time called for uncertainty to be better reflected in the collision risk estimates which are presented as part of EIAs (Box 5.1), and recent extensions of the Band model enable this to be achieved (Masden 2015; McGregor *et al.* 2018). Including a measure of uncertainty in collision estimates moves CRM towards a risk-based framework, providing information not only on the magnitude of an event, but also on the probability of occurrence. Irrespective of how and whether uncertainty is incorporated, it is important to remember that CRMs are only a tool to aid in the assessment of impacts and management of wind farms. Because of the limitations of CRMs and the assumptions made within each, the model estimates provide a means of comparison between different development scenarios, but the estimates should only be considered indicative and never absolute. Acknowledging this, CRM outputs can then be used to assess the likely changes to population trends due to wind-farm losses.

Predicting population-scale consequences of collision

As part of the impact assessment process, it is important to determine what the population-level consequences of any collisions may be. This means that collision estimates must be considered in the context of the populations concerned. In many cases, the outputs from the collision models themselves are sufficient to allow stakeholders to conclude that there are unlikely to be any significant effects on the population(s) concerned. However, where predicted collision rates are high, further analyses are required to predict the likely population-level consequences of any collisions. A wide variety of tools have been proposed to do this in both an onshore and an offshore context. These range from very simple approaches to much more complex ones (Freeman *et al.* 2014; Humphreys *et al.* 2016; Cook & Robinson 2017; O'Brien *et al.* 2017) (Figure 5.3). These approaches are discussed below.

Simple summation

At the simplest level, the population-level consequences of collisions with wind turbines may be assessed using a straightforward summation of the predicted number of collisions. While this approach has been widely used in onshore assessments, more recent offshore assessments have typically taken more complex approaches (Humphreys *et al.* 2016). Simple summation offers a clear, and easily interpretable, measure of the total number of birds liable to be affected by collisions at any given time. This can then be compared to

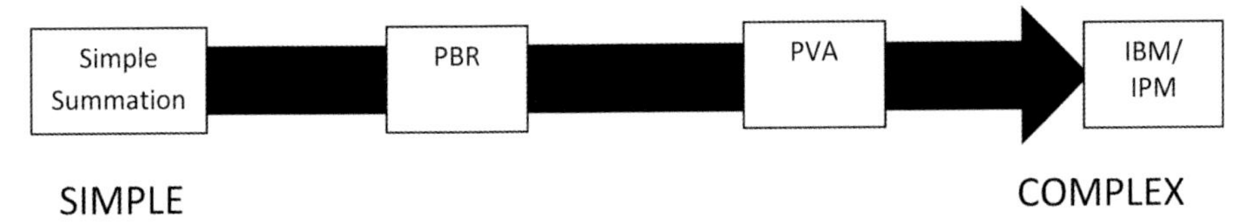

Figure 5.3 Approaches used to assess the population-level consequences of collisions range from very simple, for example summation of the number of affected birds, to more complex approaches such as potential biological removal (PBR), population viability analysis (PVA), individual-based model (IBM) and integrated population model (IPM).

the total population size or, if the data allow, some assessment of baseline mortality in the absence of any wind farm. However, predicting the population-level consequences over the lifetime of a wind farm becomes more challenging, as simply projecting estimated collisions into the future ignores recruitment and natural mortality in the population concerned.

Potential biological removal

PBR was initially developed for use in relation to the harvesting of marine mammal populations for commercial purposes (Wade 1998), and was further developed for use in relation to anthropogenic impacts on bird populations (Niel & Lebreton 2005; Dillingham & Fletcher 2008). PBR provides an estimate of the maximum level of mortality, in addition to that expected to occur naturally, which a population can experience and still remain viable. The PBR algorithm is as follows:

$$\text{PBR} = \frac{1}{2} R_{\max} N_{\min} f \tag{1}$$

where R_{max} is the maximum intrinsic growth rate of a population, N_{min} is a conservative estimate of population size, and f is a recovery factor between 0.1 and 1 (Wade 1998). As R_{max} can be difficult to estimate, Niel and Lebreton (2005) suggested supplementing it with an estimate of the maximum annual population growth rate, λ_{max}, as follows:

$$R_{\max} = \lambda_{\max} - 1 \tag{2}$$

Using the principles of life-history theory (Charnov 1993), Niel & Lebreton (2005) estimated the maximum annual population growth rate using its predictable relationship with age at first breeding and annual adult survival probability:

$$\lambda_{\max} \approx \frac{(s\alpha + s + \alpha + 1) + \sqrt{(s - s\alpha - \alpha - 1)^2 - 4s\alpha^2}}{2\alpha} \tag{3}$$

where s is annual adult survival probability and α is age at first breeding. These equations enable estimation of PBR using only estimates of minimum population size (N_{min}), adult survival (s) and age at first breeding (α). The limited data requirements and relative simplicity of the formulae make this an attractive tool with which to assess the population-level consequences of collisions with OWFs.

Developers seeking to make the case for the Triton Knoll OWF and Phase II of the London Array OWF were amongst the first to make use of PBR in relation to wind-farm EIA (Box 5.2). Subsequently, the approach was widely adopted in the assessments of other UK wind farms, including Beatrice, Hornsea and the Forth and Tay Developments (Cook *et al.* 2015). PBR has also been suggested as a tool for the assessment of OWF effects on birds elsewhere in Europe (Busch & Garthe 2016; 2017). However, significant criticisms of this approach have emerged (Green *et al.* 2016; O'Brien *et al.* 2017).

In applying PBR to bird populations, Niel & Lebreton (2005) state that it can be used to estimate a level of mortality above which extinction is likely. In the assessment of OWFs, it is typically used to set a threshold below which any level of mortality is deemed sustainable. However, Niel & Lebreton (2005) explicitly advise against this interpretation as PBR values are derived using maximum growth rates. Furthermore, PBR makes a number of assumptions about density dependence and population trajectories that may not be supported in the populations concerned (O'Brien *et al.* 2017). Both the shape and strength of any density-dependent relationships can determine whether a given level of mortality may be sustainable. However, in many seabird populations, such relationships

Box 5.2 A brief history of the use of potential biological removal in wind-farm assessment in the UK

Mark L. Tomlinson

Following Dillingham & Fletcher's (2011) application of potential biological removal (PBR) to albatrosses and petrels in New Zealand, the wind-farm industry in the UK was quick to utilise what is a relatively straightforward means of calculating sustainable thresholds of additional mortality for seabird populations. PBR analysis is thought to have been used first in the ornithology technical report that underpinned the ornithology chapters of the Environmental Statement (ES) and Habitats Regulations Assessment (HRA) for Triton Knoll (NIRAS Consulting 2012a; 2012b). In their response to the Secretary of State regarding the Triton Knoll application, the Written Representations of Natural England (NE) and the Joint Nature Conservation Committee (JNCC) accepted the use of PBR, provided necessary precaution was applied. With agreed amendments to the model parameters, PBR was used to produce sustainable threshold limits for Black-legged Kittiwake *Rissa tridactyla* (Figure 5.4) and

Figure 5.4 The highly colonial Black-legged Kittiwake *Rissa tridacytla*; a key subject of potential biological removal analysis during impact assessment in the UK. (Martin Perrow)

Northern Gannet *Morus bassanus* at Flamborough Head and Bempton Cliffs Special Protection Area (a European designation under the EU Birds Directive, 2009/147/EC) as well as the wider populations of Lesser Black-backed Gull *Larus fuscus* and Great Black-backed Gull *Larus marinus* that were of concern (Natural England 2012).

During the process, NE sought the advice of Peter Dillingham to gain a greater understanding of the outputs in order to conclude whether the simple model was suitable for decisions regarding key protected populations at colonies. Subsequently, the Statutory Nature Conservation Bodies (SNCBs) used PBR to inform their responses to offshore wind-farm applications in tandem with its use by developers and their consultants at numerous sites such as Burbo Bank Extension (NIRAS Consulting 2013), Project One of Hornsea (SMartWind Ltd/RPS 2013) and Seagreen Alpha and Bravo (Seagreen 2013; 2017).

The appeal of PBR as an assessment tool was largely a result of its simplicity. The data required, in the form of survival rates, age of first breeding and population size, were generally readily available from the literature and published colony censuses, which could be quickly agreed upon between the developers and the SNCBs. However, reaching agreement on the coefficient of variation (*CV*) around the population (reflective of the accuracy of the population estimate and the reproductive biology of the species concerned), and especially the 'recovery factor' (*f*), often proved to be more difficult, as the SNCBs required greater levels of precaution.

The values of *CV* and *f* are highly influential in the output of the model. Initially, *f* was suggested to represent the conservation status of the species. To this end, Dillingham & Fletcher (2008) utilised the International Union for Conservation of Nature and Natural Resources (IUCN) criteria, with values of 0.5 for least concern species, 0.3 for near-threatened and 0.1 for threatened species. Black-legged Kittiwake is a clear example of how this can be open to interpretation. As the world's most numerous gull species (Coulson 2011), a value of 0.5 may be applicable, although the species' rapid decline in the UK (JNCC 2015) could imply a value <0.3 in a national or local context. However, the SNCBs advised a very precautionary approach to *f* (and *CV*), and opted for a value of 0.1 in all assessments (Natural England 2014). For other species associated with SPA colonies, a maximum of 0.3 was applied except in the case of the rapidly expanding Northern Gannet population at Flamborough Head & Bempton Cliffs SPA, where 0.4 was used.

Table 5.1 Sustainable additional mortality thresholds (numbers of individuals) derived from potential biological removal analysis for a fictional Black-legged Kittiwake colony of 5,000 pairs (10,000 individuals), with a range of *f* and *CV* values (0.1–0.5)

	***f* factor**				
CV	**0.1**	**0.2**	**0.3**	**0.4**	**0.5**
0.1	71	143	214	285	356
0.2	66	131	197	262	328
0.3	60	121	181	241	301
0.4	55	111	166	222	277
0.5	51	102	153	204	255

The importance of the selection of both *f* and *CV* in assessment is shown in Table 5.1 using a fictional Black-legged Kittiwake colony with a breeding population of 10,000 and applying a survival rate of 0.82 (derived from North Shields and Marsden from 1955 to 1997) (Coulson 2011), and 4 years as the age of first breeding (Wernham *et al.* 2002). As the *CV* value decreases from 0.5 in intervals of 0.1, the sustainable threshold increases by only around 9%, but more or less doubles with each 0.1 reduction of *f*. Ultimately, this meant that a project of relatively low predicted impact at a metapopulation scale could easily be judged to be unacceptable, especially given that the accuracy of the level of predicted mortality, typically through collision risk modelling, was also subject to intense debate. In this, the requirement of an avoidance rate of 98% rather than 99% coincidentally also leads to a doubling or halving of predicted losses depending on the value selected.

The subjectivity of the parameters incorporated in PBR was at least partly responsible for the Statutory Nature Conservation Bodies in the UK now advising against its use to assess the population-level impacts of offshore wind farms on birds.

are poorly understood (Horswill *et al.* 2017). Similarly, analyses show that in situations where populations are declining, and even in some situations where populations are stable, additional mortality at the level allowable under PBR may not be sustainable (O'Brien *et al.* 2017). In addition, there are question marks over the need to incorporate sources of anthropogenic mortality other than that associated with wind farms, and whether ensuring a population remains sustainable meets the requirements of the European Union (EU) Birds and Habitats Directives (2009/147/EC and 92/43/EEC) (European Commission 2000; Zydelis *et al.* 2009; Green *et al.* 2016). Finally, the choice of recovery factor, which heavily influences the outcome of the models, is often highly subjective (Cook & Robinson 2015; Green *et al.* 2016; O'Brien *et al.* 2017) (Box 5.2). In light of these criticisms, the UK Statutory Nature Conservation Bodies (SNCB) now advise against the use of PBR in the assessment of the population-level consequences of impacts from OWFs on birds.

Population viability analysis using matrix models

There are a range of forms of PVA although, in the context of OWFs, most involve the use of matrix models to derive population projections (Figure 5.5) (WWT Consulting 2011; 2012; Humphreys *et al.* 2016; Cook & Robinson 2017). At their most basic level, these models may be deterministic, with fixed values for all parameters. However, additional complexity can be incorporated, for example demographic and/or environmental stochasticity and density-dependent regulation of demographic parameters.

$$\begin{bmatrix} 0 & 0 & . & . & P_1 \\ S_1 & 0 & . & . & 0 \\ 0 & S_2 & . & . & 0 \\ . & . & . & . & . \\ 0 & 0 & 0 & S_2 & S_{AD} \end{bmatrix} \times \begin{bmatrix} N_1 \\ N_2 \\ . \\ . \\ N_{AD} \end{bmatrix}$$

Figure 5.5 An example of a Leslie matrix model that may be used to elucidate population effects in wind-farm assessment. P_1 reflects productivity of breeding adults, S_1 is first year survival, S_2 is sub-adult survival, S_{AD} is adult survival, N_1 is number of first years, N_2 number of second years and N_{ad} number of adult birds.

These approaches have been widely used in the assessment of the population-level consequences of collisions between seabirds and turbines (Cook & Robinson 2015). The widespread use of these approaches led to the development of industry-wide guidance as part of the Strategic Ornithological Support Services (SOSS) process (WWT Consulting 2011; 2012), which used the Northern Gannet *Morus bassanus* in the North Sea as a case study. This guidance highlighted the need for a clear understanding of limitations of the data available for the population concerned. It also argues that, in cases where data are limited, simpler models may reflect a more honest assessment of the population concerned than more complex ones. In particular, it is argued that density dependence should only be incorporated into models where there is clear evidence for it in the population concerned. This advice is echoed by Green *et al.* (2016), who argued that density-independent models are likely to reflect a more precautionary approach to assessing the risk associated with OWFs. Cook & Robinson (2017) further argue that while deterministic models may reflect a simpler approach, where the data allow, stochastic models should be used as these result in predictions which are inherently more conservative. Populations should be modelled with and without the impact associated with a wind farm, and the outputs from these models should be compared. A matched-runs approach (Green *et al.* 2016; Cook & Robinson 2017), whereby impacted and unimpacted populations are assumed to have identical demographic parameters, save for the impact associated with the wind farm, should be used. This both reduces the uncertainty in the final outputs and ensures that any difference between the final populations reflects the impact associated with the wind farm only, and not random variation in the underlying demographic parameters.

Alternative modelling approaches

While matrix models are the most widespread approach used for PVA in relation to offshore wind-farm assessment, they are by no means the only approach. Approaches such as individual-based models (IBMs) have been applied to the assessment of the effects of other renewable energy developments, such as tidal barrages, on birds (Burton *et al.* 2010). Such approaches scale up individual behaviour to the population level, enabling a more accurate representation of demographic process than is possible for matrix models, which summarise these data across a population. Mackenzie *et al.* (2010) used an IBM to assess the impact of the Docking Shoal, Race Bank and Dudgeon OWFs on the North Norfolk Coast Sandwich Tern *Thalasseus sandvicensis* population (Box 5.3). A key advantage of this approach was that it enabled greater consideration of the importance of different model parameters. This meant that it was possible to put collision-related mortality estimates into the context of other sources of mortality such as severe weather events or predation by Red Fox *Vulpes vulpes* and other predators. Similarly, Warwick-Evans *et al.* (2017) developed an IBM to investigate the impact of wind farms in the English Channel on the breeding population of Northern Gannets on Alderney. The model used tracking data to investigate how the impact of birds either colliding with turbines or avoiding the planned developments would affect the birds at a population level. However, it is noted that, in addition to tracking data, such models require information about the physiology of the species concerned (Warwick-Evans *et al.* 2017).

To assess the potential impact of wind farms in the Firth of Forth and Tay, the Centre for Ecology and Hydrology developed an integrated population model to judge the potential impacts of four wind farms proposed at the same time upon six species of seabirds in the area (Freeman *et al.* 2014); one of the most important for breeding seabirds in Europe (Mitchell *et al.* 2004). These models integrated data describing the abundance,

Box 5.3 Use of a risk-based approach towards the assessment of the population-level consequences of predicted collision mortality of a breeding seabird

Richard Caldow, Aulay Mackenzie, Sophy Allen & Martin R. Perrow

In the early 2000s, proposals to develop several offshore wind farms (OWFs) in the Greater Wash were being considered. Each of these developments had the potential to lead to collisions of Sandwich Terns *Thalasseus sandvicensis* breeding at two colonies within the North Norfolk Coast Special Protection Area (SPA) (Figure 5.6). Accordingly, the Habitats Regulations Assessment (HRA) process required the government regulators [then the Department of Energy and Climate Change (DECC)] to assess the likelihood of this additional mortality leading to an adverse effect on the integrity of the SPA. At the time, Maclean *et al.* (2007) had recently reviewed the potential use of population viability analysis (PVA) to assess the impact of OWFs on bird populations. As a result of that, Natural England (NE) requested work by the developers to determine whether PVA could be used in this particular case. This work, conducted by ECON Ecological Consultancy Ltd. and the University of Essex, concluded that while there was a lack of definitive, site-specific information on key demographic rates, PVA was the only tool available to evaluate the potential population-level impact of collision mortality as a result of the Greater Wash wind farms (Mackenzie *et al.* 2009; 2010).

Discussions between the interested parties considered the fundamental advantages of 'individual-based' models rather than the matrix-based approach employed in many available software packages. As a result of these discussions, a

Figure 5.6 Sandwich Terns *Thalasseus sandvicensis* breeding at Blakeney Point, which, along with the Scolt Head colony, comprises the internationally important population designated within the North Norfolk Coast Special Protection Area. (Martin Perrow)

Table 5.2 Baseline parameters for the North Norfolk Sandwich Tern population incorporated into the individual-based population model (ViaPop)

Parameter	Subparameter	Value (mean ± SD)	Source and notes
Initial population	Size (number of individuals)	6,914 ± 250	Mean population size from colony counts (1970–2007)
	Sex ratio	0.5	Assumed
	Age distribution	As determined by mortality rate	
Mortality (except 'harvest')	Adult mortality (individuals/year) (excludes catastrophes)	10 ± 1%	Robinson (2008)
	Juvenile (individuals/ year)	23 ± 2%	Robinson (2008)
	Sex-specific rate	Equality	Assumed
Mortality through 'harvest'	Fixed harvest (/year)	Variable (0–800)	
Harvest period	Start year, end year	Year 10, year 35	Allows equilibration followed by maximum life of wind farm
Metapopulation	Metapopulation links	None incorporated	
Reproductive system and rates	Monogamy/other	Monogamy	
	Age of first reproduction (years): females	3	BWPi (2004)
	Age of first reproduction (years): males	3	BWPi (2004)
	Maximum progeny/ female/year	2	Stienen (2006): 1.6 eggs on average at Griend, 2% fledge, 2 chicks; no data from North Norfolk
	Distribution of progeny/female/year	Not highly skewed	
	Sex ratio at birth	0.5	
	Proportion of adult females breeding	100%	Non-breeders excluded from model
	Density-dependent impact on proportion of adult females breeding	Independent: 100% breeding	
	Allee effect in operation?	No	
Reproduction rate	Mean chicks fledged/ female/year	0.68 ± 0.32	Derived from North Norfolk data 1966–2007
Carrying capacity	Adult population	6,914 ± 250 individuals	

bespoke population model called ViaPop was developed. Population representation was individual based; that is, each live individual was explicitly represented within the model. This individual representation avoided structural oversimplification and allowed stochastic processes to act at an individual level; features which gave greater population realism. Population processes were simulated as discrete sequential events, which acted in a probabilistic manner on individuals (or pairs). Full details of the model structure and operation are provided in Mackenzie *et al.* (2009; 2010) and Mackenzie (2011a; 2011b).

Where possible, population parameters were determined directly from the relatively well-studied North Norfolk colonies (Table 5.2) Survival rates were determined by analysis of British Trust for Ornithology ring recovery data, and estimates of other parameter values were derived from published literature. A board of experts approved the parameter values used.

Mortality likelihoods were allocated to individuals based on the baseline mortality rates with the incorporation of Gaussian error. Baseline mortality rates were age-class specific, with two age classes specified: adults and juveniles (incorporating juveniles of age 0 and immature birds to age <3 years). Mortality due to wind-farm collisions was expressed by removing at random a fixed number of individuals from the population on every annual cycle within the 'harvesting' period, of (by default) 25 years.

The level of reproductive output, prior to any density-dependent effects, was allocated to individual females, with Gaussian error. Females could only reproduce if there was an available unmated male in the model population. Field observations indicated a considerable variation between the reproductive outputs of individual colonies within the North Norfolk Coast SPA in different years, owing to the varying impact of weather events and predator impacts. The model thus included as the default option a 'cohort' effect that allocated all individual females within one annual cycle a fixed reproductive output generated by random selection from the probability density function of observed colony reproduction rates. Density dependence is an inevitable and necessary element in the dynamics of every population process. Without a negative-feedback loop, a population will necessarily either climb to infinity or descend to extinction (albeit possibly with a transient period of unstable equilibrium). For the purposes of this PVA, negative density dependence at high densities operated through an influence on productivity alone.

Structurally, the model followed a simple flow pattern. An annual carrying capacity was allocated from the mean value with Gaussian error. Under the default parameter set, this kept the mean population size close to the recorded size. The next step was reproduction, which was modulated by density dependence at higher densities. This was succeeded by natural background mortality and then wind-farm-based mortality. At each step, the model considered the parameter (reproduction or mortality) probability of each individual in turn, as described above. At this point, the annual population size was recorded, all individuals had their age incremented by a year, and the cycle restarted.

Early versions of the model explored the population impacts of particular wind-farm configurations at Docking Shoal and Race Bank, both separately and in combination, with consideration of different combinations of number and size of turbines, and at a range of collision avoidance rates of 95–99.6%. This approach was subsequently replaced by a separation between the PVA and the collision risk model;

the final version of ViaPop did not consider wind-farm configurations or avoidance rates, but simply considered a wide range of different mortality levels that might be expected to arise from a variety of assumed wind-farm configurations and avoidance rates.

Given the stochastic nature of the population model, and its use to simulate a wide range of possible levels of collision mortality, ViaPop generated a distribution of predicted population sizes after a period of 25 years under both the baseline scenario and each of the impact scenarios (Figure 5.7). These basic outputs were also expressed in a number of other ways which acknowledged the stochasticity within the predictions under each scenario, and allowed comparisons to be made of: (1) the absolute probability that the population under each scenario would fall below a range of different threshold levels, and (2) the change in those probabilities between each impact scenario and the baseline scenario. These results were presented in a variety of graphical forms, such as Figure 5.8 and Figure 5.9.

NE and the Joint Nature Conservation Committee (JNCC), the Statutory Nature Conservation Bodies advising DECC, adopted an approach to interpreting PVA outputs which sought to identify the upper threshold to the predicted number of collision fatalities that would keep within certain limits: (1) the absolute probability of the population ending up below certain threshold sizes, and (2) the increase in those probabilities between the unimpacted and impacted scenarios. The intention was to identify a level of collision mortality that, given the stochasticity in the system, could be assessed as posing an acceptably low risk or an acceptably low increase in the risk of the notified population of the SPA declining. This goal is conceptually different from identifying an absolute level of population decline that might be considered acceptable.

In the light of the model outputs, and the way in which these were interpreted, the SNCBs advised DECC regarding the number of predicted collisions that, if exceeded, would in their view mean that an adverse effect on the integrity of the SPA could not be ruled out beyond reasonable scientific doubt. In the light of this advice, DECC conducted a strategic appropriate assessment of the predicted impact of four proposed wind-farm developments in the Greater Wash area (in combination with the existing Sheringham Shoal wind farm). The result was a decision that the

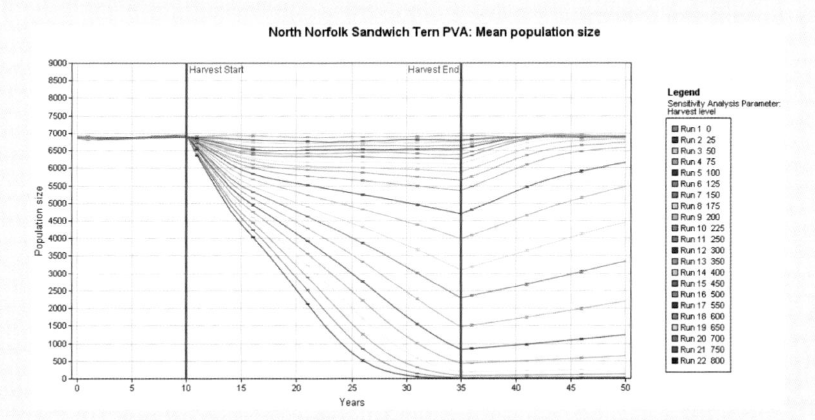

Figure 5.7 Mean population size (±SE) response to increasing levels of additional collision mortality of adult Sandwich Terns, from 0 (baseline scenario) to 800 individuals per year.

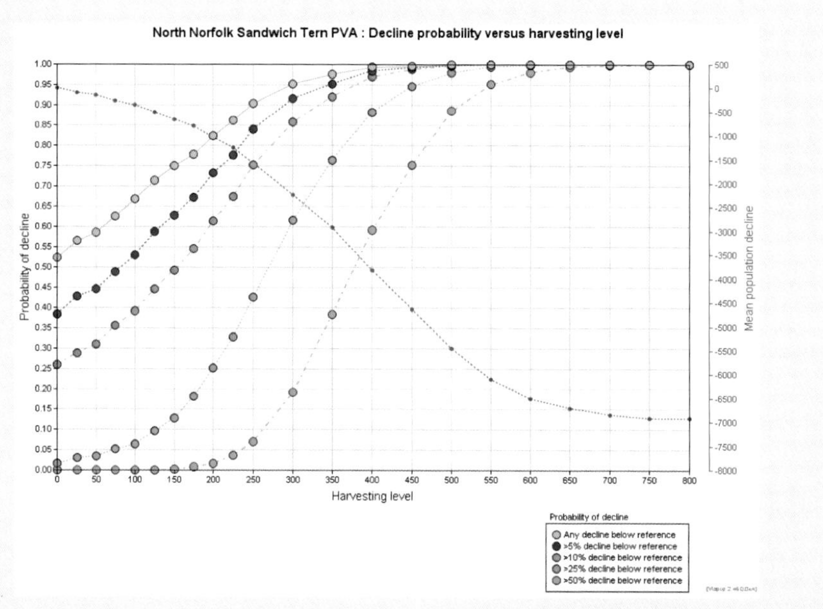

Figure 5.8 Left-hand axis: the probability of the Sandwich Tern population declining by more than a number of five differing percentage points relative to the reference population size (i.e. the median population under the baseline scenario) when subjected to levels of additional collision mortality ranging from 0 (baseline scenario) to 800 individuals per year. Right-hand axis: the magnitude of the mean population decline after 25 years of additional collision mortality as that mortality increases from 0 to 800 individuals per year.

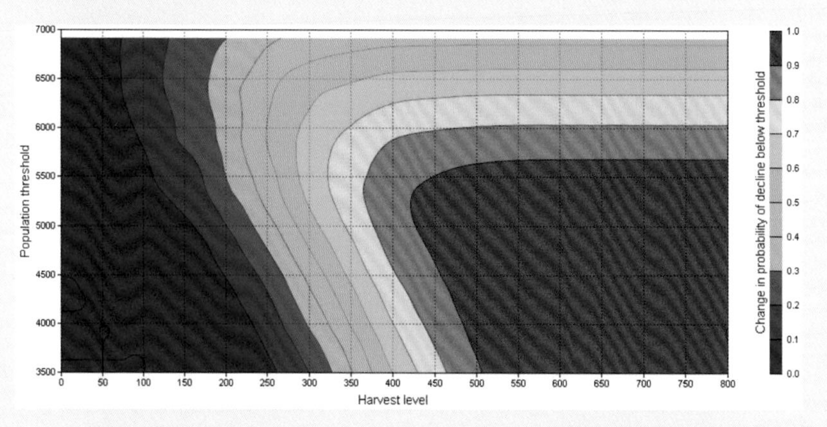

Figure 5.9 Change (relative to the baseline scenario) in the probability of the Sandwich Tern population declining below a range of threshold values at levels of additional collision mortality ranging from 0 (baseline scenario) to 800 individuals per year.

development closest to the Sandwich Tern colonies, and so with the largest per turbine predicted mortality rate, namely Docking Shoal, could not be consented without the level of predicted collision mortality exceeding that at which the risk of the population declining was considered to be acceptably low.

This case study set several precedents. First, this was the first use within the field of wind-farm impact assessments in the UK of an individual-based simulation PVA to gauge the likelihood of an adverse effect on the integrity of an SPA. Secondly, this was the first example of the use of a risk-based approach to the interpretation of the outputs of a stochastic PVA, and so the first application of the fundamental concept of scientific judgement (i.e. that of probability), to a science-based assessment of a predicted impact within an HRA. Thirdly, this was the first example of a strategic approach to making wind-farm consenting decisions in which the predicted contributions of several developments to a cumulative effect were assessed and differing consent decisions reached in each case. Finally, even though dependent upon a stochastic population model, a risk-based approach to interpretation of probabilistic model outputs, and the use of density dependence in that modelling, all of which have subsequently been subject to much criticism on the grounds of being 'poor science' and lacking precaution (Green *et al.* 2016), this was the first case in which a proposed OWF development in UK waters was refused consent on the basis of the level of risk it was predicted to pose to the well-being of a population of seabirds. To this day, as far as the authors are aware, it remains the only such case.

The modelling studies outlined were funded by Centrica Renewable Energy Ltd. on behalf of Centrica Race Bank Wind Ltd. and Centrica Docking Shoal Wind Ltd. Material presented here is reproduced from unpublished reports to Centrica Renewable Energy Ltd. with permission from Centrica and Ørsted. Aulay Mackenzie was then at the University of Essex and Sophy Allen at the Joint Nature Conservation Committee

survival and productivity for all species from multiple sources. A key advantage of this approach was that it enabled both environmental stochasticity and observation error to be modelled simultaneously. By combining these data sources using a Bayesian framework, it was possible to make probabilistic predictions about future population changes under different scenarios such as impact versus no impact and for impacts of different magnitude.

In many ways, the strength of approaches such as these is also their weakness. They enable extremely detailed predictions about the likely consequences of impacts associated with OWFs. However, at the same time, they also require incredibly detailed data, which are unavailable for many species, particularly at a site-specific level. Freeman *et al.* (2014), for example, could not include demographic data for all species at all sites in their analysis. However, as the sites were close to each other, they were thought likely to exhibit similar temporal patterns in demographic parameters (Cook *et al.* 2011). As seabirds often exhibit strong regional patterns in demography (Frederiksen *et al.* 2005a; 2005b), where no data exist to support these models, it is unlikely to be appropriate to simply apply demographic data from colonies in other regions. However, it is to be hoped that with the rapid growth of seabird tracking studies that establish connectivity between colonies and particular wind farms (see Thaxter & Perrow, Chapter 4), and other advances in technology that allows the automated monitoring of seabirds, the data necessary for such powerful analytical approaches will become more widely available.

Interpretation of effects upon populations

Regardless of the approach used to assess any population-level consequences of the impacts associated with OWFs, careful consideration needs to be given to the interpretation of their outputs. Ultimately, these outputs are going to be used by decision makers to determine whether or not the impacts associated with a wind farm can be deemed acceptable. For this reason, despite the issues listed in the previous section, approaches such as PBR, which offer a clearly defined threshold of acceptability, are attractive to decision makers and developers alike. However, a variety of metrics are available with which to quantify the outputs from PVAs and communicate likely wind-farm impacts (e.g. Green *et al.* 2016; Cook & Robinson 2017). Ideally, models used to derive these metrics should incorporate the impacts associated with all effects associated with the wind farm, including displacement (Vanermen & Stienen 2019), and not relate solely to collision. However, it is recognised that determining the consequences of some of these effects, and quantifying them, can be difficult (Humphreys *et al.* 2015). For example, the extent to which displacement contributes to increased energetic costs, reduced survival and/or reduced productivity is unknown and can only be estimated, such as by modelling changes according to the distribution of important resources such as prey (Freeman *et al.* 2014).

Acceptable biological change

Acceptable biological change (ABC) was set out as a method for assessing the population impact of an OWF by Marine Scotland (2015). Using terminology from the Intergovernmental Panel on Climate Change (IPCC), ABC allows for a change of up to one-third in the probability of a defined target being achieved as a result of the impact of a management intervention, which is classed by the IPCC as representing an outcome that is 'as likely as not' (Mastrandrea *et al.* 2010). This was applied to different population model metrics, for example, population size (ABCn) or the probability of the population growth rate (λ) being less than 1 (ABC$\lambda_{<1}$). This is binary, with impacts assessed as 1 (acceptable) or 0 (not acceptable). However, this approach has been widely criticised (e.g. Cook & Robinson 2015; Green *et al.* 2016) and it seems unlikely that it will be used in the future.

A key flaw in ABC is that whether or not the impacts associated with a wind farm may be deemed acceptable can be driven by uncertainty surrounding the demographic parameters used in the underlying models. This uncertainty may be influenced by sampling variance (Gould & Nichols 1998) or may not have been estimated over a sufficient period to reflect the true variability of the population concerned (Lande *et al.* 2003). Consequently, any assessment made using ABC may reflect limitations in the available data, rather than the biology of the species concerned. It is further argued that thresholds determined by ABC are applied inappropriately (Green *et al.* 2016).

Decline probability difference

Using the decline probability difference (DPD), the probability of a population declining is estimated for models with and without a management intervention. This probability could be expressed in relation to growth rate [e.g. probability that growth rate <1 (DPD$\lambda_{<1}$)] or population size [e.g. probability that population declines by 25% (DPD$_{\Delta 25}$)]. This metric is then assessed in relation to the intervention and non-intervention scenarios. These metrics criteria are assessed over a scale from 0 (no impact) to 1 (severe impact). Criticisms of DPD have been made that are similar to those made in relation to ABC (Green *et al.* 2016), because the baseline probability of decline can be largely determined by the choice of demographic parameters, which may be subject to significant uncertainty.

Counterfactuals

Population models can be used to estimate the size, or growth rate, of a population through time both with and without the impact of a management intervention. The metric is then assessed in relation to either the counterfactual of the impacted to unimpacted population size (CIUn) or growth rate (CIUλ). These counterfactuals could be estimated at a fixed point in time or at a series of intervals throughout the lifetime of a project. These metrics are assessed over a scale from 0 (severe impact) to 1 (no impact). These approaches appear less susceptible to the problems associated with ABC and DPD, although they do retain some (limited) sensitivity to the chosen demographic parameters (Cook & Robinson 2017).

As alluded to above, regardless of the approach used to quantify the population-level consequences of wind-farm impacts, there is a need to determine whether or not these consequences are acceptable. However, it is important to take a step back and question whether such a delineation is appropriate in the first place. Population trends vary both between species and between colonies (JNCC 2015). Similarly, in conservation terms, not all populations may be of equal importance. Consequently, at a societal level, people may be willing to accept a greater population-level impact where that population is increasing, or where it involves a geographically widespread and/or abundant species. Ultimately, it may be necessary for decisions to be made about what is or is not acceptable on a case-by-case basis. This then becomes a question about what is important at a societal level, and decisions should be made in consultation with a range of relevant stakeholders and legislation.

Cumulative impacts

As individual projects become operational, a better understanding about the impacts they have on species and populations can be gained (e.g. Dierschke *et al.* 2016). This shifts the focus of assessment towards considers the cumulative impact of multiple developments (Maclean *et al.* 2014; Goodale & Milman 2016; Thaxter *et al.* 2017), particularly given that many species face pressure from multiple anthropogenic activities (Burthe *et al.* 2014). Often, assessments only consider developments of a similar type; for example, a wind-farm cumulative impact assessment (CIA) will consider other wind farms or renewable energy developments at most, but not other anthropogenic activities such as fisheries or oil and gas installations (Masden *et al.* 2010). In addition, the assessments are often restricted in spatial and temporal extent and with regard to the species or receptors included. As a consequence, the practices and existing approaches for CIA so far have been largely found to be inadequate (Ma *et al.* 2012; Willsteed *et al.* 2017).

Despite a proliferation in potential CIA approaches for a broad range of taxa including birds, turtles, marine mammals and bats (Maxwell *et al.* 2013; Brabant *et al.* 2015; Sanz-Aguilar *et al.* 2015; Voigt *et al.* 2015; Bastos *et al.* 2016; Busch & Garthe 2017), there is a significant mismatch between the relatively simplistic approaches employed by industry and the latest modelling techniques that better capture biological processes (Humphreys *et al.* 2016). Given the building evidence about how density dependence (Horswill *et al.* 2017), dispersal (Grüebler *et al.* 2015) and source-sink dynamics (Kirol *et al.* 2015) may affect population responses to anthropogenic pressure, widely adopted approaches to CIA are likely to be limited to producing realistic predictions of impact only for a narrow range of circumstances. The key challenge is to scale-up potentially multiple impacts on individual animals to the level of the population(s) concerned, by predicting changes to demographic rates (May *et al.* 2019). This may require more data than are often available for impacted populations, but it is unclear how a failure to account for such processes may affect the conclusions and ultimately the decisions made by consenting bodies.

By identifying the circumstances under which different approaches and assumptions may be applied, and the consequences for doing so, it may be possible to balance biological realism with the requirements and realities of the assessment process (Getz *et al.* 2017; Milner-Gulland & Shea 2017). In some circumstances, an investment in the collection of additional demographic data may be required. There is an urgent need to address the mismatch between regulatory requirements and biological realism to minimise both the risk of negative cumulative impacts affecting vulnerable species and the risk of poor regulatory decision making hampering renewable energy deployment and economic development.

Concluding remarks

The use of both CRMs and population models has become a key part of the consenting process for the offshore wind industry. However, there remains relatively little consensus over some of the approaches that are used. This can contribute to significant costs for all stakeholders involved in the process (Box 5.1). Key challenges relate to ensuring that uncertainty is adequately captured by the consenting process and that the models and parameters used are evidence based and biologically realistic.

There is recognition about the value of guidance and standardised parameters for use in models. However, while they may represent the best available data, there are question marks about how representative many of these representative values are. For example, the recommended speed of 13.1 m/s for Black-legged Kittiwake, viewed as one of the key species in relation to collision risk (Garthe & Hüppop 2004; Furness *et al.* 2013), is based on two tracks recorded using radar over a period of 11 minutes in Sweden (Alerstam *et al.* 2007). In contrast, a study using laser rangefinders estimated a much slower mean flight speed for Black-legged Kittiwake, at 8.71 m/s, based on short tracks of 296 birds (Skov *et al.* 2018), although this is based on data from a single location in favourable weather conditions. Such examples highlight the uncertainties in the data used at present. The widespread use of technology, such as GPS tags (see Thaxter & Perrow, Chapter 4), is likely to enable more representative measurements to be used in the future (e.g. Fijn & Gyimesi 2018). The use of more empirical data such as obtained from tracking data is likely to be welcomed by industry and other stakeholders (Box 5.1). The move towards incorporating estimates of uncertainty into CRMs is likely to be welcomed. However, consideration needs to be given to whether it is possible to develop these models further, for example with the use of tracking data to gain more accurate assessments of collision risk.

As the assessment process for the offshore wind industry has developed from that applied onshore, there has been recognition of the need to move beyond simple metrics, such as a simple summation of the number of birds predicted to be affected by a wind farm, and consider the population-level consequences using approaches such as population modelling. While ever more complex approaches to population modelling are developed, the wider applicability of the different models is hindered by the availability of the necessary data. Until better data are more widely available, it is thought to be most appropriate to fall back on less complex approaches (e.g. Cook & Robinson 2017) while acknowledging the uncertainties in the outputs from these models.

In relation to both CRMs and population models, it is important to make use of the best available data. There is a welcome move towards making assessments more biologically realistic. However, it is important that these efforts are based on solid evidence, for example, not making assumptions about density-dependent processes operating on a population where these assumptions are not supported by empirical data. As with any models, it is important that efforts are made to validate those used in the assessment process. A failure to do this is likely to lead to problems as the offshore wind industry develops and stakeholders repeatedly return to the same questions about the accuracy and precision of model outputs during the consenting process.

Acknowledgements

This chapter is a summary of work that has been funded by and carried out on behalf of NERC, Marine Scotland Science, JNCC, The Scottish Wind Farm Bird Steering Group, The Crown Estate and BTO. We are also grateful to Alison Johnston, Liz Humphreys, Chris Thaxter, Niall Burton, Rob Robinson, James Pearce Higgins (BTO), Lucy Wright, Aly McCluskie, Rowena Langston (RSPB), Orea Anderson, Julie Black, Sue O'Brien (JNCC), Tim Frayling, Melanie Kershaw, Richard Caldow (Natural England), Alex Robbins, Glen Tyler (SNH), Mark Trinder (MacArthur Green), Chris Pendlebury (Natural Power), Jared Wilson, Finlay Bennet, Ian Davies (Marine Scotland Science), Bill Band and Sue King for various discussions about different aspects of this work. Thanks also to Martin Perrow for his useful comments on a previous version of this chapter.

References

Alerstam, T., Rosén, M., Bäckman, J., Ericson, P.G.P. & Hellgren, O. (2007) Flight speeds among bird species: allometric and phylogenetic effects. *PLoS Biology* 5(8): e197. doi: 10.1371/journal.pbio.0050197.

Band, W. (2012) Using a collision risk model to assess bird collision risks for offshore windfarms. Strategic Ornithological Support Services. Project SOSS-02. Thetford: British Trust for Ornithology. Retrieved 23 August 2017 from https://www.bto.org/sites/default/files/u28/downloads/Projects/Final_Report_SOSS02_Band1ModelGuidance.pdf

Band, W., Madders, M. & Whitfield, D.P. (2007) Developing field and analytical methods to assess avian collision risk at wind farms. In de Lucas, M., Janss, G.F.E. & Ferrer, M. (eds) *Birds and Wind Farms: Risk assessment and mitigation.* Madrid: Quercus/Servicios Informativos Ambientales. pp. 259–275.

Barrios, L. & Rodríguez, A. (2004) Behavioural and environmental correlates of soaring-bird mortality at on-shore wind turbines. *Journal of Applied Ecology* 41: 72–81.

Bastos, R., Pinhanços, A., Santos, M., Fernandes, R.F., Vicente, J.R., Morinha, F., Honrado, J.P., Travassos, P., Barros, P. & Cabral, J.A. (2016) Evaluating the regional cumulative impact of wind farms on birds: how can spatially explicit dynamic modelling improve impact assessments and monitoring? *Journal of Applied Ecology* 53: 1330–1340.

Bowgen, K. & Cook, A.S.C.P. (2018) Bird collision avoidance: empirical evidence and impact assessments. JNCC Report No. 614. Peterborough: Joint Nature Conservation Committee. Retrieved 15 April 2019 from http://jncc.defra.gov.uk/pdf/Report_614_FINAL_WEB.pdf

Box, G.E.P., Hunter, J.S. & Hunter, W.G. (2005) *Statistics for Experimenters: Design, innovation and discovery*, 2nd edn. Hoboken, NJ: John Wiley & Sons.

Brabant, R., Vanermen, N., Stienen, E.W.M. & Degraer, S. (2015) Towards a cumulative collision risk assessment of local and migrating birds in North Sea offshore wind farms. *Hydrobiologia* 756: 63–74.

Burthe, S.J., Wanless, S., Newell, M.A., Butler, A. & Daunt, F. (2014) Assessing the vulnerability of the marine bird community in the western North Sea to climate change and other anthropogenic impacts. *Marine Ecology Progress Series* 507: 277–295.

Burton, N.H.K., Cook, A.S.C.P., Thaxter, C.B., Austin, G.E. & Clark, N.A. (2010) Severn tidal power – SEA Topic Paper: Waterbirds. BTO (PB/BV Consortium) Paper to DECC. Retrieved 15 April 2019 from https://assets.publishing.service.gov.uk/government/uploads/system/uploads/attachment_data/file/69895/36._Waterbirds.pdf

Busch, M. & Garthe, S. (2016) Approaching population thresholds in presence of uncertainty: assessing displacement of seabirds from offshore wind farms. *Environmental Impact Assessment Review* 56: 31–42.

Busch, M. & Garthe, S. (2017) Looking at the bigger picture: the importance of considering annual cycles in impact assessments illustrated in a migratory seabird species. *ICES Journal of Marine Science* 75: 690–700.

BWPi (2004) *Birds of the Western Palearctic Version 1.0 (DVD)*. Oxford: Birdguides and Oxford University Press.

Camphuysen, C.J., Fox, A.D., Leopold, M.F. & Petersen, I.K. (2004) Towards standardised seabirds at sea census techniques in connection with environmental impact assessments for offshore wind farms in the UK: a comparison of ship and aerial sampling methods for marine birds, and their applicability to offshore wind farm assessments. NIOZ Report No. BAM-02-2002 to COWRIE. Texel: Royal Netherlands Institute for Sea Research (NIOZ). Retrieved 15 April 2019 from http://jncc.defra.gov.uk/PDF/Camphuysenetal2004_COWRIEmethods.PDF

Charnov, E.L. (1993) *Life History Invariants. Some explanations of symmetry in evolutionary ecology*. Oxford: Oxford University Press.

Collier, M., Dirksen, S. & Krijgsveld, K. (2012) A review of methods to monitor collisions or micro-avoidance of birds with offshore wind turbines. Report for The Crown Estate. Strategic Ornithological Support Services Project SOSS-03a. Retrieved 15 April 2019 from https://www.bto.org/sites/default/files/u28/downloads/Projects/Final_Report_SOSS03A_Part1.pdf

Cook, A.S.C.P. & Robinson, R.A. (2015) The scientific validity of criticisms made by the RSPB of metrics used to assess population level impacts of offshore wind farms on seabirds. BTO Research Report No. 665. Thetford: British Trust for Ornithology. Retrieved 15 April 2019 from https://doi.org/https://www.bto.org/sites/default/files/publications/rr665.pdf

Cook, A.S.C.P. & Robinson, R.A. (2017) Towards a framework for quantifying the population-level consequences of anthropogenic pressures on the environment: the case of seabirds and windfarms. *Journal of Environmental Management* 190: 113–121.

Cook, A.S.C.P., Parsons, M., Mitchell, I. & Robinson, R.A. (2011) Reconciling policy with ecological requirements in biodiversity monitoring. *Marine Ecology Progress Series* 434: 267–277.

Cook, A.S.C.P., Humphreys, E.M., Masden, E.A. & Burton, N.H.K. (2014) The avoidance rates of collision between birds and offshore turbines. BTO Research Report 656. Thetford: British Trust for Ornithology. Retrieved 15 April 2019 from https://www.bto.org/research-data-services/publications/research-reports/2014/avoidance-rates-collision-between-birds-an

Cook, A.S.C.P., Humphreys, E.M., Bennet, F., Masden, E.A. & Burton, N.H.K. (2018a) Quantifying avian avoidance of offshore wind turbines: current evidence and key knowledge gaps. *Marine Environmental Research* 140: 278–288.

Cook, A.S.C.P., Ward, R.M., Hansen, W.S. & Larsen, L. (2018b) Estimating seabird flight height using LiDAR. *Scottish Marine and Freshwater Science* 9(14). Retrieved 15 April 2019 from http://marine.gov.scot/data/estimating-seabird-flight-height-using-lidar

Coulson, J.C. (2011) *The Kittiwake*. London: T. & A.D. Poyser.

Dahl, E.L., Bevanger, K., Nygård, T., Røskaft, E. & Stokke, B.G. (2012) Reduced breeding success in white-tailed eagles at Smøla windfarm, western Norway, is caused by mortality and displacement. *Biological Conservation* 145: 79–85.

de Lucas, M., Janss, G.F.E., Whitfield, D.P. & Ferrer, M. (2008) Collision fatality of raptors in wind farms does not depend on raptor abundance. *Journal of Applied Ecology* 45: 1695–1703.

de Lucas, M. & Perrow, M.R. (2017) Birds: collision. In Perrow, M.R. (ed.) *Wildlife and Wind Farms, Conflicts and Solutions. Volume 1. Onshore: Potential effects*. Exeter: Pelagic Publishing. pp. 155–190.

Department of Energy and Climate Change (DECC) (2012) Decision letter in relation to Docking Shoal Offshore Wind Farm. WWW Document. Retrieved 18 September 2017 from https://www.og.decc.gov.uk/EIP/pages/projects/DockingDecision.pdf

Desholm, M. (2006) Wind farm related mortality among avian migrants – a remote sensing study and model analysis. PhD thesis, University of Copenhagen. Retrieved 15 April 2019 from http://www.dmu.dk/Pub/PHD_MDE.pdf

Dierschke, V., Furness, R.W. & Garthe, S. (2016) Seabirds and offshore wind farms in European waters: avoidance and attraction. *Biological Conservation* 202: 59–68.

Dillingham, P.W. & Fletcher, D. (2008) Estimating the ability of birds to sustain additional human-caused mortalities using a simple decision rule and allometric relationships. *Biological Conservation* 141: 1783–1792.

Dillingham, P.W. & Fletcher, D. (2011) Potential biological removal of albatrosses and petrels with minimal demographic information. *Biological Conservation* 144: 1885–1894.

Drewitt, A.L. & Langston, R.H.W. (2006) Assessing the impact of wind farms on birds. *Ibis* 148: 29–42.

Eichhorn, M., Johst, K., Seppelt, R. & Drechsler, M. (2012) Model-based estimation of collision risks of predatory birds with wind turbines. *Ecology and Society* 17(2): 1. doi: 10.5751/ES-04594-170201.

European Commission (2000) Managing Natura 2000 sites: The Provisions of Article 6 of the 'Habitats' Directive 92/43/EEC. Luxembourg: Office for Official Publications of the European Communities. Retrieved 15 April 2019 from http://ec.europa.eu/environment/nature/natura2000/management/docs/art6/provision_of_art6_en.pdf

Everaert, J. (2014) Collision risk and micro-avoidance rates of birds with wind turbines in Flanders. *Bird Study* 61: 220–230.

Ferrer, M., de Lucas, M., Janss, G.F.E., Casado, E., Muñoz, A.R., Bechard, M.J. & Calabuig, C.P. (2012) Weak relationship between risk assessment studies and recorded mortality in wind farms. *Journal of Applied Ecology* 49: 38–46.

Fijn, R.C. & Gyimesi, A. (2018) Behaviour related flight speeds of sandwich terns and their implications for wind farm collision rate modelling and impact assessment. *Environmental Impact Assessment Review* 71: 12–16.

Frederiksen, M., Harris, M.P. & Wanless, S. (2005a) Inter-population variation in demographic parameters: a neglected subject? *Oikos* 111: 209–214.

Frederiksen, M., Wright, P.J., Harris, M.P., Mavor, R.A., Heubeck, M. & Wanless, S. (2005b) Regional patterns of kittiwake *Rissa tridactyla* breeding success are related to variability in sandeel recruitment. *Marine Ecology Progress Series* 300: 201–211.

Frederiksen, M., Moe, B., Daunt, F., Phillips, R.A., Barrett, R.T., Bogdanova, M.I., Boulinier, T., Chardine, J.W., Chastel, O., Chivers, L.S., Christensen-Dalsgaard, S., Clément-Chastel, C., Colhoun, K., Freeman, R., Gaston, A.J., González-Solís, J., Goutte, A., Grémillet, D., Guilford, T., Jensen, G.H., Krasnov, Y., Lorentsen, S.-H., Mallory, M.L., Newell, M., Olsen, B., Shaw, D., Steen, H., Strøm, H., Systad, G.H., Thórarinsson, T.L. & Anker-Nilssen, T. (2012) Multicolony tracking reveals the winter distribution of a pelagic seabird on an ocean basin scale. *Diversity and Distributions* 18: 530–542.

Freeman, S., Searle, K., Bogdanova, M., Wanless, S. & Daunt, F. (2014) Population dynamics of Forth & Tay breeding seabirds: review of available models and modelling of key breeding populations. Report No. MSQ-0006 to Marine Scotland Science. Retrieved 15 April 2019 from https://www2.gov.scot/Topics/marine/marineenergy/Research/SeabirdsForthTay/FinalReport

Furness, R.W., Wade, H.M. & Masden, E.A. (2013) Assessing vulnerability of marine bird populations to offshore wind farms. *Journal of Environmental Management* 119: 56–66.

Garthe, S. & Hüppop, O. (2004) Scaling possible adverse effects of marine wind farms on seabirds: developing and applying a vulnerability index. *Journal of Applied Ecology* 41: 724–734.

Getz, W.M., Marshall, C.R., Carlson, C.J., Giuggioli, L., Ryan, S.J., Romañach, S.S., Boettiger, C., Chamberlain, S.D., Larsen, L., D'Odorico, P. & O'Sullivan, D. (2017) Making ecological models adequate. *Ecology Letters* 21: 153–166.

Gibson, L., Wilman, E.N. & Laurance, W.F. (2017) How green is 'green' energy? *Trends in Ecology & Evolution* 32: 922–935.

Goodale, M.W. & Milman, A. (2016) Cumulative adverse effects of offshore wind energy development on wildlife. *Journal of Environmental Planning and Management* 59: 1–21.

Gould, W.R. & Nichols, J.D. (1998) Estimation of temporal variability of survival in animal populations. *Ecology* 79: 2531–2538.

Green, R.E., Langston, R.H.W., McCluskie, A., Sutherland, R. & Wilson, J.D. (2016) Lack of sound science in assessing wind farm impacts on seabirds. *Journal of Applied Ecology* 53: 1635–1641.

Grüebler, M.U., Schuler, H., Spaar, R. & Naef-Daenzer, B. (2015) Behavioural response to anthropogenic habitat disturbance: indirect impact of harvesting on whinchat populations in Switzerland. *Biological Conservation* 186: 52–59.

Holmstrom, L.A., Hamer, T.E., Colclazier, E.M., Denis, N., Verschuyl, J.P. & Ruché, D. (2011) Assessing avian–wind turbine collision risk: an approach angle dependent model. *Wind Engineering* 35: 289–312.

Horswill, C., O'Brien, S.H. & Robinson, R.A. (2016) Density dependence and marine bird populations: are wind farm assessments precautionary? *Journal of Applied Ecology* 54: 1406–1414.

Humphreys, E.M., Cook, A.S.C.P. & Burton, N.H.K. (2015) Collision, displacement and barrier effect concept note. BTO Research Report No. 669. Thetford: British Trust for Ornithology. Retrieved 15 April 2019 from https://www.bto.

org/sites/default/files/shared_documents/publications/research-reports/2015/rr669.pdf

Humphreys, E.M., Masden, E.A., Cook, A.S.C.P. & Pearce-Higgins, J.W. (2016) Review of Cumulative Impact Assessments in the context of the onshore wind farm industry. Scottish Windfarm Bird Steering Group Commissioned Report No. 1505. Retrieved 15 April 2019 from http://www.swbsg.org/images/1505_Research_Cumulative_Impact_Assessment.pdf

Jessopp, M.J., Cronin, M., Doyle, T.K., Wilson, M., McQuatters-Gollop, A., Newton, S. & Phillips, R.A. (2013) Transatlantic migration by post-breeding puffins: a strategy to exploit a temporarily abundant food resource? *Marine Biology* 160: 2755–2762.

Johnston, A. & Cook, A.S.C.P. (2016) How high do birds fly? Development of methods and analysis of digital aerial data of seabird flight heights. BTO Research Report No. 676. Thetford: British Trust for Ornithology. Retrieved 15 April 2019 from https://www.bto.org/file/337905/download?token=bAbLpuAG

Johnston, A., Cook, A.S.C.P., Wright, L.J., Humphreys, E.M. & Burton, N.H.K. (2014) Modelling flight heights of marine birds to more accurately assess collision risk with offshore wind turbines. *Journal of Applied Ecology* 51: 31–41.

Joint Nature Conservation Committee (JNCC) (2015) Seabird population trends and causes of change: 1986–2013 report. Peterborough: JNCC. Retrieved 15 April 2019 from http://jncc.defra.gov.uk/page-3201

Katzner, T.E., Nelson, D.M., Braham, M.A., Doyle, J.M., Fernandez, N.B., Duerr, A.E., Bloom, P.H., Fitzpatrick, M.C., Miller, T.A., Culver, R.C.E., Braswell, L. & DeWoody, J.A. (2017) Golden eagle fatalities and the continental-scale consequences of local wind-energy generation. *Conservation Biology* 31: 406–415.

King, S. (2019) Seabirds: collision. In Perrow, M.R. (ed.) *Wildlife and Wind Farms, Conflicts and Solutions. Volume 3. Offshore: Potential effects*. Exeter: Pelagic Publishing. pp. 206–234.

Kirol, C.P., Beck, J.L., Huzurbazar, S.V., Holloran, M.J. & Miller, S.N. (2015) Identifying greater sage-grouse source and sink habitats for conservation planning in an energy development landscape. *Ecological Applications* 25: 968–990.

Kleyheeg-Hartman, J.C., Krijgsveld, K.L., Collier, M.P., Poot, M.J.M., Boon, A.R., Troost, T.A. & Dirksen, S. (2018) Predicting bird collisions with wind turbines: comparison of the new empirical Flux Collision Model with the SOSS Band model. *Ecological Modelling* 387: 144–153.

Lande, R., Engen, S. & Saether, B. (2003) *Stochastic Population Dynamics in Ecology and Conservation*. Oxford: Oxford University Press.

Ma, Z., Becker, D.R. & Kilgore, M.A. (2012) Barriers to and opportunities for effective cumulative impact assessment within state-level environmental review frameworks in the United States. *Journal of Environmental Planning and Management* 55: 961–978.

Macalister, T. Wind farm scrapped over fears for birds. *Guardian*, 6 July 2012. Retrieved 20 Jine 2019 from https://www.theguardian.com/environment/2012/jul/06/wind-farm-scrapped-fear-birds

Mackenzie, A. (2011a) Population viability analysis of the North Norfolk Sandwich Tern *Sterna sandvicensis* population. Unpublished report to Centrica Renewable Energy Ltd.

Mackenzie, A. (2011b) Technical report: ViaPop population viability analysis model. Unpublished report to Centrica Renewable Energy Ltd.

Mackenzie, A., Perrow, M.R., Gilroy, J.J. & Skeate, E. (2009) Population viability analysis of the North Norfolk Sandwich Tern *Sterna sandvicensis* population. Unpublished report to Centrica Renewable Energy Ltd./AMEC Power & Process (Europe).

Mackenzie, A., Perrow, M.R., Gilroy, J. & Skeate, E.R. (2010) Population viability analysis of the North Norfolk Sandwich Tern *Sterna sandvicensis* population. Final Revised Report. Unpublished report to Centrica Renewable Energy Ltd./AMEC Power & Process (Europe).

Maclean, I.M.D., Frederiksen, M. & Rehfisch, M.M. (2007) Potential use of population viability analysis to assess the impact of offshore windfarms on bird populations. BTO Research Report No. 480 to COWRIE. Thetford: British Trust for Ornithology. Retrieved 15 April 2019 from https://www.bto.org/sites/default/files/shared_documents/publications/research-reports/2007/rr480.pdf

Maclean, I.M.D., Inger, R., Benson, D., Booth, C.G., Embling, C.B., Grecian, W.J., Heymans, J.J. Plummer, K.E., Shackshaft, M., Sparling, C., Wilson, B., Wright, L.J., Bradbury, G., Christen, N., Godley, B.J., Jackson, A., McCluskie, A., Nichols-Lee, R. & Bearhop, S. (2014) Resolving issues with environmental impact assessment of marine renewable energy installations. *Frontiers in Marine Science* 75(1): 1–5.

Marine Scotland (2015) Appropriate Assessment: Marine Scotland's consideration of a proposal affecting designated Special Areas of Conservation ('SACs') or Special Protection Areas ('SPAs'). Marine Scotland. Retrieved 15 April 2019 from https://www2.gov.scot/Resource/0044/00446505.pdf

Masden, E.A. (2015) Developing an avian collision risk model to incorporate variability and uncertainty. Edinburgh: Scottish Government. *Scottish Marine and Freshwater Science* 6(14). doi: 10.7489/1659-1.

Masden, E.A. & Cook, A.S.C.P. (2016) Avian collision risk models for wind energy impact assessments. *Environmental Impact Assessment Review* 56: 43–49.

Masden, E.A., Fox, A.D., Furness, R.W., Bullman, R. & Haydon, D.T. (2010a) Cumulative impact assessments and bird/wind farm interactions: developing a conceptual framework. *Environmental Impact Assessment Review* 30: 1–7.

Masden, E.A., McCluskie, A., Owen, E. & Langston, R.H.W. (2015) Renewable energy developments in an uncertain world: the case of offshore wind and birds in the UK. *Marine Policy* 51: 169–172.

Mastrandrea, M.D., Field, C.B., Stocker, T.F., Edenhofer, O., Ebi, K., Frame, D.J., Held, H., Kriegler, E., Mach, K.J., Matschos, P.R., Plattner, G., Yohe, G.W. & Zweirs, F.W. (2010) Guidance note for lead authors of the IPCC fifth assessment report on consistent treatment of uncertainties. Intergovernmental Panel on Climate Change (IPCC). Retrieved 15 April 2019 from https://wg1.ipcc.ch/AR6/documents/AR5_Uncertainty_Guidance_Note.pdf

Maxwell, S.M., Hazen, E.L., Bograd, S.J., Halpern, B.S., Breed, G.A., Nickel, B., Teutschel, N.M., Crowder, L.B., Benson, S., Dutton, P.H., Bailey, H., Kappes, M.A, Kuhn, C.E., Weise, M.J., Mate, B., Shaffer, S.A., Hassrick, J.L., Henry, R.W., Irvine, L., McDonald, B.I., Robinson, P.W., Block, B.A. & Costa, D.P. (2013) Cumulative human impacts on marine predators. *Nature Communications* 4: 2688.

May, R., Masden, E.A., Bennet, F. & Perron, M. (2019) Considerations for upscaling individual effects of wind energy development towards population-level impacts on wildlife. *Journal of Environmental Management* 230: 84–93.

McAdam, B.J. (2005) A Monte-Carlo model for bird/wind turbine collisions. MSc thesis, University of Aberdeen.

McGregor, R., King, S., Donovan, C.R., Caneco, B. & Webb, A. (2018) A stochastic collision risk model for seabirds in flight. Report for Marine Scotland. Retrieved 15 April 2019 from https://www2.gov.scot/Topics/marine/marineenergy/mre/current/StochasticCRM/fullreport

Milner-Gulland, E.J. & Shea, K. (2017) Embracing uncertainty in applied ecology. *Journal of Applied Ecology* 54: 2063–2068.

Ministry of Economic Affairs (2015) Framework for assessing ecological and cumulative effects of offshore wind farms. Part A: Methods. Ministry of Economic Affairs. Retrieved 15 April 2019 from https://www.noordzeeloket.nl/publish/pages/122212/framework_for_assessing_ecological_and_cumulative_effects_of_offshore_wind_farms_-_part_a_methods_46.pdf

Mitchell, P.I., Newton, S.F., Ratcliffe, N. & Dunn, T.E. (2004) *Seabird populations of Britain and Ireland*. London: T. & A.D. Poyser..

Natural England (2012) The Planning Act 2008. The infrastructure planning (Examination procedure) Rules 2010. Triton Knoll Wind Farm Order Application. Annex C: Expert Report on Ornithology by Dr. Alex Banks. Retrieved 24 February 2016 from http://infrastructure.planninginspectorate.gov.uk/projects/east-midlands/triton-knoll-offshore-wind-farm/?ipcsection=docs&stage=4&filter=Written+Representations

Natural England (2014) The Planning Act 2008. The infrastructure planning (Examination procedure) Rules 2010. Hornsea Offshore Wind Farm – Project One Application. Annex F: Expert on offshore ornithology (HRA matters) by Melanie Kershaw. Retrieved 24 February 2016 from http://infrastructure.planninginspectorate.gov.uk/wp-content/ipc/uploads/projects/EN010033/2. Post-Submission/Representations/Written Representations/Natural England.pdf

Newton, I. & Little, B. (2009) Assessment of wind farm and other bird casualties from carcasses found on a Northumbrian beach over an 11-year period. *Bird Study* 56: 158–167.

Niel, C. & Lebreton, J.D. (2005) Using demographic invariants to detect overharvested bird populations from incomplete data. *Conservation Biology* 19: 826–835.

NIRAS Consulting (2012a) Triton Knoll Offshore Wind Farm Limited. Report to inform the Habitats Regulations Assessment. Retrieved 15 April 2019 from https://infrastructure.planninginspectorate.gov.uk/wp-content/ipc/uploads/projects/EN010005/EN010005-000305-0402%20Habitats%20Regulations%20Assessment%20report.pdf

NIRAS Consulting (2012b) Triton Knoll Offshore Wind Farm Limited. Volume 2: Chapter 6 – Ornithology. Retrieved 15 April 2019 from https://infrastructure.planninginspectorate.gov.uk/wp-content/ipc/uploads/projects/EN010005/EN010005-000258-0501%2002%20ES%20V2%20C6%20Bird%20ecology.pdf

NIRAS Consulting (2013) DONG Energy Burbo Extension (UK) Ltd. Clarification Note: Potential biological removal analysis for Kittiwake. Response to Natural Resources Wales written representation dated 28 October 2013. Retrieved 15 April 2019 from https://infrastructure.planninginspectorate.gov.uk/wp-content/ipc/uploads/projects/EN010026/EN010026-001036-Clarification%20Note%20-%20Response%20to%20NRW%20Potential%20Biological%20Removal%20Analysis%20for%20Kittiwake.pdf

Obama, B. (2017) The irreversible momentum of clean energy. *Science (New York, NY)* 355: 126–129.

O'Brien, S.H., Cook, A.S.C.P. & Robinson, R.A. (2017) Implicit assumptions underlying simple harvest models of marine bird populations can mislead environmental management decisions. *Journal of Environmental Management* 201: 163–171.

Patrick, S.C., Bearhop, S., Grémillet, D., Lescroël, A., Grecian, W.J., Bodey, T.W., Hamer, K.C., Wakefield, E., Le Nuz, M. & Votier, S.C. (2014) Individual differences in searching behaviour and spatial foraging consistency in a central place marine predator. *Oikos* 123: 33–40.

Perrow, M.R., Harwood, A.J.P., Berridge, R. & Skeate, E.R. (2017) The foraging ecology of sandwich terns in north Norfolk. *British Birds* 110: 257–277.

Pettersson, J. (2005) The impact of offshore wind farms on bird life in southern Kalmar Sound, Sweden: a final report based on studies. Retrieved 15 April 2019 from https://tethys.pnnl.gov/sites/default/files/publications/The_Impact_of_Offshore_Wind_Farms_on_Bird_Life.pdf

Plonczkier, P. & Simms, I.C. (2012) Radar monitoring of migrating Pink-footed Geese: behavioural responses to offshore wind farm development. *Journal of Applied Ecology* 49: 1187–1194.

Podolsky, R. (2008) Method of and article of manufacture for determining probability of avian collision. US Patent 7,315,799: 1.

Robinson, R.A. (2008) Survival of sandwich terns in Britain and Ireland. BTO Research Report No. 509. Thetford: British Trust for Ornithology.

Ross-Smith, V.H., Thaxter, C.B., Masden, E.A., Shamoun-Baranes, J., Burton, N.H.K., Wright, L.J., Rehfisch, M.M. & Johnston, A. (2016) Modelling flight heights of lesser black-backed gulls and great skuas from GPS: a Bayesian approach. *Journal of Applied Ecology* 53: 1676–1685.

Sanz-Aguilar, A., Sánchez-Zapata, J.A., Carrete, M., Benítez, J.R., Ávila, E., Arenas, R. & Donázar, J.A. (2015) Action on multiple fronts, illegal poisoning and wind farm planning, is required to reverse the decline of the Egyptian vulture in southern Spain. *Biological Conservation* 187: 10–18.

Schaub, M. (2012) Spatial distribution of wind turbines is crucial for the survival of red kite populations. *Biological Conservation* 155: 111–118.

Scottish Courts and Tribunals (2016) Royal Society for Protection of Birds, Scotland (the RSPB) v The Scottish Ministers and Inch Cape Offshore Limited [2016] CSOH 103 P28/15. Petition of the Royal Society for the Protection of Birds for Judicial Review. Retrieved 15 April 2019 from https://www.scotcourts.gov.uk/search-judgments/judgment?id=d69419a7-8980-69d2-b500-ff0000d74aa7

Scottish Natural Heritage (SNH) (2014) Recommended bird survey methods to inform impact assessment of onshore wind farms. Inverness: Scottish Natural Heritage. Retrieved 15 April 2019 from https://www.nature.scot/sites/default/files/2018-06/Guidance%20Note%20-%20Recommended%20bird%20survey%20methods%20to%20inform%20impact%20assessment%20of%20onshore%20windfarms.pdf

Seagreen Wind Energy (2013) Seagreen Phase 1 Offshore Project. Habitats Regulations Appraisal information to inform appropriate assessment. Seagreen Phase 1 Offshore Project. Addendum. Retrieved 15 April 2019 from http://marine.gov.scot/datafiles/lot/SG_FoF_alpha-bravo/Seagree_Phase1_Offshore_Project_Addendum/Part%202/A4MR-SEAG-Z-DEV275-SRP-233%20Part%202%20-%20Seagreen%20Phase%201%20Offshore%20Project%20HRA.pdf

Seagreen Wind Energy (2017) Chapter 8: Ornithology, Environmental Statement. Seagreen Alpha and Bravo Offshore Project. Consent Application Documentation. Retrieved 15 April 2019 from http://marine.gov.scot/sites/default/files/chapter_8_ornithology.pdf

Skov, H., Leonhard, S.B., Heinänen, S., Zydelis, R., Jensen, N.E., Durinck, J., Johansen, T.W., Jensen, B.P., Hansen, B.L., Piper, W. & Grøn, P.N. (2012) Horns Rev 2 monitoring 2010–2012. Migrating birds. Orbicon, DHI, Marine Observers and Biola. Report commissioned by DONG Energy.

Retrieved 15 April 2019 from https://tethys.pnnl.gov/sites/default/files/publications/Horns_Rev_2_Migrating_Birds_Monitoring_2012.pdf

Skov, H., Heinänen, S., Norman, T., Ward, R., Méndez-Roldán, S. & Ellis, I. (2018) ORJIP bird collision and avoidance study. Final Report April 2018. London: Carbon Trust. Retrieved 15 April 2019 from https://www.carbontrust.com/media/675793/orjip-bird-collision-avoidance-study_april-2018.pdf

Smales, I., Muir, S., Meredith, C. & Baird, R. (2013) A description of the biosis model to assess risk of bird collisions with wind turbines. *Wildlife Society Bulletin* 37: 59–65.

Smallwood, K.S. & Thelander, C. (2008) Bird mortality in the Altamont Pass wind resource area, California. *Journal of Wildlife Management* 72: 215–223.

Smallwood, K.S., Rugge, L. & Morrison, M.L. (2009) Influence of behavior on bird mortality in wind energy developments. *Journal of Wildlife Management* 73: 1082–1098.

SMartWind Ltd/RPS (2013) Hornsea Offshore Wind Farm Project One. Habitats Regulations Assessment Report: Information to support the Appropriate Assessment for Project One. Retrieved 15 April 2019 from https://infrastructure.planninginspectorate.gov.uk/wp-content/ipc/uploads/projects/EN010033/EN010033-000665-12.6%20Habitats%20Regulation%20Assessment.pdf

Statutory Nature Conservation Bodies (SNCBs) (2014) Joint Response from the Statutory Nature Conservation Bodies to the Marine Scotland Science Avoidance Rate Review. Retrieved 15 April 2019 from https://www.nature.scot/sites/default/files/2018-02/SNCB%20Position%20Note%20on%20avoidance%20rates%20for%20use%20in%20collision%20risk%20modelling.pdf

Stienen, E.W.M. (2006) Living with gulls: trading off food and predation in the sandwich tern *Sterna sandvicensis*. PhD Thesis, Rijksuniversiteit, Groningen. Retrieved 15 April 2019 from https://www.rug.nl/research/portal/publications/living-with-gulls(51438613-1ca3-4f4c-8cf1-5933785422a7)/export.html

Stienen, E.W.M., Courtens, W., Everaert, J. & Van de Valle, M. (2008) Sex-biased mortality of common terns in wind farm collisions. *Condor* 110: 154–157.

Thaxter, C.B., Buchanan, G.M., Carr, J., Butchart, S.H.M., Newbold, T., Green, R.E., Tobias, J.A., Foden, W.B., O'Brien, S. & Pearce-Higgins, J.W. (2017) Bird and bat species' global vulnerability to collision mortality at wind farms revealed through a trait-based assessment. *Proceedings of the Royal Society B: Biological Sciences* 284(1862): 20170829.

Toke, D. (2011) The UK offshore wind power programme: a sea-change in UK energy policy? *Energy Policy* 39: 526–534.

Tucker, V.A. (1996) A mathematical model of bird collisions with wind turbine rotors. *Journal of Solar Energy Engineering* 118: 253–262.

US Fish and Wildlife Service (2013) Eagle conservation plan guidance. Module 1 – Land-based wind energy. Retrieved 15 April 2019 from https://www.fws.gov/migratorybirds/pdf/management/eagleconservationplanguidance.pdf

Vanermen, N. & Stienen, E.W.M. (2019) Seabirds: displacement. In Perrow, M.R. (ed.) *Wildlife and Wind Farms, Conflicts and Solutions. Volume 3. Offshore: Potential effects*. Exeter: Pelagic Publishing. pp. 174–205.

Voigt, C.C., Lehnert, L.S., Petersons, G., Adorf, F. & Bach, L. (2015) Wildlife and renewable energy: German politics cross migratory bats. *European Journal of Wildlife Research* 61: 213–219.

Wade, P.R. (1998) Calculating limits to the allowable human-caused mortality of cetaceans and pinnipeds. *Marine Mammal Science* 14: 1–37.

Warwick-Evans, V., Atkinson, P.W., Walkington, I. & Green, J.A. (2017) Predicting the impacts of wind farms on seabirds: an individual-based model. *Journal of Applied Ecology* 55: 503–515.

Wernham, C.V., Toms, M.P., Marchant, J.H., Clark, J.A., Siriwardena, G.M. & Baillie, S.R. (2002) *The Migration Atlas: Movements of the birds of Britain and Ireland*. London: T. & A.D. Poyser.

Willsteed, E., Gill, A.B., Birchenough, S.N.R. & Jude, S. (2017) Assessing the cumulative environmental effects of marine renewable energy developments: establishing common ground. *Science of the Total Environment* 577: 19–32.

WindEurope (2018) Offshore wind in Europe: key trends and statistics 2017. February 2018. Brussels: WindEurope. Retrieved 15 April 2019 from https://windeurope.org/wp-content/uploads/files/about-wind/statistics/WindEurope-Annual-Offshore-Statistics-2017.pdf

WWT Consulting (2011) Project SOSS-04: Gannet population viability analysis: demographic data, population model, and outputs. Report commissioned by The Crown Estate Strategic Ornithological Support Services. Thetford:

British Trust for Ornithology. Retrieved 15 April 2019 from https://www.bto.org/sites/default/files/u28/downloads/Projects/Final_Report_SOSS04_GannetPVA.pdf

WWT Consulting (2012) Developing guidelines on the use of population viability analysis for investigating bird impacts due to offshore wind farms. Report commissioned by The Crown Estate Strategic Ornithological Support Services. Thetford: British Trust for Ornithology. Retrieved 15 April 2019 from https://www.bto.org/sites/default/files/u28/downloads/Projects/Final_Report_SOSS04_PVAGuidelines.pdf

Zydelis, R., Bellebaum, J., Osterblom, H., Vetemaa, M., Schirmeister, B., Stipniece, A., Dagys, M., van Eerden, M. & Garthe, S. (2009) Bycatch in gillnet fisheries – an overlooked threat to waterbird populations. *Biological Conservation* 142: 1269–1281.

CHAPTER 6

Measuring bird and bat collision and avoidance

MARKUS MOLIS, REINHOLD HILL, OMMO HÜPPOP, LOTHAR BACH, TIMOTHY COPPACK, STEVE PELLETIER, TOBIAS DITTMANN and AXEL SCHULZ

Summary

This chapter collates the scope and limitations of technology and methods to quantify the density, identity, flight height and behaviour of bats and birds in the offshore environment, including both seabirds and migratory landbirds such as passerines and waterfowl. Such information is needed because offshore wind-farm development has reached the industrial stage in European waters and is now rapidly increasing in many countries around the world such as in Southeast Asia and the USA. Development poses direct and indirect threats to wildlife, particularly in a cumulative context. Many aspects of animal–turbine interactions appear to be site, season and species specific, so that uncertainties about the magnitude of threats and potential impacts remain. Quantification of bat and bird activity and risk-associated behaviour is especially challenging in the offshore environment for practical reasons, such as technical constraints on measurements (e.g. wave impact on acoustic and radar detection) as well as for reasons of remoteness and limited accessibility, thus demanding the use of elaborate remote-sensing techniques rather than observer-based visual observations. The practicability of a range of methods and techniques to achieve these aims is introduced and described, with a focus on multisensor systems. These systems intend to maximise the quality of bird and bat data needed as input for collision risk models. The quality of existing risk models would greatly benefit from (1) improved input data through technical advances of both well-established and more recently developed methods, such as advanced radars, thermal imaging devices, video cameras (visual and short-wave infrared light), radio and satellite telemetry, and acoustic analysis software, and (2) adjusted parameterisation, such as the effects of avoidance or attraction of wind turbines and variability in these input parameters. A recommendation for what is considered to be the best available practical solution to quantify interactions of birds and bats with offshore wind farms is provided.

Introduction

Growing concerns over the effects of climate change, tied with a growing recognition of the vast potential of using offshore wind power and its immediate proximity to major human population centres around the globe, have created worldwide markets and increasing demand for this clean and renewable energy source. Following the lead of countries such as the UK, with just under 36% of installed capacity in 2016, and Germany with 29%, new offshore projects in countries such as China, Japan, South Korea, Taiwan and the USA are quickly coming on line (GWEC 2016; Bundesverband WindEndergie 2018; Jameson *et al.* 2019). Offshore wind energy could conceivably contribute to population declines in birds and bats as a result of the industry's spatial scale and potential for expansion (Hüppop *et al.* 2019), as shown or suspected for some vulnerable species groups onshore (Perrow 2017). For migratory landbirds and all bats, crossing the open sea poses a higher risk than flying over land simply because of the almost total lack of landing opportunities at sea. Anecdotal reports (e.g. Aumüller *et al.* 2011) and long-term data on bird activity around illuminated, anthropogenic offshore structures (Hüppop *et al.* 2016) also illustrate the potential attraction and collision risk for migratory birds in particular, but also resident seabirds (reviewed in Ballasus *et al.* 2009; Orr *et al.* 2013; Ronconi *et al.* 2015). About one-third of the latter, as well as many migratory bird species, are considered globally threatened or even critically endangered by the International Union for Conservation of Nature and Natural Resources (IUCN) (Croxall *et al.* 2012; Vickery *et al.* 2014; Bairlein 2016). Conservation of bird and bat populations is also considered mandatory for a number of reasons: (1) as an ethical obligation, (2) for the maintenance of ecological function, as a result of the pivotal role that bats and birds play in, for instance, nutrient cycling, preservation of ecosystem health, and the diversity and dispersal of organisms (e.g. Kunz *et al.* 2011; Whelan *et al.* 2015), and (3) for significant socio-economic benefits (Kunz *et al.* 2011); for example, the foraging activity of pest-controlling insectivorous species is worth 3.7 billion USD each year in North America (Boyles *et al.* 2011).

Population-level effects of offshore wind farms (OWFs) on bat and bird species may result *inter alia* from: (1) collision-induced mortality (Kunz *et al.* 2007; Ahlén *et al.* 2009; Masden *et al.* 2010; Furness *et al.* 2013; Masden & Cook 2016; Horswill *et al.* 2017), (2) distortion of flight routes causing detours around wind farms that can impose energetic debts (Masden *et al.* 2010), and (3) fragmentation or loss of seabird foraging grounds (Garthe *et al.* 2017), but possibly also (4) from the creation of new foraging and roosting opportunities (Ronconi *et al.* 2015). Of these, collision effects have attracted considerable empirical, but also theoretical research interest. Empirical evidence on bird mortality at offshore structures, such as platforms and rigs, corroborates the notion that the probability of bird collisions increases with sudden changes to adverse weather when birds are already *en route* across the open sea (e.g. Aumüller *et al.* 2011; reviewed in Schuster *et al.* 2015). Under these conditions, nocturnally migrating songbirds comprise the majority of collision victims, as shown in Figure 6.1 and documented in numerous studies conducted on platforms (e.g. Hansen 1954; Aumüller *et al.* 2011; Schulz *et al.* 2011; Hüppop *et al.* 2016; 2019). For offshore wind turbines, however, such information is still missing. Theoretical research includes the development of models to predict collision risk of birds and bats with wind turbines (reviewed in Masden & Cook 2016; Peterson 2016; Cook & Masden, Chapter 5). However, the quality of bird count and flight height data is thought to be insufficient for precise estimates of collision risk (Furness *et al.* 2013; Green *et al.* 2016). For example, most data on the interactions between birds and oil and gas platforms are qualitative (Ronconi *et al.* 2015). Despite the documented empirical evidence of avifauna

Figure 6.1 Nocturnal long-distance migrants found as victims of collision with structures on the research platform FINO 2 in the German Baltic Sea after a foggy night in August. Eleven Willow Warblers *Phylloscopus trochilus*, one Whinchat *Saxicola rubetra* and one Spotted Flycatcher *Muscicapa striata* are shown. (IfAÖ)

interactions with wind farms, including fatal collisions (King 2019), accurate estimates of the impetus of wind turbines on demographic rates of birds and bats are missing, owing to inadequate data collection and modelling (Green *et al.* 2016) as well as poorly developed collision risk assessment (Hüppop *et al.* 2016). Thus, for effective bird and bat conservation it is essential first to gather sufficient quantitative data to support advanced population models, which enable detailed estimates of wind-farm effects at the population level.

Offshore, the collection of adequate, species-specific data to improve parameterisation of collision models poses several methodological challenges.

- Species-specific behaviour at sea (see selection of behavioural seabird traits in Garthe and Hüppop 2004) and responses to anthropogenic structures (Hill *et al.* 2014; Ronconi *et al.* 2015) clearly require adequate taxonomic identification of the birds recorded, although this may be no easy task considering the high number of species potentially occurring at OWFs (Hüppop *et al.* 2006; Hill *et al.* 2014).
- An ability to detect birds and bats under conditions of poor visibility (rain, fog or darkness) is needed as these conditions may match with maximum collision risk (e.g. Aumüller *et al.* 2011) and, at night, with maximum migratory traffic (Hüppop *et al.* 2006; Hill *et al.* 2014).
- As bats and most migratory terrestrial birds are relatively small (<100 g body mass), sensors with high optical resolution or limited weight (e.g. telemetry technology) are required (e.g. Bridge *et al.* 2011).

- Critical components of collision models include flight altitude (Fijn *et al.* 2015), the proportion of birds at risk of collision with turbine blades (Fijn *et al.* 2015) and assessment of avoidance rates (Chamberlain *et al.* 2005; Masden & Cook 2016), as well as attraction rates, which have not yet been described. All the above require tracking and quantification of individual movements relative to a wind farm over time and spatial scales (e.g. Desholm & Kahlert 2005).
- While much of the activity of many seabirds occurs during daylight hours, migration of many terrestrial landbirds also takes place at night, which thereby requires methods that enable continuous recording.
- The remoteness of most OWFs demands the use of remote-sensing techniques, including technologies requiring high data storage, and offering only very restricted maintenance and data-transfer opportunities.
- Offshore conditions provide no or very limited opportunity for ground-truthing collision events as a result of the considerable difficulty or even impossibility of recovering carcasses of birds and bats at sea.

Scope

This chapter introduces the main methods used to monitor collisions and avoidance/attraction behaviour of birds and bats at operational offshore wind turbines. The methods considered here encompass visual daytime and night-time observation, thermal imaging, radar, acoustic recording and telemetry. Some of these are included in this chapter for completeness, but are explained in more detail in other chapters of this volume, such as Thaxter & Perrow (Chapter 4) in the case of telemetry and tracking and Cook & Masden (Chapter 5) in relation to modelling of collision risk. The methods outlined are assessed and compared with each other in terms of their feasibility to operate effectively under offshore conditions. The intention is to provide an overview on the strengths and limitations of the various methods and technologies used for monitoring birds and bats at operational sites. As the focus of offshore wind energy has been in European waters, most examples originate from north-western Europe, although development-related studies from the USA are also included. The information collated in this chapter has been extracted from scientific articles published in peer-reviewed journals, publicly available reports, conference presentations and websites. The chapter concludes with what the authors consider to be the best practical approach to monitor the interactions of seabirds, migratory terrestrial landbirds and bats with operational wind farms in a range of diverse environmental conditions offshore.

Themes

Visual observations and optical devices

Visual surveys generally require clear vision and can be carried out either by experienced observers with binoculars, spotting scopes and/or rangefinders to quantify the distances involved (Skov *et al.* 2018) from stable, wind-protected vantage points (platforms or anchored ships) or by means of high-resolution aerial digital imagery (Buckland *et al.* 2012; Williams *et al.* 2015; Kemper *et al.* 2016). Digital aerial surveys involve post-flight visual analysis of birds in the resulting imagery (Kemper *et al.* 2016) and provide spatial information on the distribution of diurnal birds relative to offshore wind turbines (see Webb & Nehls, Chapter 3). But as with observer-based survey techniques, digital aerial surveys primarily depend on high visibility conditions.

With observer-based surveys using spotting scopes (≤60× magnification), identification of many flying birds passing at a distance of up to 6 km is possible, but size dependent (Dierschke *et al.* 2005). Under favourable conditions, a few larger species, for example Great Cormorant *Phalacrocorax carbo*, may be identified even up to a distance of 15 km compared to georeferenced buoys (Dierschke 1991; Dierschke *et al.* 2005). The majority of songbirds, however, can only be detected at most to a few hundred metres even under good conditions (Kramer 1931; Dierschke *et al.* 2005). If carried out systematically, observer-based surveys will yield (semi-)quantitative data on encountered birds including their identity and behaviour (e.g. foraging, resting, actively migrating, avoidance, attraction and collision), migratory phenology and altitudinal distribution. As with most surveillance methods, visual monitoring of birds is restricted by fog, precipitation, high seas and strong winds, thus missing migratory activity at times of potentially enhanced collision risk. In general, vision-based survey techniques demand a lot of human resources (field observers or personnel for screening imagery), incur high logistical costs and are limited by the fact they cannot be conducted on a continuous basis. Consequently, the risk of missing sporadic incidents such as mass collision events (Hüppop *et al.* 2019) increases with the spacing of observation periods.

Nocturnal visual surveys are generally challenged by visible light limitation, which may be resolved to some degree by using ambient or amplified natural as well as artificial light. The description of the ceilometer technique using spotlights (e.g. portable version by Gauthreaux 1969) has been omitted from this review because spotlights tend to attract birds and bats either directly or indirectly through enhanced food supply from light-attracted insects. As the potential detection bias has not been rigorously quantified, the ceilometer technique is no longer considered relevant for quantifying nocturnal bird activity. In contrast, the moon provides the strongest ambient, night-time light source and moon-watching, used in ornithological research to quantify bird migration since the early 1950s (Lowery 1949), is still viewed as a valuable technique. In this, the silhouettes of bats and birds are visible with a spotting scope as they pass in front of the illuminated disc of the moon, providing a means of detecting and generally quantifying the animals present under pristine environmental conditions such as clear skies (Lowery 1949; Liechti *et al.* 1995). However, range is restricted to a few hundred metres owing to the difficulties in discriminating bat from bird silhouettes beyond 150 m (Kunz *et al.* 2007). Moon-watching can therefore yield passage rates, expressed as number of animals per time. Estimates on flight height are also possible, but are rather crude and strongly observer dependent. The advantages of moon-watching compared to more automated methods include low equipment cost, a lack of artefacts caused by artificial lights and no enhanced insect contamination of samples. The primary disadvantage of this labour-intensive method comprises limited opportunities for observations, set by astronomical and weather parameters (Kunz *et al.* 2007). Furthermore, Liechti *et al.* (1995) report on a conspicuous observer bias, with beginners spotting at most half of the bats and birds seen by experienced observers and a distance-dependent spotting efficacy among the latter. A key constraint is also that the passage rates derived from the analysis are only relative to other moon-watching surveys, and while they may help to document high- versus low-activity nights, they are limited in providing a quantitative estimate of migration intensity. Conditions for moon-watching are astronomically restricted to about four sufficiently moonlit nights per month and further constrained in meteorological terms by the need for cloudless conditions and good weather. The opportunity to directly observe bird collisions through moon-watching is thus extremely limited, which renders moon-watching unsuitable for assessing collisions of birds or bats with turbines.

The feasibility of fluorescent light detection and ranging (lidar) to serve in nocturnal bird classification has been explored by Lundin *et al.* (2011) and Brydegaard *et al.* (2010), who found that birds could be crudely classified up to 100 m in darkness. Alternative approaches to quantify nocturnal bird and bat activity include the use of night-vision equipment that amplifies available natural visible light (i.e. moon and stars) or equipping infrared-light sensitive video cameras with infrared-light emitting sources (Figure 6.2). Short-wave infrared radiation is invisible to birds (Hecht & Pirenne 1940; Goldsmith 2006), minimising potential attraction influences due to the effects of visible light, which are known to be an issue in ceilometer surveys (Bruderer *et al.* 1999). Individual observers may carry night-vision goggles or use scope-supported options to directly spot birds and bats. So-called third-generation devices are available with 30,000–50,000-fold light amplification capabilities. Night-vision devices as well as video cameras with active infrared illumination allow discrimination between bats and small birds at distances up to 150 m, whereas larger birds can be identified up to a distance of several hundred metres (Schulz *et al.* 2014). Measures obtained by these types of devices include the relative abundance of passing bats and birds per unit time as well as their flight direction and behaviour, while taxonomic identification is mainly based on behavioural traits. A rough assessment of flight height is possible, but its accuracy is strongly observer dependent.

The advantage of night-vision devices is a rather high degree of detail concerning species composition and behaviour of animals moving in the surroundings of the observer (Kirkwood & Cartwright 1993; Allison & Destefano 2006; Kunz *et al.* 2007; Horn *et al.* 2008). The disadvantages of night-vision devices include the relatively high costs for qualitatively acceptable equipment, weather dependency, susceptibility to light-pollution and overexposure (Kunz *et al.* 2007). Furthermore, recording nocturnal flight activity near turbines with night-vision equipment or video cameras requires mounting devices on or close to a turbine.

Thermal imaging

Thermal imaging devices detect the long-wave (2–15 μm) infrared radiation emitted by any bodies, which are, in the case

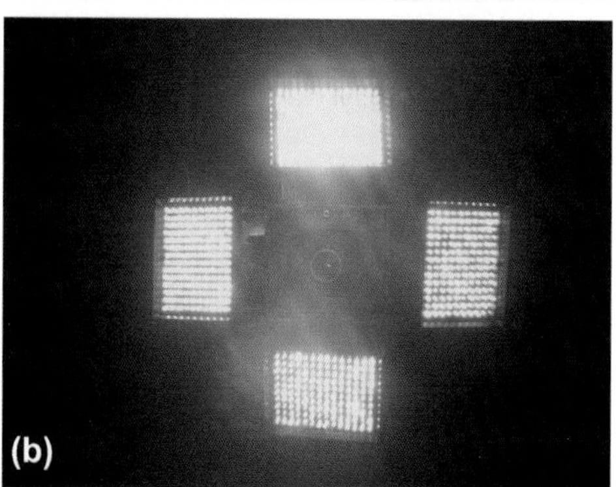

Figure 6.2 (a) VARS infrared-sensitive camera system mounted on the nacelle of a wind turbine at Alpha Ventus in the German Bight; (b) an infrared-sensitive photograph of the front of the camera system showing the infrared spotlights surrounding the camera that enable bird detection at night. (IfAÖ)

of birds and bats, usually significantly warmer than their environment. As the detectability of infrared radiation increases with the temperature difference between a body and its background, birds and bats are generally readily perceived by thermal imaging devices offshore, at least in temperate and higher latitudes where water and air temperatures are lower, year-round, than the body temperature of homoeothermic animals. This thermal disparity is further enhanced at night, improving the accuracy of thermal imaging for quantification of nocturnal animal movements at sea. As a result, thermal imaging is particularly appealing for monitoring the interactions between offshore wind turbines and bats or migratory birds because bats are nocturnal and avian migration passage rates tend to be higher at night than during daylight hours (e.g. Hüppop *et al.* 2006; Welcker *et al.* 2017). Nevertheless, thermal imaging devices have a lower spatial resolution than visible light-sensitive video cameras of comparable focal length and, hence, operate at shorter ranges. They are also unable to detect bats and smaller birds that are still detectable by visible-light sensitive video cameras.

Thermal imaging has been regularly used to record bird and bat collisions with wind turbines since the late 1980s (Winkelmann 1992; Horn *et al.* 2008). Sensors have also continually developed, for example to allow 360° detection (Figure 6.3) or the replacement of actively cooled detectors with lower cost, passively cooled units (Desholm *et al.* 2006). The sensors are also being tailored to monitor offshore avian and bat collisions in conjunction with other remote-sensing techniques, providing more comprehensive and robust detection systems (see *Collision detection systems*, below). Similar to infrared video cameras, thermal imaging can track flights, although with limited spatial resolution (but see Gauthreaux & Livingston 2006), and quantify flock size and migration intensity (Desholm *et al.* 2006; Collier *et al.* 2012). Thermal imaging has four major advantages: (1) full-time day and night recordings are achieved, (2) taxonomic identification of animals is often possible to the species-group or species level, although highly dependent on the experience of the analyst as identification is more often based on behavioural rather than morphological traits, (3) direct detection of collisions is possible with some systems even under conditions of poor visibility (Collier *et al.* 2012), although cloud cover generally lowers the ability to detect birds and bats with thermal imaging (Horton *et al.* 2015), and (4) thermal imaging enables detection of passerines, which as a group experience high numbers of collisions offshore (e.g. Aumüller *et al.* 2011; Hüppop *et al.* 2016; see also Figure 6.1). Bats and smaller songbirds are detected at a lower range compared to thrushes. There are, however, also several disadvantages associated with thermal imaging. Compared to simultaneous weather radar measurements, Horton *et al.* (2015) reported that thermal imaging detections decreased with the presence of cloud cover and flight height, but increased with mean ground flight speed of animals.

Figure 6.3 Rotating panoramic (360°) thermal imaging camera near the helicopter deck of the FINO 1 platform near the Alpha Ventus wind farm in the German North Sea. (FuE-Zentrum FH Kiel GmbH)

For thermal imaging as well as for video systems, the maximum detection range is generally dependent on animal size, sensor dimensions and resolution, and camera focal length. The detection of bats and small birds over the full length of a rotor blade, for instance, usually requires a large focal length, although this is also generally dependent on the camera mounting location, such as from structures on the lower platform or nacelle. A longer or large focal length, however, yields a small field of view (as computable with online calculators such as Fulton 2018) and a smaller depth of field (Matzner *et al.* 2015) (Figure 6.4). Consequently, multiple infrared cameras are generally necessary for detection of bats and small birds within the rotor-swept area, coupled with a generally significant increase in equipment power and data storage requirements. Furthermore, detection of avoidance responses with thermal imaging is generally restricted to the rotor-swept area (i.e. 'micro-avoidance'), but permits neither the registration of three-dimensional positions of birds nor avoidance responses at a larger spatial scale, such as 'meso-avoidance' or 'macro-avoidance' (see Cook *et al.* 2018).

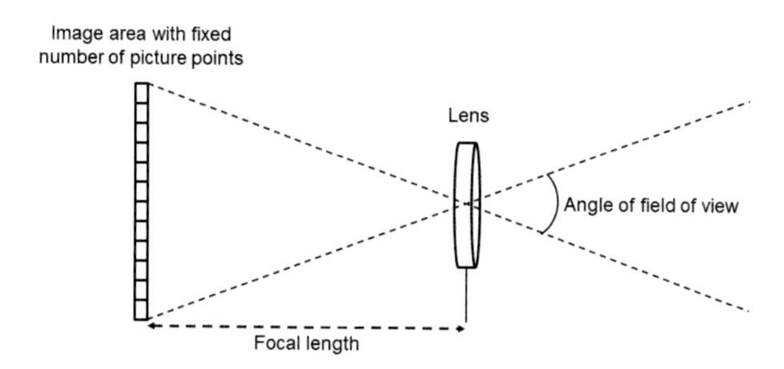

Figure 6.4 The physical principles that determine camera characteristics, such as resolution of images and detection limits of objects. (Modified after Matzner *et al.* 2015)

Radar

Radio detection and ranging (radar) uses electromagnetic radio waves ranging from a few millimetres up to 2 m for locating, tracking and identifying objects (Bruderer 1997). Radar technology has been used in bird research since the 1940s, but began to be more widely employed in ornithological research from the 1960s (Eastwood 1967; Bruderer 1997; Shamoun-Baranes *et al.* 2014; Hüppop *et al.* 2018). Subsequent methodological improvement was slow, however, and emerging evidence on the complexity of radar measurements prevented its wider application until the 1990s (Bruderer *et al.* 1999; Shamoun-Baranes *et al.* 2014). Since that period, the algorithms for radar image processing have improved and new software tools for image analysis have become available (Stepanian *et al.* 2014).

Radio waves are classified into bands by their length. Bird studies often use either an 'X band', with a range of wavelengths from 2.5 to 3.75 cm (8–12 GHz), or an 'S band' with 8–15 cm (2–4 GHz) wavelengths. The higher frequency X band typically provides a higher resolution and can detect smaller targets, while the S band has a more extended range and is less affected by rain and fog. A radar device emits either continuous or pulsed electromagnetic radiation waves and receives echoed radiation, of which only a very small quantity of energy is reflected from targets, such as birds, bats and insects, back to the radar system. These radio waves travel at light speed and the time lapse between emission and echo perception is a measure of the distance between the radar device and its target (Eastwood 1967; Bruderer 1997; Kunz *et al.* 2007; Drake & Reynolds 2012). Reflectivity is dependent on the relative difference in dielectric properties between two media. In the case of bat and bird detection, the difference in electric conductivity between air, a dielectric medium, and water, a conductive medium, in organic tissue causes reflections of radio waves from target bodies. For birds, the maximum detection range of radar is from a few kilometres to about 240 km, depending on antenna characteristics, power of emitted radiation, radar wavelength and target radar cross-section (Desholm *et al.* 2006; Drake & Reynolds 2012). Detection improves when target size is relatively large compared to the operating wavelength. Using shorter wavelengths (X band <3 cm), however, does not necessarily improve bird and bat detection because atmospheric attenuation increases with decreasing wavelength, and smaller objects, such as rain or insects, begin to blur or clutter radar images as 'insect contamination'. In such instances, it is possible to distinguish most bird and bat targets from the background clutter provided wind speed and directions are taken into consideration. Optimal radar wavelengths range between 3 and 10 cm, which generally allows more consistent detection of larger bats and smaller to medium-sized birds (10–500 g), such as most passerines and birds up to the size of ducks. Conversely, wavelengths between 10 and 15 cm enhance detection of larger (≥1,000 g) birds, but may miss smaller individuals (Eastwood 1967; Bruderer 1997; Gauthreaux & Belser 2003).

The development of wind-energy utilities and the resulting requirement to quantify the abundance of bats and birds has advanced the application of radar in wind-farm studies both onshore and offshore. The longest continuous and still ongoing radar-based monitoring of offshore bird migration has provided extremely valuable insight into offshore bird activity since 2003 (Hüppop *et al.* 2006; Hill *et al.* 2014). Radar provides several advantages for monitoring the activities of most birds and bats at wind turbines, including far-range detection, full-time operation, presumed bird imperceptibility to radio waves (Bruderer *et al.* 1999) and extensive airspace samples.

Three major types of radar systems are distinguished, with each providing different target characteristics: (1) Doppler radar systems, which quantify radial velocity of a moving target through detection of shifts in the frequency of reflecting signals, as well

Figure 6.5 Horizontal radar for quantification of bird and bat migration intensity at the FINO 1 platform near the Alpha Ventus wind farm in the German Bight, North Sea. (FuE-Zentrum FH Kiel GmbH)

as providing distance and direction, (2) 'true' tracking radar, which provides three-dimensional information about position, speed, direction and, to some extent, identity of targets based on wingbeat patterns (Bruderer 1997), and (3) marine/ship radar (Figure 6.5), which typically utilises a rotating T-shaped fan-beam antenna to detect distance and direction of moving targets and programmed software to display their two-dimensional trajectories (Desholm *et al.* 2006). Each of the three radar types provides differing 'target' signatures and thus diverging biological perspectives. Of the three, marine/ship radar is comparatively less costly and, hence, most commonly used for ornithological monitoring programmes (Kunz *et al.* 2007).

There are also different types of radar antennae for different purposes. Fan-beams have a narrow horizontal and a wide vertical opening angle. They provide precise images on the horizontal distribution of target objects, but very limited altitudinal information (or *vice versa* when the antenna rotates vertically) (Harmata *et al.* 1999; Hüppop *et al.* 2006; Kunz *et al.* 2007). Alternatively, a parabolic dish antenna with a scanning pencil beam (Figure 6.6) delivers steric polar coordinates, that is, images of high horizontal and vertical resolution. Separation of different species groups through characteristic wingbeat patterns, described as the 'target echo signature' (Figure 6.6), is possible when the antenna is in a fixed position or follows the target (Bruderer 1997; Bruderer *et al.* 1999; 2010; Bruderer & Popa-Lisseanu 2005; Zaugg *et al.* 2008; Hill *et al.* 2014). Information on flight altitude of birds can also be gathered by selectively sweeping different altitudinal strata with a pencil beam (Dokter *et al.* 2011) or tilting the rotation axis of a fan-beam antenna in 90° increments. Moreover, electromagnetic modelling techniques are emerging that can estimate three-dimensional scatter properties of anisotropic objects, such as birds and bats, which may improve

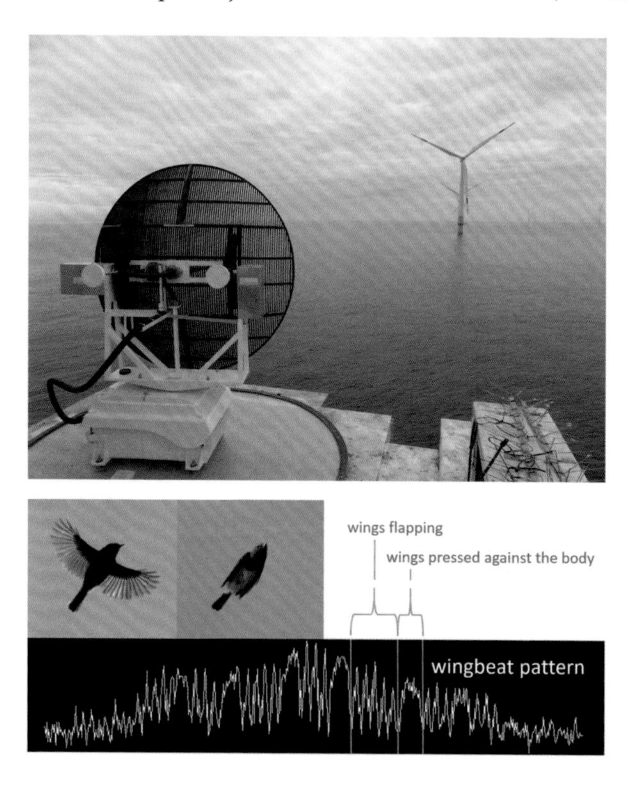

Figure 6.6 Pencil-beam radar (with cover removed) for measuring migration intensity (top) and recorded wingbeat pattern of a bird (bottom), that allows identification of the species group; in this case a small passerine such as the European Robin *Erithacus rubecula* shown. (IfAÖ)

their taxonomic identification by radar (Mirkovic *et al.* 2016). Nevertheless, separation of wingbeat patterns of birds and bats is generally difficult in fixed antenna mode to impossible in radar signals during scanning mode.

There are, however, several crucial limitations of radar technology for studying the interactions between turbines and birds and bats. Most importantly, radar cannot detect collisions because of its spatial resolution. Even modern radar devices do not differentiate between birds, bats and rotor blades because the spatial resolution of radar signals is still too coarse (Gauthreaux & Belser 2003; Urmy *et al.* 2017). Secondly, relatively high radio-wave reflectivity by the wind turbine itself masks bird or bat signals owing to overlapping radar echoes in the near-turbine airspace as well as the formation of radar shadows on the far side of the wind turbine relative to the radar. Hence, the disappearance of an echo after a bat or bird has entered the rotor-swept airspace may be due to collision, to the animal's entrance into the turbine's radar shadow or to its departure from the area covered by the radar beam. Thirdly, when using a pencil-beam radar for quantification of altitudinal bird distribution, detectability is limited at very low altitudes because of the extremely small sampling volume near the source of a pencil beam. For this reason, Gauthreaux & Livingston (2006), for instance, were unable to quantify bird migration at altitudes below 25 m. Fourthly, radar tracking is dependent on target course and velocity. Echoes of animals flying at low radial speed, for example in headwind conditions, or tangentially to a radar beam are frequently undetected by Doppler radar systems. Kelly *et al.* (2009) mention wave clutter, interference with ship radar and difficulties in detecting birds on or directly above the sea surface as specific problems for radar application in offshore environments. Nevertheless, Kelly *et al.* (2009) also suggest solutions to these problems, such as the use of: (1) vertically scanning or profiling radar systems or installation of an artificial radar horizon to avoid the effects of wave clutter, (2) 'constant false alarm rate' processing or interference rejection algorithms to minimise interference with a ship's radar, and (3) Doppler processing, scan-to-scan integration correlation or frequency diversity processing to enhance the detection of bats and birds close to or on the sea surface.

Finally, rather than generating assessments based on scattered, localised information on aerial animal movements, radar may be used at a broader landscape scale incorporating wind farms. For example, Doppler radar network systems such as ENRAM (European Network for the Radar surveillance of Animal Movement) or NEXRAD (Next-Generation Radar) synchronise individual regional European and US weather radars, respectively, to obtain broad, large-scale aerial movements of birds, bats and insects (Shamoun-Baranes *et al.* 2014; Stepanian *et al.* 2016b). The first ENRAM study revealed continental patterns of nocturnal autumn migration dynamics (Nilsson *et al.* 2018). In a similar vein, radar systems that allow tracking of flights of individuals over longer distances by software are being improved. Flyways of birds equipped with a radar beacon can be elucidated with weather radar over long distances (Bridge *et al.* 2011). This, however, involves catching (and manipulating) birds and bats, as undertaken in telemetry studies (see Thaxter & Perrow, Chapter 4).

Acoustic monitoring

Vocalisations of birds and bats cover distinct frequencies, for which different types of microphones are available. Information is therefore presented separately for birds and for bats.

Flight calls of birds

Many bird species express flight calls within a narrow frequency band ranging between 1 and 10 kHz (Ronconi *et al.* 2015) and typically of 50–300 ms duration (Farnsworth 2005). Recordings of these calls can be used to identify bird species (Evans 2005; Farnsworth 2005) and have been used to assess migration phenology and intensity (Farnsworth & Russell 2007; Hüppop *et al.* 2012; Ronconi *et al.* 2015). These signals are also potentially detectable within the rotor-swept area of turbines (Buxton & Jones 2012; Hüppop & Hilgerloh 2012) and offer the only available monitoring method for ground-truthing nocturnally migrating bird species (Stepanian *et al.* 2016a).

The first publication using vocalisation to quantify bird migration dates back about 120 years (see Evans 2005 for reviews on the history of acoustic bird research; Farnsworth 2005). The advent of digital recorders has since accelerated the science of bioacoustics by enabling development of analytical software that visualises acoustic signals, and allows collection of vast quantities of date- and time-stamped data. More recently, the combination of decreasing equipment costs and the increased power of analytical software has spurred deployments of multiple bio-microphones in arrays (Blumstein *et al.* 2011; Mennhill 2011; Stepanian *et al.* 2016a).

The use of spatially dispersed groups of microphones also makes it possible to monitor soundscapes by which individual birds can be tracked in space and time (Kirschel *et al.* 2011), yielding data of generally similar, or even enhanced, quality compared to that obtained from observer-based studies (Mennhill 2011; see *Visual observations and optical devices*, above). Farnsworth & Russell (2007), for instance, used a microphone array to monitor flight calls of migratory birds around an offshore oil platform in an area covering 600×1,000 m. Other acoustic bird migration monitoring studies have also been successfully conducted offshore (Hüppop *et al.* 2006; 2012; Hüppop & Hilgerloh 2012; Hill *et al.* 2014), with the longest record of offshore bird flight calls collected on the FINO 1 platform near the Alpha Ventus wind farm in the German part of the North Sea (Hüppop *et al.* 2012; 2019; Hill *et al.* 2014). The main advantages of acoustic bird monitoring are the continuous quantification of species-specific calls and the ability to synchronise call activity with location, temporal (i.e. seasonal and diurnal/nocturnal) and weather-dependent variables. The accuracy of this method has also been corroborated through radar-based measurements (Farnsworth *et al.* 2004). With respect to OWFs, flight call quantification can provide evidence of bird presence in the rotor-swept zone (Hüppop & Hilgerloh 2012; Hill *et al.* 2014) and, thus, is expected to improve species-specific estimates of bird collision risk.

However, acoustic monitoring also has several disadvantages. First, not all bird species call during migration, and for those that do, intraspecific variation in flight call intensity has been assumed (Farnsworth 2005), suggesting that the number of flight calls only partly reflects actual bird density or species diversity. Secondly, flight calls do not allow discrimination of gender or age. Thirdly, the number of flight calls is often higher at night and in deteriorating weather (Hüppop & Hilgerloh 2012). This may reflect an increase in per-capita call rate as a result of a behavioural response to existing conditions that causes individual birds to call more frequently (i.e. a limited proxy of the number of individuals present), an increased number of birds at lower altitude, or a combination of both. Fourthly, despite the expanding use of flight call algorithms and automated analytical software, the analysis of acoustic data remains generally labour intensive owing to the need for verification of sound files by trained staff (Bardeli *et al.* 2010). According to Ross & Allen (2014), automated analytical software has been made available in an R package (flightcallr), which reduces the manual workload by over 80%. Yet, a significant error probability remains associated with the use of this software. Automated analysis of bird sounds has also been implemented for real-time identification of songbird species in remote monitoring networks, such as the Automated Remote Biodiversity Monitoring Network (ARBIMON) (Aide *et al.* 2013). The rate of automated false positives or wrongly identified detections, however, is generally higher than in manual analyses (Hill & Hüppop 2011). Random sampling of acoustic data by trained personnel is another approach used to ensure the quality of accurate species identification and to increase the efficiency of acoustic surveys (e.g. Wimmer *et al.* 2013). In all cases, acoustic surveys should include a rigorous post-assessment quality review to maximise accuracy. Finally, although analytical programs are becoming more effective in terms of segregating calls from background noise, recordings of avian flight calls are frequently contaminated by noise generated by the wind turbines, as well as by wind and wave action, endemic in offshore environments. Farnsworth & Russell (2007), for instance, were unable to use sound recordings of frequencies below 6 kHz because of noise from waves. The use of dedicated software, such as Automatic Recording of Migrating Aves (AROMA), which separates bird calls from background noise, allows the analysis of species composition and migration phenology in combination with weather effects (Hüppop *et al.* 2012; Hüppop & Hilgerloh 2012; Hill *et al.* 2014).

Overall, the sole use of acoustic data is insufficient to monitor flight activity. However, acoustic data can supplement monitoring schemes that rely on other remote-sensing technologies and observer-based approaches (Ronconi *et al.* 2015). In this way, the predictive power of collision risk models (CRMs) can improve considerably.

Echosounding of bats

Bats use ultrasound echolocation calls for short-distance orientation (Griffin 1958). Following the development of ultrasound detectors in the 1980s, acoustic surveys have become the most common method to study the ecology of bats, including in relation to wind farms (Hein 2017). Today, several types of bat detectors of four different transformation systems (heterodyning, frequency division, time expansion and direct recording) are in use (Parsons & Szewczak 2009) and several authors discuss the use of bat detectors and the identification of bats by their sound (e.g. Fenton & Bell 1981; Ahlén 1990; Parsons & Szewczak 2009; Russ *et al.* 2012; Barataud 2015). The method does have some obvious constraints: (1) the detection range of bat echolocation calls is species-specifically limited to a range of 30–100 m (e.g. Barataud 2015), (2) as with acoustic bird studies, recorded events can arise either from one individual passing by several times or from several different individuals, and (3) echolocation calls do not provide information on the age and sex of bats. Nevertheless, the approach of finding and identifying bats by recording their calls is generally convenient since the technique is both affordable and easy to deploy (Hein 2017). In addition, intraspecific modification of echolocation calls can provide direct evidence on the presence and behaviour of bats, such as exploring or feeding at offshore structures (Hüppop & Hill 2016).

Bat detector systems have been used to study bat activity and migration in onshore (Ahlén 1997; Ahlén *et al.* 2009) and offshore environments (Walter *et al.* 2007; Ahlén *et al.* 2009; Pelletier *et al.* 2013; Peterson 2016) (Box 6.1). Automated recording devices are especially useful along coastlines, islands (Cryan & Brown 2007; Bach *et al.* 2009; Rydell *et al.* 2014) and offshore (Ahlén *et al.* 2009; Hüppop & Hill 2016), and have been regularly and successfully used to specifically monitor bat activity at both onshore (Kunz *et al.* 2007; Brinkmann *et al.* 2011; Pelletier *et al.* 2013; Rodrigues *et al.* 2015) and offshore wind turbines (Lagerveld *et al.* 2014; Hüppop & Hill 2016). Box 6.1 provides a particularly informative example of detector systems being usefully combined with radar and thermal imaging systems to document behaviour and flight characteristics. Automated acoustic bat monitoring can be considered the most efficient method available so far for assessing the collision risk for bats at wind turbines (Brinkmann *et al.* 2011; Behr *et al.* 2015; Rodrigues *et al.* 2015; Peterson *et al.* 2016; Hayes *et al.* 2019).

Box 6.1 Surveys for bats migrating through Kalmar Sound in the Swedish Baltic Sea

Several species of migratory bats cross the open sea, but relatively little is known about their interactions with wind turbines (Hüppop *et al.* 2019). Starting in 2005, a 2 year project investigated the possible impact on migrating bats of wind turbines positioned 14 km offshore in the Kalmar Sound between the Swedish mainland and the island of Öland (Ahlén *et al.* 2009). The study addressed two questions: (1) Do bats cross the Kalmar Sound (including the offshore wind farms Utgrunden and Ytre Stengrund) during migration? and, if so, (2) Is there any potential impact, such as collisions, of bats with wind turbines? This study exemplifies the usefulness of multiple methods for the quantification of encounter frequencies and interactions between offshore wind turbines and bats. Five methods were simultaneously applied to measure bat activities: (1) a land-based manual detector survey was conducted of bat departure from two sites on Öland (Eckelsudde and Ottenby); (2) a manual detector survey of flight activity was carried out from a vessel at sea for 14 nights (total 29 hours) up to 4 hours after sunset; (3) bat abundance was recorded with marine radar positioned on the Utgrunden lighthouse whereby bats were distinguished from birds by their flight behaviour and bat classification was verified through vessel-based manual detector identification during the first night; (4) automated bat detector systems were placed at the bottom of several wind turbines to record bat activity simultaneously; and (5) an infrared thermal imaging camera (Palm IR PRO, Raytheon Co., Waltham, MA, USA) was used to record bat behaviour around wind turbines.

Land-based manual detector surveys documented up to seven species leaving Öland towards the sea, mainly from Ottenby. The most common migrating species were Nathusius' Pipistrelle *Pipistrellus nathusii*, Soprano Pipistrelle *P. pygmaeus* and Noctule *Nyctalus noctula* (Figure 6.7). Most bats left the island at a wind speed of around 4 m/s, but were also regularly observed migrating with winds up to 9 m/s. Most departures from stopover sites were recorded during the first 3 hours after sunset in August and September.

Altogether, vessel-based detector surveys recorded ten bat species migrating over or foraging in Kalmar Sound. While most bats simply passed by the ships, some bats started to circle and investigate the vessel as soon as it was encountered. Bats were also observed to forage on insects that had drifted from shore. Non-migrating species Daubenton's Bat *Myotis daubentonii* and Pond Bat *M. dasycneme* were also observed to occur offshore. Thermal imaging recordings of bats, such as *N. noctula* and *P. nathusii*, revealed flight heights of 10–40 m above the sea surface, although they would frequently change elevations and fly around the nacelle and rotor blades when they encountered wind turbines. *Nyctalus noctula*, in particular, showed a typical investigative behaviour when approaching a nacelle, which suggested that it was searching for roost possibilities.

Automated bat detector systems recorded bats at all offshore wind turbines. Although the mean number of bats recorded by automated bat detector systems was low compared to radar observations, there were occasionally nights of higher numbers of detector recordings on the turbine platforms. At Ytre Stengrund, 79 bat passes of 10 species (5 wind turbines, total 49 hours) were documented, while only 20 bat passes of 6 species (7 wind turbines, total 60.5 hours) were recorded at Utgrunden in 2006. The main species were *N. noctula* and Particoloured Bat *Vespertilio murinus*, but also *P. nathusii* and *P. pygmaeus* and more rarely *M. dasycneme*.

Figure 6.7 A migrating Noctule *Nyctalus noctula* spotted in the hours of daylight. (John Larsen)

A total of 2,553 bats were observed with radar during 37.5 hours on Kalmar Sound in 2006. All bats checked by manual detector identification belonged to *N. noctula*. Although smaller bat species were also observed during these sea trips, they were not identified by radar. Seventy-six per cent of all observed *N. noctula* flew at wind speeds up to 2.5 m/s, but a few were found flying up to 9 m/s. Although most bats appeared to migrate, radar data indicated that *N. noctula* went out to sea for foraging and then returned to the same point at land.

In conclusion, a variety of bats, including both migratory and presumably non-migratory species, have been documented offshore. Moreover, bats have been observed foraging at sea while migrating. Upon encountering wind turbines, bats were also frequently observed investigating and foraging around the nacelle and rotor blades, suggesting a possible attraction behaviour, which renders collisions of bats at offshore wind turbines possible.

Impact noise

As a step beyond recording direct vocalisations, the vibrations caused by collisions of larger birds, and potentially, smaller birds and bats with wind turbines, can be recorded acoustically with contact microphones mounted on the rotor blades. While the intrinsic vibrations of the turbine can mask the acoustic signals of collision events (Flowers *et al.* 2014), the background noise of the turbine can potentially be separated from the sound of collisions, although the intensity and frequency of vibrations will differ in accordance with the mass of the colliding object. Contact microphones can be calibrated by simulated impact events using artificial objects of defined mass that are propelled towards operating wind turbines, such as empty or water-filled tennis balls that match the masses of small birds and bats (Flowers *et al.* 2014; Suryan *et al.* 2016). Algorithms have been developed that perform wavelet analyses that can distinguish vibrations caused by collisions from background noise (Jiang *et al.* 2011; Suryan *et al.* 2016).

Individual tracking

The methods described thus far, apart from visual daytime and night-time observations, principally use stationary receivers, which enable detection of flying animals over technically limited scales, from several kilometres to more than 100 km in the case of radar, to less than 100 m in the case of visual and thermal imaging and bioacoustics. Individual tracking follows a different approach by attaching devices on to individuals to trace their movements in space and time (Burger & Shaffer 2008; Exo *et al.* 2013; McKinnon & Love 2018). Animal tracking gives insight into individual behaviour, including habitat use, migration and responses to environmental cues. The technological development of animal tracking went through three stages. First, animal localisation was realised with radio (or other) beeper transmitters that were fitted to animals. The emitted signals were initially detected with hand-held and later also with automated antennae. To improve animal tracking, signals were simultaneously detected with directional antennae deployed at different sites to localise individuals via triangulation. At the second stage, data-loggers were deployed that stored different parameters, such as flight directions and durations, light levels (in geolocators) or Global Positioning System (GPS) positions. For data retrieval, however, animal recapture is mandatory. The latter drawback was solved with the use of devices of the third stage that transmit data either continuously, or in data packages upon availability of an internet connection, or on demand to special-purpose receivers.

Detail on the use of telemetry and tracking in the study of bird and bat interactions with wind farms is provided by Hein (2017) and Smallwood (2017) in relation to onshore, and Thaxter & Perrow (Chapter 4) specifically in relation to birds offshore. Further information on the use of available technology, particularly in relation to bats and small migratory landbirds, as opposed to seabirds and larger migrant waterbirds (Thaxter & Perrow, Chapter 4), is provided below. The suitability of tracking devices for studying wind farm and bird/bat interactions depends on two major aspects. First, adequate tag characteristics, such as mass, attachment mode and placement, are needed to minimise deleterious effects on the performance and fitness of birds and bats. The size of a radio transmitter is a compromise between operational tag life, transmitter power (and hence detection range) and length between signal bursts (i.e. signal resolution) on the one hand, and tag mass (mainly battery) on the other hand. The ongoing technological progress, however, will spur tag miniaturisation, which should lead to adequate tagging of even the smallest migratory animals prone to interacting with wind turbines, such as passerines, bats and insects (Box 6.2).

The meta-analysis by Bodey *et al.* (2018) indicates that tags exceeding 1% of bird body mass may already negatively affect bird survival and reproduction. Besides the risks associated with excess load, the tag attachment mode may also have an impact on bird survival. Herein, the use of harnesses, which is the most frequently applied method, significantly increased bird mortality (Bodey *et al.* 2018). Furthermore, inappropriate placement of reasonably weighted devices on animal bodies may have detrimental consequences such as excessive energy expenditure from modified flight patterns, as in Northern Gannet *Morus bassanus* (Vandenabeele *et al.* 2014). In addition, target positioning accuracy and frequency need to be high enough to assess individual avoidance behaviour. Detection of avoidance of entire wind farms demands less accurate positioning, over tens of metres, than detection of avoidance of individual turbines at less than 5 m, or a location relative to the rotor-swept area.

Automated wildlife tracking systems based on radio-telemetry are potentially useful for detecting wind-farm effects on the activities of individual small birds and bats. Receiving accurate and precise positions and flight bearings is, however, a logistical challenge with

Box 6.2 The Motus Wildlife Tracking System

The Motus Wildlife Tracking System is a research approach established by Bird Studies Canada in partnership with Acadia University, among others, and involves development of an international, collaborative radio-tracking network (Taylor *et al.* 2017). The programme network allows individual researchers to study individual movements of small birds, bats and larger insects via a direct, open source access to coordinated automated radio-telemetry arrays. Spatially deployed automated receivers are used in conjunction with individually digitally coded NanoTags (a tradename from Lotek Wireless), allowing thousands of individuals to be tracked continuously and simultaneously across broad regional landscapes. Six tag models are available, with the smallest model (11×5×3 mm; L×W×H) weighing 0.29 g and the largest one (23×9×7 mm) 2.6 g. Depending on burst interval and tag model, the manufacturer-specified range of operational tag lifetime varies between 10 and 928 days (Lotek Wireless Inc. 2018). The system allows tracking of a single radiofrequency across multiple receiving stations, and then systematically coordinates, disseminates and archives detections and associated metadata in a central repository. Over 350 receiving stations are active and more than 9,000 individuals of over 87 species of birds, bats and insects have been tracked (Taylor *et al.* 2017). More than 250 million date- and time-stamped animal detections have been recorded since the inception of the programme in 2012 (Taylor *et al.* 2017). The data set includes detections of individuals during all phases of the annual cycle (breeding, migration and non-breeding). The research value of the network will continue to grow as spatial coverage of stations and number of partners and collaborators increases, offering a framework for global collaboration and coordinated approach to better understand animal movements over a broad scale (Loring 2014; Crysler *et al.* 2016; Loring *et al.* 2016; Peterson *et al.* 2016; Taylor *et al.* 2017).

respect to configuration and spacing of receiver stations as well as computational power to analyse the collected data (Smolinsky *et al.* 2013; Mitchell *et al.* 2015; Taylor *et al.* 2017). Given a high vantage point of an antenna, the approximate position of radio-tagged birds and bats may be detected within a range of up to several kilometres (Burger & Shaffer 2008; Taylor *et al.* 2017). One example is the rapidly growing ground-based automated wildlife-tracking system Motus, which uses radio receivers and coded tags weighing between 0.29 and 2.6 g (Box 6.2) (Taylor *et al.* 2017; Lotek Wireless Inc. 2018; McKinnon & Love 2018). The maximum accuracy of positioning of tagged animals within the Motus network is *ca.* 1 km, which is sufficient for describing migratory flyways and corridors at larger than local scales (Taylor *et al.* 2017). Motus is, however, not eligible for assessing post-construction collision risk at micro- and meso-scale levels. By installing additional arrays of ground-based automated receiver stations for VHF radio signals with multiple antennae on or near wind turbines, flight paths of animals weighing less than 10 g, including many migratory birds, could be traced with higher resolution in time and space (Taylor *et al.* 2011; Smolinsky *et al.* 2013; Sjöberg *et al.* 2015). Installing such arrays of receivers at a more regional scale and tagging significant samples of birds or bats seems to be a very promising approach to track residence times as well as timing, direction, speed and types of movements of animals as small as insects (Taylor *et al.* 2017). As such, initiatives like Motus will most probably enable the assessment of animal interactions with wind-energy facilities at the level of wind farms, for example in documenting macro-avoidance.

GPS loggers or transmitters are a type of tracking system by which data of high temporal and spatial resolution, including altitude, are gathered (Corman & Garthe 2014). The mass of contemporary GPS devices is as low as 1 g, and they have an accuracy of about 10 m and operational life of 1–2 years (McKinnon & Love 2018). They can be used for localisation of birds with a body mass of more than 100 g, such as waders, ducks, geese or raptors. Data readout of GPS devices is possible after animal recapture or automated data transmission to nearby receiving facilities. In the case of the International Cooperation for Animal Research Using Space (ICARUS) project, the receiving antenna is mounted on the international space station or on airplanes (Wikelski *et al.* 2007).

Geolocators are devices that represent the third stage of development in animal tracking technology. They record light levels over time to estimate longitude and latitude (Stutchbury *et al.* 2009; Lisovski *et al.* 2012). The advantage of geolocators is their extremely low mass (0.35 g), which comes at the costs of rendering at most one position per day and a low positioning accuracy, particularly around the vernal and autumnal equinoxes (Lisovski *et al.* 2012; McKinnon & Love 2018). Moreover, data export requires retrieval of geolocators (McKinnon & Love 2018).

Tracking technology has at least three major advantages over other remote-sensing techniques in the quantification of offshore avian collision risk. First, individual behavioural responses of birds and bats to offshore wind turbines can, theoretically, be recorded at all relevant spatiotemporal scales (Thaxter & Perrow, Chapter 4), providing modellers with urgently needed information on avoidance rates (Green *et al.* 2016). Secondly, tracking will enhance the quality of environmental impact assessments during the pre-construction period through data-based determinations of distance limits of wind farms around, for instance, bird colonies (Perrow *et al.* 2006; Thaxter & Perrow, Chapter 4). Thirdly, tagging studies may be relatively inexpensive in the context of an intensive programme of digital aerial or boat-based surveys (Webb & Nehls, Chapter 3) or an intensive study of bird collision risk using multisensor systems (see *MUSE*, below; Skov *et al.* 2018). Limitations of using tracking devices involve bird capture and tag mounting, which may impose stresses on birds, as may non-ergonomically shaped or placed harnesses and the extra weight imposed by tags (Barron *et al.* 2010; Exo *et al.* 2013; Vandenabeele *et al.* 2014). Even though technology continues to advance with smaller and lighter devices enabling a greater range of birds to be tagged, some species may still respond unfavourably to tagging (Thaxter *et al.* 2016). Finally, tags that archive data, as opposed to those able to transmit data, require the recapture of tagged individuals, which restricts their use to animals with high site fidelity and return rate (Exo *et al.* 2013).

Carcass collection

The most straightforward method for quantifying wind turbine-induced fatalities of birds or bats would be searching for carcasses beneath turbines, typically known as fatality monitoring (Huso *et al.* 2017). This approach is, with some limitations, applicable onshore but is unfeasible offshore because carcasses retrieval is virtually impossible at sea. Hüppop *et al.* (2016), for instance, estimated that up to 90% of carcasses are likely to be missed at an offshore research platform with an 80 m high lattice tower, such as FINO 1. In contrast to many types of wind turbines, FINO 1 is equipped with a 256 m^2 platform 20 m a.s.l., on which Aumüller *et al.* (2015) determined the average retention time of bird carcasses to be 27 days. Nevertheless, the ratio of carcass catchment area to collision-relevant area is much larger for research platforms such as FINO 1 than for wind turbines. The large size of the rotor blades in combination with higher wind speeds at sea further complicates the retrieval of carcasses as they scatter over a larger area than on land. Hence, it seems quite unlikely that carcass searches around offshore wind turbines recapture birds and bats injured or killed by the turbine. Moreover, as the time of collision is unknown, it is impossible to reconstruct the environmental conditions that prevailed at the time of collision (Korner-Nievergelt *et al.* 2013; Hüppop *et al.* 2016). Such knowledge, for instance on weather conditions, would provide hints on possible processes underlying collision events. The use of floating barriers or nets, which may prevent fatal casualties from being lost, seems likely to be constrained by the technical issues of securing these structures against waves and currents, particularly during storm events. Yet, the few available areas in a wind farm where carcasses can be retrieved, such as converter stations and working or landing decks, can provide qualitative information on species identity of the victims and the circumstances surrounding collision, such as time of day or night and weather conditions at the time (e.g. Baerwald *et al.* 2008; Hüppop *et al.* 2016). Consequently, carcass recovery yields highly variable, erroneous estimates of collision risk and impact, but in combination with weather information, radar and/or acoustic measurements, carcass counts can complement collision risk assessment (Aumüller *et al.* 2011; Hüppop *et al.* 2016).

Collision detection systems

No single method can comprehensively assess the impact of wind turbines on bird or bat populations (Hüppop *et al.* 2006; Kunz *et al.* 2007). To achieve this goal, monitoring methods must allow taxonomic identification of all species during daytime and even more so at night, which normally is the period of highest migratory activity, and under adverse weather conditions, such as rain and fog. These demands have led to the design of multidetector systems in which the advantages of different remote-sensing methods are combined to optimise spatiotemporal resolution and maximise the empirical database for CRMs. Several multisensor systems exist in different stages of development (Dirksen 2017). A brief overview on (1) three systems using multiple devices of the same type of sensor and (2) six systems using devices of different types of sensors is presented in Box 6.3, while Table 6.1 compares the performance and configuration of the (five) referenced systems of the latter group. The following paragraphs provide more detail on eight of the systems (excluding the MultiBird system listed in Box 6.3).

ID Stat

As a continuously operating acoustic collision detector system, ID Stat can be used to obtain patterns of wind-farm effects at broader spatial scales as well as on the timing of collisions (Collier *et al.* 2012) if used at multiple turbines. One directional microphone is attached at the base of each rotor blade inside the hub of a wind turbine. Sound spectra of collision events are logged and transmitted by the Global System for Mobile Communications (GSM) network. To calibrate ID Stat, objects weighing as little as 2.5 g were shot at the blades of a wind turbine (Delprat & Alcuri 2011). In addition, filtering software eliminates ambient noise from rain and the operational noise from the turbines. Confirmation of collision events was not possible from images or species identification of fatally collided individuals (Collier *et al.* 2011; 2012; Ahmad *et al.* 2013). ID Stat has been tested with one onshore wind turbine in western France, although to the best of the authors' knowledge the results of this study have not been reported. Likewise, information on its commercial application is not available.

Box 6.3 A brief comparison of existing multisensor systems

Multisensor systems fall into two broad categories. The first comprises systems with multiple devices of the same sensor type that strive for maximum spatial coverage in the detection of visual or acoustic signals. This group comprises ID Stat as a microphone array, and the video arrays of the Thermal Animal Detection System (TADS) and the Visual Automatic Recording System (VARS). The second category includes systems that are combinations of different types of sensors integrated into one system. At the time of writing, there are six existing systems, but in different stages of development. (1) WT-Bird is a collision-triggered system combining infrared cameras, acoustic and contact microphones, with the objective of detecting and registering bird collisions. A prototype was tested onshore and offshore in Dutch waters (Wiggelinkhuizen & den Boon 2010). (2) DTBird combines acoustic with optical (visual video cameras) devices and allows the addition of thermal imaging cameras. Its objective is the detection and deterrence of near-flying birds, quantification of bird collisions and mitigation of collisions through automated shutdown when birds underrun a critical distance from a wind turbine. Technical additions for nocturnal measurements have been developed. The accuracy of a visually triggered prototype was tested over a 5-year period (May *et al.* 2012) and is commercially available. (3) The Acoustic Thermographic Offshore Monitoring (ATOM) system, a prototype equipped with thermal imaging, infrared and visual cameras, as well as two types of microphones (acoustic and ultrasound), was tested in the USA in a 1-month onshore and 15-month offshore environment (Willmott *et al.* 2015). Its objective is to provide information on abundance, size, identity and flight parameters (altitude, direction, velocity) of birds and bats to mitigate collision mortality. (4) The MultiSensor wildlife detection system (MUSE) uses pan/tilt visual and thermal cameras together with surveillance radar. Its objective is to combine quantification of avoidance behaviour and collision events with mitigation procedures. The system was tested offshore during the Offshore Renewables Joint Industry Programme (ORJIP) Bird Collision Avoidance study from 2014 to 2016 (Box 6.4). (5) The Wind Turbine Sensor Unit constitutes of accelerometers, contact microphones, stereovisual and stereo-infrared cameras and acoustic and ultrasonic microphones with the objective of providing contact-triggered monitoring of bird and bat collisions. A prototype was developed and tested in the USA in two 1 month runs (Suryan *et al.* 2016). (6) MultiBird is the multisensor extension of DTBird, comprising a rotating 360° thermal imaging camera with automatic bird detection and recording software, four video cameras from DTBird and a Robin Radar 3D system consisting of a horizontally and a vertically rotating radar. The objective of the system is to provide information on abundance, size, identity and flight parameters (altitude, direction, velocity) of birds during night and day at the highest possible range and also during poor weather conditions. All sensors are mounted at the FINO 1 research platform in the south-eastern North Sea (unpublished data R. Hill).

In conclusion, the currently available multisensory detection systems show limited success in the detection of collision events. Moreover, they do not allow estimation of either micro- or macro-avoidance of small birds at the species-group or species level. Furthermore, only data on false positives are available, but the number of birds or bats missed at the sensor level by the different systems cannot be assessed based on the published information.

Table 6.1 Comparison of the configuration and performance of all published multisensor systems presented in this chapter used to quantify the activity of birds and bats around and their collisions with offshore structures

Feature		WT-Bird	DTBird/Bat	ATOM	MUSE	Wind Turbine Sensor Unit
Deployed sensor types		Accelerometer, IR and visual video camera, contact microphone	Visual video, audio acoustic, thermal images	Thermal images, visual video camera, audio acoustic, ultrasound	Radar, thermal images, visual video camera	Accelerometer, contact microphone, stereo IR and visual video cameras, audio and ultrasonic acoustic microphones
Record trigger		Acoustic	Video image	N/A	Radar	Accelerometer
Avoidance	Micro (RSA)	IR video (RSA single turbine)	Visual video (RSA multiple turbines)	Visual video (but not specified)	IR and visual video (RSA single turbine), thermal imaging	IR and visual video (RSA single turbine)
	Macro	No	Visual video (600 m for >150 cm wingspan)	≤180 m range	Surveillance radar	No
Collision detection		Accelerometers, contact microphones, visual video	Visual video	Visual video	IR and visual video camera, thermal imaging	Accelerometer, contact microphone
Nocturnal operation	Acoustic	N/A	Audio	Audio and ultrasonic	N/A	Audio and ultrasonic
	Visual	IR video	No	Thermal images	IR video, thermal imaging	IR video

Table 6.1 – *continued*

Feature		WT-Bird	DTBird/Bat	ATOM	MUSE	Wind Turbine Sensor Unit
Bird identification	Acoustic	N/A	Species level	Group to species	N/A	species level
	Thermal	N/A	No	Group level passerine (<20 cm) vs non-passerine (>30 cm)	Group to species level	N/A
	Visual	Species level (only daylight)	Species level from above passerine level	Species level	Group to species level (only daylight)	Yes, unspecified level
Bat detection		No	Ultrasonic	Ultrasonic	Radar	Ultrasonic
Bat identification	Acoustic	N/A	Species level	Species level	N/A	Species level
Project status		Offshore tested prototype	Commercially available onshore and offshore systems	Offshore tested prototype	Commercially available onshore and offshore systems	Provisional patented, seeking commercial partner
References		Wiggelinkhuizen & den Boon (2010)	Puente *et al.* (2017; Collier *et al.* (2011)	Willmott *et al.* (2015)	Skov & Jensen (2015)	Suryan *et al.* (2016)

IR: infrared; N/A: not applicable; RSA: rotor-swept area.

TADS

The Thermal Animal Detection System (TADS) was initially developed in the early 2000s (Desholm 2003). This system aims at an automated monitoring of nocturnal bird behaviour and involves an array of three or six thermal imaging cameras targeting the rotor-swept area or airspace around a wind turbine (Petersen *et al.* 2006). Video sequences are automatically downloaded to a hard disc after an object whose temperature exceeds that of a user-defined temperature level enters the camera field. TADS records the number of collisions and flight altitude of birds, although potentially all types of activity and behaviour can be captured by the cameras. Bats and smaller birds, including passerines, may potentially be identified under appropriate conditions, but this is dependent on distance and camera quality. Old-generation thermal imaging cameras with a resolution of only 320×240 pixels had a potential detection range as low as approximately 50 m and a limited ability for species identification (Collier *et al.* 2011). The system was operated for several years off the Danish coast to assess the collision risk of Common Eider *Somateria mollissima* (Petersen *et al.* 2006). The proportion of false positives triggered, for example, by clouds was 99.7%. Verification of false positives is feasible by video sequences, but is time consuming.

VARS

The Visual Automatic Recording System camera system or VARS was designed to quantify the number of birds and bats passing the rotor-swept area of a wind turbine (Schulz *et al.* 2009; Collier *et al.* 2012) to assess collision risk. VARS can be positioned on a turbine nacelle, monitoring a radial sector behind the rotor parallel to the plane of the rotor-swept area. The opening angle can be adapted to the size of the target species and to the intended detection range, such as the length of rotor blades. Because of its installation on the nacelle, the position of the camera's field of view remains constant relative to the surveyed plane of the rotor-swept zone. The full-time operating camera uses a charge-coupled device (CCD) sensor that is sensitive to infrared radiation which, in combination with infrared spotlights, makes observations during daylight and in darkness possible (Schulz *et al.* 2014). Real-time recognition and recording of birds and bats passing through the camera field is automated by motion analysis software. Animals that move at a predefined minimum speed within the camera's field of view cause specific changes in pixels, which triggers the storage of video streams for as long as movements take place (Collier *et al.* 2012). With VARS, the identification of small-bodied birds is possible to the group level, such as warblers, thrushes, terns and swifts, while larger birds are often identified to species level. The distance at which birds are detected ranges from 60 m to several hundred metres for passerines and larger birds, respectively (Collier *et al.* 2012). In a 3-year survey at Alpha Ventus OWF in the North Sea, bird traffic measurements with VARS within the rotor-swept area was compared to that obtained from fixed-beam radar measurements taken in the immediate surroundings. This allowed micro-avoidance rates to be empirically determined for the first time at an OWF at night, when the majority of recorded flights occurred (Schulz *et al.* 2014; Coppack *et al.* 2015). A bird collision rate for the turbine was calculated using the VARS-measured traffic rate through the rotor-swept area. Two further comparable VARS units have been running at the FINO 2 research platform in the south-western Baltic Sea since 2007 and have continuously recorded the nocturnal light attraction of birds at the platform. A high-definition version with colour videos has also been developed, leading to improved species identification and extended detection ranges in daylight conditions.

WT-Bird

This system was developed to automatically record bird collisions at wind turbines and inform users about individual incidents (Ahmad *et al.* 2013). The system consists of microphones and two video cameras equipped with CCDs sensitive to infrared radiation (Wiggelinkhuizen *et al.* 2006). Cameras are mounted together with infrared light-emitting diode (LED) floodlights on the lower part of a turbine monopole facing towards the rotor-swept area. Two acceleration sensors are fixed to each blade of the rotor for collision detection, which triggers storage of video and acoustic signals during daytime and night-time (Wiggelinkhuizen *et al.* 2006; Ahmad *et al.* 2013). Moreover, an email is sent to inform the user upon system activation. To minimise the rate of false positives, the system continuously analyses and verifies all incoming data (see Figure 1 in Collier *et al.* 2011).

WT-Bird records the number and timing of collision events and enables taxonomic identification of larger birds (Collier *et al.* 2011). Monitoring of additional parameters by WT-Bird, such as birds in the immediate vicinity of the wind turbine, rotor position and weather conditions, is intended; and WT-Bird could subsume turbine maintenance costs through detection of rotor-blade damage. Field testing of a WT-Bird prototype in the Netherlands commenced in late 2005, indicating moderate accuracy in detecting collisions with airborne dummy objects that represented the smallest abundant bird species (Wiggelinkhuizen *et al.* 2006). Technical testing appears to be ongoing as a commercial version is not yet available.

DTBird

DTBird is a modular system developed by LIQUEN in Spain to detect and respond to bird presence with the aim of protecting birds from collision with wind turbines (e.g. Hanagasioglu *et al.* 2015). For bird detection, DTBird uses up to eight high-definition video cameras that fully cover the airspace around a wind turbine, which is autonomously and continuously monitored at light levels above 50 lux (May *et al.* 2012; Hanagasioglu *et al.* 2015). In addition, the system uses thermal cameras and acoustic recordings at night (DTBird 2018).

No detailed information about system performance, such as the range of detection of nocturnal birds, is available. However, in principle the detection range is dependent on bird size and weather conditions. In daylight, larger birds, such as raptors and vultures, for which the system was originally developed, are detected at a distance between 400 and 1,000 m (Figure 6.8), while smaller birds, such as passerines, are recorded under favourable weather conditions if they are less than 50 m away from a DTBird unit. Taxonomic identification of birds is possible to the group level (Ahmad *et al.* 2013). Following bird detection, DTBird software monitors birds approaching a wind turbine in real time. If monitored birds cross into a predefined safety zone, DTBird's 'Dissuasion Control' module is designed to trigger the emission of deterrent calls to encourage threatened birds to move out of the hazardous zone. If birds continue flying towards the rotor-swept airspace, DTBird software will automatically shut down the respective turbine with the mitigation module 'Stop Control'. Finally, the 'Collision Control' module is designed to register bird collisions with rotor blades. DTBird also stores sound records of each bird detected visually. All recorded data can be reviewed, for instance, to check for false positives, that is videos without a bird. On average, the daily rate of false positives is less than 1.5, but this increases to up to four per hour during unfavourable weather. Moreover, large remotely distant objects, such as aircraft, also trigger video recording (Hanagasioglu *et al.* 2015). Such false positives will

Figure 6.8 Example of automatic bird detection (birds marked with red frames) by a video camera as part of the DTBird multisensor system.

erroneously initiate automatic shutdown of wind turbines, suggesting that this technology needs further research and testing (Gartman *et al.* 2016).

DTBird has been commercially available for several years and, according to DTBird (2018), 131 units of DTBird and DTBat are installed in '47 existing/projected, onshore/offshore wind farms in 13 countries (Austria, France, Germany, Greece, Italy, Norway, Poland, Spain, Sweden, Switzerland, The Netherlands, United Kingdom and the United States)'. In a 7 month field test, the reliability of DTBird was tested in a wind farm on the Norwegian Island Smøla by comparing DTBird output to data obtained from the 'Merlin' mobile avian radar system (DeTect) and tracking data of resident White-tailed Eagles *Haliaeetus albicilla* fitted with GPS transmitters (May *et al.* 2012). The total operation period at Smøla spanned 5 years. For radii of 300 m and 150 m around a wind turbine, DTBird registered, respectively, 76% and 96% of all radar-recorded birds. Based on a 24 hour recording period, DTBird detected 59–80% of all flight movements around the site within a 300 m radius. In contrast, small birds were only rarely recorded, although it remains debatable as to whether this is due to the low sensitivity of DTBird or low ambient flight traffic of small birds during the study period. The rate of false positives was 1.2 per day after fine-tuning the system. No birds were detected at night (<200 lux). Comparison with GPS data indicates that DTBird repeatedly underestimated the distance of tracked eagles to the test turbine, resulting in more events where DTBird's mitigation modules were activated.

ATOM

The Acoustic Thermographic Offshore Monitoring system (ATOM) was developed by Normandeau Associates, Inc. (Gainesville, USA) in 2009 to provide baseline information to assess and mitigate offshore wind-park effects on the avifauna (Normandeau Associates Inc. 2018). The original ATOM used acoustic and ultrasound sensors to record calls of birds and bats, respectively, along with two forward-facing infrared thermal cameras (Willmott *et al.* 2015). The detection range of the thermal cameras was less than 180 m, with microphones achieving lower ranges, depending on environmental conditions and species involved, while the detection range for bat calls was less than 20 m (Collier *et al.* 2011). An updated version of ATOM also includes a visual light camera (Willmott *et al.* 2015). Simultaneous operation of both thermal-imaging cameras allows the calculation of the flight altitude of birds and bats. ATOM also generates data on the abundance, flight velocity and flight direction of birds and bats, but does not record collision events. Data are stored on site, but data transfer via satellite connections to a remote station can also be achieved. A 1 month field test onshore as well as a 15 month test off the Atlantic US coast demonstrated the feasibility of ATOM for gathering baseline data (Willmott *et al.* 2015).

MUSE

The intention of the MUlti SEnsor wildlife detection system (MUSE), formerly known as the Thermal & Visual Animal Detection System (TVADS) (Skov & Jensen 2015), is to generate a comprehensive estimate of bird and bat collision and avoidance behaviour at OWFs from the scale of individual wind turbines to multiple wind farms (Skov & Jensen 2015). MUSE combines horizontal surveillance and vertical radar with pan-tilt digital cameras in visual and thermal modes (see Figure 6.10 in Box 6.4) for optimal spatial and temporal coverage. The system is designed to deliver species-specific information about collision events, flight altitude and three-dimensional flight tracks both day and night to continuously depict lateral and vertical movements of birds and bats approaching wind turbines. This system enables users to quantify species-specific avoidance rates of larger birds, such as gulls and Northern Gannet, to improve the data basis of CRMs. MUSE has been tested under operational conditions at Thanet OWF in the UK (Box 6.4) (Skov *et al.* 2018) and is being used in the Block Island OWF in the USA.

Box 6.4 Development of an automated monitoring system to detect and track bird movements in offshore wind farms

Henrik Skov, Stefan Heinänen, Sara Méndez-Roldán, Tim Norman & Robin Ward

Compliance with regulatory requirements increasingly demands clear evidence on how birds behave within and around offshore wind farms (OWFs). To quantify bird collision risk, collision risk models are used, and to obtain realistic risk estimates, the collision risk modelling is subsequently corrected to take account of behavioural responses of birds to the presence of turbines; that is, avoidance. However, there is considerable uncertainty over the avoidance rates of most groups of birds owing to the relatively few monitoring studies so far undertaken in OWFs that have gathered empirical evidence. To meet the demand for further robust evidence on the avoidance behaviour of seabirds to OWFs, the Offshore Renewables Joint Industry Programme (ORJIP) Bird Collision Avoidance (BCA) Study coordinated by the Carbon Trust, UK, was undertaken at the Thanet OWF in the UK, developed and managed by Vattenfall, between July 2014 and June 2016 (Skov *et al.* 2018). The three main aims of the project were: (1) development of a bird monitoring system, which allows detecting and tracking bird movements at the species level in and around an operational OWF; (2) monitoring of seabird behaviour by deploying a multiple sensor monitoring system at the macro- (response to wind-farm perimeter), meso-

(response to individual turbines) and micro- (response to rotor and individual turbine blades) scales; and (3) development of an appropriate methodology for quantifying empirical avoidance rates (EARs).

So far, recording of meso- and micro-avoidance behaviour of birds has been constrained by the access of observers to platforms and turbines being limited to benign weather conditions. In the ORJIP project, an automated multisensor monitoring system (MUSE) based on integrated radar and digital camera was developed and applied from two turbine platforms to provide data on seabird behaviour within the wind farm. In addition to meso- and micro-avoidance behaviour, the monitoring system recorded actual collisions. The system integrated automated tracking with identification of flying birds based on a combination of radar and digital camera recordings. The system made it possible to monitor four turbines from one unit and generate documentation for flight paths both as visual and thermal video and geographic information system (GIS) tracks. Examples of thermal video frames are shown in Figure 6.9. The MUSE system is based on digital communication between a fan-beam radar and a digital camera, in which the camera receives original direction and distance coordinates from the radar, and subsequently records the event by video. The integrated radar and camera unit is remotely controlled, and is subdivided into a Data Acquisition and Pre-processing System (DAPS) and a software package DAPSControl, for controlling the DAPS (Figure 6.10). The DAPS samples at 100 MHz and performs real-time median filtering of data from the radar. Multiple options for interfacing and data connections between the sensors exist; at Thanet OWF the interface format was TCP/IP. Remote access facilitated the transfer of data to a server onshore and the ability to watch live videos through a web plug-in.

The classification of bird objects by the radar is facilitated by the high temporal resolution of the signal acquisition, which makes it possible to filter radar echoes against the known signature of a bird signal when in flight. Once the camera receives the position of the flying bird from the radar, it will zoom on the bird while recording the

Figure 6.9 Thermal image of a European Herring Gull *Larus argentatus* approaching the spinning rotor of a wind turbine, but avoiding collision by flying along the plane of the rotor (recorded on the test site in Denmark). (DHI)

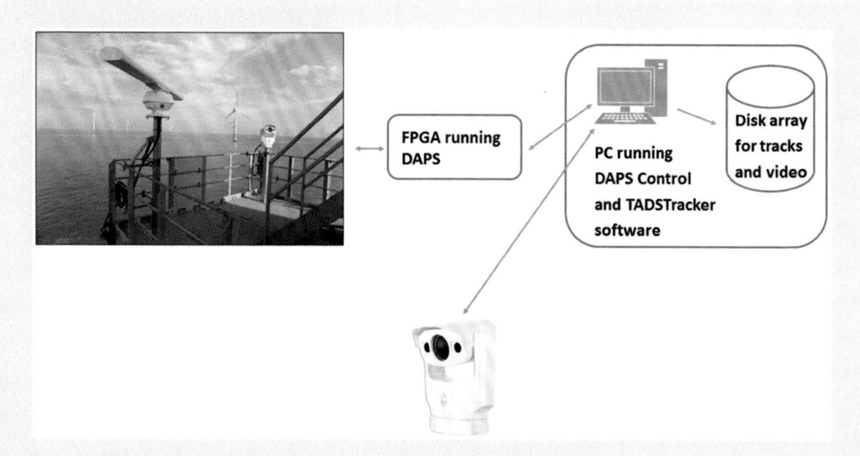

Figure 6.10 Overview of the flow of information in the MUSE unit used in the ORJIP Bird Collision Avoidance study. DAPS: Data Acquisition and Pre-processing System; FPGA: field programmable gate array. (Inset photograph: Thomas W. Johansen)

event on video. In this project, the system was applied with a standard marine radar (magnetron X-band) and a combined visual–thermal camera, but it is capable of running with both solid-state and magnetron-based radars, as well as with a range of digital pan-tilt cameras with visual and thermal channels. The number of turbines (rotors) that can be monitored will depend on the target species and the distance between turbines. As a rule of thumb, the radar can detect a standard seabird at 4 km range, and a digital camera with high-resolution motion control can follow a seabird at approximately 1 km distance.

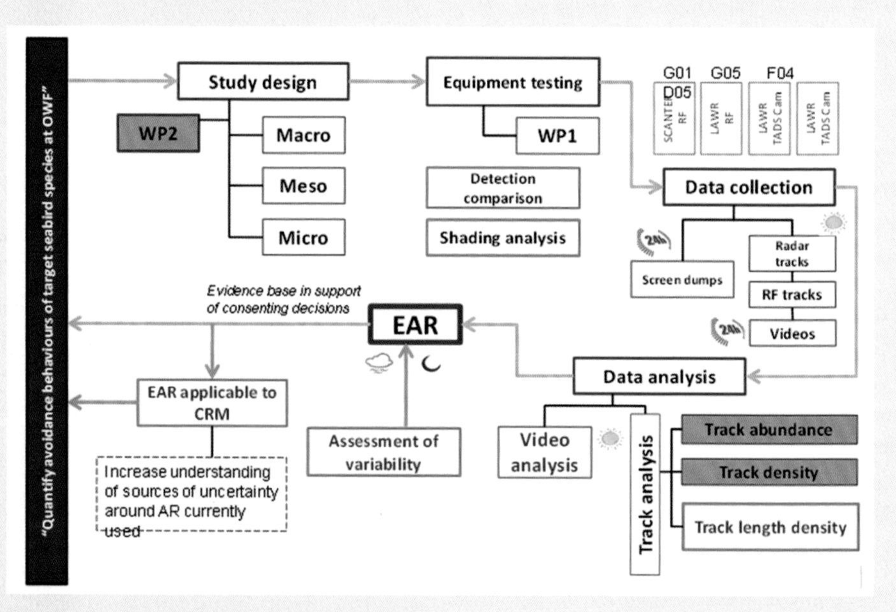

Figure 6.11 Analytical framework applied to determine of empirical avoidance rates in the ORJIP Bird Collision Avoidance study. OWF: offshore wind farm; EAR: empirical avoidance rate; CRM: collision risk model; RF: laser rangefnder; WP1: Work Package 1; WP2: Work Package 2.

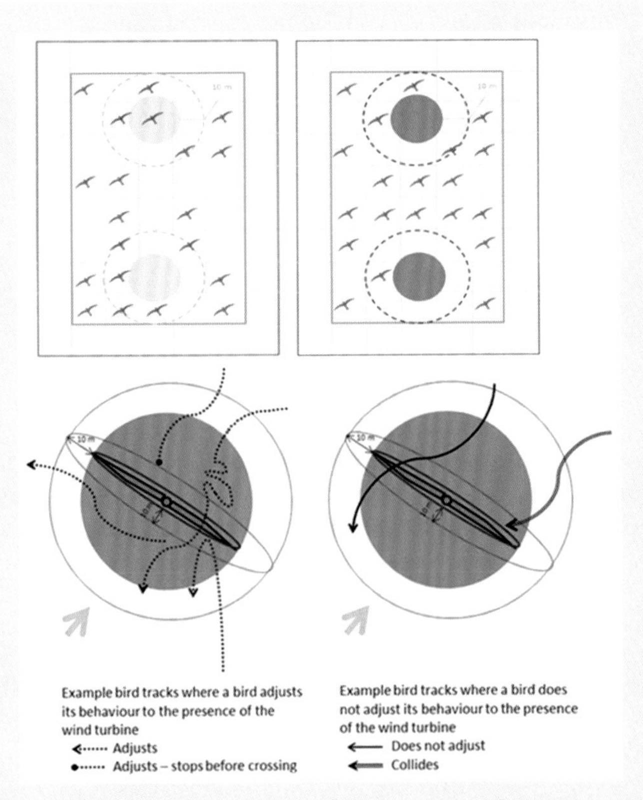

Figure 6.12 Sketch of the calculation of empirical meso-avoidance (top) and micro-avoidance (bottom) in the ORJIP Bird Collision Avoidance study. For meso-avoidance, the right panel describes observed track length density, while the left panel describes the hypothetical scenario with the same total track length as observed, but without the wind turbines. For micro-avoidance (bottom), arrows represent bird movement in relation to the rotor, shown as a dark blue ellipse, with the addition of a 10 m buffer, shown as a red ellipse. The light blue arrow represents wind direction. Seabirds have been observed to adjust their flight to avoid individual blades (dotted arrow), or not to adjust, but cross the rotor and survive (solid arrow), or, on rare occasions, collide (double arrow).

An overview of the analytical framework applied during ORJIP for determination of empirical avoidance rates is shown in Figure 6.11. Meso-avoidance was quantified by comparing the observed density of bird tracks (measured as track length per unit area) inside the rotor-swept zone (with a 10 m buffer) of monitored turbines with a hypothetical situation in the absence of the wind farm, in which the same total track length would have been observed in the wind-farm area (Figure 6.12). The formula for meso-avoidance was thus:

Meso-avoidance = 1 – (Track length density in rotor-swept zone/
Mean track length density in entire wind farm)

Micro-avoidance was quantified by calculating the proportion of birds adjusting their flight path to the presence of blades to all birds entering the rotor-swept zone. Adjustments are assessed by analysing bird movements on the video documentation in relation to the rotor, represented as a dynamic ellipse, surrounded by a 10 m buffer that changes its orientation with the wind direction.

The development of the MUSE system during ORJIP has facilitated the possibility to perform continuous and automated monitoring of bird behaviour in OWFs at the species level. Thus, the development of this technology will potentially benefit both research, by allowing data collection in a wide range of weather conditions, and industry, by enhancing compliance with regulatory requirements.

Wind Turbine Sensor Unit

The Wind Turbine Sensor Unit was developed by R. Albertani and colleagues at Oregon State University and consists of three sensor components (Suryan *et al.* 2016). The system includes accelerometers and contact microphones mounted at the base of each blade that are designed to continuously register collisions of animals with turbines. The system also includes a stereovisual camera and an infrared camera, both mounted on the nacelle, for tracking birds and bats, that record target distance and size. Finally, two nacelle-mounted microphones (one acoustic, one ultrasound) monitor bird and bat calls. Data are managed with the aid of a buffer ring set-up that, when triggered by a presumed collision event registered by any of the sensors, automatically records all systems data occurring 10 seconds before and 10 seconds after the event. The complete system was tested in two field campaigns during which empty and water-filled tennis balls were shot under different conditions against an experimental wind turbine in Colorado, USA. Depending on the conditions, between 35% (normal operation) and 100% (idle rotation) of tennis-ball collisions were detected against background vibrations.

Concluding remarks

Predicting the probability of collisions of birds and bats with wind turbines is important for assessing the extent to which wind farms may affect their survival and ultimately population dynamics (e.g. Horswill *et al.* 2017). The assessment of the interaction of birds and bats with wind turbines is based on theoretical assumptions rather than on solid empirical evidence (Green *et al.* 2016). There are several reasons why this is the case; above all, the limited technical possibilities to continuously record collisions. Additional reasons include: (1) the infrequent and stochastic nature of collisions, (2) limited knowledge on environmental and biological input data for population models, and (3) formal requirement to predict site-specific collision rates before the consenting and construction of a wind farm. Therefore, predictive CRMs are the method of choice for quantitatively estimating collision risk (see Cook & Masden, Chapter 5).

To the authors' knowledge, none of the models accounts for the attraction of night-migrating birds or bats (i.e. negative avoidance) by artificial light, which has been identified as the key factor influencing bird presence near the rotor-swept zone (Hill *et al.* 2014). Moreover, already slightly inaccurate estimates of model parameters can strongly modify model output. Chamberlain *et al.* (2006) demonstrated that small changes in avoidance rate can lead to inflationary deviations in predicted mortality rates. As avoidance rates have rarely been empirically determined at different spatial scales during day and night or in relation to the operational status of a wind farm, in-depth measurement of avoidance/attraction behaviour of birds and bats at different spatial scales seems mandatory to improve the output of CRMs. Furthermore, causal detachment of input values introduces considerable uncertainty to model output (Green *et al.* 2016). For example, many Environmental Impact

Assessments merge site-specific pre-construction data on bird activity with avoidance rates determined elsewhere or in other taxa. Last but not least, the majority of CRMs neglect environmental, intraspecific and interspecific variability, which may at least partly explain the weak match between the mortality predicted by a model and the observed mortality in the field (Green *et al.* 2016; Masden & Cook 2016, and references therein).

Progress in the field of collision risk assessment is hampered by a mismatch between rigid model assumptions and the quality of the empirical input data. A model's output can only be as good as its input. Thus, the validation of model-based hypotheses on collision rates and the population consequences of added mortality for birds and bats will remain difficult to assess as long as the density and quality of data have not improved. Empirical model validation, and if necessary, refinement is decisive owing to the lack of appropriate methods to quantify actual collision events.

The following three steps will help to improve impact assessments of offshore wind turbines for birds and bats:

- First, the accuracy in the quantification of avoidance/attraction rates needs to be raised, as small changes alter CRM output tremendously (Chamberlain *et al.* 2006; Cook & Masden, Chapter 5). In terms of macro-avoidance, tracking with radio-tags seems to be a practical solution for small avifauna, including landbirds and waterbirds as well as migratory bats. For most seabird species, GPS tags with high spatiotemporal resolution can be used in combination with other remote-sensing technology, such as radar, video or thermal imaging. In terms of micro-avoidance, the turbine structure, and in particular, the rotor-swept area, need to be continuously monitored with stereovisual and infrared or thermal cameras of high optical resolution strategically mounted at the base of each rotor blade (Suryan *et al.* 2016) or turbine nacelle (Schulz *et al.* 2014). Additional cameras are needed to record barotraumatic bat (and possibly small bird) fatalities potentially caused in the wake of the turbine outside the rotor-swept area.
- Secondly, the ability to directly register bird and bat collisions with wind turbines needs improvement because fatality monitoring cannot be carried out offshore to quantify collisions or to validate CRMs. The most appropriate available methodology to achieve this goal could involve accelerometers fixed to each turbine blade.
- Thirdly, priority should be given to the development of techniques that enable robust identification of bird and bat species, as it has become evident that turbine–animal interactions are highly species specific. This can be achieved using high-resolution cameras, but also through recording the calls of bats and birds with microphones fixed to the nacelle or the monopile of a wind turbine. Effort should focus on those species that seem most threatened by collisions with wind turbines, such as nocturnally migrating birds, migrating bats and, perhaps, foraging seabirds.

Each of these three steps needs to be directly associated with date- and time-stamped observations so that a conditions-based assessment (including seasonal, day/night and weather conditions) of risk factors can be developed. Care must also be taken regarding the design of impact studies in relation to clearly defined study objectives and intended outputs. Preferably, comparable data should be obtained from surveys before, during and after turbine construction. For example, taking pre-construction measurements of bat activity at ground level while post-construction measurements are taken at rotor-swept altitude is very likely to confound the assessment of activity levels (Lintott *et al.* 2016). However, practical constraints often preclude comparable sample taking, particularly offshore, which has to be considered in the interpretation of study results.

Multisensor systems that assemble components of the remote-sensing technologies described in this chapter provide the best solution to quantify interactions between turbines and birds and bats offshore. Ideally, such systems would be fixed to each individual wind turbine of a wind farm. As a minimal solution, the authors recommend the installation of one multisensor system in each corner as well as in the centre of a wind farm (a total of five systems) to ensure balanced sampling of birds and bats entering the wind farm and inside the wind farm, irrespective of flight direction. However, the diversity of available systems renders comparability among different systems almost impossible. To gain a better understanding of the spatiotemporal variation of the effects of a wind farm and ultimately cumulative impacts of winds farms on the phenology and dynamics of bird and bat populations, coordinated monitoring of multiple wind farms using the sampling design outlined should occur simultaneously. In this way, estimates on macro-avoidance can be better obtained, as reported, for example, for migratory geese (Plonczkier & Simms 2012), through which energetic costs of circumventing wind farms could be assessed. In the longer term, such a concerted empirical approach would improve model parameterisation, validate current CRMs and improve their predictive performance.

Acknowledgements

We are grateful for financial support from the German Federal Agency for Nature Conservation through the project BIRDMOVE (grant no. 3515822100).

References

Ahlén, I. (1990) *Identification of Bats in Flight*. Stockholm: Swedish Society for Conservation of Nature.

Ahlén, I. (1997) Migratory behaviour of bats at south Swedish coasts. *International Journal of Mammalian Biology* 62: 375–380.

Ahlén, I., Baagøe, H.J. & Bach, L. (2009) Behavior of Scandinavian bats during migration and foraging at sea. *Journal of Mammalogy* 90: 1318–1323.

Ahmad, N., O'Nils, M. & Lawal, N. (2013) A taxonomy of visual surveillance systems. Sundsvall: Research Report in Electronics of the Mid Sweden University. Retrieved 26 March 2019 from http://www.diva-portal.org/smash/record.jsf?pid=diva2%3A618452&dswid=-553

Aide, T.M., Corrada-Bravo, C., Campos-Cerqueira, M., Milan, C., Vega, G. & Alvarez, R. (2013) Real-time bioacoustics monitoring and automated species identification. *PeerJ* 1: e103. doi: 10.7717/peerj.103.

Allison, N.L. & Destefano, S. (2006) Equipment and techniques for nocturnal wildlife studies. *Wildlife Society Bulletin* 34: 1036–1044.

Aumüller, R., Boos, K., Freienstein, S., Hill, K. & Hill, R. (2011) Beschreibung eines Vogelschlagereignisses und seiner Ursachen an einer Forschungsplattform in der Deutschen Bucht. *Vogelwarte* 49: 9–16.

Aumüller, R., Hill, R., Rebke, M., Hill, K. & Weiner, C. (2015) Weiterführende Messungen zur Vogelzugforschung auf der Forschungsplattform FINO3 zeitgleich mit dem Bau eines großen Offshore-Windparks in der nördlichen Deutschen Bucht. Offshore birds. Final Report No. 11083 submitted to Bundesministerium für Umwelt, Naturschutz & Reaktorsicherheit. Osterholz-Scharmbeck: Avitec-Research. Retrieved 26 March 2019from https://www.fino3.de/files/forschung/vogelzug/Schlussbericht%20Avitec%20Research%200327533C.pdf

Bach, L., Bach, P., Helge, A., Maatz, K., Schwarz, V., Teuscher, M. & Zöller, J. (2009) Fledermauszug auf Wangerooge – erste Ergebnisse aus dem Jahr 2008. *Zeitschrift Mellumrat* 8(1): 8–10.

Baerwald, E.F., D'Amours, G.H., Klug, B.J. & Barclay, R.M.R. (2008) Barotrauma is a significant cause of bat fatalities at wind turbines. *Current Biology* 18(16): R695–R696.

Bairlein, F. (2016) Migratory birds under threat. *Science* 354(6312): 547–548.

Ballasus, H., Hill, K. & Hüppop, O. (2009) Gefahren künstlicher Beleuchtung für ziehende Vögel und Fledermäuse. *Berichte zum Vogelschutz* 46: 127–157.

Barataud, M. (2015) *Acoustic Ecology of European Bats: Species identification, study of their habitats and foraging behaviour*. Mèze: Biotope Editions; Museúm national d'histoire naturelle, Paris.

Bardeli, R., Wolff, D., Kurth, F., Koch, M., Tauchert, K.-H. & Frommolt, K.-H. (2010) Detecting bird sounds in a complex acoustic environment and application to bioacoustic monitoring. *Pattern Recognition Letters* 31: 1524–1534.

Barron, D.G., Brawn, J.D. & Weatherhead, P.J. (2010) Meta-analysis of transmitter effects on avian behaviour and ecology. *Methods in Ecology and Evolution* 1: 180–187.

Behr, O., Brinkmann, R., Korner-Nievergelt, F., Nagy, M., Niermann, I., Reich, M. & Simon, R. (2015) *Reduktion des Kollisionsrisikos von Fledermäusen an Onshore-Windenergieanlagen (RENEBAT II)*. Göttingen: Cuvillier.

Blumstein, D.T., Mennill, D.J., Clemins, P., Girod, L., Yao, K., Patricelli, G., Deppe, J.L., Krakauer, A.H., Clark, C., Cortopassi, K.A., Hanser, S.F., McCowan, B., Ali, A.M. & Kirschel, A.N.G. (2011) Acoustic monitoring in terrestrial environments using microphone arrays. Applications, technological considerations and prospectus. *Journal of Applied Ecology* 48: 758–767.

Bodey, T.W., Cleasby, I.R., Bell, F., Parr, N., Schultz, A., Votier, S.C., Bearhop, S. & Paradis, E. (2018) A phylogenetically controlled meta-analysis of biologging device effects on birds. Deleterious effects and a call for more standardized reporting of study data. *Methods in Ecology and Evolution* 9: 946–955.

Boyles, J.G., Cryan, P.M., McCracken, G.F. & Kunz, T.H. (2011) Economic importance of bats in agriculture. *Science* 332(6025): 41–42.

Bridge, E.S., Thorup, K., Bowlin, M.S., Chilson, P.B., Diehl, R.H., Fléron, R.W., Hartl, P., Kays, R., Kelly, J.F., Robinson, W.D. & Wikelski, M. (2011) Technology on the move: recent and forthcoming innovations for tracking migratory birds. *BioScience* 61: 689–698.

Brinkmann, R., Behr, O., Niermann, I. & Reich, M. (eds) (2011) *Entwicklung von Methoden zur Untersuchung und Reduktion des Kollisionsrisikos von Fledermäusen an Onshore-Windenergieanlagen. Ergebnisse eines Forschungsvorhabens*. Göttingen: Cuvillier.

Bruderer, B. (1997) The study of bird migration by radar. Part 1: The technical basis. *Naturwissenschaften* 84(1): 1–8.

Bruderer, B. & Popa-Lisseanu, A.G. (2005) Radar data on wing-beat frequencies and flight speeds of two bat species. *Acta Chiropterologica* 7: 73–82.

Bruderer, B., Peter, D. & Steuri, T. (1999) Behaviour of migrating birds exposed to x-band radar and bright light beam. *Journal of Experimental Biology* 202: 1015–1022.

Bruderer, B., Peter, D., Boldt, A. & Liechti, F. (2010) Wing-beat characteristics of birds recorded with tracking radar and cine camera. *Ibis* 152: 272–291.

Brydegaard, M., Guan, Z., Runemark, A., Akesson, S. & Svanberg, S. (2010) Feasibility study: fluorescence lidar for remote bird classification. *Applied Optics* 49: 4531–4544.

Buckland, S.T., Burt, M.L., Rexstad, E.A., Mellor, M., Williams, A.E. & Woodward, R. (2012) Aerial surveys of seabirds. The advent of digital methods. *Journal of Applied Ecology* 49: 960–967.

Bundesverband WindEnergie (2018) Windenergie Factsheet Deutschland 2017. Berlin: Bundesverband WindEnergie. Retrieved 26 March 2019 from https://www.wind-energie.de/themen/zahlen-und-fakten/

Burger, A.E. & Shaffer, S.A. (2008) Application of tracking and data-logging technology in research and conservation of seabirds. *Auk* 125: 253–264.

Buxton, R.T. & Jones, I.L. (2012) Measuring nocturnal seabird activity and status using acoustic recording devices. Applications for island restoration. *Journal of Field Ornithology* 83: 47–60.

Chamberlain, D.E., Freeman, S., Rehfisch, M.R., Fox, A. & Desholm, M. (2005) Appraisal of Scottish Natural Heritage's wind farm collision risk model and its application. Research Report 401. Thetford: British Trust for Ornithology. Retrieved 26 March 2019 from https://www.bto.org/research-data-services/publications/research-reports

Chamberlain, D.E., Rehfisch, M.R., Fox, A.D., Desholm, M. & Antony, S.J. (2006) The effect of avoidance rates on bird mortality predictions made by wind turbine collision risk models. *Ibis* 148: 198–202.

Collier, M.P., Dirksen, S. & Krijgsveld, K.L. (2011) A review of methods to monitor collisions or micro-avoidance of birds with offshore wind turbines. Part 1: Review. Final report submitted to The Crown Estate. Culemborg: Bureau Waardenburg. Retrieved 26 March 2019 from https://www.bto.org/science/wetland-and-marine/soss/projects

Collier, M.P., Dirksen, S. & Krijgsveld, K.L. (2012) A review of methods to monitor collisions or micro-avoidance of birds with offshore wind turbines. Part 2: Feasibility study of systems to monitor collisions. Final report submitted to The Crown Estate. Culemborg: Bureau Waardenburg. Retrieved 26 March 2019 from https://www.bto.org/science/wetland-and-marine/soss/projects

Cook, A.S.C.P., Humphreys, E.M., Bennet, F., Masden, E.A. & Burton, N.H.K. (2018) Quantifying avian avoidance of offshore wind turbines: Current evidence and key knowledge gaps. *Marine Environmental Research* 140: 278–288.

Coppack, T., Dittmann, T. & Schulz, A. (2015) Avian collision risk and micro-avoidance rates determined at an existing offshore wind farm. In Köppel, J. & Schuster, E. (eds) *What Have We Learnt So Far? A synoptical perspective on wind energy's wildlife implications. Book of Abstracts. Conference on Wind energy and Wildlife impacts (CWW 2015).* Berlin, Germany, 10–12 March. Technische Universität Berlin: 25.

Corman, A.-M. & Garthe, S. (2014) What flight heights tell us about foraging and potential conflicts with wind farms. A case study in Lesser Black-backed Gulls (*Larus fuscus*). *Journal of Ornithology* 155: 1037–1043.

Croxall, J.P., Butchart, S.H.M., Lascelles, B., Stattersfield, A.J., Sullivan, B., Symes, A. & Taylor, P. (2012) Seabird conservation status, threats and priority actions: a global assessment. *Bird Conservation International* 22: 1–34.

Cryan, P.M. & Brown, A.C. (2007) Migration of bats past a remote island offers clues toward the problem of bat fatalities at wind turbines. *Biological Conservation* 139: 1–11.

Crysler, Z.J., Ronconi, R.A. & Taylor, P.D. (2016) Differential fall migratory routes of adult and juvenile Ipswich sparrows (*Passerculus sandwichensis princeps*). *Movement Ecology* 4: 3.

Delprat, B. & Alcuri, G. (2011) ID Stat: Innovative technology for assessing wildlife collisions with wind turbines. Conference on Wind energy and Wildlife impacts (CWW 2011), Trondheim, Norway, 5 May 2011. Norwegian Institute for Nature Research. Retrieved 26 March 2019 from http://cww2011.nina.no/

Desholm, M. (2003) Thermal Animal Detection System (TADS). Development of a method for estimating collision frequency of migrating birds at offshore wind turbines. Technical Report No. 440. Aarhus: National Environmental Research Institute. Retrieved 26 March 2019 from https://www.dmu.dk/1_viden/2_publikationer/3_fagrapporter/

Desholm, M. & Kahlert, J.A. (2005) Avian collision risk at an offshore wind farm. *Biology Letters* 1: 296–298.

Desholm, M., Fox, A.D., Beasley Patrick, D.L. & Kahlert, J.A. (2006) Remote techniques for counting and estimating the number of bird–wind turbine collisions at sea: a review. *Ibis* 148: 76–89.

Dierschke, J., Dierschke, V. & Krüger, T. (2005) Anleitung zur Planbeobachtung des Vogelzugs über dem Meer ('Seawatching'). *Seevögel* 26: 2–13.

Dierschke, V. (1991) Seawatching auf Helgoland. *Ornithologischer Jahresbericht Helgoland* 1: 49–53.

Dirksen, S. (2017) Review of methods and techniques for field validation of collision rates and avoidance amongst birds and bats at offshore wind turbines. Final report phase 1 submitted to Rijkswaterstaat WVL for the Dutch Governmental Offshore Wind Ecological Programme. Utrecht: Sjoerd Dirksen Ecology research, consultancy & management., Retrieved 26 March 2019 from https://www.noordzeeloket.nl/functies-gebruik/windenergie/ecologie/wind-zee-ecologisch/documenten-wozep-0/vogels/@166999/review-methods/

Dokter, A.M., Liechti, F., Stark, H., Delobbe, L., Tabary, P. & Holleman, I. (2011) Bird migration flight altitudes studied by a network of operational weather radars. *Journal of the Royal Society, Interface* 8: 30–43.

Drake, V.A. & Reynolds, D.R. (2012) *Radar Entomology Observing Insect Flight and Migration.* Wallingford: CABI.

DTBird (2018) DTBird System Brochure September 2018: Bird monitoring & reduction of collision risk with wind turbines. Madrid: DTBird. Retrieved 26 March 2019 from https://dtbird.com/index.php/dtbird-dtbat-document-downloads

Eastwood, E. (1967) *Radar Ornithology*. London: Methuen & Co.

Evans, W.R. (2005) Monitoring avian night flight calls. The new century ahead. *The Passenger Pigeon* 67(1): 15–24.

Exo, K.-M., Fiedler, W. & Wikelski, M. (2013) On the way to new methods: a lifetime of round the clock monitoring. In Schäffer, N. (ed.) *Der Falke – Journal für Vogelbeobachter. Bird migration.* Wiebelsheim: AULA. pp. 20–25.

Farnsworth, A. (2005) Flight calls and their value for future ornithological studies and conservation research. *Auk* 122: 733–746.

Farnsworth, A. & Russell, R.W. (2007) Monitoring flight calls of migrating birds from an oil platform in the northern Gulf of Mexico. *Journal of Field Ornithology* 78: 279–289.

Farnsworth, A., Gauthreaux, J.S.A. & van Blaricom, D. (2004) A comparison of nocturnal call counts of migrating birds and reflectivity measurements on Doppler radar. *Journal of Avian Biology* 35: 365–369.

Fenton, M.B. & Bell, G.P. (1981) Recognition of species of insectivorous bats by their echolocation calls. *Journal of Mammalogy* 62: 233–243.

Fijn, R.C., Krijgsveld, K.L., Poot, M.J.M. & Dirksen, S. (2015) Bird movements at rotor heights measured continuously with vertical radar at a Dutch offshore wind farm. *Ibis* 157: 558–566.

Flowers, J., Albertani, R., Harrison, T., Polagye, B. & Suryan, R.M. (2014) Design and initial component tests of an integrated avian and bat collision detection system for offshore wind turbines. In Proceedings of the 2nd Marine Energy Technology Symposium, Seattle, 15 April 2014. Retrieved 26 March 2019 from https://vtechworks.lib.vt.edu/handle/10919/49197

Fulton, W. (2018) Calculator for field of view (FOV) of a camera and lens. Retrieved 26 March 2019 from https://www.scantips.com/lights/fieldofview.html#top

Furness, R.W., Wade, H.M. & Masden, E.A. (2013) Assessing vulnerability of marine bird populations to offshore wind farms. *Journal of Environmental Management* 119: 56–66.

Garthe, S. & Hüppop, O. (2004) Scaling possible adverse effects of marine wind farms on seabirds. Developing and applying a vulnerability index. *Journal of Applied Ecology* 41: 724–734.

Garthe, S., Markones, N. & Corman, A.-M. (2017) Possible impacts of offshore wind farms on seabirds. A pilot study in northern gannets in the southern North Sea. *Journal of Ornithology* 158: 345–349.

Gartman, V., Bulling, L., Dahmen, M., Geißler, G. & Köppel, J. (2016) Mitigation measures for wildlife in wind energy development, consolidating the state of knowledge – Part 2. Operation, decommissioning. *Journal of Environmental Assessment Policy and Management* 18(3): 1650014.

Gauthreaux, S.A. (1969) A portable ceilometer technique for studying low level nocturnal migration. *Journal of Field Ornithology* 40: 309–319.

Gauthreaux, S.A. & Belser, C.G. (2003) Radar ornithology and biological conservation. *Auk* 120: 266–277.

Gauthreaux, S.A. & Livingston, J.W. (2006) Monitoring bird migration with a fixed-beam radar and a thermal-imaging camera. *Journal of Field Ornithology* 77: 319–328.

Goldsmith, T.H. (2006) What birds see. *Scientific American* 295(1): 68–75.

Green, R.E., Langston, R.H.W., McCluskie, A., Sutherland, R. & Wilson, J.D. (2016) Lack of sound science in assessing wind farm impacts on seabirds. *Journal of Applied Ecology* 53: 1635–1641.

Griffin, D.R. (1958) *Listening in the Dark: The acoustic orientation of bats and men.* New Haven, CT: Yale University Press.

Global Wind Energy Council (GWEC) (2016) Europe's push for off-shore wind. *Nature* 536(7617): 378–379.

Hanagasioglu, M., Aschwanden, J., Bontadina, F. & de la Puente Nilsson, M. (2015) Investigation of the effectiveness of bat and bird detection of the DTBat and DTBird systems at Calandawind turbine. Final report submitted to Swiss Federal Office of Energy (SFOE), Federal Office for the Environment (FOEN). Zürich: Interwind AG, Swiss Ornithological Institute, SWILD – Urban Ecology & Wildlife Research & DTBird. Retrieved 26 March 2019 from https://www.aramis.admin.ch/Default.aspx?DocumentID=3928&Load=true

Hansen, L. (1954) Birds killed at lights in Denmark 1886–1939. *Videnskabelige Meddelelser fra Dansk Naturhistorisk Forening* 116: 269–368.

Harmata, A.R., Podruzny, K.M., Zelenak, J.R. & Morrison, M.L. (1999) Using marine surveillance radar to study bird movements and impact assessment. *Wildlife Society Bulletin* 27(1): 44–52.

Hayes, M.A., Hooton, L.A., Gilland, K.L., Grandgent, C., Smith, R.L., Lindsay, S.R., Collins, J.D., Schumacher, S.M., Rabie, P.A., Gruver, J.C. & Goodrich-Mahoney, J. (2019) A smart curtailment approach for reducing bat fatalities and curtailment time at wind energy facilities. *Ecological Applications*, e01881.

Hecht, S. & Pirenne, M.H. (1940) The sensibility of the nocturnal long-eared owl in the spectrum. *Journal of General Physiology* 23: 709–717.

Hein, C.D. (2017) Monitoring bats. In Perrow, M.R. (ed.) *Wildlife and Wind Farms, Conflicts and Solutions. Volume 2. Onshore: Monitoring and mitigation*. Exeter: Pelagic Publishing. pp. 31–57.

Hill, R. & Hüppop, O. (2011) Zugrufe über der Nordsee – welche Erkenntnisse lassen sich aus einer automatisierten Erfassung gewinnen? *Vogelwarte* 49: 318–319.

Hill, R., Hill, K., Aumüller, R., Schulz, A., Dittman, T., Kulemeyer, C. & Coppack, T. (2014) Of birds, blades and barriers: detecting and analysing mass migration events at Alpha Ventus. In BSH & BMU (eds) *Ecological Research at the Offshore Windfarm Alpha Ventus: Challenges, results and perspectives*. Wiesbaden: Springer Spektrum. pp. 111–131.

Horn, J.W., Arnett, E.B. & Kunz, T.H. (2008) Behavioral responses of bats to operating wind turbines. *Journal of Wildlife Management* 72: 123–132.

Horswill, C., O'Brien, S.H. & Robinson, R.A. (2017) Density dependence and marine bird populations. Are wind farm assessments precautionary? *Journal of Applied Ecology* 54: 1406–1414.

Horton, K.G., Shriver, W.G. & Buler, J.J. (2015) A comparison of traffic estimates of nocturnal flying animals using radar, thermal imaging, and acoustic recording. *Ecological Applications* 25: 390–401.

Hüppop, K., Dierschke, J., Hill, R. & Hüppop, O. (2012) Jahres- und tageszeitliche Phänologie der Vogelrufaktivität über der Deutschen Bucht. *Vogelwarte* 50: 93–114.

Hüppop, O. & Hilgerloh, G. (2012) Flight call rates of migrating thrushes. Effects of wind conditions, humidity and time of day at an illuminated offshore platform. *Journal of Avian Biology* 43: 85–90.

Hüppop, O. & Hill, R. (2016) Migration phenology and behaviour of bats at a research platform in the south-eastern North Sea. *Lutra* 59: 5–22.

Hüppop, O., Dierschke, J., Exo, K.-M., Fredrich, E. & Hill, R. (2006) Bird migration studies and potential collision risk with offshore wind turbines. *Ibis* 148: 90–109.

Hüppop, O., Hüppop, K., Dierschke, J. & Hill, R. (2016) Bird collisions at an offshore platform in the North Sea. *Bird Study* 63: 73–82.

Hüppop, O. , Ciach, M. , Diehl, R. , Reynolds, D.R., Stepanian, P.M. and Menz, M.H. (2018) Perspectives and challenges for the use of radar in biological conservation. *Ecography* 42: 1–19, doi:10.1111/ecog.04063

Hüppop, O., Michalik, B., Bach, L., Hill, R. & Pelletier, S.K. (2019) Migratory birds and bats. In Perrow, M.R. (ed.) *Wildlife and Wind Farms, Conflicts and Solutions. Volume 3. Offshore: Potential effects*. Exeter: Pelagic Publishing. pp. 142–173.

Huso, M.M.P., Dalthorp, D. & Korner-Nievergelt, F. (2017) Statistical principles of post-construction fatality monitoring. In Perrow, M.R. (ed.) *Wildlife and Wind Farms, Conflicts and Solutions. Volume 2. Onshore: Monitoring and mitigation*. Exeter: Pelagic Publishing. pp. 84–102.

Jameson, H., Reeve, E., Laubek, B. & Sittel, H. (2019) The nature of offshore wind farms. In Perrow, M.R. (ed.) *Wildlife and Wind Farms, Conflicts and Solutions. Volume 3. Offshore: Potential effects*. Exeter: Pelagic Publishing. pp. 1–29.

Jiang, Y., Tang, B., Qin, Y. & Liu, W. (2011) Feature extraction method of wind turbine based on adaptive Morlet wavelet and SVD. *Renewable Energy* 36: 2146–2153.

Kelly, T.A., West, T.E. & Davenport, J.K. (2009) Challenges and solutions of remote sensing at offshore wind energy developments. *Marine Pollution Bulletin* 58: 1599–1604.

Kemper, G., Weidauer, A. & Coppack, T. (2016) Monitoring seabirds and marine mammals by georeferenced aerial photography. *International Archives of the Photogrammetry, Remote Sensing and Spatial Information Sciences* XLI-B8: 689–694.

King, S. (2019) Seabirds: collision. In Perrow, M.R. (ed.) *Wildlife and Wind Farms, Conflicts and Solutions. Volume 3. Offshore: Potential effects*. Exeter: Pelagic Publishing. pp. 206–234.

Kirkwood, J.J. & Cartwright, A. (1993) Comparison of two systems for viewing bat behavior in the dark. *Proceedings of the Indiana Academy of Science* 102: 133–137.

Kirschel, A.N.G., Cody, M., Harlow, Z.T., Promponas, V.J., Vallejo, E.E. & Taylor, C.E. (2011) Territorial dynamics of Mexican ant-thrushes *Formicarius moniliger* revealed by individual recognition of their songs. *Ibis* 153: 255–268.

Korner-Nievergelt, F., Brinkmann, R., Niermann, I. & Behr, O. (2013) Estimating bat and bird mortality occurring at wind energy turbines from covariates and carcass searches using mixture models. *PLoS ONE* 8(7): e67997. doi: 10.1371/journal.pone.0067997.

Kramer, G. (1931) Zug in grosser Höhe. *Vogelzug* 2: 69–71.

Kunz, T.H., Arnett, E.B., Cooper, B.M., Erickson, W.P., Larkin, R.P., Mabee, T., Morrison, M.L., Strickland, M.D. & Szewczak, J.M. (2007) Assessing impacts of wind-energy development on nocturnally active birds and bats. A guidance document. *Journal of Wildlife Management* 71: 2449–2486.

Kunz, T.H., Braun de Torrez, E., Bauer, D., Lobova, T. & Fleming, T.H. (2011) Ecosystem services provided by bats. *Annals of the New York Academy of Sciences* 1223: 1–38.

Lagerveld, S., Poerink, B.J., Haselager, R. & Verdaat, H. (2014) Bats in Dutch offshore wind farms in autumn 2012. *Lutra* 57: 61–69.

Liechti, F., Bruderer, B. & Paproth, H. (1995) Quantification of nocturnal bird migration by moonwatching: comparison with radar and infrared observations. *Journal of Field Ornithology* 66: 457–468.

Lintott, P.R., Richardson, S.M., Hosken, D.J., Fensome, S.A. & Mathews, F. (2016) Ecological impact assessments fail to reduce risk of bat casualties at wind farms. *Current Biology* 26: R1135-R1136.

Lisovski, S., Hewson, C.M., Klaassen, R.H.G., Korner-Nievergelt, F., Kristensen, M.W. & Hahn, S. (2012) Geolocation by light. Accuracy and precision affected by environmental factors. *Methods in Ecology and Evolution* 3: 603–612.

Loring, P.H. (2014) Evaluating digital VHF technology to monitor shorebird and seabird use of offshore wind energy areas in the Western North Atlantic. PhD thesis, University of Massachusetts Amherst, MA.

Loring, P.H., Sievert, P.R., Griffin, C.R., Janaswamy, R. & Johnston, S. (2015) Tracking offshore movements of common terns across the southern New England continental shelf using Nanotags and automated radio telemetry stations. Presentation at British Ornithologists' Union conference, Leicester, UK, 31 March–2 April 2015. Retrieved 26 March 2019 from https://www.youtube.com/watch?v=xr_0pXEXx3c

Lotek Wireless Inc. (2018) Bird & Bat NanoTag. Retrieved 26 March 2019 from http://www.lotek.com/avian-nanotags.htm

Lowery, G.H., Jr (1949) A quantitative study of the nocturnal migration of birds. PhD thesis, University of Kansas.

Lundin, P., Samuelsson, P., Svanberg, S., Runemark, A., Akesson, S. & Brydegaard, M. (2011) Remote nocturnal bird classification by spectroscopy in extended wavelength ranges. *Applied Optics* 50: 3396–3411.

Masden, E.A. & Cook, A.S.C.P. (2016) Avian collision risk models for wind energy impact assessments. *Environmental Impact Assessment Review* 56: 43–49.

Masden, E.A., Haydon, D.T., Fox, A.D. & Furness, R.W. (2010) Barriers to movement: modelling energetic costs of avoiding marine wind farms amongst breeding seabirds. *Marine Pollution Bulletin* 60: 1085–1091.

Matzner, S., Cullinan, V.I. & Duberstein, C.A. (2015) Two-dimensional thermal video analysis of offshore bird and bat flight. *Ecological Informatics* 30: 20–28.

May, R., Hamre, Ø., Vang, R. & Nygård, T. (2012) Evaluation of the DTBird video-system at the Smøla wind-power plant. Detection capabilities for capturing near-turbine avian behavior. NINA Report 910. Trondheim: Norwegian Institute for Nature Research. Retrieved 26 March 2019 from https://www.researchgate.net/profile/Roel_May/publication/294823923_Evaluation_of_the_DTBird_video-system_at_the_Smola_wind-power_plant_Detection_capabilities_for_capturing_near-turbine_avian_behaviour/links/56c4336208ae8a6fab5b2394/Evaluation-of-the-DTBird-video-system-at-the-Smola-wind-power-plant-Detection-capabilities-for-capturing-near-turbine-avian-behaviour.pdf

McKinnon, E.A. & Love, O.P. (2018) Ten years tracking the migrations of small landbirds. Lessons learned in the golden age of bio-logging. *Auk* 135: 834–856.

Mennhill, D.J. (2011) Individual distinctiveness in avian vocalizations and the spatial monitoring of behaviour. *Ibis* 153: 235–238.

Mirkovic, D., Stepanian, P.M., Kelly, J.F. & Chilson, P.B. (2016) Electromagnetic model reliably predicts radar scattering characteristics of airborne organisms. *Scientific Reports* 6: 35637.

Mitchell, G.W., Woodworth, B.K., Taylor, P.D. & Norris, D.R. (2015) Automated telemetry reveals age specific differences in flight duration and speed are driven by wind conditions in a migratory songbird. *Movement Ecology* 3(1): 19.

Nilsson, C., Dokter, A.M., Verlinden, L., Shamoun-Baranes, J., Schmid, B., Desmet, P., Bauer, S., Chapman, J., Alves, J.A., Stepanian, P.M., Sapir, N., Wainwright, C., Boos, M., Górska, A., Menz, M.H.M., Rodrigues, P., Leijnse, H., Zehtindjiev, P., Brabant, R., Haase, G., Weisshaupt, N., Ciach, M. & Liechti, F. (2018) Revealing patterns of nocturnal migration using the European weather radar network. *Ecography* 47, https://doi.org/10.1111/ecog.04003.

Normandeau Associates Inc. (2018) Acoustic Thermographic Offshore Monitoring (ATOM). Retrieved 26 March 2019 from https://www.normandeau.com/environmental-specialists-consultant-atom-technology/

Orr, T.L., Herz, S.M. & Oakley, D.L. (2013) Evaluation of lighting schemes for offshore wind facilities and impacts to local environments. Submitted to US Department of the Interior, Bureau of Ocean Energy Management, Office of Renewable Energy Programs, Herndon, VA. Retrieved 26 March 2019 from https://www.boem.gov/ESPIS/5/5298.pdf

Parsons, S. & Szewczak, J. (2009) Detecting, recording and analysing the vocalisations of bats. In Kunz, T.H. & Parsons, S. (eds) *Ecological and Behavioral Methods for the Study of Bats*. Baltimore, MD: Johns Hopkins University Press. pp. 91–111.

Pelletier, S.K., Omland, K.S., Watrous, K.S. & Peterson, T.S. (2013) Information synthesis on the potential for bat interactions with offshore wind facilities. Final Report. Herndon, VA: US Department of the Interior, Bureau of Ocean Energy Management. Retrieved 26 March 2019 from https://www.boem.gov/ESPIS/5/5289.pdf

Perrow, M.R. (2017) A synthesis of effects and impacts. In Perrow, M.R. (ed.) *Wildlife and Wind Farms, Conflicts and Solutions. Volume 1. Onshore: Potential effects*. Exeter: Pelagic Publishing. pp. 241–276.

Perrow, M.R., Skeate, E.R., Lines, P., Brown, D. & Tomlinson, M.L. (2006) Radio telemetry as a tool for impact assessment of wind farms. The case of little terns *Sterna albifrons* at Scroby Sands, Norfolk, UK. *Ibis* 148: 57–75.

Petersen, I.K., Christensen, T.K., Kahlert, J.A., Desholm, M. & Fox, A.D. (2006) Final results of bird studies at the offshore wind farms at Nystedand Horns Rev, Denmark. Report submitted to DONG Energy and Vattenfall A/S. Aarhus: National Environmental Research Institute. Retrieved 26 March 2019 from https://dtbird.com/images/Downloads/DTBird_System_Specifications._November_2017.pdfPeterson, T.S. (2016) Bats in the rotor zone – managing risk with acoustics. Presentation at the National Wind Coordination Collaborative (NWCC) Wildlife Wind and Wildlife Research Meeting XI, Broomfield, CO, 2 December 2016. Retrieved 26 March 2019 from www.nationalwind.org/wp-content/uploads/2018/02/WWRM-XI-Proceedings-May-2017.pdf

Peterson, T., Pelletier, S. & Giovanni, M. (2016) Long-term bat monitoring on islands, offshore structures, and coastal sites in the Gulf of Maine, mid-Atlantic, and Great Lakes – Final Report. Retrieved 26 March 2019 from https://www.osti.gov/servlets/purl/1238337

Plonczkier, P. & Simms, I.C. (2012) Radar monitoring of migrating pink-footed geese. Behavioural responses to offshore wind farm development. *Journal of Applied Ecology* 49: 1187–1194.

Puente, M., Diaz, J. & Riopérez, A. (2017). DTBird® System Specifications for Wind Turbines. Day & Night, On & Offshore. DTBird Bird & Bat Protection brochure. Retrieved 26 March 2019 from https://dtbird.com/images/Downloads/DTBird_System_Specifications._November_2017.pdf

Rodrigues, L., Bach, L., Dubourg-Savage, M.-J., Karapandža, B., Kovač, D., Kervyn, T., Dekker, J., Kepel, A., Bach, P., Collins, J., Harbusch, C., Park, K., Micevski, B. & Mindermann, J. (2015) Guidelines for consideration of bats in wind farm projects, Revision 2014. EUROBATS Publication Series No. 6. Bonn: UNEP/EUROBATS Secretariat. Retrieved 26 March 2019 from http://www.eurobats.org/sites/default/files/documents/publications/publication_series/pubseries_no6_english.pdf

Ronconi, R.A., Allard, K.A. & Taylor, P.D. (2015) Bird interactions with offshore oil and gas platforms: review of impacts and monitoring techniques. *Journal of Environmental Management* 147: 34–45.

Ross, J.C. & Allen, P.E. (2014) Random Forest for improved analysis efficiency in passive acoustic monitoring. *Ecological Informatics* 21: 34–39.

Russ, J., Barlow, K.E., Briggs, P.A. & Sowler, S. (2012) *British Bat Calls: A guide to species identification*. Exeter: Pelagic Publishing.

Rydell, J., Bach, L., Bach, P., Diaz, L.G., Furmankiewicz, J., Hagner-Wahlsten, N., Kyheröinen, E.-M., Lilley, T., Masing, M., Meyer, M.M., Pētersons, G., Šuba, J., Vasko, V., Vintulis, V. & Hedenström, A. (2014) Phenology of migratory bat activity across the Baltic Sea and the south-eastern North Sea. *Acta Chiropterologica* 16: 139–147.

Schulz, A., Kube, J., Kellner, T., Schleicher, K. & Sordyl, H. (2009) Entwicklung und Einführung eines automatischen Erfassungssystems für die Ermittlung des Vogelschlages unter Praxisbedingungen auf FINO2. Final Report. Neu Broderstorf: Forschungsvorhaben des Bundesministeriums für Umwelt, Naturschutz und Reaktorsicherheit. Retrieved 26 March 2019 from https://doi.org/10.2314/GBV:636092496

Schulz, A., Kulemeyer, C., Röhrbein, V. & Coppack, T. (2011) Die Forschungsplattform FINO 2 – eine automatisierte Vogelwarte inmitten der Ostsee. *Seevögel* 32: 99–101.

Schulz, A., Dittmann, T. & Coppack, T. (2014) Erfassung von Ausweichbewegungen von Zugvögeln mittels Pencil Beam Radar und Erfassung von Vogelkollisionen mit Hilfe des Systems VARS. Ökologische Begleitforschung am Offshore-Testfeldvorhaben alpha ventus zur Evaluierung des Standarduntersuchungskonzeptes des BSH (StUKplus). Final Report submitted to Bundesamt für Seeschifffahrt und Hydrographie (BSH), Hamburg. Neu Broderstorf: Institut für Angewandte Ökosystemforschung. Retrieved 26 March 2019 from https://doi.org/10.2314/GBV:845722808

Schuster, E., Bulling, L. & Köppel, J. (2015) Consolidating the state of knowledge: a synoptical review of wind energy's wildlife effects. *Environmental Management* 56: 300–331.

Shamoun-Baranes, J., Alves, J.A., Bauer, S., Dokter, A.M., Hüppop, O., Koistinen, J., Leijnse, H., Liechti, F., van Gasteren, H. & Chapman, J.W. (2014) Continental-scale radar monitoring of the aerial movements of animals. *Movement Ecology* 2(1): article 9.

Sjöberg, S., Alerstam, T., Åkesson, S., Schulz, A., Weidauer, A., Coppack, T. & Muheim, R. (2015) Weather and fuel reserves determine departure and flight decisions in passerines migrating across the Baltic Sea. *Animal Behaviour* 104: 59–68.

Skov, H. & Jensen, N.E. (2015) Multi-sensor monitoring system for determination of bird behaviour in marine wind farms. In Proceedings of the Rave R&D Offshore Wind Conference, Fraunhofer Institute, Bremerhaven, 2015. Retrieved 26 March 2019 from https://www.google.com/url?sa=t&rct=j&q=&esrc=s&source=web&cd=1&ved=2ahUKEwjZiILM3KnfAhU4UxUIHXJWALwQFjAAegQICxAC&url=http%3A%2F%2Fwww.rave-offshore.de%2Ffiles%2Fdownloads%2Fkonferenz%2Fkonferenz-2015%2Fposter-session%2F16_Skov.pdf&usg=AOvVaw3SJ-clyqZIGt5YD34Qclgk

Skov, H., Heinänen, S., Norman, T., Ward, R., Méndez-Roldán, S. & Ellis, I. (2018) ORJIP Bird Collision and Avoidance Study. Final Report. London: The Carbon Trust. Retrieved 26 March 2019 from https://www.carbontrust.com/media/675793/orjip-bird-collision-avoidance-study_april-2018.pdf

Smallwood, K.S. (2017) Monitoring birds. In Perrow, M.R. (ed.) *Wildlife and Wind Farms, Conflicts and Solutions. Volume 2. Onshore: Monitoring and mitigation*. Exeter: Pelagic Publishing. pp. 1–30.

Smolinsky, J.A., Diehl, R.H., Radzio, T.A., Delaney, D.K. & Moore, F.R. (2013) Factors influencing the movement biology of migrant songbirds confronted with an ecological barrier. *Behavioural Ecology and Sociobiology* 67: 2041–2051.

Stepanian, P.M., Chilson, P.B., Kelly, J.F. & Tatem, A. (2014) An introduction to radar image processing in ecology. *Methods in Ecology and Evolution* 5: 730–738.

Stepanian, P.M., Horton, K.G., Hille, D.C., Wainwright, C.E., Chilson, P.B. & Kelly, J.F. (2016a) Extending bioacoustic monitoring of birds aloft through flight call localization with a three-dimensional microphone array. *Ecology and Evolution* 6: 7039–7046.

Stepanian, P.M., Horton, K.G., Melnikov, V.M., Zrni, D.S. & Gauthreaux, S.A. (2016b) Dual-polarization radar products for biological applications. *Ecosphere* 7(11): e01539.

Stutchbury, B.J.M., Tarof, S.A., Done, T., Gow, E., Kramer, P.M., Tautin, J., Fox, J.W. & Afanasyev, V. (2009) Tracking long-distance songbird migration by using geolocators. *Science* 323(5916): 896.

Suryan, R.M., Albertani, R. & Polagye, B. (2016) A synchronized sensor array for remote monitoring of avian and bat interactions with offshore renewable energy facilities. Oregon State University and University of Washington. Retrieved 26 March 2019 from https://tethys.pnnl.gov/sites/default/files/publications/Suryan-et-al-2016.pdf

Taylor, P.D., Mackenzie, S.A., Thurber, B.G., Calvert, A.M., Mills, A.M., McGuire, L.P. & Guglielmo, C.G. (2011) Landscape movements of migratory birds and bats reveal an expanded scale of stopover. *PLoS ONE* 6(11): e27054. doi: 10.1371/journal.pone.0027054.

Taylor, P.D., Crewe, T.L., Mackenzie, S.A., Lepage, D., Aubry, Y., Crysler, Z., Finney, G., Francis, C.M., Guglielmo, C.G., Hamilton, D.J., Holberton, R.L., Loring, P.H., Mitchell, G.W., Norris, D.R., Paquet, J., Ronconi, R.A., Smetzer, J.R., Smith, P.A., Welch, L.J. & Woodworth, B.K. (2017) The Motus Wildlife Tracking System. A collaborative research network to enhance the understanding of wildlife movement. *Avian Conservation and Ecology* 12(1): article 8.

Thaxter, C.B., Ross-Smith, V.H., Clark, J.A., Clark, N.A., Conway, G.J., Masden, E.A., Wade, H.M., Leat, E.H.K., Gear, S.C., Marsh, M., Booth, C., Furness, R.W., Votier, S.C., Burton, N.H.K. & Daunt, F. (2016) Contrasting effects of GPS device and harness attachment on adult survival of lesser black-backed gulls *Larus fuscus* and great skuas *Stercorarius skua*. *Ibis* 158: 279–290.

Urmy, S.S., Warren, J.D. & Parrini, F. (2017) Quantitative ornithology with a commercial marine radar. Standard-target calibration, target detection and tracking, and measurement of echoes from individuals and flocks. *Methods in Ecology and Evolution* 8: 860–869.

Vandenabeele, S.P., Grundy, E., Friswell, M.I., Grogan, A., Votier, S.C. & Wilson, R.P. (2014) Excess baggage for birds. Inappropriate placement of tags on gannets changes flight patterns. *PLoS ONE* 9(3): e92657. doi: 10.1371/journal.pone.0092657.

Vickery, J.A., Ewing, S.R., Smith, K.W., Pain, D.J., Bairlein, F., Škorpilová, J., Gregory, R.D. & Fox, T. (2014) The decline of Afro-Palaearctic migrants and an assessment of potential causes. *Ibis* 156: 1–22.

Walter, G., Matthes, H. & Joost, M. (2007) Fledermauszug über Nord- und Ostsee – Ergebnisse aus Offshore-Untersuchungen und deren Einordnung in das bisher bekannte Bild zum Zuggeschehen. *Nyctalus (Neue Folge)* 12: 221–233.

Welcker, J., Liesenjohann, M., Blew, J., Nehls, G. & Grünkorn, T. (2017) Nocturnal migrants do not incur higher collision risk at wind turbines than diurnally active species. *Ibis* 159: 366–373.

Whelan, C.J., Şekercioğlu, Ç.H. & Wenny, D.G. (2015) Why birds matter. From economic ornithology to ecosystem services. *Journal of Ornithology* 156(S1): 227–238.

Wiggelinkhuizen, E.J., Rademakers, L.W.M.M., Barhorst, S.A.M., den Boon, H.J., Dirksen, S. & Schekkerman, H. (2006) WT-Bird. Collision recording for offshore wind farms. Research Report No. 06-60. Petten: Energy Research Centre of The Netherlands. Retrieved 26 March 2019 from https://publicaties.ecn.nl/PdfFetch.aspx?nr=ECN-RX--06-060

Wiggelinkhuizen, E.J. & den Boon, H.J. (2010) Monitoring of bird collisions in wind farm under offshore-like conditions using WT-BIRD system Final report. Research Report No. 09-033. Petten: Energy Research Centre of the Netherlands. Retrieved 26 March 2019 from https://publicaties.ecn.nl/PdfFetch.aspx?nr=ECN-E--09-033

Wikelski, M., Kays, R.W., Kasdin, N.J., Thorup, K., Smith, J.A. & Swenson, G.W., Jr (2007) Going wild: what a global small-animal tracking system could do for experimental biologists. *Journal of Experimental Biology* 210: 181–186.

Williams, K.A., Stenhouse, I.J., Connelly, E.E. & Johnson, S.M. (2015) Mid-Atlantic wildlife studies: distribution and abundance of wildlife along the eastern seaboard 2012–2014. Portland, ME: Biodiversity Research Institute. Retrieved 26 March 2019 from http://roa.midatlanticocean.org/wp-content/uploads/2016/01/williams-et-al-2015.pdf

Willmott, J.R., Forcey, G.M. & Hooton, L.A. (2015) Developing an automated risk management tool to minimize bird and bat mortality at wind facilities. *Ambio* 44(Suppl 4): 557–571.

Wimmer, J., Towsey, M., Roe, P. & Williamson, I. (2013) Sampling environmental acoustic recordings to determine bird species richness. *Ecological Applications* 23: 1419–1428.

Winkelmann, J.E. (1992) The impact of the Sep wind park near Oosterbierum (Fr.), the Netherlands, on birds, 1: Collision victims. Research Report 92/2. Arnhem: DOL-Instituut voor Bos-en Natuuronderzoek, Rijksinstituut voor Naturbeheer. Retrieved 26 March 2019 from http://edepot.wur.nl/385853

Zaugg, S., Saporta, G., van Loon, E., Schmaljohann, H. & Liechti, F. (2008) Automatic identification of bird targets with radar via patterns produced by wing flapping. *Journal of the Royal Society, Interface* 5: 1041–1053.

CHAPTER 7

Mitigating the effects of noise

FRANK THOMSEN and TOBIAS VERFUSS

Summary

The construction, operation and decommissioning of offshore wind farms involve activities that generate underwater sound. Most concern has been raised about the impacts of noise during construction using impact pile-driving. Pile-driving noise can lead to displacement of marine mammals and some marine fish species. Physiological effects such as temporary hearing loss are likely if the acoustic dose at the receiver exceeds certain thresholds. In this chapter, the methods to mitigate sound during offshore wind-farm construction are divided into three categories: (1) source mitigation, comprising methods that reduce the sound directly at the source; (2) channel mitigation, comprising methods that reduce the emitted underwater sound in the water column; and (3) receiver mitigation, comprising methods that prevent the receiver from being close to the sound source. Both source- and channel-mitigation methods are confounded by the need to achieve a certain amount of hammer energy to penetrate the seabed. Source mitigation can involve adjustment of piling energy, pulse prolongation and use of alternative installation technologies such as vibrodriving, 'BLUE Piling' or suction buckets. Channel mitigation involves bubble curtains, casings and resonators including hydro-sound dampers. Bubble curtains and casings have proven to be very effective during piling. However, as they dampen higher frequencies more effectively than lower frequencies, this makes these methods more effective in reducing impacts on marine mammals than on fish. Receiver mitigation involves acoustic management devices aimed at deterring animals out of the danger zone, as well as safety zones, soft-start and temporary piling restrictions. Several European countries that are developing offshore wind prescribe the application of one or several mitigation methods. It is suggested that the selection of the most appropriate mitigation measures in a particular circumstance should be based on a risk-based approach.

Introduction

The construction, operation and decommissioning of offshore wind farms (OWFs) lead to the emission of underwater sound (Box 7.1). Adverse impacts on marine life triggered by unwanted sound, typically termed 'noise', have become a focus issue in marine environmental research and policy in the European Union (EU) and elsewhere (OSPAR 2009; Tasker *et al.* 2010; NMFS 2016; Nehls *et al.* 2019). This is because water is an excellent medium for sound transmission. Many forms of marine life such as fish and marine mammals, and perhaps even some invertebrates, rely on sound as their primary mode of communication, and for orientation and navigation (reviewed by Tyack & Clark 2000; Ladich *et al.* 2006; Edmonds *et al.* 2016; Hawkins & Popper 2016). Human-generated underwater sound can affect marine life by leading to a variety of behavioural responses, masking important sounds and even causing hearing loss in the form of temporary or permanent hearing threshold shifts (TTS and PTS, respectively). At the very high levels received, sound can injure or kill marine life (see reviews by Richardson *et al.* 1995; Popper *et al.* 2004; Southall *et al.* 2007; McGregor *et al.* 2013). Offshore wind turbines are typically installed on substructures that are connected to the seabed. The most common substructure design is the monopile, a large-diameter pile that is driven tens of metres into the seabed. Other design options include jackets, tripods, and gravity-based and floating foundations (Jameson *et al.* 2019). The installation of monopiles, jackets and tripods involves in most cases impact pile-driving. This method is associated with significant noise emissions in the form of shock waves that are induced by the hammer and propagate away from the pile. Thus, the construction phase is most critical when it comes to environmental impact and management (Madsen *et al.* 2006; Thomsen *et al.* 2006; 2015; Bailey *et al.* 2014). There is enough evidence to conclude that underwater sound from impact pile-driving can lead to large-scale displacement of cetaceans such as Harbour Porpoises *Phocoena phocoena*, pinnipeds such as Harbour Seals *Phoca vitulina* and several marine fish species (Brandt *et al.* 2011; Thomsen *et al.* 2012; Dähne *et al.* 2013; Hawkins *et al.* 2014; Russell *et al.* 2016). Several studies have shown that the displacement effects are, however, mostly temporary. For porpoises and seals, there is a risk of physiological response in the form of TTS or even PTS closer to the sound source and/or at relatively high acoustic doses (Hastie *et al.* 2015; Kastelein *et al.* 2016). Physiological effects of impact pile-driving on fish are also likely, and mortality very close to the source cannot be ruled out (Thomsen *et al.* 2006; Popper & Hastings 2009). Indeed, there are anecdotal records of large aggregations of birds seemingly scavenging injured or disorientated fish during or immediately after piling events (Perrow 2019).

Government agencies regulate marine industries as part of the planning process for new developments or activities within the seascape. In Europe, most regulation derives from the national implementation of EU-wide directives such as the Environmental Impact Assessment (EIA) Directive (1985, updated 2011) and the Habitats Directive (1992). In the USA, the Marine Mammal Protection Act (MMPA) plays an important role. In any of these jurisdictions, the Environmental Impact Statement (EIS) or Environmental Statement (ES) generally involves a description of the local environment and often includes an inventory of species and other components that could be affected, in what are generally termed 'baseline' conditions. The EIS also comprises a description of the maximum extent of the wind-farm development and a detailed assessment of the possible effects of the development on the local environment. This often includes proposed mitigation to reduce those effects. Regulators then examine the EIS and decide whether the residual effects are acceptable; if so, a permit (consent) can then be issued. Any granted permit may come

Box 7.1 A brief background to underwater sound

Sound is an alteration in pressure that propagates as a wave through an elastic medium. It involves local compression and expansion of the medium. Sound pressure is expressed in pascals (Pa) and the sound pressure level (SPL) in decibels (dB). The decibel is a relative measure of sound pressure that is of logarithmic nature. To derive absolute levels, the measured pressure needs to be related to a reference pressure value. In water, the reference value is 1 μPa, whereas in air it is 20 μPa. To compare decibel levels in air to decibel levels under water, 61.5 dB must be added to the in-air values. An addition of 25.5 dB is due to the differences in the reference pressure values. Another 36 dB must be added owing to the higher acoustic impedance of water compared to air. Thus, an SPL of 100 dB re 20 μPa in air is comparable to an SPL of 161.5 dB re 1 μPa under water.

Different metrics exist to describe continuous and impulsive sounds. In the underwater sound literature, the terms and definitions of these metrics are not used consistently. In this chapter, the nomenclature of ISO 18405 is followed (ISO 2017).

The SPL (given in dB re 1 μPa) is the main metric to describe continuous sounds such as the noise emissions that are associated with drilling or vibrodriving operations. It is derived from the root-mean-square (RMS) value of the sound pressure averaged over a stated time interval, compared to the reference value of 1 μPa.

Impulsive sounds, such as pile-driving strikes, are commonly characterised by the sound exposure level (SEL) (given in dB re 1 μPa^2 s), and their peak SPL (given in dB Re 1 μPa). The SEL is a measure of the total sound energy of an impulse, normalised to 1 second (1 s). The SEL is often used to describe the sound energy received by an organism. The SEL(cum) describes the cumulative sound exposure level which takes into account both received level and duration of exposure in relation to individual activities/sources (NMFS 2018). The peak SPL for impulsive sounds is given as zero-to-peak sound pressure level ($L_{p,0\text{–}pk}$), derived by the maximum sound pressure of the impulse.

Although SEL and SPL vary in the underlying definitions, they have the same value for the integration time interval of 1 s. For pulses shorter than 1 s down to 100 ms, which typically emerge from impact pile-driving, the SPL value is always higher than the corresponding SEL. Caution should be taken when comparing sounds that are described with different metrics.

Travelling through the water column from the sound source, sound becomes weaker and distorted with distance. The transmission loss is described by:

$$\mathrm{RL} = \mathrm{SL} - \mathrm{TL}$$

where RL is the received level, SL is the source level and TL is transmission loss.

The source level is expressed as dB re 1 μPa m. It can be understood as dB re 1 μPa referred to 1 m as a measure of the acoustic output of a source. It is important to note that the source level is back-calculated to 1 metre using measurements from distances farther away from the source.

Once the source level and sound propagation are known, the basis of any sound modelling is all about identifying the transmission loss to arrive at the received level.

In a simplified approach, transmission loss can be described by geometric spreading. Geometric spreading can involve spherical spreading, which happens near the source or in deep water. Here, the sound can propagate uniformly in all

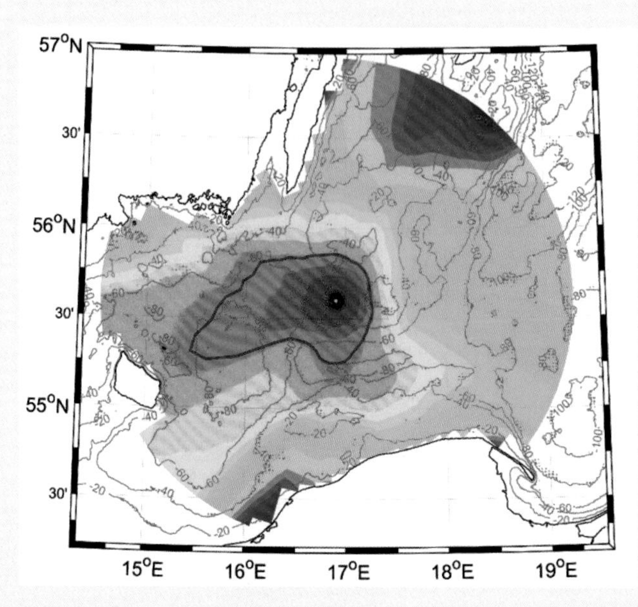

Figure 7.1 A sound map showing the sound propagation from a hypothetical impact pile-driving event in the Baltic Sea. The solid blue, black and green lines indicate impact ranges for behavioural response, temporary threshold shift and permanent threshold shift, respectively, for Harbour Seals when using a bubble curtain as mitigation. Such maps are usually used in sound impact assessments in conjunction with data on the distribution and abundance of marine animals. (Uwe Stöber, DHI)

directions. The TL is given by 20 $\log_{10}$ (R/1m). This means that for each doubling of distance, the sound becomes 6 dB weaker. The other form of geometric spreading is cylindrical spreading, which happens in shallow water. Here, the sound can propagate only as a cylindrical wave bound by the sea floor and sea surface, with TL given by 10 $\log_{10}$ (R/1m). This means that for each doubling of distance, the sound becomes 3 dB weaker. TL is often found to lie between 20 and 10 log (R) and thus 15 log (R) seems to describe TL quite well in some cases.

More advanced studies use numerical sound-propagation models in the determination of TL. All of the models available solve the wave equation under appropriate boundary conditions. Ray-based models are widely used and are particularly useful at high frequencies. Wave-number integration and normal modes are good at lower frequencies and solve range-independent problems. Alternatively, parabolic equation models are popular for range-dependent problems and are good for low-frequency modelling. Parabolic equations are suitable for modelling transmission loss from pile driving.

Noise modelling results are often shown in sound maps (Figure 7.1), which are used in conjunction with data on the distribution and abundance of fish or marine mammals to provide an estimate of the number of individuals affected. The impact is then extrapolated using empirical data on hearing sensitivities and demonstrated effects.

For more background and detailed information on the standards for underwater noise measurements, see JASCO (2009), TNO (2011), Robinson *et al.* (2014) and Verfuß *et al.* (2014).

with the obligation to carry out mitigation measures (see early reviews in relation to noise by Thomsen *et al.* 2006; Nehls *et al.* 2007). Thus, new sound-mitigation technologies have been developed, partly in response to the establishment of noise thresholds and other regulatory measures in some parts of the EU (BSH 2013a).

Scope

This chapter describes the current state of knowledge in technologies and approaches to mitigate underwater noise that arises during the installation of OWFs. It builds on earlier reviews (Caltrans 2009; OSPAR 2016; Verfuß 2014) and covers a wide range of peer-reviewed papers and grey literature in the form of reports and other publicly available information from press releases and websites. Experiences are mainly drawn from the North Sea region in north-western Europe, as this has been the focal area for offshore wind to date. However, theoretical considerations that may be relevant for sites being developed in other markets across the globe, such as in the USA, Taiwan or Japan, are also provided. To structure the review, mitigation measures were divided into three main categories: (1) *Source mitigation*, whereby sound is reduced directly at source, (2) *Channel mitigation*, which refers to any method that reduces the sound emitted into the water column; that is, in the channel between the sound source and the receiver (marine life), and (3) *Receiver mitigation*, categorised as any other methods that prevent the receiver from being close to the sound source, including acoustic management devices aimed at deterring animals from the danger zone, establishment of safety zones, soft-start for piling operations and temporal restrictions. The chapter then reviews relevant low-noise foundation installation technologies as an alternative to impact pile-driving, before closing with a review of other foundation types such as gravity-based foundations and floating designs.

Although what is considered to be best practice of sound mitigation is described, the decision about the exact methodology (source, channel and/or receiver mitigation) used to manage the risk of wind-farm related sounds lies with the specific wind-farm developer, installation contractor and regulator. Regulators will eventually base their approval decisions on site-specific EIAs. Any noise-mitigation measure has its cost implications, and the costs can vary hugely based on the chosen mitigation strategy and site specificities (see examples by Nehls *et al.* 2007; Diederichs *et al.* 2014; Schorcht 2014). In the event that the deployment of noise-reduction technologies (NRTs) affects the construction schedule, this may add significantly to the installation costs. A detailed cost discussion is, however, beyond the scope of this chapter.

For acoustic terminology, ISO 18405 *Underwater acoustics – terminology* (ISO 2017) is followed throughout (see also Box 7.1).

Themes

Source mitigation

Methods involving source mitigation are primary measures aimed at modifying the piling process, the piling equipment or the pile properties.

Adjustment of piling energy

The most obvious measure to reduce wind-farm construction noise is applying less energy to the hydraulic hammers that are used for impact pile-driving. Measurements during wind-farm construction have shown that cutting the piling energy by half leads to a noise reduction of 2–3 dB (Bellmann 2014; Gündert *et al.* 2014). The soft-start procedure is important in this context (see *Receiver mitigation*, below). However, there are limitations in the use of a reduction of piling energy, as, for example, a certain level is required for effective pile penetration into the seabed. The minimum energy required varies with different soil properties. Moreover, at lower energy levels, the duration of a piling event will be extended, and a higher number of strikes will be needed to install the pile (e.g. Koschinski & Lüdemann 2013). It needs to be considered that there is a common understanding that physiological effects are related to the dose of exposure, which includes the duration of impact and the number of strikes for impulsive sounds (Southall *et al.* 2007; NMFS 2016). Accordingly, the received acoustic dose may not be much different between low-energy, high-frequency piling and high-energy, low-frequency piling.

Pulse prolongation

The longer the contact time between hammer and pile, the lower the developing noise, particularly the peak sound pressure levels ($L_{p,0-pk}$). Thus, any measure prolonging the piling impulse and thereby altering the shape of the produced sound wave is desirable. In conventional impact pile-driving, a dynamic layer or 'piling cushion' between the hydraulic hammer and the pile can be inserted. As Elmer *et al.* (2007a) have determined by model calculations, a doubled impulse length at constant piling energy reduces the $L_{p,0-pk}$ by about 9 dB. Measurements during the installation of a 3.2 m diameter monopile for the FINO 2 research platform, using a steel rope as a piling cushion, verified this theoretically derived value (Elmer *et al.* 2007b). Impulse prolongation can, however, negatively affect pile drivability, especially when combined with higher seabed resistances. Hence, offshore engineering companies have not yet employed this method. Practical experience is still limited to some research and demonstration projects (e.g. Rustemeier *et al.* 2012).

BLUE Piling technology

BLUE Piling is a relatively new approach for driving large offshore piles, which was invented by Fistuca BV, a spin-off from Eindhoven University of Technology, and is now being further developed by IHC IQIP. Instead of a hydraulically operated steel anvil, the BLUE Hammer utilises the acceleration and weight of a massive water column, which is contained in a steel housing. The first generation BLUE Hammer consists of an expandable combustion chamber that is placed underneath the water column. By combusting a gas mixture, the gas volume expands and the water column is accelerated upwards while the downward force drives the pile into the seabed. By falling back into its original position, the water column creates a second blow on the pile. After the release of the exhaust gases, the cycle is repeated (Figure 7.2). Tests with the BLUE 25M Hammer, a full-scale prototype, demonstrated that the blow of a BLUE Hammer differs substantially from that of a hydraulic hammer, as it delivers a blow with a comparatively long duration of typically more than 100 ms. This is much longer than the contact times of hydraulic pile drivers, which range around 4 ms. The gradually increasing and decreasing impact force of BLUE Hammers significantly reduces the acceleration levels inside the monopile, which, in turn, reduces

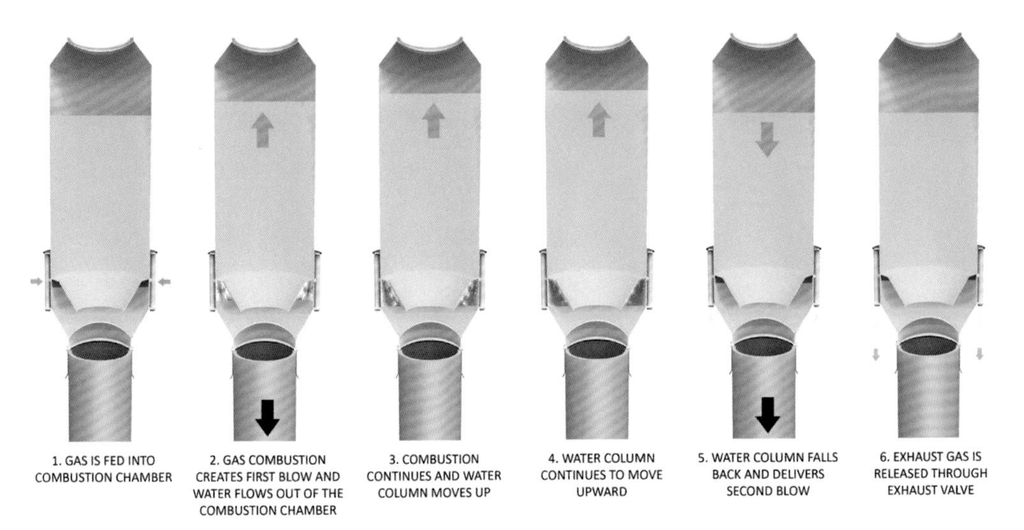

Figure 7.2 Working principle (piling sequence) of the BLUE 25M, a BLUE Hammer prototype with internal combustion chamber. (Jasper Winkes, Fistuca BV)

the vibrations and thereby the level of underwater noise. The Netherlands Organisation for Applied Scientific Research (TNO) conducted measurements in a confined harbour area during pile driving with a BLUE Piling machine on a test pile at refusal (Jansen & de Jong 2016, unpublished). The measured sound exposure levels (SELs), under selected piling conditions at about 400 m distance in a water depth of 24 m, ranged between 140 and 162 dB re 1 μPa^2 s ($L_{p,0\text{-}pk}$: 160–186 dB re 1 μPa^2). The corresponding energies transmitted to the pile in single strikes ranged from 4,000 to 7,500 kJ. Most of the acoustic energy was concentrated in the mid-frequency range from 250 Hz to 1 kHz. The recorded noise levels are lower than typical SELs of piling sequences for conventional hydraulic hammers operated with energies up to 1,000 kJ (165–175 dB re 1 μPa^2 s at 750 m range and valid for water depths of 25–30 m). The TNO study thus suggests that BLUE Piling is more silent than conventional impact pile-driving (Jansen & de Jong 2016, unpublished). Modelling indicates that the technology has the potential to reduce the SEL by more than 20 dB compared to hydraulic impact pile-driving and to keep the SEL below the noise threshold criterion which has been established in Germany (Trimoreau & Smidt-Luetzen 2017, unpublished; see *Concluding remarks*, below). A first full-scale prototype of the BLUE Hammer, the BLUE 25M, was designed to drive state-of-the-art monopiles with diameters up to 10 m. After an inshore test at Rotterdam Maasvlakte, a full-scale offshore test in the Dutch part of the North Sea was executed in mid-2018 (Figure 7.3). The tests, supported by the Carbon Trust's Offshore Wind Accelerator (OWA) initiative (Carbon Trust 2018a; 2018b), revealed that further adjustments need to be made before BLUE Piling technology can be commercially used. It can be expected that the design and working principle of the next generation BLUE Hammer will differ from the prototype tested in 2018.

Channel mitigation

Channel mitigation, whereby the sound emitted into the water column is reduced, is achieved by applying noise reduction technologies (NRTs). Any NRT involves placing a sound-dampening obstacle between the pile and the water body. The obstacle can either be air or any other sound-absorbing, reflecting or shielding material. The development of NRTs was mainly triggered by the implementation of threshold criteria for underwater

Figure 7.3 BLUE 25M Hammer hooked up to an installation vessel 'Svanen', driving a 6.5 m diameter monopile during a full-scale test campaign off the coast of the Netherlands in August 2018. (Jasper Winkes, Fistuca BV)

noise in Germany (BMU 2013; BSH 2013b). A variety of methods and technical solutions have since been developed, tested and qualified for the offshore wind sector. An extensive overview on NRTs is provided in OSPAR (2016). The most relevant technologies to date are bubble curtains, pile casings and modular systems consisting of resonators such as hydro-sound dampers (HSDs). All of these have proven their effectiveness during hundreds of pile installations (Table 7.1) and are commercially available. Detailed information for the OWF projects that are discussed in this chapter are provided in Table 7.2. Based on existing experience, NRT can be regarded as a mature technology for water depths of up to 30–40 m. However, deployment in deeper waters remains a challenge.

Bubble curtains

At the time of writing in 2018, bubble curtains had been the most frequently used NRT for the installation of offshore wind turbine foundations. Bubble curtains are formed by pumping compressed air through perforated hoses with nozzles. For the predominant design, the big bubble curtain (BBC), the hoses are deployed on the seabed in a radius of

Table 7.1 Number of deployments for different noise-reduction technologies in the context of offshore wind-farm construction up to March 2018

Water depth (m)	Noise-reduction technology			
	Big bubble curtains	Small bubble curtains	Pile casings	Hydro-sound dampers
0–20	~80	–	~70	1
20–30	>500	1	>230	~140
30–40	>400	2	>100	~90
>40	~50	–	–	~20
Total	>1,000	3	>400	~250

Table 7.2 Deployment of noise-reduction technologies (NRTs) during selected offshore wind farm (OWF) construction from 2009–2016

Year	NRT	OWF name	WD (m)	Foundation	*n*	PD (m)	PL (m)	EL (m)
2009	SBC	Alpha Ventus	29	Tripod	3	2.6	40	30–35
2009	BBC	FINO 3[a]	23	Monopile	1	4.8	55	30
2011/12	(D)BBC	Trianel Windpark Borkum[b]	27–33	Tripod	93	2.4	31–40	20–28
2012	NMS-6900	Riffgat	18–23	Monopile	30	5.7–6.5	53–72	30–48
2012	HSD	London Array	15	Monopile	1	5.7	59–64	40
2014	NMS-6900	Borkum Riffgrund 1	23–29	Monopile	77	6.0	39–59	16–30
2014/15	HSD & BBC	Amrumbank West	19–24	Monopile	48	6.0	53–60	30
2016	NMS-IMI & BBC	Nordsee One	26–29	Monopile	54	6.7	~70	N/A
2016	HSD & BBC	Veja Mate	39–41	Monopile	67	7.9	85	44

[a]Research platform; [b]former project name: Borkum West II offshore wind farm.
WD: water depth; *n*: number of piles installed with NRT; PD: pile diameter; PL: pile length; EL: pile embedment length; SBC: small bubble curtain; (D) BBC: (double) big bubble curtain; NMS, Noise Mitigation Screen; IMI, Integrated Monopile Installer; HSD: hydro-sound damper.
For details regarding the achieved noise reduction, see text.

Figure 7.4 Use of a double big bubble curtain in combination with a hydro-sound damper system around the pile during installation of one of the turbine foundations at Wikinger offshore wind farm in the Baltic Sea in 2016. (Hydrotechnik Lübeck GmbH)

50–80 m around the piling site. Ascending air bubbles act as a barrier and extenuate the sound propagation by reflection, scattering and resonance effects. To boost the efficiency, one can deploy two (a double big bubble curtain or DBBC) or more bubble curtain layers instead of one (Figure 7.4). In a few instances, small bubble curtains (SBCs) have been directly attached to the pile or the foundation itself, but these designs have shown to be vulnerable to currents and are therefore less effective. Therefore, so far, only the BBC design has played a role. The first serial test of a BBC layout was executed at the Trianel Windpark Borkum wind farm (Diederichs *et al.* 2014). Several BBC set-ups, mainly differing in the size of the nozzles and the quantity of supplied air, were successfully tested during the installation of 31 (out of 40) tripod foundations that each were anchored with three pin piles in the seabed. The best BBC configuration regarding both effectiveness and practicality delivered an average noise reduction of 11 dB (SEL) and 14 dB ($L_{p,0\text{-}pk}$), resulting in a drastically reduced noise impact area (Diederichs *et al.* 2014; Dähne *et al.* 2017). Since then, similar BBC set-ups have proven their effectiveness at numerous other OWF projects in German waters. Experience gained from these installations has shown that the attainable noise reduction may vary owing to differing soil properties, local currents and water depth. Over the past few years, larger pile diameters and higher piling energies have required multilayer BBCs or combined systems (near-field NRT plus BBC or DBBC) to stay within the statutory noise limits (Figure 7.4).

Casings

Casings are especially well suited to install monopiles. Design concepts for casings range from flexible pile sleeves made of different fabrics to hollow steel tubes. The most widely

Figure 7.5 Deployment of the Integrated Monopile Installer (IHC) Noise Mitigation Screen NMS-8000 at NordseeOne offshore wind farm in the North Sea in 2016. The NMS-8000 stands upright on the installation vessel next to the hydraulic hammer and in front of two 6.5 m diameter monopiles. (Henk van Vessem, IHC IQIP)

used application is the Noise Mitigation Screen (NMS), which was developed and designed by the Dutch company IHC (Figure 7.5). The NMS is a double-walled steel cylinder with sound-insulated connections between the inner and outer walls and an air-filled cavity. A confined bubble curtain creates an air–water environment inside the cylinder to reduce the sound coupling. This system is also known as the Integrated Monopile Installer.

The Riffgat OWF demonstrated the first serial application of the technology in 2012. A total of 30 monopile foundations were first vibrated into the seabed and then driven to target depth by impact pile-driving shielded by an NMS-6900 system. The casing had a length of 30 m, an outer diameter of 10 m and a weight of 360 t. Noise measurements carried out at 750 m for five piling events revealed an SEL (single-strike, 50% percentile) of 163 dB re $1\ \mu Pa^2\ s$ (Fischer *et al.* 2013). The project clearly demonstrated that the noise could be kept below the statutory thresholds, but as a result of the lack of reference measurements, the extent of the achieved noise reduction remained unclear. In 2014, the NMS-6900 was used for another 80 pile-driving operations at the Borkum Riffgrund 1 OWF. In comparison to the reference condition without NRT, the noise reduction in SEL was 14 dB. The $L_{p,0\text{–}pk}$ was reduced by 16 dB (pile length=58 m, target depth=27 m, water depth=27 m) (Gündert *et al.* 2015). The NMS-6900 performed best around 1 kHz ,where the SEL was reduced by more than 30 dB (Figure 7.6). The results of the noise mitigation achieved using a BBC, a DBBC and the NMS-6900 system at both Trianel Windpark Borkum (Diederichs *et al.* 2014) and Borkum Riffgrund 1 OWF (Schiedek *et al.* unpublished) are compiled in Figure 7.6.

More advanced systems designed for pile diameters up to 8.0 m (NMS-8000, NMS-8200) are thought to be even more efficient (H. van Vessem, personal communication 2018). The use of casings is dependent on sufficient storage and crane capacity on the installation vessel. Logistically, it may not be possible to transport very long sleeves in

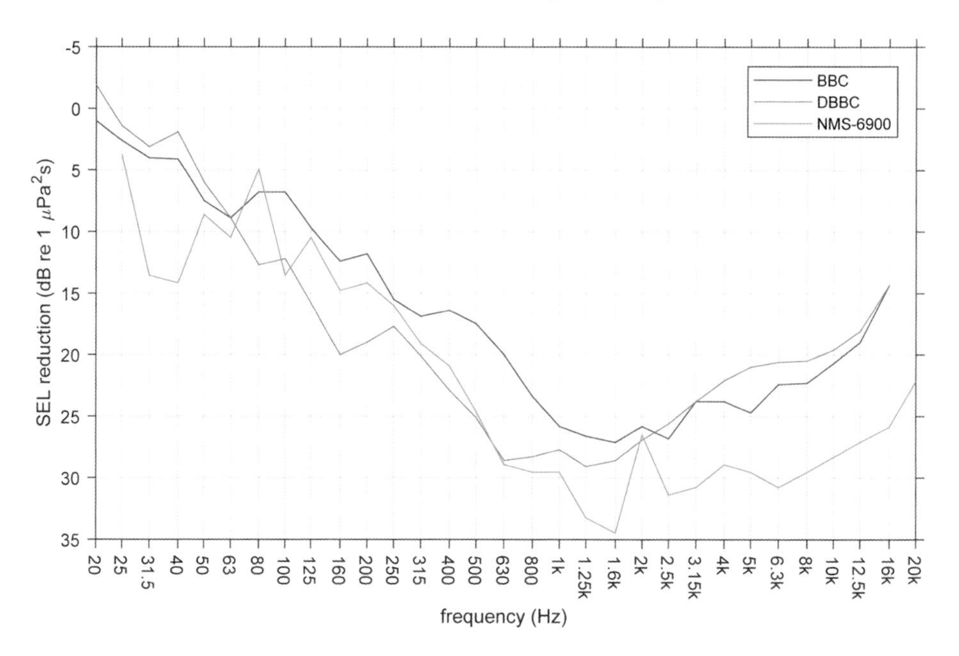

Figure 7.6 Frequency-dependent reduction in sound exposure level (SEL) for three different noise-reduction technologies: big bubble curtain (BBC), double big bubble curtain (DBBC) and IHC Noise Mitigation Screen (NMS-6900). BBC and DBBC data were read off Figure 1-38 in Diederichs *et al.* (2014), with the NMS-6900 data obtained from Schiedek *et al.* (unpublished). (Uwe Stöber, DHI)

a vertical position, which would otherwise take up most of the deck space if they were transported horizontally. As the weight of a casing increases with length, the applicability of the method is limited to water depths up to approximately 40 m. The deployment of large casings from floating vessels and at construction sites with varying water depths between the individual turbine positions remains challenging.

It can be seen that, in general, the DBBC and the NMS resulted in greater sound dampening compared to the BBC system. It should be noted that the sound reduction is frequency dependent, as between 20 Hz and approximately 1 kHz, the sound reduction increases with frequency. Above 1 kHz, the sound reduction is less, but still higher compared to the lower frequencies. This has implications for the impact assessment. Harbour Porpoises and Harbour Seals hear better at higher than at lower frequencies (Møhl 1968; Kastelein *et al.* 2002) and thus, for these species, the bubble curtain and the NMS are very efficient. As fish have a much smaller bandwidth of hearing, mainly below 1 kHz (see review by Gill *et al.* 2012), the reduction for them is less pronounced. It is therefore important to consider this frequency dependency of mitigation measures when trying to establish the effectiveness of NRTs (see Dähne *et al.* 2014 for a discussion on the topic).

Resonators

The most widely used resonators in offshore pile driving are HSDs. In this approach, stationary air-filled balloons or robust foam elements of different sizes reduce underwater sound. The balloons cause sound reflections, resonant effects with high scattering and absorption and material damping effects. Typically, the HSD elements are arranged either in nets or in boxes and placed in the vicinity of the noise source (Kuhn & Bruns 2013; Lee *et al.* 2013; Elmer & Savery 2014). A winch frame above the water and a basket that can be lowered as ballast weight keep the elements in place around the pile (Figure 7.7). In

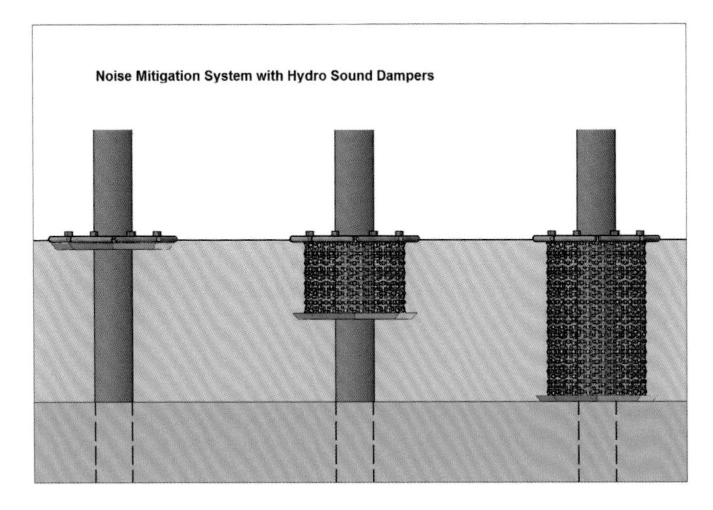

Figure 7.7 Working principle of the hydro-sound damper system, with the system first fixed under the pile gripper in a compact arrangement (left), before it is released (centre) and then lowered to the seafloor (right). (Karl-Heinz Elmer, OffNoise-Solutions GmbH; Elmer *et al.* 2011).

contrast to the BBC concept, no compressed air is needed, and there is nearly no influence by currents. The sound mitigation strictly depends on the number, distribution and type of HSD elements used. The main advantage of the HSD system is that the technicians can adjust the sound mitigation to a predefined frequency range. Laboratory and small-scale tests with different HSD set-ups around 110 Hz resonance frequency indicate a noise-reduction potential in the near field of up to 23 dB between 100 Hz and 600 Hz (Kuhn & Bruns 2013).

The London Array wind-farm construction team carried out the first full-scale test of HSDs during installation works in August 2012. The crew deployed a self-expanding prototype (Figure 7.8) with an outer diameter of 9 m and a total weight of 17 t. The system consisted of three main components: a buoyancy ring, the HSD net and a ballast box.

Figure 7.8 Deployment of a hydro-sound damper (HSD) prototype at London Array wind farm in the Thames estuary, UK. The HSD system is folded in the frame that is lifted above the monopile. (Karl-Heinz Elmer, OffNoise-Solutions GmbH)

Once lifted over the pile, the system was released and recovered automatically. Noise measurements were performed during the driving of a 5.7 m diameter monopile (Remmers & Bellmann 2013). Compared to a reference case without noise reduction, a reduction in SEL of 7–13 dB and in $L_{p,0-pk}$ of 7–15 dB was determined. The best damping efficiency in the range of 15 dB was recorded at the peak frequencies of the ramming noise at 200–500 Hz (Remmers & Bellmann 2013).

In 2013–2014, an improved HSD system was deployed during the installation of 47 (out of 80) monopiles at the Amrumbank West OWF. With an average noise reduction of 8–13 dB (SEL) (Bellmann 2014) and a 20 dB noise reduction in the frequency range between 100 and 800 Hz (Bruns *et al.* 2014), the HSD system showed a good constant performance (Figure 7.9). However, the handling of the commercial prototype required two crane operations, which caused delays of 3–4 hours per pile during the installation process (S. Schorcht, personal communication 2014). Therefore, all following projects have used improved HSD systems where a winch frame with an HSD box that may be easily lowered is already positioned below the pile gripper. Several projects, including Sandbank and Veja Mate OWFs, both located in the German Bight, have used such an advanced system.

In a variation of the HSD concept, Wochner *et al.* (2016) developed arrays (trays) of underwater air-filled Helmholtz resonators, the AdBm system, which surround the noise source and are tuned to optimally attenuate noise in a frequency band of interest. Offshore tests with a full-scale prototype executed in the autumn of 2018 at a wind farm in the Belgian part of the North Sea, showed a 7–8 dB reduction in sound pressure level compared to an unmitigated scenario (Elzinga *et al.* 2019). The best efficiency is reached around the design frequency (100 Hz) of the damping elements, whilst the reduction in higher frequencies is lower to some degree. Dähne *et al.* (2017) have stated that this frequency response is not of benefit for high-frequency cetaceans such as Harbour Porpoise. The system would, however, work very well for fish and other cetaceans such as baleen whales that are more sensitive at lower frequencies (Southall *et al.* 2007; Gill *et al.* 2012), and a combined deployment with a bubble curtain can achieve a good overall performance in both the low and the high-frequency band (Elzinga *et al.* 2019). In general, it is possible to design the individual resonators to achieve reduction at a whole range of frequencies (M. Wochner, personal communication 2017).

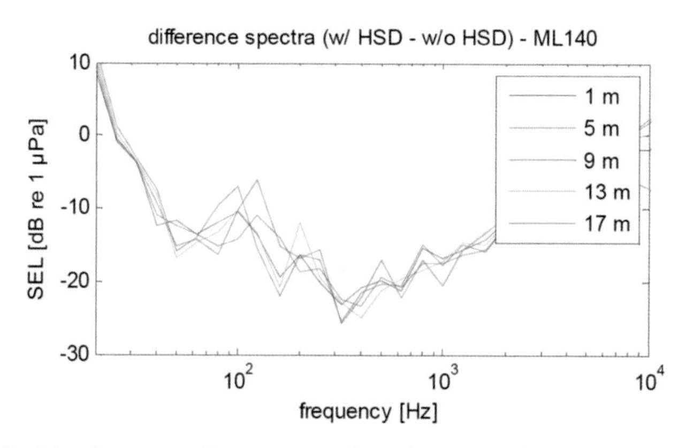

Figure 7.9 Sound mitigation at a piling energy of 800 kJ using a hydro-sound damper system at Amrumbank West offshore wind farm (from Bruns *et al.* 2014) in a water depth of approximately 22 m. ML140: measurement location at 140 m from the piling site. Hydrophone recordings were taken at 1, 5, 9, 13 and 17 m above the seabed. SEL: sound exposure level. (From Bruns *et al.* 2014)

Receiver mitigation

Mitigation measures that address the receiver can be divided into acoustic deterrent devices (ADDs) and acoustic harassment devices (AHDs), monitoring of safety zones, soft-start procedures for pile driving to encourage sensitive animals to leave the danger zone before certain noise or energy levels are exceeded, and temporal restrictions on piling.

Acoustic deterrent and harassment devices

An ADD is an underwater sound emitter that is intended to deter marine mammals from dangerous objects or areas such as fishing nets or marine construction activities. Similarly, an AHD is a sound emitter intended to keep marine mammals, usually pinnipeds, away from particular areas such as fish farms, leading to alternative names such as 'seal scarers' and 'seal scrammers'. The distinction between ADDs and AHDs in the context of offshore wind is arbitrary, as both devices are being used to exclude marine mammals from areas of exposure to high-intensity noise such as impact pile-driving. Yet, it has to be pointed out that AHDs have much higher estimated source sound pressure level (SPL) at >185 dB re 1 μPa at 1 m than ADDs, with <185 dB re 1 μPa at 1 m (Gordon *et al.* 2007; OSPAR 2009).

The evidence base regarding the efficacy of seal scarers in deterring seals, for which they were designed, is still inconclusive (Götz & Janik 2013). However, seal scarers have been shown to cause aversive behaviour in Harbour Porpoises to a distance of several kilometres (Brandt *et al.* 2012; 2013; Dähne *et al.* 2013; 2017). Despite this, the deterrence of porpoises by seal scarers is not 100% effective as a small number of individuals may remain in the area afterwards. From a conservation perspective, it should be noted that the use of seal scarers could become an issue as a result of the potential to lead to large-scale habitat exclusion, at least in porpoises and perhaps also in other odontocetes (Götz & Janik 2013; McGregor *et al.* 2013).

Safety zones

Safety zones can be established to avoid marine mammals being exposed to noise doses where injury would otherwise be caused. For example, the Joint Nature Conservation Committee (JNCC) in the UK advises an exclusion zone of 500 m for pile-driving operations (JNCC 2010). Trained marine mammal observers are required to detect marine mammals within the safety zone by sight. Passive acoustic monitoring (PAM) is used in addition to detect marine mammals at night or during averse sighting conditions (JNCC 2010; Verfuß *et al.* 2016; Scheidat & Porter, Chapter 2). PAM technology is to likely become more advanced in identifying species and perhaps even individuals, and also in supporting real-time monitoring. However, it can only monitor safety zones if the species of interest produces sound most of the time, which may not be the case for all such species (Weilgart 2007).

Soft-start

'Soft-start' or 'ramp-up' is a gradual increase in hammer impact energy and strike rate. This is often done at the beginning of impact pile-driving to penetrate the seabed without damage to the equipment. Soft-start allows marine mammals to swim away while being exposed to lower levels of sound compared to the full impact. Although widely used across various European countries, it is remains unclear how efficient this technique really is (OSPAR 2009; Boyle & New 2018).

Temporal–spatial restrictions

Temporal–spatial restrictions have the potential to be very effective because the sound-producing activity is permitted in a certain period or season in a certain area. Several European states have imposed seasonal restrictions on piling, mainly due to fish spawning times. For example in the UK, a number of OWF projects have seen Atlantic Herring *Clupea harengus*-related temporal–spatial consenting restrictions or other mitigation requirements associated with marine licences. Piling restrictions are applied in these cases because of the potential effect of piling upon spawning adult herring and/or their behaviour (MMO 2014; Boyle & New 2018). Perrow *et al.* (2011) suggest that the negative impact of pile driving upon recruitment of Atlantic Herring at one site, with indirect effects on a protected breeding seabird, could have readily been avoided with a short temporal restriction on piling activity to avoid their main spawning period. In the USA, several wind-farm developers have signed an agreement to phase out impact pile-driving at planned projects sites on the East Coast during critical times (e.g. during the spring migration) for North Atlantic Right Whales *Eubalaena glacialis* (Deepwater Wind 2014).

The principal issue with any temporal restriction is that the cost of such measures could be relatively high if this would lead to a delay in wind-farm construction; for example, when construction activities are moved to a period with adverse weather that prevents a continuous installation process. This is especially true when NRTs are available to mitigate sound to a level that leads to non-significant impacts (see Dähne *et al.* 2017 for a case study).

Low-noise foundation installation technologies

The alternatives to impact pile-driving are generally less noisy and thus can be considered 'low-noise' foundation installation technologies. These have only been used occasionally, although efforts are being made to increase the market share of these alternatives, not only for environmental, but also for economic reasons. Here, information is presented on three methods: vibratory piling, drilling and the suction bucket concept.

Vibratory piling

Vibratory piling, also known as vibropiling or vibrodriving, is a low-noise foundation installation technology in which a vibratory 'hammer' unit placed on top of a pile replaces the conventional impact pile-driver. The working principle of a vibrodriver is based on balanced pairs of counter-rotating eccentric masses that generate a resultant downward force. The pile is then vibrated into the soil at a specific frequency along its vertical axis. Vibropiling is suitable for medium to dense seabed substrates, including medium to densely bedded sands and cohesive materials. Employment of modern technology permits the installation of very large piles. In extreme cases, piles with diameters up to 30 m have been driven with several vibratory units that were combined to make one bigger system (APE 2017). Experience in offshore wind installation covers a larger number of pin piles for tripod foundations, and monopiles with diameters of up to 6.5 m (de Neef *et al.* 2013).

Vibratory piling offers several benefits over impact pile-driving, including a faster installation time, lower demand for auxiliary energy and lower noise emissions. The technology has been used at a number of wind-farm sites in the German part of the North Sea, including the BARD Offshore 1 (80 tripiles), Riffgat (30 monopiles) and Global Tech 1 (80 tripods) wind farms. However, owing to some unresolved aspects with regard to the bearing capacity of vibrodriven piles, and uncertainties regarding predictions of the 'vibrability' of piles, the technology has still not broken through in the industry. As existing design guidelines that specify requirements for axial proof loads are only valid for impact-driven piles, where these requirements exist, vibratory piling is restricted to the upper metres of a piling process. Thus, a pre-vibrating and post-hammering approach was followed in all of the aforementioned OWF projects, which means that after initial vibratory piling, the piles needed to be impact driven to target depth to establish the needed bearing capacity (DFI & GDG 2015). The downside of this two-stage approach is that the installation vessel needs to be fitted with both a vibrodriver and an impact pile-driver. In addition, the application of noise-mitigation systems during post-hammering is mandatory in some European countries to keep the noise below certain thresholds (Table 7.3).

To close existing knowledge gaps regarding the bearing capacity of vibrodriven piles, a pilot project led by RWE and supported by the Carbon Trust's OWA initiative was conducted at an onshore test site at Cuxhaven in northern Germany in 2014. The objective was to demonstrate that for sandy soils, the lateral load capacity of vibrated piles is equivalent to that of hammered piles. Three 4.3 m diameter test piles (length 21.5 m, wall thickness 40–45 mm) were installed with a hydraulic PVE 500M vibrodriving unit and another three with a hydraulic impact hammer to an embedment depth of 18.5 m (Carbon Trust 2015b; Dieseko Group 2016). This field experiment provided additional evidence that vibratory piling is considerably faster and less intense compared to impact pile-driving. Three years after installation, a follow-up study assessed the soil set-up effect that had developed over time. The results suggest that, under consideration of several installation aspects, the bearing capacity of the installed vibrated and hammered piles is comparable to a large extent (Innogy 2017).

Calculations based on field measurements indicate a 15–20 dB lower noise level for vibropiling compared to undamped impact pile-driving (Elmer *et al.* 2007a; Betke & Matuschek 2011). However, on refusal or in harder substrate, noise levels may be higher (J. Kringelum, personal communication 2014). In addition, one must be cautious when comparing sound signals of different metrics (se also Box 7.1). During vibropiling, the emitted frequency spectrum comprises peaks occurring at the vibration frequency (usually 20–40 Hz) and its harmonics. At frequencies above 500 Hz, the spectrum is more even in nature and can extend to several kilohertz (Spence *et al.* 2007).

Several wind-farm projects have provided field experience with the sound emissions of vibratory piling. However, vibropiling was not used to drive the piles to target depth in any of these cases. At the German pilot wind farm Alpha Ventus for example, 18 pin piles for six tripod foundations were vibrated 9 m down into the sandy seabed before hydraulic impact piling commenced. The measured SPL (averaged over 5 seconds, at 750 m) in vibratory piling was 142–157 dB re 1 μPa while the measured SEL (single-strike, at 750 m) in impact piling was 171–175 dB re 1 μPa^2 s (Betke & Matuschek 2011; Koschinski & Lüdemann 2013). SPLs in the 100 Hz to 1 kHz frequency range were significantly reduced for the vibratory hammer compared to the impact pile-driver (Betke & Matuschek 2011). At Riffgat OWF, the CAPE Holland Super Quad Kong, a combination of four in-phase vibrodrivers, was used to install 30 large monopiles 15–30 m deep into the seabed. After that, construction work involved impact pile-driving with an IHC S-1800 hammer to take the piles to target depth. The attainable penetration depth for each vibrated monopile varied with local seabed substrate properties. At 750 m, an SPL (50% percentile, averaged over 30

Table 7.3 Overview of mitigation measures for offshore wind farm (OWF) noise used across the market leaders of offshore wind in European waters

Measure	Belgium	Denmark	Germany	Netherlands	UK
Literature	OD Nature (2015)	ENS (2015; 2017), Skjellerup *et al.* (2015)	BSH (2008; 2011), BMU (2013)	RVO (2017)	JNCC (2010), MMO (2014), Boyle & New (2018)
Mitigation plan		Concession holder is required to prepare a 'Prognosis for Underwater Noise' and conduct a control measurement programme	Permit holder is obliged to prepare a 'Noise Mitigation Concept' that includes noise prognosis, deterrence, noise mitigation and monitoring/control of efficiency	Permit holder obliged to prepare a 'Piling Plan' detailing how the foundations will be installed, which sound-mitigation measures will be used and how sound levels will be measured	Noise risk assessment and 'Marine Mammal Mitigation Plan' to be developed as part of EIA. For wind-farm construction in European protected areas, an 'Appropriate Assessment' may be required
Marine mammal monitoring during construction	Marine mammal observers around the piling site (observation radius 200–500 m). If marine mammals are observed, the piling operations must be delayed		PAM (*post hoc* analysis of collected data)		Marine mammal observers are mandatory around the piling site (observation radius 500 m). PAM (real-time monitoring) is regarded as a useful supplement to visual monitoring. If a marine mammal is observed in the mitigation zone, the commencement of piling operations must be delayed. If a marine mammal enters the mitigation zone during soft-start, the energy level should not be increased. Ongoing piling operations at full power can, however, be continued until the pile reaches target depth

Table 7.3 – *continued*

Measure	Belgium	Denmark	Germany	Netherlands	UK
Noise monitoring during construction	Concession holder must inform authorities about location, start and stop of pile-driving activities. Reported noise levels must relate to a distance of 750 m from piling site	Mandatory noise measurements to be executed for first foundations at 750 m from piling site (reference distance) and other distances	Mandatory noise measurements at 750 m and 1.5 km from any installation site	Mandatory noise measurements	Mandatory noise measurements on first foundations (minimum of four); further noise monitoring may be required (to be determined on a case-by-case basis)
Seasonal–spatial restrictions for pile driving	Pile-driving activities that exceed the noise threshold may not take place between 1 January and 30 April		No wind-farm construction in Natura 2000 areas. Cumulative acoustic disturbance must be managed (for details see text)	Roll-out schedule defined in 'Offshore Wind Energy Roadmap' ensures limited amount of parallel wind-farm construction activities	Piling may be on a case-by-case basis restricted in periods of peak fish spawning (condition of the consent)
Threshold for underwater noise; measured at 750 m from noise source (piling site)	$L_{p,0\text{-}pk}$: 185 dB	Thresholds for SEL(cum) may apply on a case-by-case basis. Examples include SEL(cum) thresholds of 183 dB (Horns Rev 3 OWF) and 190 dB (Nissum Bredning OWF)	SEL (5% percentile): 160 dB; peak SPL: 190 dB	SEL (dependent on number of wind turbines to be installed): 163–167 dB (January to May); 169–173 dB (June to August); 171–175 dB (August to December)	

Table 7.3 – *continued*

Measure	Belgium	Denmark	Germany	Netherlands	UK
Use of noise-reduction technologies	Mandatory, if expected noise levels from pile-driving would exceed the defined threshold		Mandatory, if expected noise levels from pile-driving would exceed the defined thresholds	Mandatory, if expected noise levels from pile-driving would exceed the defined thresholds	
Soft-start at beginning of pile-driving activities	Mandatory for first 10 minutes of pile-driving operations	Mandatory	Mandatory	Mandatory	Mandatory
Duration of piling activities			Limited to max. 180 minutes per monopile; max. 140 minutes per pile of jacket foundation	Permit holder is asked to execute pile-driving activities in time-frame 'as short as possible'	Concurrent piling of adjacent piles recommended to reduce spatial and temporal footprint of disturbance
Use of ADDs in the context of construction works	Mandatory use 30 minutes before the commencement of piling activities	Mandatory before and during piling operations	Mandatory use 15–40 minutes before the commencement of piling activities	Mandatory use 30 minutes before the commencement of, and during, piling activities	Deployment judged on a case-by-case basis (condition of the consent)

ADD: acoustic deterrent device; EIA: Environmental Impact Assessment; EEZ: Exclusive Economic Zone; OWF: offshore wind farm; PAM: passive acoustic monitoring; SEL: sound exposure level; SEL(cum): cumulative sound exposure level; SPL: sound pressure level.

seconds) of 145 dB re 1 μPa was recorded for the vibratory piling (Fischer *et al.* 2013). In the absence of reference measurements without a noise-mitigation system, this value must be compared against the predicted SEL (single-strike, at 750 m) for impact pile-driving at Riffgat, which was 180 dB re 1 μPa^2 s (Koschinski & Lüdemann 2013). However, the noise levels during vibrodriving were also well below the measured SEL (during the impact pile-driving operations shielded by an IHC NMS-6900 casing (see *Casings*, above).

At the Danish Anholt wind farm, two 5.3 m diameter monopiles were vibrated 17–18 m into a sandy seabed using three synchronised PTC 200HD vibropiling units (PTC 2012). Owing to insurmountable resistance in hard soil, for one of the piles, the installation process used impact piling to achieve the last metres of the required pile depth. This allowed comparative noise measurements to be obtained. The average measured SPL (averaged over 30 seconds) was 153 dB re 1μPa for vibrodriving compared to an average SEL associated with impact piling of 172 dB re 1 μPa^2 s (reference measurements taken across four piles). As noted in Box 7.1, while sounds of different metrics should be compared with caution, the measured peak SPLs at Anholt were reduced by 26–28 dB (J. Kringelum, personal communication 2014).

Drilling

Offshore pile installation by drilling is also considered to be less noisy than impact pile-driving. Two main approaches to drilling technology may be distinguished: (1) the 'drive–drill–drive' concept and (2) 'pure' drilling solutions. In drive–drill–drive operations, the installation process involves impact pile-driving at the start of the process until the hammer meets resistance in a hard bottom layer. Then, the construction team inserts a drilling system in the pile, lowers it to the bottom and uses it to cut a shaft through the hard layers. Finally, the engineers switch back to the impact pile-driver to bring the pile to target depth. In pure drilling solutions, construction involves only a drilling device to accomplish the pile-installation process. Thus, the authors regard only pure drilling as a 'low-noise' foundation installation technology. However, to reflect the current state of technology, experience with drive–drill–drive operations is also discussed.

The application of pure drilling solutions is limited. Although several companies have worked on adapting existing onshore solutions for the serial installation of offshore piles, none of these technologies is commercially available. Concepts include Ballast Nedam's Drilled Concrete Monopile approach (Saleem 2011), which was finally abandoned, and Hochtief and Herrenknecht's Offshore Foundation Drilling (Koschinski & Lüdemann 2013), where a drill with a cutterhead, operated from a mobile offshore platform, is inserted in the pile and cuts its way down through the seabed regardless of geological properties. Assuming that installation reaches cutting speeds of 3.0 m/h (Jung 2013), this would enable monopile installation within one day.

A first estimate of the noise emissions that are associated with this technology was derived in 2008, when a Herrenknecht vertical shaft machine was used to drill a 5 m wide shaft 39 m into a water-saturated soil in Naples, Italy. Noise recordings suggest that when transferred to offshore conditions, an SPL of ≤117 dB re 1 μPa and a peak SPL of ≤122 dB re 1 μPa can be expected at a distance of 750 m from the drilling site (Rustemeier *et al.* 2012).

Despite these promising results, the development of pure drilling solutions was not taken further owing to cost considerations. Because of its high cost relative to other installation methods, wind-farm developers use drive–drill–drive technology only occasionally, for example, when piles need to be installed in bedrock or substrates with hard bottom layers. Examples include the North Hoyle (2003), Barrow (2005), Teesside

(2012/13) and Gwynt y Môr (2013/14) OWFs. At North Hoyle, piles were drilled in underlying bedrock after they were first driven through the upper sand and sediment layers. A strong fundamental component at 125 Hz and harmonics up to 1 kHz dominated the emitted frequency spectrum (Nedwell *et al.* 2003). Unfortunately, the authors could not establish the source level and transmission loss. At Barrow, a pile top-drilling system provided by Bauer was used to install three monopiles with a diameter of 4.75 m. The drill cut through various layers of dense sand, silt, clay, mudstone/siltstone and siltstone/sandstone. Thereafter, piling was undertaken until target depth. Rates of progress varied between 0.35 and 1.0 m/h (Beyer & Brunner 2006). Developers gained further experience with drive–drill–drive operations at Gwynt y Môr, where the world's largest reverse circulation drill rig provided by LDD was used for several 6.5 m diameter monopiles. Reported progress rates were 0.6 m/h in mudstone and up to 2.2 m/h in softer ground (Fugro Seacore 2014). Although a significant cost reduction in drilling technology has been achieved, the method is still not cost competitive to impact pile-driving and, apart from being used at sites with challenging soil conditions, does not play a major role in offshore pile installation.

Suction buckets

Suction buckets, also known as suction cans or caissons, are a proven low-noise installation technology for anchoring offshore structures in the seabed. The installation process starts with the placement of the suction bucket foundation on to the seabed. Owing to the deadweight of the structure, the bucket penetrates the substrate to a certain depth. Then, pre-installed pumps are switched on to evacuate water and loosen sediment from the interior of the bucket, thereby creating a negative pressure on the inside. With ongoing evacuation, the hydrostatic pressure difference and the structure's deadweight cause the bucket to penetrate farther into the seabed. Levelling is accomplished by alternating pumping operations between several suction chambers or individual suction cans. The only noise associated with the installation process comes from the suction pumps and the installation vessels.

The installation of suction buckets is possible in a variety of sediments such as sands, clay, silt, other water-saturated soils and layered compositions of these materials. Large boulders, or other hard inclusions such as cemented layers or coral outcrops can be problematic, however, and areas with large sand waves or high seabed mobility may exclude the applicability of suction buckets. Finally, as suction bucket installation is a function of water depth (Houlsby & Byrne 2005), shallow waters limit the amount of driving force that is available for suction bucket penetration.

To date, more than 2,000 suction-assisted installations have been performed in oil and gas applications. However, until recently, all experience in offshore wind was limited to near-shore sites, met masts (Figure 7.10) or the anchoring of floating foundations (Ibsen *et al.* 2005; Zhang *et al.* 2012; Nielsen *et al.* 2017; Equinor 2017). Two principal design concepts can be discriminated: (1) suction bucket jackets, with comparatively small suction buckets at the bottom of each jacket leg, and (2) the Mono Bucket, a monopile with one big suction bucket at its bottom. Suction bucket jackets have a larger footprint and shallower seabed embedment than other foundation types. This is due to the loading conditions of large offshore wind turbines and the limited pressure that can be applied in relatively shallow waters during the installation process. Significant lateral loads resulting from the wind and waves are transferred to the soil via the jacket foundation through a push–pull mechanism, predominantly resulting in vertical foundation loads (Shonberg *et al.* 2017).

Figure 7.10 Two Mono Bucket foundations on board an installation vessel in February 2013. The foundations served as substructures for meteorological met masts installed at Dogger Bank in the North Sea. (Kristian Ascanius Jacobsen, Universal Foundation A/S)

The load transferred to the ground is a function of the distance between the suction buckets such that an increase in the footprint of the jacket reduces the load on the foundations.

To close existing knowledge gaps, in particular with regard to the penetration ability of suction buckets in layered and relatively hard soils, the Carbon Trust initiated a trial installation campaign at different projected wind-farm sites in the North Sea (Nielsen & Ibsen 2015). Further practical experience was gained when a full-scale suction bucket jacket prototype carrying a 4.0 MW wind turbine was installed at Borkum Riffgrund 1 wind farm in August 2014 (DONG 2013; Carbon Trust 2014) in a water depth of 24.4 m (Figure 7.11). During the installation process, an SPL (averaged over 5 seconds; 5% percentile) of 138–141 dB re 1μPa and an $L_{p,0-pk}$ of 149–166 dB re 1μPa were measured at a distance of 750 m (Remmers & Bellmann 2015, unpublished). It is of note that the ship propeller noise dominated the sound recordings and that the operation of the suction pumps could not be identified against this background.

Building on the gathered experience with suction buckets, Vattenfall and Ørsted have decided to install a larger number of suction bucket jackets at the Aberdeen (Scottish North Sea) and Borkum Riffgrund 2 (German Bight) wind farms, respectively (Vattenfall 2017; ST3 Offshore 2018). The projects are the first commercial-scale deployments of suction bucket technology in offshore wind. At Aberdeen, 11 jackets with 9.5 m diameter and 13 m tall suction buckets on each leg were successfully anchored to the seabed between March and May 2018. Another 20 suction bucket jackets were installed at Borkum Riffgrund 2 wind farm between July and August 2018. At both sites, the foundations will carry 8.4 MW offshore wind turbines.

Regarding the Mono Bucket concept, two full-scale prototypes of 18.5 m diameter and up to 18.5 m height, topped by two 8.4 MW offshore wind turbines, are scheduled for installation adjacent to the Deutsche Bucht wind farm (German Bight) in mid-2019. The removal of two meteorological met masts, which were installed on smaller Mono Bucket foundations (15 m diameter, 7 m height) at Dogger Bank in December 2013, has already

Figure 7.11 Installation of a suction bucket jacket at Borkum Riffgrund 1 in the North Sea in August 2014. (Ørsted Wind Power A/S / SPT Offshore)

demonstrated that Mono Bucket foundations can be fully decommissioned without significant noise emissions by reversing the installation process (Carbon Trust 2013; Universal Foundation 2017). This is also important from the perspective of marine life.

Other foundation types

In principle, other foundation types, such as gravity (base) foundations and floating OWFs (see overviews in Carbon Trust 2015a; Jameson *et al.* 2019), may be installed with far less noise than those using piles, especially monopiles. Gravity foundations are large box-like structures that are typically made of concrete and steel. Their stability is achieved by the deadweight of the structure, supplemented by additional ballast. The available models differ in shape and production details. Developers have installed gravity-based foundations in several projects, especially in the Danish part of the Baltic Sea and the Belgian part of the North Sea, in water depths no more than 25 m.

Floating wind farms exist as several conceptual designs, divided into spar-buoys, such as Hywind, semi-submersible platforms, such as WindFloat, and tension-leg platforms, such as the BLUE H concept. All of these include a floating turbine moored to the seabed using taut-leg or catenary mooring systems. The anchoring and mooring solution depends on the seabed conditions and the holding capacity required. Catenary moorings often involve drag-embedded anchors or gravity anchors. Taut-leg moorings are typically anchored with small-diameter driven piles, suction piles or gravity anchors. The size of the anchor is variable, with larger and heavier anchors able to generate a greater holding capacity, and the associated noise of installation varies with the method used. The full potential of floating wind farms will be realised in locations that are challenging or not suitable for monopiles or jackets; for instance, sites with water depths beyond 50 m or very hard bottom substrates.

Figure 7.12 A substructure for the Hywind Scotland floating wind-farm project being floated off a heavy transport vessel at Stord in Norway, before being towed 30 km to the east coast of Scotland. (Olaf Nagelhus, Equinor ASA)

The first commercial operational floating wind farm, Equinor's Hywind Scotland project, was commissioned in 2017. Five 6.0 MW turbines on spar-boys each 91 m long and weighing 2,300 t were installed at Buchan Deep, approximately 25 km off the east coast of Scotland, in water depths of 95–120 m. Each floating substructure, constructed by Navantia in Spain (Figure 7.12), is anchored to the seabed with three suction anchors, being 16 m tall and 5 m in diameter (Equinor 2017). The wind turbines were pre-assembled on land, before being lifted on their spar-buoy substructures and towed to the site. The noise emissions associated with the installation of floating foundations are much lower than for piled foundations, with the noise from the installation of the suction anchors being comparable to that emitted during the installation of suction bucket foundations (Koschinski & Lüdemann 2013).

Concluding remarks

To ensure that the expansion of offshore wind energy takes place in a sustainable and environmentally sound manner, most European countries developing offshore wind have undertaken action to mitigate noise. However, apart from a few common measures such as the soft-start procedure at the beginning of pile-driving operations, national approaches are highly diverse (Table 7.3), as also highlighted in a review of the practices in the UK and Germany for use in the developing market in the USA (BOEM 2018). In the UK, for example, the approach is simple in the sense that pile-driving operations are subject to on-site marine mammal monitoring by trained observers. The piling team has to postpone the process if observers spot a marine mammal within a 500 m radius around the piling site. In addition, on a few occasions piling has been restricted in periods of peak fish spawning, as applied to the construction of Gwynt y Môr OWF (MMO 2014).

In Germany, the first step was the establishment of legislation that prohibits the installation of OWFs in Special Areas of Conservation (BMVBS 2009). In addition, an administrative regulation to prevent auditory injuries in marine mammals was first introduced in 2004, based on scientific background given by the Federal Environment Agency (UBA 2011), and became effective in 2008 (BSH 2008). The regulation defines noise thresholds for the single-pulse SEL and the peak SPL. Consequently, the use of noise-reduction measures is mandatory for pile-driving works in the German parts of the North and Baltic Seas in the conjunction of activities that would exceed the thresholds (Schorcht 2014). To avoid cumulative effects on marine mammals from noise disturbance, a Concept for the Protection of Harbour Porpoises from Offshore Wind Farm Construction Noise was adopted and introduced by the Federal Ministry for the Environment (BMU 2013). This concept prescribes that, with respect to certain defined disturbance radii that depend on the measured SEL at 750 m from any piling site, the cumulative acoustic disturbance in the German Exclusive Economic Zone (EEZ) of the North Sea must be limited to a maximum of 10% of the entire (EEZ) area between September and April, and a maximum of 1% of a main concentration area around the Sylt Outer Reef between May and August. The tightening of the limit between May and August corresponds to the Harbour Porpoise mating and nursing season. The Harbour Porpoise is regarded as an indicator species in this context. As a result, parallel construction works need to be managed in a spatial–temporal way to keep displacement effects under a tolerable level.

The timely roll-out of offshore wind developments in the Netherlands follows the Offshore Wind Energy Roadmap (Routekaart Wind op zee 2030, EZK 2018), which ensures a gradual extension of the installed offshore wind capacity. For the designated wind-farm developments, the regulations are summarised in individual Wind Farm Site Decisions (Kavelbesluits), which contain a number of provisions to prevent physical harm and/or disturbance to marine mammals and fish, such as noise thresholds for the SEL that must not be exceeded in the construction phase. The applicable noise threshold is dependent on the final number of wind turbines and the season in which pile-driving operations will be executed (MEZ 2016; RVO 2017). As for Germany, the Harbour Porpoise serves as an indicator species for the establishment of noise-mitigation practices.

Notwithstanding any differences in approach between countries, it is clear that considerable progress has been made in developing technologies for the mitigation of effects of OWF noise on marine life. This is especially true for developments in source mitigation such as BLUE Piling, and channel mitigation in the form of casings, bubble curtains and resonators. The application of the precise method depends on the site-specific conditions and is strongly governed by the applicable regulatory framework as well as the project economics. It is important to note that in order not to exceed the noise thresholds required in German waters in particular, there has been an increasing trend towards the use of several noise-mitigation systems in combination with each other (Table 7.2). This includes the use of far-field systems in combination with near-field systems, such as the combined deployment of a BBC and an HSD system (see *Bubble curtains* and *Resonators*, above) at Wikinger OWF (Figure 7.4).

Finally, the authors suggest that the selection of the appropriate noise-mitigation method or combination of methods should follow a risk-based approach (Boyd *et al.* 2008; WODA 2013). A risk assessment framework results in a more systematic approach to sound impact studies. It involves a stepwise procedure, including: (1) risk identification, (2) exposure assessment, (3) dose–response assessment, and (4) the overall characterisation of the risk, which leads to (5) risk management and the selection of appropriate mitigation measures (Figure 7.13). Applying this approach means that mitigation measures would be applied only when risks are high, such that sound levels are likely to lead to significant

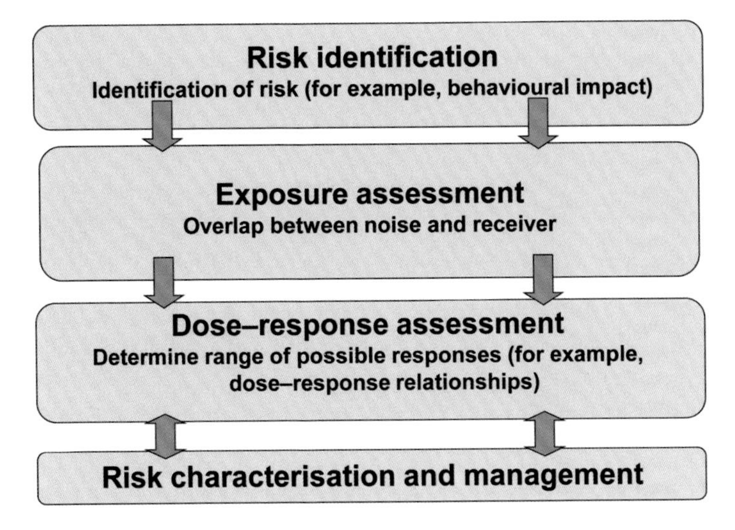

Figure 7.13 Overview of a risk-based approach to select the most appropriate course of action for mitigating noise impacts upon marine life. (Adapted from WODA 2013)

effects on marine life. This can be investigated using acoustic modelling (Farcas *et al.* 2016; Nehls *et al.* 2019) and a thorough analysis of the abundance and distribution of marine life in the project area through specific surveys and research (see Dahlgren *et al.*, Chapter 1; Scheidat & Porter, Chapter 2).

Acknowledgements

We would like to thank Ørsted Wind Power for the permission to use results from the vibrodriving test at Anholt OWF and the NMS-6900 tests at Borkum Riffgrund 1, respectively. Furthermore, we thank the following people and companies for the permission to reproduce images for the illustration of this chapter: Jasper Winkes (Fistuca BV) for Figure 7.2 and Figure 7.3, Cay Grunau (Hydrotechnik Lübeck GmbH) for Figure 7.4, Henk van Vessem (IHC IQIP) for Figure 7.5, Karl-Heinz Elmer (OffNoise-Solutions GmbH) for Figure 7.7 and Figure 7.8, Kristian Ascanius Jacobsen (Universal Foundation A/S) for Figure 7.10, Ørsted Wind Power A/S for Figure 7.11 and Equinor ASA for Figure 7.12. Uwe Stöber of DHI produced Figure 7.1 and Figure 7.6. We thank G. Nehls of BioConsult SH GmbH & Co. for the permission to use data for Figure 7.6. Figure 7.13 was adapted from WODA 2013 with permission.

References

American Piledriving Equipment (APE) (2017, 12 January) World's largest pile driven by APE Dodecakong. American Piledriving Equipment. Retrieved 14 December 2018 from http://www.pilebuck.com/industry-news/worlds-largest-pile-driven-ape-dodecakong/

Bailey, H., Brookes, K.L. & Thompson, P.M. (2014) Assessing environmental impacts of offshore wind farms: lessons learned and recommendations for the future. *Aquatic Biosystems* 10: 1–13.

Bellmann, M.A. (2014) Overview of existing noise mitigation systems for reducing pile-driving noise. In *Proceedings of Inter.noise 2014: 43rd International Congress on Noise Control Engineering*, 16–19 November 2014, Melbourne, Australia. Retrieved 14 December 2018 from https://www.acoustics.asn.au/conference_proceedings/INTERNOISE2014/papers/p358.pdf

Betke, K. & Matuschek, R. (2011) *Messungen von Unterwasserschall beim Bau der Windenergieanlagen im Offshore-Testfeld Alpha Ventus. Abschlussbericht zum Monitoring nach StUK3 in der Bauphase.* Oldenburg: Institute of Technical and Applied Physics (ITAP).

Beyer, M. & Brunner, W.G. (2006) Drilling monopiles at Barrow Offshore Windfarm, UK. *Tiefbau* 6: 346–350.

Boyd, I., Brownell, B., Cato, D., Clarke, C., Costa, D., Evans, P.G.H., Gedamke, J., Genrty, R., Gisiner, B., Gordon, J., Jepson, P., Miller, P., Rendell, L., Tasker, M., Tyack, P., Vos, E., Whitehead, H., Wartzok, D. & Zimmer, W. (2008) The effects of anthropogenic sound on marine mammals – a draft research strategy. Ostend: European Science Foundation and Marine Board. Retrieved 19 December 2018 from http://archives.esf.org/fileadmin/Public_documents/Publications/MBpp13.pdf

Boyle, G. & New, P. (2018) ORJIP: impacts from piling on fish at offshore wind sites: collating population information, gap analysis and appraisal of mitigation options. Final report, June 2018. London: The Carbon Trust. Retrieved 14 December 2018 from https://www.carbontrust.com/media/676435/orjip-piling-study-final-report-aug-2018.pdf

Brandt, M.J., Diederichs, A., Betke, K. & Nehls, G. (2011) Responses of harbour porpoises to pile driving at the Horns Rev II offshore wind farm in the Danish North Sea. *Marine Ecology Progress Series* 421: 205–216.

Brandt, M.J., Höschle, C., Diederichs, A., Betke, K., Matuschek, R., Witte, S. & Nehls, G. (2012) Far-reaching effects of a seal scarer on harbour porpoises, *Phocoena phocoena*. *Aquatic Conservation Marine and Freshwater Ecosystems* 23:222–232.

Brandt, M.J., Höschle, C., Diederichs, A., Betke, K., Matuschek, R. & Nehls, G. (2013) Seal scarers as a tool to deter harbour porpoises from offshore construction sites. *Marine Ecology Progress Series* 475: 291–302.

Bruns, B., Stein, P., Kuhn, C., Sychla, H. & Gattermann, J. (2014) Hydro sound measurements during the installation of large diameter offshore piles using combinations of independent noise mitigation systems. In *Proceedings of Inter.noise 2014: 43rd International Congress on Noise Control Engineering*, 16–19 November 2014, Melbourne, Australia. Retrieved 14 December 2018 from https://www.acoustics.asn.au/conference_proceedings/INTERNOISE2014/papers/p864.pdf

Bundesamt für Seeschifffahrt und Hydrographie (BSH) (2008) Genehmigung Offshore Windenergiepark 'Borkum West II' (Consenting Decision Offshore Wind Farm 'Borkum West II'). Hamburg: Federal Maritime and Hydrographic Agency (BSH). Retrieved 17 April 2019 from https://www.eib.org/attachments/pipeline/20080237_nts_de.pdf

Bundesamt für Seeschifffahrt und Hydrographie (BSH) (2011) Offshore wind farms. Measuring instruction for underwater sound monitoring: current approach with annotations – application instructions. Hamburg: Federal Maritime and Hydrographic Agency (BSH). Retrieved 14 December 2018 from https://www.bsh.de/DE/PUBLIKATIONEN/_Anlagen/Downloads/Offshore/Anlagen-EN/Measuring-instruction-for-underwater-sound-monitoring.pdf

Bundesamt für Seeschifffahrt und Hydrographie (BSH) (2013a) Offshore-Windparks. Prognosen für Unterwasserschall: Mindestmaß an Dokumentation. Hamburg: Federal Maritime and Hydrographic Agency (BSH). Retrieved 14 December 2018 from https://www.bsh.de/DE/PUBLIKATIONEN/_Anlagen/Downloads/Offshore/Anlagen-DE/Ergaenzung-Offshore-Windparks.pdf

Bundesamt für Seeschifffahrt und Hydrographie (BSH) (2013b) Messvorschrift für die quantitative Bestimmung der Wirksamkeit von Schalldämmmaßnahmen. Hamburg: Federal Maritime and Hydrographic Agency (BSH). Retrieved 14 December 2018 from https://www.bsh.de/DE/PUBLIKATIONEN/_Anlagen/Downloads/Offshore/Anlagen-DE/Ergaenzung-Messvorschrift-quantitative-Bestimmung-Schalldaempfmassnahmen.pdf

Bundesministerium für Umwelt, Naturschutz und Reaktorsicherheit (BMU) (2013) Konzept für den Schutz der Schweinswale vor Schallbelastungen bei der Errichtung von Offshore-Windparks in der deutschen Nordsee (Schallschutzkonzept). Berlin: Federal Ministry for the Environment, Nature Conservation and Nuclear Safety (BMU). Retrieved 14 December 2018 from https://www.bfn.de/fileadmin/BfN/awz/Dokumente/schallschutzkonzept_BMU.pdf

Bundesministerium für Verkehr, Bau und Stadtentwicklung (BMVBS) (2009) Verordnung über die Raumordnung in der deutschen ausschließlichen Wirtschaftszone in der Nordsee (AWZ Nordsee-ROV) vom 21. September 2009. Federal

Ministry of Transport, Building and Urban Affairs (BMVBS). In *Bundesgesetzblatt (Federal Law Gazette)*, Bonn, Part I (61): 25.

Bureau of Ocean Energy Management (BOEM) (2018) *Summary Report: Best Management Practices Workshop for Atlantic Offshore Wind Facilities and Marine Protected Species (2017)*. OCS Study BOEM 2018-015. Sterling, VA: US Department of the Interior, Bureau of Ocean Energy Management, Atlantic OCS Region.

Caltrans (2009) Technical guidance for assessment and mitigation of the hydroacoustic effects of pile driving on fish. Sacramento, CA: California Department of Transportation. Retrieved 14 December 2018 from http://www.dot.ca.gov/hq/env/bio/files/Guidance_Manual_2_09.pdf

Carbon Trust (2013, 15 September) Second meteorological mast installed at Dogger Bank. London: The Carbon Trust. Retrieved 14 December 2014 from https://www.carbontrust.com/news/2013/09/second-meteorological-mast-installed-at-dogger-bank/

Carbon Trust (2014, 28 August) Innovative offshore wind turbine foundation deployed. London: The Carbon Trust. Retrieved 14 December 2018 from https://www.carbontrust.com/news/2014/08/innovative-offshore-wind-turbine-foundation-deployed/

Carbon Trust (2015a) Floating offshore wind: market and technology review. Report prepared for the Scottish Government, June 2015. London: The Carbon Trust. Retrieved 14 December 2018 from https://www.carbontrust.com/media/670664/floating-offshore-wind-market-technology-review.pdf

Carbon Trust (2015b, 25 November) Positive vibrations for new monopile installation methodology for offshore foundations. London: The Carbon Trust. Retrieved 14 December 2018 from https://www.carbontrust.com/news/2015/11/positive-vibrations/

Carbon Trust (2018a, 13 March) Carbon Trust Offshore Wind Accelerator launches new project to reduce costs and underwater noise in offshore wind construction. London: The Carbon Trust. Retrieved 14 December 2018 from https://www.carbontrust.com/news/2018/03/offshore-wind-accelerator-blue-pilot/

Carbon Trust (2018b, 20 August) Carbon Trust Offshore Wind Accelerator announces successful execution of offshore test of innovative new installation technology to reduce underwater noise and costs. London: The Carbon Trust. Retrieved 14 December 2018 from https://www.carbontrust.com/news/2018/08/carbon-trust-offshore-wind-accelerator-announces-successful-execution-of-offshore-test-of-innovative-new-installation-technology-to-reduce-underwater-noise-and-costs/

Dähne, M., Adler, S., Krügel, K., Siebert, U., Gilles, A., Lucke, K., Peschko, V. & Sundermeyer, J. (2013) Effects of pile-driving on harbour porpoises (*Phocoena phocoena*) at the first offshore wind farm in Germany. *Environmental Research Letters* 8. doi: 10.1088/1748-9326/8/2/025002.

Dähne, M., Peschko, V., Gilles, A., Lucke, K., Adler, S., Ronneberg, K. & Siebert, U. (2014) Marine mammals and windfarms: effects of Alpha Ventus on harbour porpoises. In BSH & BMU (eds) *Ecological research at the offshore windfarm Alpha Ventus: Challenges, results and perspectives.* Federal Maritime and Hydrographic Society of Germany (BSH) and Federal Ministry for the Environment, Nature Conservation and Nuclear Safety (BMU). Wiesbaden: Springer Spektrum. pp. 133–149.

Dähne, M., Tougaard, J., Carstensen, J., Rose, A. & Nabe-Nielsen, J. (2017) Bubble curtains attenuate noise from offshore wind farm construction and reduce temporary habitat loss for harbour porpoises. *Marine Ecology Progress* Series 580: 221–237.

de Neef, L., Middendorp, P. & Bakker, J. (2013) Installation of monopiles by vibrohammers for the Riffgat project. In *Pfahlsymposium 2013, Proceedings*. Braunschweig: Institute for Soil Mechanics and Foundation Engineering, Technische Universität Braunschweig. Retrieved 14 December 2018 from http://www.cape-holland.com/upload/files/Riffgat_Pfahlsymposium_2013_de_Neef.pdf

Deep Foundations Institute (DFI) & Gavin & Docherty Geo Solutions (GDG) (2015) Comparison of impact versus vibratory driven piles: with a focus on soil-structure interaction. Report 14007-01, Rev 2. Deep Foundations Institute (DFI) and Gavin & Docherty Geo Solutions (GDG). Retrieved 14 December 2018 from http://www.dfi.org/update/Comparison%20of%20impact%20vs%20vibratory%20driven%20piles.pdf

Deepwater Wind (2014) Leading environmental and conservation groups and deepwater wind announce agreement to protect endangered whales at Deepwater ONE offshore wind farm site. Retrieved 14 December 2018 from http://dwwind.com/press/leading-environmental-conservation-groups-deepwater-wind-announce-agreement-protect-endangered-whales-deepwater-one-offshore-wind-farm-site/

Diederichs, A., Pehlke, H., Nehls, G., Bellmann, M., Gerke, P., Oldeland, J., Grunau, C., Witte, S. & Rose, A. (2014) Entwicklung und Erprobung des Großen Blasenschleiers zur Minderung der Hydroschallemissionen bei Offshore-Rammarbeiten. BMU Project No. 0325309A/B/C. Husum: BioConsult. Retrieved 14 December 2018 from http://142px.com/hydroschall/wp-content/uploads/2014/10/BWII_ENDVERSION_20130624_final_2-1.pdf

Dieseko Group (2016) Offshore vibratory hammers – product range. Retrieved 14 December 2018, http://www.diesekogroup.com/wms/fm/userfiles/content/909DB8AC-A551-8459-4C31-BA3A7E69CB4E.pdf

DONG Energy (2013) Jacket on suction bucket. Full scale test in the German North Sea. Document No. 1636880. DONG Energy.

Edmonds, N.J., Firmin, C.J., Goldsmith, D., Faulkner, R.C. & Wood, D.T. (2016) A review of crustacean sensitivity to high amplitude underwater noise: data needs for effective risk assessment in relation to UK commercial species. *Marine Pollution Bulletin* 108: 5–11.

Elmer, K.-H. & Savery, J. (2014) New hydro sound dampers to reduce underwater pile driving noise emissions In *Proceedings of Inter.noise 2014: 43rd International Congress on Noise Control Engineering,* 16–19 November 2014, Melbourne, Australia. Retrieved 14 December 2018 from https://tethys.pnnl.gov/sites/default/files/publications/Elmer_and_Savery_2014.pdf

Elmer, K.H., Betke, K. & Neumann, T. (2007a) Standardverfahren zur Ermittlung und Bewertung der Belastung der Meeresumwelt durch die Schallimmission von Offshore-Windenergieanlagen – Schall II. BMU Project Reference 0329947. Hannover: Leibniz Universität Hannover. Retrieved 14 December 2018 from https://www.tib.eu/de/suchen/id/TIBKAT%3A532200500/Standardverfahren-zur-Ermittlung-und-Bewertung/

Elmer, K.H., Gerasch, W.J., Neumann, T., Gabriel, J., Betke, K. & Schultz von Glahn, M. (2007b) Measurement and reduction of offshore wind turbine construction noise. *DEWI Magazine* 30: 33–38.

Elmer, K.-H., Betke, K. & Neumann, T. (2011) Standardverfahren zur Ermittlung und Bewertung der Belastung der Meeresumwelt durch die Schallimmission von Offshore-Windenergieanlagen – Untersuchung von Schallminderungsmaßnahmen an FINO 2. Final Report. BMU Project Reference 0329947A. Hannover: Leibniz Universität Hannover. Retrieved 14 December 2018 from https://www.tib.eu/en/search/id/TIBKAT%3A546614620/Standardverfahren-zur-Ermittlung-und-Bewertung/

Elzinga, J., Mesu, A., van Eekelen, E., Wochner, M., Jansen, E. & Nijhof, M. (2019) Installing offshore wind turbine foundations quieter: a performance overview of the first full-scale demonstration of the AdBm underwater noise abatement system. Paper OTC-29613-MS. Offshore Technology Conference 2019, Houston, TX. Retrieved 11 June 2019 from https://www.onepetro.org/conference-paper/OTC-29613-MS

Energistyrelsen (ENS) (2015) Betingelser for udbud af etablering af havmølleparken Horns Rev 3 – Endelige udbudsbetingelser revideret 6. februar 2015. Copenhagen: Danish Energy Agency (ENS). Retrieved 18 December 2018 from https://ens.dk/sites/ens.dk/files/Vindenergi/hr_3_udbudsmateriale_final_feb15.pdf

Energistyrelsen (ENS) (2017) Tilladelse til etablering af elproduktionsanlæg i Nissum Bredning, 28. februar 2017. Copenhagen: Danish Energy Agency (ENS). Retrieved 18 December 2018 from https://ens.dk/sites/ens.dk/files/Vindenergi/etableringstilladelse_-_nissum_bredning.pdf

Equinor (2017) Hywind Scotland: the world's first commercial floating wind farm. Fact Sheet. Equinor. Retrieved 14 December 2018 from https://www.equinor.com/content/dam/statoil/documents/newsroom-additional-documents/news-attachments/brochure-hywind-a4.pdf

Farcas, A., Thompson, P.M. & Merchant, N.D. (2016) Underwater noise modelling for environmental impact assessment. *Environmental Impact Assessment Review* 57(Suppl C): 114–122.

Fischer, J., Sychla, H., Bakker, J., de Neef, L. & Stahlmann, J. (2013) A comparison between impact driven and vibratory driven steel piles in the German North Sea. Conference on Maritime Energy (COME), Hamburg University of Technology. Retrieved 14 December 2018 from https://www.researchgate.net/publication/281901773_A_comparison_between_impact_driven_and_vibratory_driven_steel_piles_in_the_German_North_Sea

Fugro Seacore (2014) Developments in XL drilling for monopiles/shafts used in the offshore energy sector. Presentation at the Offshore Site Investigation and Geotechnics (OSIG) South West Geoforum, June 2014. Retrieved 14 December 2018 from https://docplayer.net/44003948-Fugro-seacore-ltd-developments-in-xl-drilling-for-monopiles-shafts-used-in-the-offshore-energy-sector-june.html

Gill, A.B., Bartlett, M. & Thomsen, F. (2012) Potential interactions between diadromous teleosts of UK conservation importance and electromagnetic fields and subsea noise from marine renewable energy developments. *Journal of Fish Biology* 81: 664–695.

Gordon J., Thompson, D., Gillespie, D., Lonergan, M., Calderan, S., Jaffey, B. & Todd, V. (2007) Assessment of the potential for acoustic deterrents to mitigate the impact on marine mammals of underwater noise arising from the construction of offshore windfarms. COWRIE Report DETER-01-07. Newbury: COWRIE. Retrieved 19 December 2018 from https://www.osc.co.uk/userfiles/Gordon_etal_2007_AcousticDeterrentsCOWRIE.pdf

Götz, T. & Janik, V.M. (2013) Acoustic deterrent devices to prevent pinniped depredation: efficiency, conservation concerns and possible solutions. *Marine Ecology Progress* Series 492: 285–302.

Gündert, S., van de Parb, S. & Bellmann, M.A. (2014) Empirische Modellierung zur Prädiktion von Hydroschallimmissionen bei Impulsrammung von Fundamentstrukturen für Offshore-Windenergieanlagen. DAGA, Oldenburg, Germany. pp. 449–450. Retrieved 14 December 2018 from https://www.itap.de/media/guendert_et_al._daga2014.pdf

Gündert S., Bellmann, M.A. & Remmers, P. (2015) Offshore Messkampagne 3 (OMK 3) für das Projekt BORA im Offshore-Windpark Borkum Riffgrund I. Hydroakustische Messungen zur Evaluierung der Wirksamkeit des eingesetzten Schallminderungssystems IHC NMS-6900 und zur Untersuchung der Schallabstrahlung einer Fundamentstruktur (Monopile). Project No. 2271-14. Version 2. Oldenburg: Institute of Technical and Applied Physics (ITAP). Retrieved 14 December 2018 from https://docplayer.org/43349792-Offshore-messkampagne-3-omk-3-fuer-das-projekt-bora-im-offshore-windpark-borkum-riffgrund-01.html

Hastie, G.D., Russell, D.J.F., McConnell, B., Moss, S., Thompson, D., Janik, V.M. & Punt, A. (2015) Sound exposure in harbour seals during the installation of an offshore wind farm: predictions of auditory damage. *Journal of Applied Ecology* 52: 631–640.

Hawkins, A.D. & Popper, A.N. (2016) Quo Vadimus – a sound approach to assessing the impact of underwater noise on marine fishes and invertebrates. *ICES Journal of Marine Science* 74: 635–651.

Hawkins, A.D., Roberts, L. & Cheesman, S. (2014) Responses of free-living coastal pelagic fish to impulsive sounds. *Journal of the Acoustical Society of America* 135: 3101–3116.

Houlsby, G.T. & Byrne, B.W. (2005) Design procedures for installation of suction caissons in clay and other materials. *Geotechnical Engineering* 158: 75–82.

Ibsen, L.B., Liingaard, M. & Nielsen, S.A. (2005) Bucket foundation, a status. Copenhagen Offshore Wind Conference 2005, Proceedings. Copenhagen: Aalborg University. Retrieved 14 December 2018 from http://vbn.aau.dk/files/57365677/Proceedings._Bucket_Foundation__a_status

Innogy (2017, 26 April) Innogy treibt die Forschung zum Thema Vibrationsrammen von Offshore-Windkraft-Fundamenten weiter voran. Essen: Innogy SE. Retrieved 14 December 2018 from https://news.innogy.com/vibrationsrammen-von-offshore-windkraft-fundamenten-innogy-treibt-forschung-weiter-voran/

International Organization for Standardization (ISO) (2017) ISO 18405. *Underwater acoustics – terminology*. Geneva: International Organization for Standardization. Retrieved 14 December 2018; https://www.iso.org/obp/ui/#iso:std:iso:18405:ed-1:v1:en

Jameson, H., Reeve, E., Laubek, B. & Sittel, H. (2019) The nature of offshore wind farms. In Perrow, M.R. (ed.) *Wildlife and Wind Farms, Conflicts and Solutions. Volume 3. Offshore: Potential effects.* Exeter: Pelagic Publishing. pp. 1–29.

Jansen, H.W. & de Jong, C.A.F. (2016) Second acoustic underwater measurements BLUE piling. TNO report 2016 R10567. The Hague: TNO, May 2016, 32 pp., unpublished.

JASCO (2009) *Underwater Acoustics: Noise and the effects on marine mammals. A pocket handbook.* Victoria, BC; Halifax, NS; Brisbane: JASCO Applied Sciences.

Joint Nature Conservation Committee (JNCC) (2010) Statutory nature conservation agency protocol for minimising the risk of injury to marine mammals from piling noise (August 2010). Aberdeen: Joint Nature Conservation Committee. Retrieved 18 December 2018 from http://jncc.defra.gov.uk/pdf/JNCC_Guidelines_Piling%20protocol_August%202010.pdf

Jung, B. (2013) Marktreife Weiterentwicklung der VSM-Technologie für die Erstellung von Fundamenten für Offshore-Windenergieanlagen. Final Report to Federal Ministry for the Environment. BMU Project No. 0325233. Schwanau: Herrenknecht AG. Retrieved 14 December 2018 from https://www.tib.eu/en/search/id/tema%3ATEMA20140101707/

Marktreife-Weiterentwicklung-der-VSM-Technologie/

Kastelein, R.A., Bunskoek, P., Hagedoorn, M. & Au, W.W.L. (2002) Audiogram of a harbour porpoise (*Phocoena phocoena*) measured with narrow-band frequency modulated signals. *Journal of the Acoustical Society of America* 112: 334–344.

Kastelein, R.A., Helder-Hoek, L., Covi, J. & Gransier, R. (2016) Pile driving playback sounds and temporary threshold shift in harbour porpoises (*Phocoena phocoena*): effect of exposure duration. *Journal of the Acoustical Society of America* 139: 2842.

Koschinski, S. & Lüdemann, K. (2013) Development of noise mitigation measures in offshore wind farm construction. Nehmten and Hamburg: Federal Agency for Nature Conservation. Retrieved 14 December 2018 from https://tethys.pnnl.gov/sites/default/files/publications/Koschinski-and-Ludemann-2013.pdf

Kuhn, C. & Bruns, B. (2013) Hydro sound dampers (HSD): a new offshore piling noise mitigation system. In Papadokis, J. & Bjørnø, L. (eds) *UA 2013: 1st Underwater Acoustics International Conference and Exhibition 2013, Corfu, Greece, Proceedings*. pp. 1303–1309. Retrieved 14 December 2018 from http://www.uaconferences.org/docs/Past_Proceedings/UACE2013_Proceedings.pdf

Ladich, F., Collin, S. & Moller, P. (2006) *Communication in Fishes*. Enfield, NH: Science Publishers.

Lee, K.M., Wilson, P.S. & Wochner, M.S. (2013) Attenuation of underwater sound through stationary arrays of large tethered encapsulated bubbles. In Papadokis, J. & Bjørnø, L. (eds) *UA 2013: 1st Underwater Acoustics International Conference and Exhibition 2013, Corfu, Greece, Proceedings*. pp. 783–791. Retrieved 14 December 2018 from http://www.uaconferences.org/docs/Past_Proceedings/UACE2013_Proceedings.pdf

Madsen, P.T., Wahlberg, M., Tougaard, J., Lucke, K. & Tyack, P. (2006) Wind turbine underwater noise and marine mammals: implications of current knowledge and data needs. *Marine Ecology Progress Series* 309: 279–295.

Marine Management Organisation (MMO) (2014) Review of post-consent offshore wind farm monitoring data associated with license conditions. MMO Project No. 1031. Report produced for the Marine Management Organisation. Retrieved 14 December 2018 from https://assets.publishing.service.gov.uk/government/uploads/system/uploads/attachment_data/file/317787/1031.pdf

McGregor, P.K., Horn, A.G., Leonhard, M.L. & Thomsen, F. (2013) Anthropogenic noise and conservation. In Brum, H. & Janik, V. (eds) *Animal Communication and Noise*. New York: Springer. pp. 409–444

Minister vor Economische Zaken (MEZ) (2016) Kavelbesluit I windenergiegebied Hollandse Kust (zuid). Staatscourant No. 67082. Minister vor Economische Zaken (MEZ). Retrieved 14 December 2018 from https://www.rvo.nl/sites/default/files/2016/12/Definitief%20kavelbesluit%20I%20Hollandse%20Kust%20zuid.pdf

Ministerie van Economische Zaken en Klimaat (EZK) (2018) Routekaart windenergie op zee 2030. Retrieved 14 December 2018 from https://www.rijksoverheid.nl/actueel/nieuws/2018/03/27/kabinet-maakt-plannen-bekend-voor-windparken-op-zee-2024-2030

Møhl, B. (1968) Auditory sensitivity of the common seal in air and water. *Journal of Auditory Research* 8: 27–38.

National Marine Fisheries Service (NMFS) (2016) Technical guidance for assessing the effects of anthropogenic sound on marine mammal hearing: underwater acoustic thresholds for onset of permanent and temporary threshold shifts. Silver Spring, MD: NOAA, US Department of Commerce. Retrieved 14 December 2018 from https://www.sprep.org/attachments/VirLib/Global/technical-guidance-assessing-effects-anthropogenic-sound-marine-mammal-hearing-noaa.pdf

National Marine Fisheries Service (NMFS) (2018). 2018 Revisions to: Technical guidance for assessing the effects of anthropogenic sound on marine mammal hearing (version 2.0): underwater thresholds for onset of permanent and temporary threshold shifts. US Department of Commerce, NOAA. NOAA Technical Memorandum NMFS-OPR-59. Retrieved 11 June 2019 from https://www.fisheries.noaa.gov/webdam/download/75962998

Nederlandse Organisatie voor Toegepast Natuurwetenschappelijk Onderzoek (TNO) (2011) Standard for measurement and monitoring of underwater noise, Part I: physical quantities and their units. TNO Report TNO-DV 2011 C235. The Hague: TNO. Retrieved 14 December 2018 from https://tethys.pnnl.gov/sites/default/files/publications/Ainslie-2011.pdf

Nedwell, J., Langworthy, J. & Howell, D. (2003) Assessment of sub-sea acoustic noise and vibration from offshore wind turbines and its impact on marine wildlife; initial measurements of underwater noise during construction of offshore windfarms, and comparison with background noise. Report No. 544 R 0424. Newbury: COWRIE. Retrieved 2 April 2019 from http://www.subacoustech.com/about-us/published-papers/

Nehls, G., Betke, K., Eckelmann, S. & Ros, M. (2007) Assessment and costs of potential engineering solutions for the mitigation of the impacts of underwater noise arising from the construction of offshore windfarms. COWRIE ENG-01-2007. Newbury: COWRIE. Retrieved 2 April 2019 from https://tethys.pnnl.gov/sites/default/files/publications/Nehls et al 2007.pdf

Nehls, G., Harwood, A.J.P. & Perrow, M.R. (2019) Marine mammals. In Perrow, M.R. (ed.) *Wildlife and Wind Farms, Conflicts and Solutions. Volume 3. Offshore: Potential effects*. Exeter: Pelagic Publishing. pp. 112–141.

Nielsen, S.A. & Ibsen L.B. (2015) The offshore bucket trial installation. Poster presentation, EWEA Offshore 2015, Copenhagen. Retrieved 14 December 2018 from http://www.ewea.org/offshore2015/conference/allposters/PO031.pdf

Nielsen, S.D., Ibsen, L.B. & Nielsen S.A. (2017) Performance of a Mono Bucket foundation – a case study at Dogger Bank. *International Journal of Offshore and Polar Engineering* 27: 326–332.

Operational Directorate Natural Environment (OD Nature) (2015) Environmental permit Northwester 2 of December 18th, 2015. Annex I: Conditions in the environmental permit (Bijlage I – De Gebruiksvoorwaarden voor de bouw en de expolitatie van den activiteit). Retrieved 18 December 2018 from https://odnature.naturalsciences.be/downloads/mumm/northwester_2/mbnorthwester2_18dec2015_voorwaarden.pdf

OSPAR (2009) Overview of the impacts of anthropogenic underwater sound in the marine environment. Biodiversity Series. London: OSPAR Commission. Retrieved 19 December 2018 from https://tethys.pnnl.gov/sites/default/files/publications/Anthropogenic_Underwater_Sound_in_the_Marine_Environment.pdf

OSPAR (2016) OSPAR Inventory of measures to mitigate the emission and environmental impact of underwater noise (2016 update). Biodiversity Series. London: OSPAR Commission. Retrieved 2 April 2019 from https://www.ospar.org/about/publications?q=&a=&y=2016&s=

Perrow, M.R. (2019) A synthesis of effects and impacts. In Perrow, M.R. (ed.) *Wildlife and Wind Farms, Conflicts and Solutions. Volume 3. Offshore: Potential effects*. Exeter: Pelagic Publishing. pp. 235–277.

Perrow, M.R., Gilroy, J.J., Skeate, E.R. & Tomlinson, M.L. (2011) Effects of the construction of Scroby Sands offshore wind farm on the prey base of little tern *Sternula albifrons* at its most important UK colony. *Marine Pollution Bulletin* 62: 1661–1670.

Popper, A.N. & Hastings, M.C. (2009) The effects of anthropogenic sources of sound on fishes. *Journal of Fish Biology* 75: 455–489.

Popper, A.N., Fewtrell, J., Smith, M.E. & McCauley, R.D. (2004) Anthropogenic sound: effects on the behavior and physiology of fishes. *Marine Technology Society Journal* 37: 35–40.

PTC Fayat Group (2012) The Trio 3x200HD driving monopiles in the Anholt offshore wind farm, Denmark. PTC – Job report – Offshore piling. PTC Fayat Group. Retrieved 14 December 2018 from https://www.scribd.com/document/116409069/ERKE-Group-Trio-3x200HD-PTC-Vibrodriving-monopiles-in-the-Anholt-Offshore-wind-farm-Denmark

Remmers, P. & Bellmann, M.A. (2013) Untersuchung und Erprobung von Hydroschalldämpfern (HSD) zur Minderung von Unterwasserschall bei Rammarbeiten für Gründungen von Offshore-Windenergieanlagen. Auswertung der Hydroschallmessungen im OWP London Array. Project No. 1918-c-bel. Oldenburg: Institute of Technical and Applied Physics (ITAP). Retrieved 14 December 2018 from https://docplayer.org/45712508-Auswertung-der-hydroschallmessungen-im-owp-london-array-dr-rer-nat-michael-a-bellmann.html

Remmers, P. & Bellmann, M.A. (2015) Offshore-Windpark Borkum Riffgrund 1 – Hydroschallmessungen während der Installation einer OWEA Gründungsstruktur mittels eines Suction Buckets. Messbericht. Version 4 – Project no. 2381-14. Oldenburg: Institute of Technical and Applied Physics Ltd (ITAP), 29 pp., unpublished.

Richardson, W.J., Malme, C.I., Green, C.R., Jr & Thomson, D.H. (1995) *Marine Mammals and Noise*. San Diego, CA: Academic Press.

Rijksdienst voor Ondernemend (RVO) (2017) Appendices Hollandse Kust (zuid) Wind Farm Sites I & II. Appendix A: Applicable law. Part of project and site description. Assen: Netherlands Enterprise Agency (RVO). Retrieved 14 December 2018 from https://offshorewind.rvo.nl/file/download/54543032

Robinson, S.P., Lepper, P.A. & Hazelwood, R.A. (2014) Good practice guide for underwater noise measurement. NPL Good Practice Guide No. 133. National Measurement Office, Marine

Scotland. The Crown Estate. Retrieved 2 April 2019 from https://www.researchgate.net/publication/263229365_Good_Practice_Guide_No_133_Underwater_Noise_Measurement

Russell, D.J.F., Hastie, G.D., Thompson, D., Janik, V.M., Hammond, P.S., Scott-Hayward, L.A.S., Matthiopoulos, J., Jones, E.L. & McConnell, B.J.J. (2016) Avoidance of wind farms by harbour seals is limited to pile driving activities. *Journal of Applied Ecology* 53: 1642–1652.

Rustemeier, J., Neuber, M., Grießmann, T., Ewaldt, A., Uhl, A., Schultz-von Glahn, M., Betke, M., Matushek, R. & Lübben, A. (2012) Konzeption, Erprobung, Realisierung und Überprüfung von lärmarmen Bauverfahren und Lärmminderungsmaßnahmen bei der Gründung von Offshore-WEA – Schall 3. BMU Project Reference 0327645. Hannover: Leibniz Universität Hannover. Retrieved 14 December 2018 from https://www.tib.eu/en/search/id/tema%3ATEMA20140305510/Konzeption-Erprobung-Realisierung-und-%C3%9Cberpr%C3%BCfung/

Saleem, Z. (2011) Alternatives and modifications of monopile foundation or its installation technique for noise mitigation. Report commissioned by the North Sea Foundation. Delft: TU Delft. Retrieved 18 December 2018 from www.vliz.be/imisdocs/publications/223688.pdf

Schorcht, S. (2014) Technischer Schallschutz in Offshore-Bauvorhaben am Beispiel der Offshore-Windparks Meerwind Süd-Ost, Global Tech I, NordseeOst, DanTysk, EnBW Baltic 2, Borkum Riffgrund 1, Amrumbank West und Butendiek. Presentation at BSH Workshop, Hamburg, 9 October 2014. Retrieved 14 December 2018 from https://docplayer.org/51501498-Technischer-schallschutz-in-offshore-wind-bauvorhaben.html

Shonberg, A., Harte, M., Aghakouchak, A., Brown, C.S.D., Pacheco Andrade, M. & Liingaard, M.A. (2017) Suction bucket jackets for offshore wind turbines: applications from in situ observations. In Shin, Y. (ed.) *Foundation Design of Offshore Wind Structures. Proceedings of TC 209 Workshop, 19th International Conference on Soil Mechanics and Geotechnical Engineering (ICSMGE)*, Seoul, 20 September 2017. pp. 65–77. Retrieved 14 December 2018 from https://www.researchgate.net/publication/327572904_Suction_bucket_jackets_for_offshore_wind_turbines_applications_from_in_situ_observations

Skjellerup, P., Maxon, C.M., Tarpgaard, E., Thomsen, F., Schack, H.B., Tougaard, J., Teilmann, J., Madsen, K.N., Mikaelsen, M.A. & Heilskov, N.F. (2015) Marine mammals and underwater noise in relation to pile driving – Working Group 2014. Report to the Danish Energy Authority. Energinet.dk: 1–20. Retrieved 19 December 2018 from https://www.researchgate.net/publication/279884643_Marine_mammals_and_underwater_noise_in_relation_to_pile_driving_-_Working_Group_2014_Report_to_the_Danish_Energy_Authority

Southall, B.L., Bowles, A.E., Ellison, W.T., Finneran, J.J., Gentry, R.L., Greene, C.R.J., Kastak, D., Ketten, D.R., Miller, J.H., Nachtigall, P.E., Richardson, W.J., Thomas, J.A. & Tyack, P. (2007) Marine mammal noise exposure criteria: initial scientific recommendations. *Aquatic Mammals* 33: 411–521.

Spence, J., Fischer, R., Bahtiarian, M., Boroditsky, L., Jones, N. & Dempsey, R. (2007) Review of existing and future potential treatments for reducing underwater sound from oil and gas industry activities.NCE Report 07-001. Noise Control Engineering, Inc. for Joint Industry Programme on E&P Sound and Marine Life. Retrieved 18 December 2018 from https://gisserver.intertek.com/JIP/DMS/ProjectReports/Cat1/JIP-Proj1.3_UnderwaterSoundFromOil_2007.pdf

ST3 Offshore (2018) First transport with ST3 Offshore's jackets on the way to Cuxhaven. ST3 Offshore. Retrieved 14 December 2018 from https://st3-offshore.com/news/article/article/first-transport-with-st3-offshores-jackets-on-the-way-to-cuxhaven-55/

Tasker, M.L., Amundin, M., Andre, M., Hawkins, T., Lang, W., Merck T., Scholik-Schlomer, A., Teilmann, J., Thomsen, F., Werner, S. & Zakharia, M. (2010) Marine Strategy Framework Directive – Task Group 11 Report – Underwater noise and other forms of energy. Luxembourg: Joint Research Centre and International Council for the Exploration of the Sea. Retrieved 19 December 2018 from http://ec.europa.eu/environment/marine/pdf/10-Task-Group-11.pdf

Thomsen, F., Lüdemann, K., Kafemann, R. & Piper, W. (2006) Effects of offshore wind farm noise on marine mammals and fish. Hamburg: Biola, on behalf of COWRIE. Retrieved 2 April from https://tethys.pnnl.gov/sites/default/files/publications/Effects_of_offshore_wind_farm_noise_on_marine-mammals_and_fish-1-.pdf

Thomsen, F., Mueller-Blenkle, C., Gill, A., Metcalfe, J., McGregor, P., Bendall, V., Amdersson, M., Sigray, P. & Wood, D. (2012) Effects of pile driving on the behavior of cod and sole. In Hawkins, A. & Popper, A.N. (eds) *Effects of Noise on Aquatic Life*. New York: Springer. pp. 387–389.

Thomsen, F., Gill, A., Kosecka, M., Andersson, M., Andre, M., Degraers, S., Folegot, T., Gabriel, J., Judd, A., Neumann, T., Norro, A., Risch, D.,

Sigray, P., Wood, D. & Wilson, B. (2015) MarVEN – environmental impacts of noise, vibrations and electromagnetic emissions from marine renewable energy. Final Study Report. Brussels: European Commission, Directorate General for Research and Innovation. Retrieved 19 December 2018 from https://publications.europa.eu/en/publication-detail/-/publication/01443de6-effa-11e5-8529-01aa75ed71a1

Trimoreau, B.& Smidt-Luetzen, R. (2017) Modelling of Blue Piling hammer radiated noise – Comparison to hydraulic hammering. Report no. 17.4362 Rev 2 for Fistuca BV. Hellerup: Lloyd's Register Consulting – Energy A/S, February 2017, 22 pp., unpublished.

Tyack, P.L. & Clark, C.W. (2000) Communication and acoustic behavior of dolphins and whales. In Au, W., Popper, A.N. & Fay, R. (eds) *Hearing by Whales and Dolphins.* Springer Handbook of Auditory Research Series. New York: Springer. pp. 156–224.

Umweltbundesamt (UBA) (2011) Empfehlung von Lärmschutzwerten bei der Errichtung von Offshore-Windenergieanlagen (OWEA), Information Unterwasserlärm. Dessau: Federal Environment Agency (UBA). Retrieved 18 December 2018 from https://www.umweltbundesamt.de/sites/default/files/medien/publikation/long/4118.pdf

Universal Foundation (2017) Dogger Bank Mono Buckets successfully decommissioned. Universal Foundation. Retrieved 18 December 2018 from https://www.windbusinessintelligence.com/news/dogger-bank-mono-buckets-successfully-decommissioned

Vattenfall (2017) Offshore wind: suction buckets reduce costs and underwater noise. Vattenfall AB. Retrieved 18 December 2018 from https://group.vattenfall.com/press-and-media/news--press-releases/newsroom/2017/offshore-wind-suction-buckets-reduce-cost-and-underwater-noise

Verfuß, T. (2014) Noise mitigation systems and low-noise installation technologies. In BSH & BMU (eds) *Ecological research at the offshore windfarm Alpha Ventus: Challenges, results and perspectives.* Federal Maritime and Hydrographic Society of Germany (BSH) and Federal Ministry for the Environment, Nature Conservation and Nuclear Safety (BMU). Wiesbaden: Springer Spektrum. pp. 181–191.

Verfuß, U.K., Andersson, M., Folegot, T., Laanearu, J., Matuschek, R., Pajala, J., Sigray, P., Tegowski, J. & Tougaard, J. (2014) BIAS standards for noise measurements. Background information and guidelines and quality assurance. Baltic Sea Information on tthe Acoustic Soundscape. Retrieved 18 December 2018 from https://biasproject.files.wordpress.com/2013/11/bias_standards_v3-2.pdf

Verfuß, U.K., Gillespie, D., Gordon, J., Marques, T., Miller, B., Plunkett, R., Theriault, J. Tollit, D., Zitterbart D., Hurbert, P. & Thomas, L. (2016) Low-visibility real-time monitoring techniques review. Report No. SMRUM-OGP2015-002. Provided to IOGP. St Andrews: SMRU Consulting. Retrieved 18 December 2018 from https://gisserver.intertek.com/JIP/DMS/ProjectReports/Cat4/Other/Verfuss2017_LowVisMonitoring.pdf

Weilgart, L.S. (2007) The impacts of anthropogenic ocean noise on cetaceans and implications for managements. *Canadian Journal of Zoology* 85: 1091–1116.

Wochner, M., Lee, K., McNeese, A. & Wilson, P. (2016) Underwater noise mitigation from pile driving using a tuneable resonator system. Paper ICA2016-0503. In Miyara, F., Accolti, E., Pasch, V. & Vechiatti, N. (eds) *2nd International Congress on Acoustics, Buenos Aires. Wind Farm Noise Buenos Aires.* Retrieved 18 December 2018 from http://www.ica2016.org.ar/ica2016proceedings/ica2016/ICA2016-0503.pdf

World Organisation of Dredging Associations (WODA) (2013) Technical guidance on: Underwater sound in relation to dredging. Delft: World Organisation of Dredging Associations. Retrieved 19 December 2018 from https://dredging.org/documents/ceda/html_page/2013-06-woda-technicalguidance-underwatersoundlr.pdf

Zhang, P., Ding, H., Le, C. & Liu, X. (2012) Test on the dynamic response of the offshore wind turbine structure with the large-scale bucket foundation. *Procedia Environmental Sciences* 12: 856–863.

CHAPTER 8

Mitigation for birds

With implications for bats

ANDREW J. P. HARWOOD and MARTIN R. PERROW

Summary

This chapter reviews the available options for reducing the potential effects of offshore wind farms (OWFs) upon birds through collision, displacement and indirect effects on their prey. Possible options to mitigate bat collision are also reviewed. There are only a few examples of mitigation, and therefore untested theoretical options and approaches used onshore that could translate to the offshore environment are included. As a first step, careful marine spatial planning to locate wind farms in less sensitive areas may avoid some potential impacts. Otherwise, measures to minimise, reduce or compensate for impacts will be required. Modifications to OWF configurations and turbine design, combined with sensitive construction methods, provide clear means of minimising impacts. Where a residual risk of collision remains, this may be reduced by making turbines more visible by painting blades before installation, and deterring birds from approaching turbines, including the use of 'green' lighting for nocturnal migrants. Curtailing turbine operation for migratory species, altering cut-in speeds for bats and providing alternative perches for seabirds, which might otherwise use OWF structures, all show promise to mitigate collision losses. However, the options to reduce displacement of sensitive species such as divers (loons) *Gavia* spp. are limited. Where impacts cannot be reduced to acceptable levels, compensatory measures should be implemented. These are likely to be site- and species-specific and for seabirds incorporate a range of strategies from habitat management and predator control at breeding sites, to reducing interactions with fisheries at sea. Repowering affords the opportunity to reduce impacts through refined spatial planning and the use of fewer larger turbines; although the 'renewables to reefs' transition may see few OWFs fully decommissioned and baseline conditions restored. Detailed monitoring of effects/impacts at both a site and cumulative scale remains crucial in order to promote understanding and to test any mitigation measures employed.

Introduction

While wind energy is a critical component of the efforts to address climate change targets, the construction and operation of offshore wind farms poses a risk to several groups of birds including seabirds (King 2019; Vanermen & Stienen 2019) and non-seabirds, such as migratory waterbirds, passerines and raptors; as well as migratory bats (Hüppop *et al.* 2019), which share some similarities with small nocturnally migrating birds. Negative effects from OWFs may result from collisions, displacement including barrier effects, and indirect effects such as reductions in prey availability and other ecosystem effects (Perrow 2019). Reducing any risk of impacts to an acceptable level constitutes a key environmental issue for consenting and public acceptance of wind farms (Drewitt & Langston 2006; Fox *et al.* 2006; Langston 2013; Schuster *et al.* 2015; Green *et al.* 2016). Reduction of negative impacts to an acceptable level must be undertaken with sufficient confidence to ensure that the development is compliant with applicable conservation legislation (Jameson *et al.* 2019). Thus, consideration of mitigation measures, during for example Environmental Impact Assessment (EIA) or Habitats Regulations Assessment (HRA) in relation to European protected sites, is crucial to ensuring that conservation objectives are met (OSPAR 2008). In some countries, such as Australia, an Environmental Management Plan (EMP) may also be required, incorporating a mitigation strategy to deal with specific impacts. Statutory bodies may also impose licence conditions that require the developer to take further steps in relation to mitigation and monitoring, particularly of predicted residual effects.

Given the global expansion in the offshore wind industry, the probability of significant cumulative effects occurring is increasing, further complicating the consenting process and placing the onus on effective mitigation measures (Gartman *et al.* 2016a; Willsteed *et al.* 2018). Consideration of mitigation at an early stage may thus reduce overall consenting risk and avoid delays in the consenting process that may otherwise affect renewable energy targets. Even where no significant impacts are likely to arise from a project, there is always scope for mitigation of possible effects to minimise or reduce risk, particularly where there is uncertainty in assessments. Mitigation also provides a pathway for engagement with critics and opponents of developments, through which opinions might be challenged, awareness raised and amicable solutions found. However, mitigation often involves trade-offs or costs for the developer and operator, and there is often uncertainty about the efficacy of specific measures, particularly when these are untested.

The adoption of an accepted 'mitigation hierarchy' during assessments and planning provides a tool for prioritising and targeting mitigation and helps to ensure that no net loss (NNL) or a net positive impact (NPI) is achieved (PricewaterhouseCoopers 2010). Some authors, such as Langston & Pullan (2003) and Marques *et al.* (2014), note a lack of consistency in the application of mitigation in relation to projects that may affect biodiversity. Nevertheless, May (2017) demonstrated the validity and application of a mitigation hierarchy approach in relation to onshore wind farms.

Scope

This chapter presents and discusses mitigation options for the reduction of collision risk, displacement and barrier effects, and indirect effects associated with OWF projects, in relation to birds with implications for bats. As the offshore equivalent to the onshore chapter provided by May (2017) in Volume 2 of this series, the themes are structured in relation to the mitigation hierarchy of: (1) avoid, (2) minimise, (3) reduce, (4) compensate and (5) restore. Thus, in simple terms, impacts should be avoided during planning, but if they cannot be, then steps can be taken in the design stage to minimise impacts and/or to reduce or compensate for them during construction and operation. If a wind farm is to be decommissioned, effort is required to restore the site to its original pre-impact condition, although this may not be desirable should the site have acquired increased biodiversity and conservation value (Dannheim *et al.* 2019). Alternatively, wind farms may be repowered with more efficient and powerful turbines, presenting an opportunity to minimise and reduce impacts.

The information presented in this chapter builds on previous reviews by a number of authors, including Langston & Pullan (2003), Hüppop *et al.* (2006), Drewitt & Langston (2006; 2008), Edkins (2008), Cook *et al.* (2011), Gove *et al.* (2013), Marques *et al.* (2014), May *et al.* (2015) and Gartman *et al.* (2016a; 2016b), as well as an extensive internet search (via Google and Google Scholar) of the most recent peer-reviewed and grey literature, and following referenced literature. In addition, the authors have direct experience of many wind farms in the UK and potential discussions surrounding mitigation of potential effects, which have not been specifically published. The focus of any OWF literature is invariably concerned with development in north-western European waters in the North, Baltic and Irish Seas, where the global wind-farm industry has been concentrated. This is now changing with burgeoning markets particularly in south-east Asia, including China, Japan and Taiwan, along the eastern seaboard of the USA, as well as in other European countries (Jameson *et al.* 2019).

Searches revealed relatively few examples of the application of targeted mitigation measures for birds in the offshore environment. As a result, the possible application of measures adopted onshore was also considered, specifically by inviting opinion in the form of boxed case studies from selected researchers. This included the case of painting blades to increase visibility to vulnerable bird species (Hodos 2003; Johnson *et al.* 2007; May *et al.* 2017) and the potential to reduce collisions from temporary turbine shutdown (Tomé *et al.* 2017a). In effect, these studies provide real examples of the application of measures or the processes involved. Moreover, many of the concepts to reduce effects, particularly in relation to displacement and indirect effects that are presented, remain hypothetical and require testing.

No literature on mitigation of effects of OWFs on bats was found. This is primarily because basic knowledge is still accumulating on the extent of the migration of bats in specific areas where OWFs are being developed, such as the Baltic Sea, as well as the more general potential for bats to occur offshore, especially in the USA (Petersen *et al.* 2016; see Hüppop *et al.* 2019 for a review of literature). As a result, an opinion piece on the topic of potential interactions of bats with offshore turbines and the ways in which any negative effects may be mitigated is provided (Box 8.1). Otherwise, birds are the primary focus of this chapter.

Box 8.1 Scope for mitigating risk to bats at offshore wind farms

Lothar Bach

In the terrestrial environment, the risk of wind farms to bats from collision or barotrauma during migration, investigation or foraging is well documented (Arnett *et al.* 2008; Baerwald & Barclay 2009; Brinkmann *et al.* 2011; Kunz *et al.* 2007; Cryan *et al.* 2014; Jameson & Willis 2014; Rydell *et al.* 2010; Voigt *et al.* 2015). Observations in the Baltic Sea and North Sea have provided evidence that bats also approach and forage within wind farms (Ahlén *et al.* 2007; 2009; Lagerveld *et al.* 2014), other offshore installations (Seebens *et al.* 2013; Hüppop & Hill 2016; Bach *et al.* 2017) and vessels (Sonntag *et al.* 2006; Walter *et al.* 2007). Bats have also been found foraging at sea, up to 2.2 km from the German Baltic coast in June (Seebens *et al.* 2013), although the importance of at-sea prey resources such as moths and other insects to bats remains unknown. It is also unclear to what extent insects are attracted to wind farms as a result of the colour of the structures (Long *et al.* 2011), or lighting, given that they may be the only refuge for many miles at sea. There is also the potential for bats to roost in wind-farm structures, particularly in relation to the larger substations (Ahlén *et al.* 2009). Pelletier *et al.* (2013) conducted a synthesis of information on the potential for bat interaction with offshore wind farms (OWFs) and concluded that there was sufficient evidence to suggest that many species migrate offshore and use islands, ships and other offshore structures as opportunistic or deliberate stopover sites, but that little is still known about the seasonality and extent of the migrations and how behaviour such as foraging patterns may differ offshore. They also found that migratory species were equally likely to be recorded at offshore, coastal or inland sites, while non-migratory species were less likely to be recorded at offshore sites relative to coastal and inland sites.

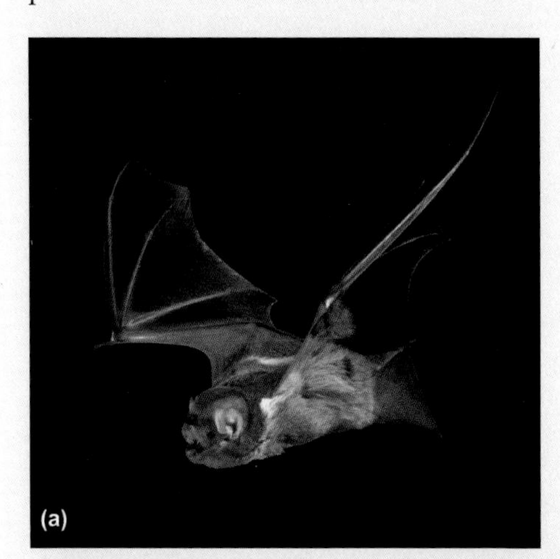

Figure 8.1 Migratory bats known to routinely cross large bodies of open water on migration in the USA include: (a) Eastern Red Bat *Lasiurus borealis* (Hatch *et al.* 2013) and (b) Silver-haired Bat *Lasionycteris noctivagans* (see McGuire *et al.* 2012). Note the Silver-haired Bat has been fitted with a MOTUS tag (see Molis *et al.*, Chapter 6). (Sherri and Brock Fenton)

A review of the characteristics of offshore migration in bats by Hüppop *et al.* (2019) outlines that activity may be focused on a specific flight for particular species, and although a range of species may be encountered, some, such as Eastern Red Bat *Lasiurus borealis* in the USA (Figure 8.1) and Nathusius' Pipistrelle *Pipistrellus nathusii* in Europe, are especially prevalent. In relation to the former, Hatch *et al.* (2013) recorded a total of 12 Eastern Red Bats 16.9–41.9 km from shore during aerial digital and boat-based surveys mainly targeted at birds (see Webb & Nehls, Chapter 3) on the Atlantic seaboard of the USA. All six of the bats for which flight height could be confidently calculated were recorded at >100 m, with five of these six at >200 m from the sea surface.

However, very little is known about the potential impacts of OWFs on bats. As with onshore wind farms, while offshore turbines may represent a significant collision risk, they are thought unlikely to cause displacement or disturbance of bats. Thus, the range of possible mitigation measures at onshore sites (Arnett 2017) seem likely to be applicable to offshore projects (see also Wilson *et al.* 2010). Many of the measures discussed in relation to birds in this chapter may therefore also be applicable to bats, although site- and species-specific knowledge is critical to the selection of appropriate measures.

As with birds, avoidance of impacts relies on ensuring that wind farms are not built in high-risk areas, including known migration routes or areas intensively used by bats. As little is yet known about offshore movements and habitat use, owing in part to difficulties surveying in the offshore environment (Petersen *et al.* 2016), this is rather difficult. Where wind farms cannot be re-sited to avoid potential impacts, there are a number of potential options for minimising and reducing risks. Deterrents, including ultrasound emissions (Arnett *et al.* 2008; 2013a; Horn *et al.* 2008) or radar (Nicholls & Racey 2007; 2009), may provide a means of reducing collision risk. However, these methods are often costly, are currently unproven at operating wind farms, and could have negative impacts on other conservation interests such as seabirds that could counter any benefits accrued for bats (Amorim *et al.* 2012; Rodrigues *et al.* 2015). Minimising attraction to the site or a particular turbine could also reduce potential collision risk. Long *et al.* (2011) found that 'pure white' and 'light grey' colours seem to attract insects and consequently result in increased insectivore activity. Thus, coloration could be modified to reduce prey attraction. Furthermore, potential roost sites could be limited at an offshore array by limiting access to internal spaces, for instance using 'scrubbers' (filters) on covers, as applied to some onshore wind turbines.

The most promising mitigation measures onshore have been the temporary shutdown of turbines when bats are likely to be present during migration (Baerwald & Barclay 2009; Arnett *et al.* 2010; 2011; 2013b; Brinkmann *et al.* 2011; Rodrigues *et al.* 2015; Arnett 2017; Behr *et al.* 2018). Research at onshore wind farms has also shown bat activity to be correlated with weather, including wind speed, temperature and rain (Horn *et al.* 2008; Bach & Bach 2009; Behr *et al.* 2011; Brinkmann *et al.* 2011; Amorim *et al.* 2012). Bats also generally prefer low wind speeds of <4 m/s, though this may be variable, and high temperatures (Ahlén *et al.* 2007; 2009; Arnett *et al.* 2008; Poerink & Haselager 2013; Lagerveld *et al.* 2014). Below the manufacturer's cut-in speed, turbines 'freewheel' without producing electricity, although the rotational speeds achieved may be more than sufficient to kill bats, and therefore Arnett (2017) suggested that raising the cut-in speed by at least 1.5 m/s (range 1.5–3 m/s) would have the effect of preventing the turbines rotating until the cut-in speed is achieved, thereby significantly reducing the number of bat mortalities. Feathering the blades, that is, pitching the blades to 90° and parallel to the wind, below cut-in speed also has the effect of reducing bat mortality. While this type of mitigation, particularly during bat migration periods, shows promise, the cut-in speeds would be best tailored to specific species and sites.

In relation to potential compensation measures, Peste *et al.* (2015) provide a review of several possible approaches to enhance assemblages of bats that have been negatively impacted. For example, they suggest management of autochthonous forest, diversification of forest and agriculture monocultures, preservation of existing roosts, provision of new roosts and creation of ponds. However, such compensation measures may be ineffective if the origin of the bats passing offshore turbines is unknown.

More research is required to inform understanding of general bat migration across large, open bodies of water and bat behaviour at OWFs. To assess the potential impact of planned or existing sites, specific at-sea bat surveys are needed during Environmental Impact Assessment (BSH 2013; Peterson *et al.* 2014; Rodriguez *et al.* 2015; Sjollema *et al.* 2014; Bach *et al.* 2017), and during and after construction (Lagerveld *et al.* 2014). Experience of monitoring at onshore sites (e.g. Hein 2017) should be used to develop offshore survey methods and assessments, with modifications to account for the conditions and size of the structures and arrays.

Besides avoiding building wind turbines in areas intensively used by bats (migrating and/or foraging), temporary shutdowns during periods of high bat activity (mainly autumn migration period) are likely to be the most effective mitigation measure, with only minor economic impacts (Brinkmann *et al.* 2011). In the European Union (EU), the Habitats Directive focuses on the protection of individual specimens of a species, with regulations that prevent the intentional killing of a bat rather than protecting specific populations (Voigt *et al.* 2015; Rodrigues *et al.* 2015). Mitigation should therefore be targeted to prevent fatalities in the first instance, but where there are unavoidable impacts, suitable methods of compensation should be identified to offset the losses (e.g. habitat creation, restoration or enhancement). However, it may be very difficult to accurately target and implement such measures for specific populations, particularly for migratory species that are of particular importance at OWFs, over large spatial scales (Voigt *et al.* 2012; Lehnert *et al.* 2014). Further research is required to quantify the impacts of the combined anthropogenic pressures on bat populations and to identify appropriate conservation measures.

Themes

Avoiding risk

Mitigation is only required to reduce any impacts to an acceptable level. Therefore, if any predicted unacceptable levels of impact can be avoided in the planning stages, there is no requirement for mitigation. It therefore makes sense to avoid areas with likely conflict with particular bird interest, including foraging areas or migratory routes (e.g. Exo *et al.* 2003; Everaert & Stienen 2007). Such areas should be considered in a holistic approach to marine spatial planning, whereby environmental aspects as well as profitability, social and security factors are taken into account (Ho *et al.* 2018; Köppel *et al.*, Chapter 9). Geographic information systems (GIS) are an important tool for mapping and visualising the multiple factors that influence site selection, such as the physical characteristics of an area, alongside economic, social and environmental metrics. A 2018 study, focusing on potential development of OWFs in the south-west coastal area of South Korea, exemplifies this approach (Choong-Ki *et al.* 2018). Similarly, Chaouachi *et al.* (2017) used a multicriteria approach to help identify appropriate areas for OWFs in relation to the Baltic States in Europe.

More detailed sensitivity mapping may then consider an individual bird species' vulnerability to potential negative effects of OWFs stemming from a wide range of species-specific factors, including flight behaviour (e.g. tendency to flock and manoeuvrability), site fidelity, avoidance responses, habitat preferences and conservation status. Such analyses have been carried out both onshore and offshore in a number of countries, including Germany (Garthe & Hüppop 2004), Scotland (Bright *et al.* 2008), Denmark (Desholm 2009), England (Bright *et al.* 2009; Bradbury *et al.* 2014), Ireland (McGuinness *et al.* 2015) and the USA (Kelsey *et al.* 2018; American Bird Conservancy 2019).

The SeaMast tool developed by Bradbury *et al.* (2014) produced sensitivity maps for a range of seabird species both in isolation and in combination (see Figure 3.1 in Webb & Nehls, Chapter 3) in English waters to a maximum of 200 nautical miles or the territorial water boundary with a neighbouring country. SeaMast used data from boat-based and aerial surveys (see Webb & Nehls, Chapter 3) spanning the period 1979–2012, coupled with species sensitivity values for seabirds developed by Garthe & Hüppop (2004) and expanded by Furness & Wade (2012) and Furness *et al.* (2013a). As well as identifying priority areas for seabirds, the intention of the tool was to inform marine spatial planning (see Köppel *et al.*, Chapter 9) and EIA in relation to a particular development, with the further possibility that it would operate as a 'sense check' of site-specific data. A key feature of the tool is that the base maps may be readily updated as more data become available and incorporated into the different data layers. It should be noted, however, that SeaMast was unable to inform development in the first three Rounds of wind-farm development in the UK as it was developed after these had been planned, although it remains available for future development rounds.

Furthermore, Loukogeorgaki *et al.* (2018) incorporated EIA into their spatial analysis for site selection of hybrid offshore wind and wave energy systems in Greece through development of an environmental performance value (EPV) to combine all potential impacts, including those for birds, as a metric alongside physical site characteristics and economic considerations. As a general point, where suitable data are unavailable for such analyses, inference through spatial modelling may be used to address any knowledge gaps (Lapeña *et al.* 2010).

Overall, any mapping exercises aim to quantify potential risk and help identify the 'best' areas for development where space is limited. However, in reality, there may actually be very few areas that 'tick all the right boxes' in multiple criteria assessments and it may be impossible to find suitable areas where all risks to a variety of birds, from seabirds to migratory waterfowl, passerines and raptors, as well as bats, can be easily avoided. This is especially true where baseline data for less well-studied groups (such as migratory passerines, raptors and bats) are lacking.

Minimising risk with careful planning

Wind-farm design and micro-siting

Most OWFs are designed as regular grids of turbines, most often in rectangular or diamond configurations. However, changing the size, shape, layout and orientation of a wind-farm array may reduce displacement, barrier and collision impacts. Such planning is often referred to as 'micro-siting' and requires a robust ecological understanding to inform decision making and provide confidence that reductions in risk will be meaningful and sustained. Site-specific surveys should provide detailed information about site utilisation by sensitive species, with the sources of risk for different seabird species from OWFs derived from expert opinion (Garthe & Hüppop 2004; Furness & Wade 2012; Furness *et al.* 2013b) and post-construction monitoring (Dierschke *et al.* 2016; King 2019; Vanermen & Stienen 2019) in particular.

To date, there is consensus that only a few species, such as Great Cormorant *Phalacrocorax carbo* and Great Black-backed Gull *Larus marinus*, are attracted to wind farms, with Common Eider *Somateria mollissima* consistently indifferent, and species such as divers (loons) (*Gavia* spp.), alcid auks and Northern Gannet *Morus bassanus* generally avoiding wind farms (Vanermen & Stienen 2019). As more information becomes available there may be surprises, such as the unforeseen avoidance of turbines under construction by Sandwich Tern *Thalasseus sandvicensis* at one site (Harwood *et al.* 2017). There is a clear

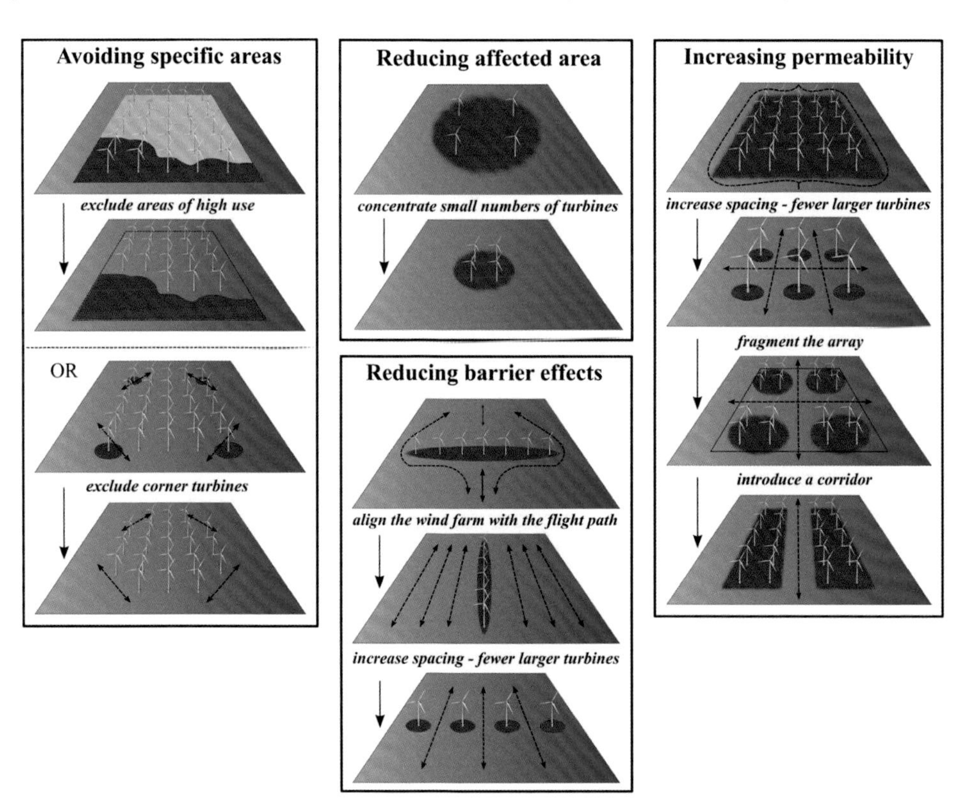

Figure 8.2 Examples of site-level spatial planning approaches that could reduce impacts on birds by avoiding development of areas associated with the highest risk, minimising the area affected, limiting barrier effects and increasing the permeability of the site to birds. See Masden *et al.* (2012) for further consideration and modelling in relation to some of these concepts

need to understand the species-specific behaviour and the drivers for the response of birds, such as the role of 'fear of the novel' even where there is little risk, relative to potentially disturbing factors such as vessels and any changes in the available resource base (Perrow 2019). Further understanding could increase the options available for reducing potential effects and impacts.

Figure 8.2 provides a conceptual overview of possible site-level spatial planning options that could be adopted for the purposes of mitigation. The options available to reduce impacts broadly divide into four areas: (1) avoiding specific areas important to birds, (2) reducing the affected area, (3) reducing barrier effects, and (4) increasing permeability. It is important to note that while the first two options may both reduce potential collision and displacement (and also perhaps indirect effects), application of the second two options may involve a trade-off between displacement and collision as these two effects are more or less mutually exclusive, as displaced birds cannot collide (Perrow 2019). Thus, reducing the level of displacement may increase collision risk for species using the airspace occupied by turbines. The trade-off between encouraging birds to use a wind farm relative to the benefits of discouraging them needs to be made on a case-by-case basis.

Dealing with the four options in turn, the simplest approach to reduce potential impacts is to avoid areas that have been identified as most important for birds within

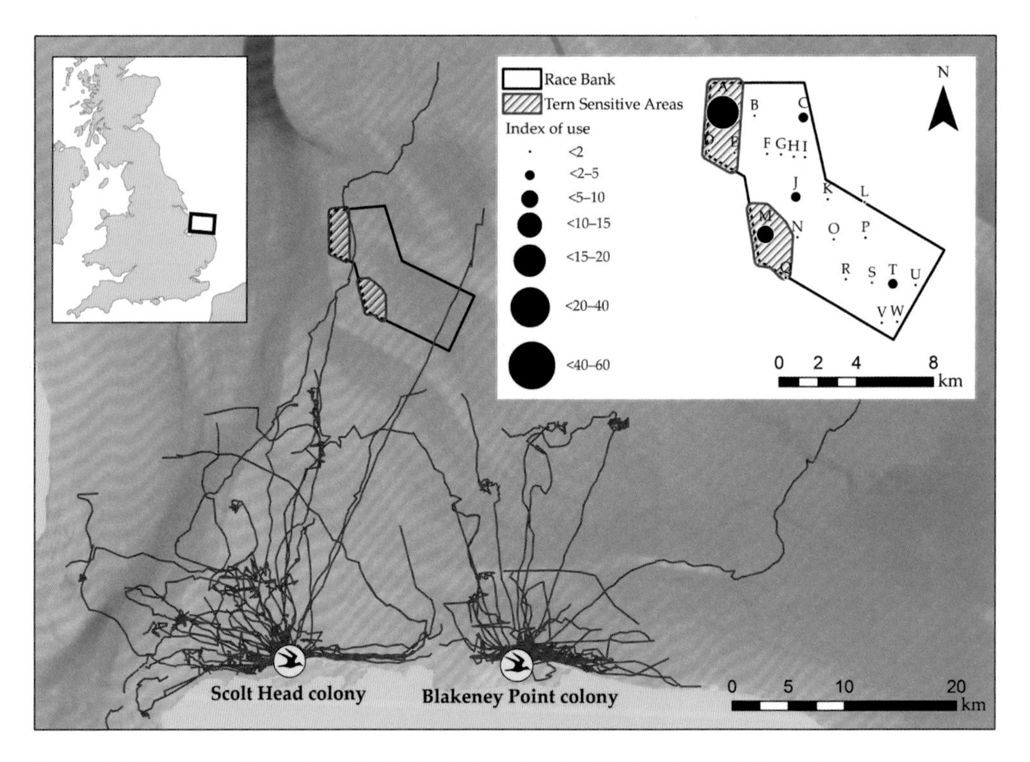

Figure 8.3 The operational Race Bank site in the Greater Wash, UK (see left inset) incorporates the exclusion of two 'tern-sensitive areas' from the original proposed site (black polygon) that exhibited the highest use (black circles in right inset) by Sandwich Terns *Thalasseus sandvicensis* recorded during boat-based surveys. Tracking from the colonies confirmed that birds breeding at two colonies within the North Norfolk Coast Special Protection Area could readily reach the site on longer foraging trips and thus all birds present at the site in the breeding season were assumed to originate from these important colonies. The exclusion of tern-sensitive areas considerably reduced predicted collision risk. [Bathymetry data from EMODnet Bathymetry Consortium 2018: EMODnet Digital Bathymetry (DTM)]

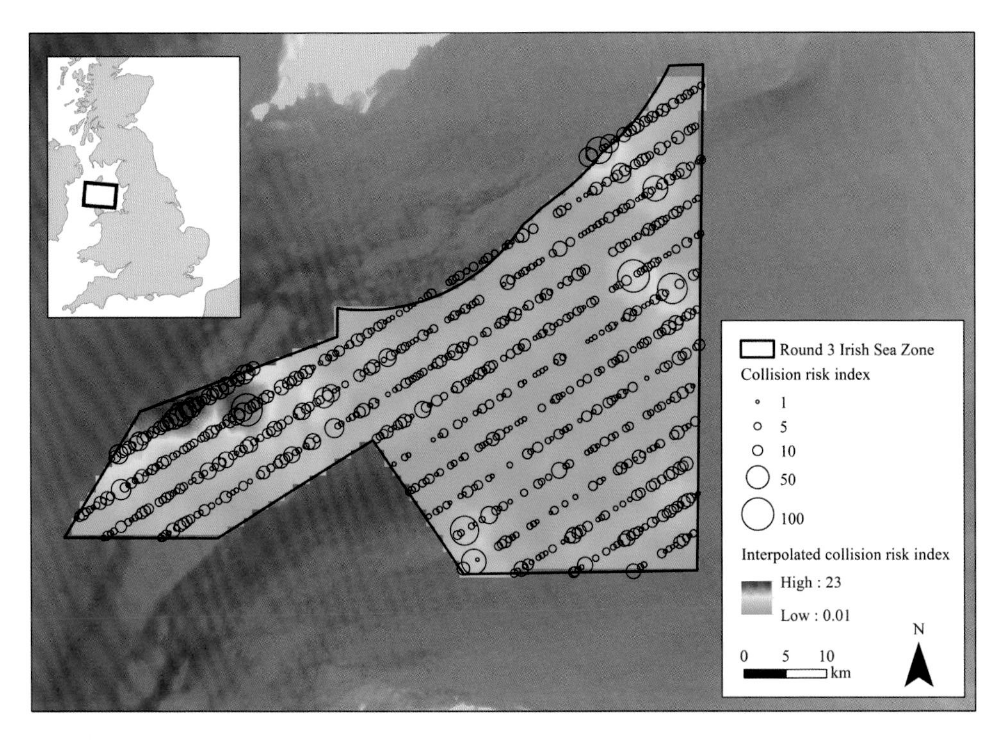

Figure 8.4 Spatial collision risk modelling applied to the Irish Sea Round 3 Zone, UK, during the Zonal Appraisal and Planning (ZAP) process. The collision risk index was derived from mortality estimates calculated using the model and mean monthly snapshot densities of flying birds in survey cells spaced at 500 m intervals. The modelling assumed that a single generic turbine was present in each cell; it used bird parameters derived from the available literature and assumed a 98% avoidance rate as a worst-case scenario. Monthly collision estimates for five sensitive species (Northern Gannet *Morus bassanus*, Black-legged Kittiwake *Rissa tridactyla*, Lesser Black-backed Gull *Larus fuscus*, European Herring Gull *Larus argentatus* and Great Black-backed Gull *Larus marinus*; see Figure 8.5) were summed to provide an overall annual collision estimate. An index for each cell was calculated as the percentage of the maximum number of collisions for any cell, capped at 100 per annum to facilitate interpretation of spatial trends during subsequent interpolation. The collision index is shown by proportional circles centred on each cell with an inverse distance weighted interpolated surface (1 km^2 resolution) to highlight the areas of greatest risk. Material presented here is reproduced from unpublished reports to Centrica Renewable Energy Ltd with permission from Centrica and Ørsted. [Bathymetry data from the EMODnet Bathymetry Consortium 2018: EMODnet Digital Bathymetry (DTM)]

the developable area. In the authors' experience, there are a number of examples where, during the planning stages, developers have decided to modify the areas available for development in order to reduce consenting risk. For example, during characterisation of the Round 3 Firth of Forth Zone using boat-based surveys, Scalp Bank was identified as being important for birds, in accordance with available literature (Wanless *et al.* 1998; Harris *et al.* 2012). As a result, Scalp Bank was excluded from the Phase 1 project development areas (Seagreen Wind Energy 2018). Similarly, at the Round 2 Race Bank in the Greater Wash, areas of particularly high use by Sandwich Terns were also excluded from development (Figure 8.3). The identification of important areas for birds at a local scale to facilitate micro-siting has previously been undertaken at terrestrial sites using a GIS-based approach in the INTACT project (see Box 6.3 in May 2017). Moreover, as well as simply using areas of high density, it is possible to generate spatial 'heat maps' of collision risk from a combination of density data and species-specific parameters that are incorporated into collision risk

Figure 8.5 A selection of key seabird species likely to be of importance in relation to any need to mitigate collision risk at wind farms in the North and Irish Seas in particular. Clockwise from upper right: Northern Gannet *Morus bassanus*, European Herring Gull *Larus argentatus* (in winter plumage), Black-legged Kittiwake *Rissa tridactyla* (in winter plumage) and Lesser Black-backed Gull *Larus fuscus*. These species, alongside Great Black-backed Gull *Larus marinus*, were the component species of the spatial collision risk mapping shown in Figure 8.4. (Martin Perrow)

modelling (see Figure 5.1 in Cook & Masden, Chapter 5), as shown in Figure 8.4. The sensitive species underpinning that exercise (with the exclusion of Great Black-backed Gull, possibly the seabird species most vulnerable to collision; see Figure 9.7 in King 2019) and that often feature in collision risk assessment and discussions of any need to mitigate collision risk are shown in Figure 8.5. In principle, a reduction of the area of the proposed site to reduce potential impacts may be achieved without compromising the capacity of the site, although this does depend on the initial proposed area effectively being surplus to requirements and the maintenance of the optimal spacing of turbines to maximise their efficiency.

The overall continuity, shape, alignment and size of a wind-farm array are likely to affect how a bird will perceive it as a potential risk and whether it is willing to enter the site or is displaced from it. In simple terms, there is evidence, particularly from onshore studies, of a species-specific avoidance distance, with 100% displacement being achieved if this exceeds the spacing between turbines (Perrow 2019). However, avoidance distance appears to vary on a site-specific basis (Hötker 2017), with this being attributable in some cases to a specific parameter such as hub height, with taller turbines increasing avoidance distance (see Figure 7.4 in Hötker 2017 in relation to Northern Lapwing *Vanellus vanellus*). Avoidance distance also appears to depend on resource availability, with birds taking more risk and coming closer to turbines where resources are limited outside the wind farm or more abundant within it (Perrow 2019).

Thus, varying the spacing between turbines in an array affects how permeable it will be to different species. The value of modifying the permeability of a site is demonstrated by Masden *et al.* (2012) using modelled flight trajectories of Common Eider based on post-construction radar data for the Danish Nysted OWF, in relation to different turbine configurations, spacing and extent to assess the likelihood of birds passing between turbines. Reducing the spacing between turbine rows to lower permeability resulted in fewer birds entering the wind farm, with associated reductions in collision risk. However, increasing the number of rows in an array resulted in increasing distances between turbines before birds would fly between them, suggesting that the response of the birds was also sensitive to the overall size of the development. A diamond array oriented with the main direction of travel produced the least permeable configuration due mainly to the alignment of the array. Arrays may also have variable permeability throughout the site, with birds more readily flying through corners of the wind farm, as there are fewer turbines to pass when transiting the risk area (Masden *et al.* 2012). Thus, removing corners and consequently producing a circular array may mean that fewer birds will enter the site, thereby again reducing collision risk (Figure 8.2).

The greatest permeability in the research by Masden *et al.* (2012) was achieved by breaking the site into four blocks in the corners of the site to provide flight corridors or undisturbed space between them (see Figure 8.2). Flight corridors do not appear to have been trialled directly offshore, perhaps partly as there is only limited evidence that birds will use them (Harwood *et al.* 2017). Any benefit of flight corridors seems most likely to be maximised where migration or foraging pathways are intersected by the development, which would otherwise operate as a barrier to movement. However, while corridors may reduce barrier effects, the funnelling and concentration of birds between the component parts of a wind farm may actually increase collision risk in some cases. As a result, the location, orientation and stability of the main flightlines should be established in detail before considering such an approach. Otherwise, barrier effects may be reduced simply by changing the orientation of the site, especially if this is a line of turbines, to align with the direction of the principal flight path (Winkelman 1992; Figure 8.2). This may go some way towards reducing any increased energetic costs that could be incurred, particularly by birds that regularly encounter one or more OWFs (Masden *et al.* 2010).

Finally, minimising any potential impacts of disturbance by construction and maintenance vessels should also be considered. In this case, the routes adopted by vessels should look to avoid important bird areas both within and outside the wind farm, keeping the footprint of the routes as small as possible. In some cases, this may involve adjustment of speed or engine noise to reduce disturbance and the risk of displacement of birds.

Modifying turbine parameters

Turbine design can strongly influence the probability of bird collisions (Anderson *et al.* 1999). The offshore industry generally uses the classic horizontal-axis three-blade monopole structures, although there are some two-bladed turbines such as the Forthwind Offshore Wind Demonstration Project (Methil, Scotland), with other designs such as the vertical-axis concept in development. Thus, while there is scope for different designs in the future, current technology is largely limited to a single design. In turn, any variability in design parameters that could affect collision risk is largely limited to the dimensions of the turbine.

A large turbine, with longer rotor blades, will have a proportionally larger swept area than a smaller turbine, thereby increasing collision risk. However, fewer larger turbines are generally required to produce as much power as more of the smaller models and it is generally accepted that the use of fewer, larger turbines within a site of specific capacity will have a lower impact on seabirds (Cook *et al.* 2011; Johnston *et al.* 2014a). Conversely, the use of larger turbines with blades reaching higher altitudes may result in greater risk for migrating birds such as passerines and waders (Krijsveld *et al.* 2011) and perhaps migratory bats (see Box 8.1). Other turbine specifications such as blade width and rotational speed can also play a role in collision risk, but will generally be proportional to the size of the turbine according to engineering requirements.

As the flight height distribution of seabirds is biased to relatively low height above the sea surface (Furness *et al.* 2013a; Johnston *et al.* 2014a), increasing the clearance of the rotor-swept area above sea level, known as the air gap, is likely to reduce the risk of seabird collisions simply through a reduction of densities of birds at risk height (Anderson *et al.* 1999; Hötker *et al.* 2006). However, increasing the air gap leads to a corresponding increase in the maximum height swept by turbine blades and the potential risk for high-flying migratory species, especially if their normal flight heights that are far in excess of turbine height are reduced by poor weather (Hüppop *et al.* 2019). Consequently, detailed data on site- and species-specific variability in flight heights and behavioural response to turbines, which are then incorporated into collision risk modelling (option 3 in the Band 2012 model), are required to determine what changes in the air gap would be needed to deliver the required impact reductions.

Box 8.2 provides an example of how modifications to rotor length and the air gap may influence collision risk estimates, and considers how collision risk may differ when using small and large turbines within a site of fixed capacity. In reality, turbine size tends to be driven by the industry, with a push towards much larger structures that can deliver more power. For example, the average size of an installed offshore wind turbine was 5.9 MW in 2017, a 23% increase on 2016 (WindEurope 2018). Increasing the air gap is dependent on engineering and financial constraints, but may be an attractive option in order to achieve consent.

Box 8.2 A theoretical demonstration of the effects of rotor blade length, air gaps and turbine size on collision risk of different seabirds

To demonstrate the effects of different rotor blade lengths and air gaps, a collision risk modelling exercise was undertaken for a hypothetical development of ten turbines populated by three common seabird species, Northern Gannet *Morus bassanus*, Lesser Black-backed Gull *Larus fuscus* and Sandwich Tern *Thalasseus sandvicensis*, which vary in morophology and flight activity parameters, as specified in Table 8.1. Option 2 of the Band model (Band 2012) was used in this exercise, in which the generic flight height distribution of the different species were derived from maximum likelihood estimates from Johnston *et al.* (2014a; 2014b). For comparative purposes, monthly density estimates of 0.5 individuals/km^2 were assumed, even though none of the species would show constant density throughout the year as a result of dispersive or migratory movements, particularly in the case of Sandwich Tern, which is a summer

Table 8.1 Indicative bird parameters for Northern Gannet *Morus bassanus*, Lesser Black-backed Gull *Larus fuscus* and Sandwich Tern *Thalasseus sandvicensis* used in collision risk estimation using the Band model (Band 2012) at a hypothetical development

Parameter	Northern Gannet	Lesser Black-backed Gull	Sandwich Tern
Bird length (m)[a]	0.94	0.58	0.38
Wingspan (m)[a]	1.72	1.42	1.00
Flight speed (m/s)	14.9[b]	13.1[b]	8.8[c]
Nocturnal activity (%)[d]	0[e]	50	0
Percentage flying above 20 m according to Johnston *et al.* (2014a; 2014b)	13%	26%	7%

Information from [a]Robinson (2005), [b]Alerstam *et al.* (2007), [c]Perrow *et al.* (2017), [d]Garthe & Hüppop (2004) and [e]Hamer *et al.* (2011).

migrant, present at breeding grounds for around 6 months. Avoidance rates were assumed to be zero to allow comparison across the different species, again despite these varying between species, although they are generally accepted to be in excess of 99% (Cook *et al.* 2012; Skov *et al.* 2018). A constant operational time of turbines for each month of 95% was adopted. Rotational speed, maximum blade width and pitch were also kept constant, at 11.5 rpm, 4.2 m and 10°, respectively. With all other parameters effectively fixed, the influence of a variety of realistic turbine rotor radii (50–80 m), given a fixed hub height of 100 m, and varying hub heights (60–100 m), given a fixed rotor radius of 50 m, upon modelled predictions was assessed.

The results illustrate the considerable variability in potential risk from simple changes in turbine parameters (Figure 8.6). For example, for Lesser Black-backed Gull, generally the most sensitive species as a result of its tendency towards higher flight height and nocturnal activity (Table 8.1), increasing hub height from 60 to 100 m (given a blade length of 50 m) resulted in a 92% reduction in collision estimates.

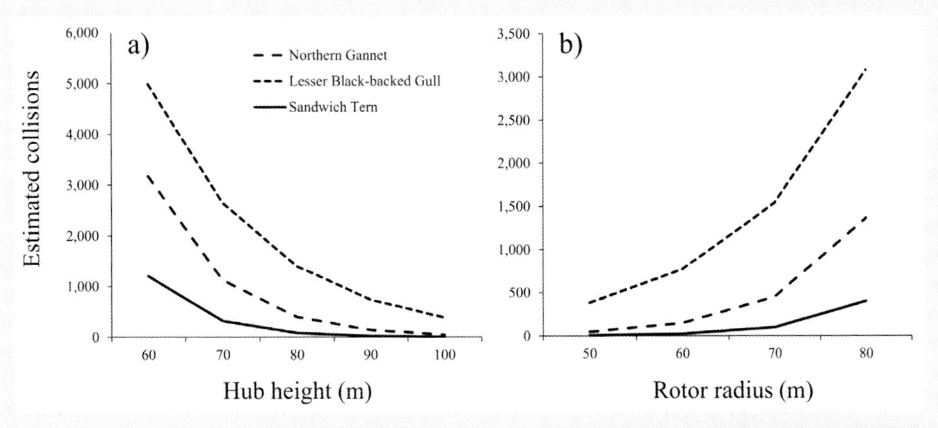

Figure 8.6 Collision risk estimates (collisions per annum) with no avoidance for Northern Gannet *Morus bassanus*, Lesser Black-backed Gull *Larus fuscus* and Sandwich Tern *Thalasseus sandvicensis* using Band model option 2 (Band 2012) according to variation in (a) hub height from 60 to 100 m and (b) rotor radius 50–80 m in isolation, while maintaining the constancy of other turbine parameters.

Conversely, increasing the rotor radius from 50 m to 80 m (maintaining a hub height of 100 m) resulted in a 702% increase in potential collisions. Thus, a single larger turbine with longer rotors will have a greater effect than a smaller one, unless this is naturally offset by increased hub height and an increasing air gap.

Similar results were noted for Northern Gannet and Sandwich Tern, despite variations in the magnitude of estimates associated with their flight attributes and height distributions. Thus, adjusting hub height so that the blades do not sweep the altitudes most frequented by the species of concern and minimising the rotor radius can help to reduce collision risk. However, the use of smaller turbines, leading to a requirement for more turbines to deliver a particular site capacity, presents a greater risk.

In addition, a simple comparison between collision risk estimates for the three species discussed (parameters from Table 8.1) based on 83×3.6 MW machines (rotor radius of 65 m) or 38×8 MW machines (rotor radius of 90 m) to provide 300 MW of capacity was also undertaken. This assumed a constant air gap of 20 m, constant operational time of 95% and constant pitch of 10°. However, the maximum blade width (4.2 m for a 3.6 MW machine and 5.4 m for an 8 MW machine) and nominal rotor speeds (12.5 m/s for a 3.6 MW machine and 10.5 m/s for an 8 MW machine) were selected based on indicative turbine model information. Under these scenarios, and given an avoidance rate of 99%, the model Option 2 results suggested 106 Northern Gannet could collide with the 3.6 MW turbine wind farm while 59 birds could collide under the 8 MW turbine scenario, a reduction of around 44%. Similar reductions were seen for Sandwich Terns (31 collisions reducing to 18) and Lesser Black-backed Gulls (243 collisions reducing to 137).

The results outlined support the notion that the best solution to deliver site capacity at the lowest risk possible for seabirds is to use the largest and thus fewest turbines possible, whilst making sure that the blades do not sweep critical flight heights of important species. Ultimately, the choices involved will, however, depend on the turbine specifications available and any engineering and financial constraints regarding how the site can be constructed.

Selection of construction methods

Impacts during construction may affect birds either directly through disturbance by associated vessels or installation works, or indirectly through potential effects upon prey such as piling-related displacement (Perrow *et al.* 2011). The direct disturbance by vessels may be minimised by limiting the numbers and size of vessels and by careful planning of the timing, routes and frequency of vessel movements (see *Wind-farm design and micro-siting*, above).

The application of noise-mitigation methods during piling largely undertaken to mitigate for impacts on marine mammals (Thomsen & Verfuβ, Chapter 7) may also help to minimise any indirect impacts on birds by reducing effects on their fish prey. Methods to reduce underwater noise using noise-mitigation systems include modifications of the piling hammer, impulse prolongation, and the use of hydro-sound dampers, bubble curtains, casings and cofferdams, especially in combination with each other (Nehls *et al.* 2007; Verfuß 2014; Thomsen & Verfuβ, Chapter 7). Noise-mitigation systems have had to be applied at all OWFs in Germany since 2012. In addition, soft-start piling procedures are widely undertaken (JNCC 2010).

Severe sound levels are also removed by the use of alternatives to piling, including gravity-base foundations (GBF), such as those used at Thornton Bank OWF (Peire *et al.* 2009), and suction caissons (Houlsby *et al.* 2005). Floating bases also remove the requirement for a physical attachment of the actual turbine structure to the seabed, but require anchor attachments with associated design considerations that generate some noise. However, to date, financial, engineering and technological constraints have limited the use of GBF and other alternatives. According to WindEurope (2018), by the end of 2017, 87% of the substructures installed in European waters, including 4,149 grid-connected turbines, had used monopiles, 9% had used using tripods and tripiles, and only 2% had used GBFs (with other technology including the first floating structures comprising the rest). Despite piling having accounted for the vast majority of the construction effort applied to date, there is almost no documented evidence of significant impacts on birds, with the exception of Perrow *et al.* (2011). This may in part be due to the short timescales involved in piling disturbance, quick recovery time or a lack of monitoring at appropriate scales to detect the impacts associated with construction, or a combination thereof.

Selecting specific construction windows may also provide a good method for minimising the impacts of construction on sensitive species. However, this is only effective for species that use the project area in a predictable manner, such as breeding birds tied to a nearby colony or perhaps those passing through in well-defined migration periods. Such an approach may be complicated by limited overlap in the seasonal distribution of important species using a site. Some less sensitive birds, especially local birds using the site regularly, may also habituate to the construction process, whereas highly sensitive or migratory species may not. The passage of migratory species once or twice a year may also typically be of little consequence. In general, restrictions on the timing of construction are likely to be costly for the developer and would need to be robustly justified on a case-by-case basis and supported with detailed behavioural information.

Phased development may also reduce the risk of acute impacts on birds and would provide the opportunity to monitor for any negative effects, which could then be addressed before considering further expansion (Gartman *et al.* 2016b). In reality, this has already happened to some degree, with extensions to many OWFs having been added or in planning. In such cases, any associated monitoring must provide sufficient information to validate impact assessments and inform the decision-making process.

Reducing impacts

Reducing displacement and attraction

Where efforts to minimise impacts have fallen short, or unanticipated impacts have arisen, measures to reduce impacts to acceptable levels are required. In relation to changing the distribution patterns of birds, this could include reducing the levels of displacement or attraction. As previously mentioned (see *Wind-farm design and micro-siting* and *Selection of construction methods*, above), vessel disturbance of sensitive species such as Red-throated Diver (Loon) *Gavia stellata* could involve modifying vessel transit routes and maintenance schedules to avoid areas of importance for birds, or times when birds might be most sensitive to disturbance. This could be achieved by real-time bird observations from the vessels themselves to establish key areas and plan alternative routes, although it should be noted that particularly sensitive groups such as divers and some seaduck (such as scoters *Melanitta* spp.) may be flushed before they can be seen and avoided.

Few bird species are attracted to wind farms (see *Wind-farm design and micro-siting*, above), with only gulls potentially attracted at the risk of increased collision. Gulls are particularly well known for scavenging from fishing vessels (Camphuysen 1994) and if commercial fishing is restricted or discouraged or is impractical due to hazards within OWFs, then fewer gulls may enter a site (Lindeboom *et al.* 2011), thereby reducing the associated risks. Similarly, if the release of any organic waste from all vessels using the wind farm can be eliminated, there may also be less reason for scavenging gulls to enter wind farms.

In relation to migratory birds in particular, lighting appears to be the single most important attractant, with nocturnal migrants such as passerines appearing to be particularly drawn to the red lights that are commonly used (Erickson *et al.* 2001; Desholm 2006; Poot *et al.* 2008). Legislation usually dictates that offshore structures must have various forms of low-intensity lighting to aid aircraft and maritime navigation. Research into the effects of lighting at a North Sea gas rig on migrating birds inferred that changing the exterior lights for ones low in spectral red, thus appearing green, reduced the numbers of birds circling for prolonged periods or landing on the rig (van de Laar 2007; Poot *et al.* 2008). Moreover, birds appear to be are less disoriented by blue and green light containing less or no visible long-wavelength radiation (Poot *et al.* 2008), and where 'green' light was used, van de Laar (2007) suggested a minimum reduction of 50% of any impacts, with up to 90% possible.

It is currently thought that turbines should use the minimum number of intermittent flashing white lights of lowest effective intensity (Hüppop *et al.* 2006). While there is potential to further develop bird-friendly artificial lighting, the approach generally remains unproven and there is scope for confounding effects on different species of birds with different visual acuity. Further research and trials are required to provide a sound basis for the adoption of a specific lighting configuration. Cook *et al.* (2011) surmised that changes to lighting could be an effective mitigation option, but noted that legislation and safety concerns limit possible application.

Luring birds away from wind farms

The use of a wind farm by birds conceivably carries an increased threat of collision should the species involved use the airspace occupied by turbine blades. Gulls and terns, and especially cormorants, will readily utilise wind-farm structures as perches. For example, gulls have routinely been seen resting on wind-farm structures within the Bligh Bank OWF (Vanermen *et al.* 2015; Figure 8.7 in Vanermen & Stienen 2019). This behaviour may be especially prevalent before the turbine is constructed, although substations may continue to be used after construction. In the study by Harwood *et al.* (2017) (see also Box 10.2 in Perrow 2019), Sandwich Terns readily used navigation buoys within 400 m of an operational wind farm (Figure 8.7), but were not recorded using the bases of the operational turbines themselves.

The tendency to perch on wind-farm structures may be exploited by the use of towers, or other structures, placed around the perimeter of a wind farm to lure birds away from the immediate wind-farm footprint (Curry & Kerlinger 2000; Larsen & Guillemette 2007). The efficacy of these 'decoys' may be enhanced by the use of anti-perching devices on structures in the wind farm itself. For example, anti-perching measures for the meteorological tower in the now abandoned Cape Wind Farm project, off Massachusetts (USA), were suggested as part of the mitigation plan to prevent raptors using the area to intercept transiting terns (ESS Group Inc. 2012).

Figure 8.7 Navigation buoys or other structures may be used as perches by a range of species, but particularly gulls and terns (all white 'dots' in this picture above the sea and on the buoy are Sandwich Terns *Thalasseus sandvicensis*), which may discourage birds from using the turbine bases within the wind farm, where they may be at risk of collision. Alternatively, such structures may attract birds to the vicinity of the wind farm and 'decoy' perches may be best installed some distance away. (Martin Perrow)

Decoy structures could, however, also attract birds to the general area of the turbines, or encourage them to linger, thereby increasing densities in the vicinity of an OWF and, accordingly, the risk of collision (Curry & Kerlinger 2000; Smallwood & Karas 2009). In theory, this would also increase the footprint of the wind farm to be considered in impact assessment. Alternatively, structures introduced some distance away from the wind farm in areas of lower risk, but within range of the birds, may be the best solution to reduce the density of birds close to the wind farm. To date, such ideas have not been specifically implemented at any wind farm.

Deterring birds from wind farms

In principle, auditory deterrents such as loud noises, alarm calls or other complex calls could be broadcast from wind-farm structures to scare birds away from the wind farm or individual turbines, thereby reducing collision risk. However, Dooling (2002) suggested that the use of acoustic deterrents may be problematic for two reasons: (1) birds may habituate and (2) any broadcast sounds will also be audible to humans because birds have a narrower range of hearing than humans. Habituation may be avoided at least in part by only using acoustic deterrents at specific times, such as during periods of bad weather or fog when turbines may be less visible to approaching birds. In relation to the use of bird distress signals, Bridges *et al.* (2004) found that these were largely ineffective when used in isolation, but could elicit evacuation of a site when coupled with a visual cue.

Despite the feasibility and relatively low cost of auditory deterrents, Cook *et al.* (2011) suggested that they were unlikely to prove effective. Nevertheless, this does not discount their successful use in some situations in a similar way to acoustic harassment devices such as pingers and seal scarers to discourage marine mammals from construction zones (Nehls *et al.* 2019; Thomsen & Verfuβ, Chapter 7). Birds would, however, need to be discouraged over the longer term, which again introduces the issue of habituation by regularly occurring individuals, even where use is restricted to specific periods.

Various forms of lighting (see *Reducing displacement and attraction*, above), such as strobes and lasers, have good potential as bird deterrents. For example, Gilsdorf *et al.* (2002) found harassment with lasers to be effective for dispersing Great Cormorant *Phalocrocarax carbo*, reducing the numbers at roosts by a minimum of 90% after one to three evenings. Cook *et al.* (2011) also note that the use of flashing lights could initiate reduced use by birds and a prospective 50–70% decrease in collision risk.

Finally, the use of microwave signals, magnets or electromagnetic waves has also been discussed for birds, but the possible risks to humans and other wildlife has generally precluded further application (Harris & Davis 1998; Gartman *et al.* 2016b).

Increasing threat awareness

Increasing the visibility of turbines, particularly the blades, to birds could reduce the potential for collision (see Johnson *et al.* 2007 for a review; May *et al.* 2017). For example, Curry & Kerlinger (2000) suggested that painting turbine blades with high-contrast and ultraviolet (UV) paint might influence flight behaviour around wind turbines. In contrast, Young *et al.* (2003) found that the use of UV-light reflective paint was not particularly successful in reducing collisions. The success of ultraviolet paint in particular is likely to be species or at least group specific, as different groups of birds such as gulls relative to other seabirds have different sensitivities in different parts of the UV spectrum (May 2017). Although the idea of painting turbine blades to make them more visible to birds had previously been attempted (Johnson *et al.* 2007), Hodos (2003) was possibly the first to do so with the specific idea of combating motion smear, whereby the brain is unable to process fast moving objects (Box 8.3). The idea is that a contrast in colour breaks up the image seen by the bird, thus making the blades more readily identifiable. Painting one turbine blade black appears to be effective, as confirmed by the study at Smøla in Norway (Box 8.3).

However, it should be noted that increasing the visibility of turbines is unlikely to be effective if birds are searching for food or roosting sites, as birds with small binocular fields of vision may not be able to identify threats in their flight path while looking downwards (Martin & Shaw 2010).

Box 8.3 The likely efficacy of contrast painting on offshore wind turbines to reduce bird collisions

Roel May

Bird collision mortality is generally perceived as a major conflict issue for wind-power development (Drewitt & Langston 2006; de Lucas & Perrow 2017; Thaxter *et al.* 2017), with the nature and magnitude of collision risk being site and species specific (May *et al.* 2017; Thaxter *et al.* 2017). In evolutionary terms, birds are adapted to a life on the wing attuned to their surrounding environment. Successfully mitigating collision risk therefore needs to be founded on a species' sensory and movement ecology (Martin 2011; May 2015; Nathan *et al.* 2008), enabling them to avoid human-made structures in due time (May *et al.* 2015; May 2017). The main differences between natural and anthropogenic obstacles are that: (1) anthropogenic structures can occur in environments where such structures are not natural, such as offshore wind turbines, and (2) anthropogenic structures can include materials and shapes that may be perceived as unnatural by birds, such as visually unobtrusive light grey smooth surfaces and rotating turbine blades. Martin (2011) suggests that birds may assume the environment ahead to be without obstructions, and rather use their lateral vision to detect movement relating to conspecifics, foraging opportunities or predators.

Most birds that collide with wind turbines fly into the rotor blades. The extent to which it is possible to increase the perception of wind turbines as 'dangerous' for birds, enabling evasive manoeuvres to come into play, has not been investigated *in situ*. What seems clear is that birds do not see the rotating blades at close range because of 'motion smear' (Hodos 2003). Motion smear occurs when the rotor blades rotate quickly (up to 250 km/h at the blade tip), making them partially transparent and blurred. The reason the birds are unable to avoid such situations may be that they are unable to perceive rotor blades in motion, lose manoeuvrability because of turbulence close to a wind turbine or fail to respond adequately in time through avoidance (May 2015). Mitigation measures to reduce collisions should therefore primarily focus on increasing the visibility of the rotor blades to birds. However, some species, such as Smøla Willow Ptarmigan *Lagopus lagopus variegatus*, are known to collide with the turbine towers (see Box 11.1 in Pedersen 2017). For these species, mitigation measures should focus on increased visibility of the towers. The rationale behind increasing the visibility of the tower is to increase the horizon, 'forcing' low-flying birds to circumvent such elevations. The main reason, however, why effective mitigation measures to reduce bird collisions have not yet been developed and implemented lies in the challenges associated with evaluating the effectiveness of such measures (May *et al.* 2015).

Figure 8.8 Turbine with one rotor part painted black as part of the experimental INTACT study at Smøla wind farm, Norway, to test the efficacy of the measure in reducing bird collisions. (Roel May)

In the INTACT project (2013–2017), contrast painting of rotor blades and turbine towers was tested experimentally at the onshore Smøla wind farm (Figure 8.8). Painting one of the three rotor blades black, according to Hodos (2003), resulted in a significant reduction in the number of bird collisions, by 70%, at four treated turbines relative to four control turbines before painting. White-tailed Eagle *Haliaeetus albicilla*, Common Snipe *Gallinago gallinago*, Meadow Pipit *Anthus pratensis* and Hooded Crow *Corvus cornix* were recorded most during collision fatality searches. Painting of the lower 10 m of the tower base at ten turbines resulted in a 30% reduction in Willow Ptarmigan collisions relative to ten control turbines before treatment.

Smøla wind farm is located on the coast of Møre og Romsdal, central Norway. The Smøla archipelago consists of a large main island and about 5,500 smaller islands and islets. The archipelago forms a habitat for a variety of birds, including White-tailed Eagle and Common Eider *Somateria mollissima*, among a range of other seaduck and diving ducks, cormorants, Northern Gannet *Morus bassanus*, a variety of *Larus* spp. gulls and Black-legged Kittiwake *Rissa tridactyla*. Some of these species are known to be vulnerable to collision with offshore wind turbines (Furness *et al.* 2013a). Although onshore, the location of the Smøla wind farm and the conditions experienced make it broadly comparable with turbines in inshore and coastal locations. More specifically, aerial foragers such as gannets and gulls spend much of their time flying at rotor-swept height often in conditions of low visibility, thus increasing their risk of collision with the rotor blades and producing similar collision vulnerability scores as White-tailed Eagle (Furness *et al.* 2013a). The soaring and circling flight behaviour of gulls and gannets during foraging, as well as social interactions, may also increase the risk of collision, as it does for raptors. Conversely, raptors migrating over the sea have been shown to be attracted to offshore wind farms, putting them at risk of colliding with turbines (Skov *et al.* 2016). Moreover, the review of seabird collisions by King (2019) notes a record of a Northern Fulmar *Fulmarus glacialis* colliding with a turbine base at Blyth Harbour in conditions of poor visibility. This opens the possibility of collision of a number of low-flying seabirds, such as auks, in a similar manner to that recorded for Willow Ptarmigan at Smøla.

In general, as shown in this study, contrast painting of rotor blades and tower bases on offshore wind turbines is expected to be effective, not least because of the small contrast in colour between the airspace and water surface relative to onshore areas. Which contrast colour would be most effective remains to be investigated, however.

Finally, should there be a trial of contrast painting in the offshore environment, experiences at Smøla show that it is important to paint rotor blades before they are installed. Even onshore, it turned out to be practically difficult and therefore costly to paint the rotor blades, as this could only be done in calm weather and from the hub downwards.

Controlling the turbines to reduce collision risk

Restricting the speed of turbine blades may reduce collision risk via two pathways: (1) increasing the visibility of blades, and (2) reducing the probability of a bird flying through the swept area being hit. In general, stationary turbine blades are likely to produce a much smaller risk to birds than rotating blades (Drewitt & Langston 2008). Therefore, temporarily shutting turbines down, or 'curtailment', when there is a clear threat is likely to be highly effective. Appropriate shutdown periods may be identified through real-time monitoring of bird flocks using radar or visual observations (e.g. de Lucas *et al.* 2012; Fijn *et al.* 2015). Temporary shutdowns can be applied to all or some of the turbines in an array, but would need to be highly targeted to specific periods such as when key species are most active, migration is occurring or weather conditions increase the risk of collision (Gartman *et al.* 2016b). Such measures have been successfully adopted at a number of onshore wind farms, with an emphasis on protecting raptors, and can be tailored to specific situations (Box 8.4).

Box 8.4 Why temporary shutdown could reduce bird fatalities through collision with turbines offshore

Filipe Canário & Ricardo Tomé

Temporary shutdown of turbines has been used successfully at some onshore wind farms as a means to prevent collisions of birds with turbine blades, with low production losses (e.g. Tomé *et al.* 2017a) (Table 8.2). Shutdown can be performed automatically, triggered by information retrieved by cameras or radar, or manually; that is, on demand by visual observers with or without the aid of radar. So far, this measure has mainly been applied for migratory soaring birds, including raptors, storks, pelicans and cranes (Figure 8.9).

Table 8.2 Two case studies of successful shutdown on demand operations in wind farms located in migratory soaring bird 'bottlenecks', in Portugal and Egypt

Parameter	Wind farm	
	Barão de São João	Gabal el Zait
Country	Portugal	Egypt
Capacity	50 MW	200 MW
Shutdown operation time	8 years	2 years
Total soaring bird fatalities	2	~15
Equivalent shutdown time/year (min–max)	8.3–104.7 hours	3.5–3.8 hours
Production loss/year (min–max)	0.1–1.2%	0.03%

Figure 8.9 Aggregation of Griffon Vultures *Gyps fulvus* on migration at the Barão S. João wind farm in south-west Portugal. (Ricardo Tomé)

Since thermal currents and updraughts, which are essential for soaring flight, do not form over water, soaring birds avoid crossing large bodies of water during migration. Consequently, they tend to concentrate in places where water-crossing between continents or land masses is shorter. In most cases, turbine shutdown on demand is being applied in wind farms located at such migratory 'bottlenecks', where hundreds or thousands of migratory soaring birds converge, for example at Tarifa, near the Strait of Gibraltar (de Lucas *et al.* 2012), Sagres in south-western Portugal (Tomé *et al.* 2017a) and Gabal el Zait on the west coast of the Gulf of Suez in Egypt (Table 8.2) (Tomé *et al.* 2017b).

In most of these locations there are currently no plans for offshore wind farms (OWFs). However, in places such as the Baltic Sea, where offshore wind energy is being developed at a fast rate, wind farms may intercept migratory routes. This may pose a serious threat to soaring birds because of their difficulties in flying over water, which force them to lose altitude and may also hamper their ability to avoid obstacles, especially under poor weather conditions. Moreover, Skov *et al.* (2016) provided evidence that soaring species, including those that do not depend so much on thermals, such as falcons and harriers, may be attracted to offshore turbines. This is probably best explained by an 'island effect', similar to that in which migratory landbirds crossing water are attracted to islands, especially during adverse weather (wind and rain), because they may provide a place to land and rest. These findings suggest that shutdown on demand may also be useful in some offshore situations to prevent soaring bird collisions.

Any procedures adopted could be implemented with similar protocols to those being used on land. In an offshore situation radar assistance would be of increased importance, since it is more difficult and expensive to deploy visual observers at sea. Also, bird assemblages in the whole 'over-water' context are typically less complex in the offshore environment, making it easier for radar software to discriminate birds from other targets as well as to identify specific groups of birds that could be predefined as shutdown criteria.

Ultimately, and for these reasons, a fully automated radar shutdown system would probably be more efficient at sea than onshore. In this case, three-dimensional information regarding birds' trajectories and altitude would be crucial, presuming the deployment of both horizontal and vertical radar. Shutdown decisions would then be based on a threshold of the number of birds moving towards the wind farm at high-risk flight altitudes.

Radar and visual observers' locations should be carefully selected to detect the birds sufficiently early, allowing the shutdown of the turbines in time. Previous studies on bird movements in the area are essential to determine such locations. Depending on the case, and on the birds' behaviour in the area, radar and observers may be located around the wind farm or 'upstream' relative to the previously established main flightline(s). Radar sensors could be placed on turbines located at the edge of the wind farm, on other supporting structures that may exist in the area, or even on land, if the wind farm is not far from the coast.

Visual observers may be essential if the shutdown criteria involve the presence of threatened species, which cannot be identified by radar alone. The use of observers could reduce the number of shutdown events, since an automatic radar-based system unable to discriminate target species would have to follow a precautionary approach and therefore trigger unnecessary shutdowns more frequently.

Although this measure is mostly being implemented for soaring birds, with a few exceptions, it could also be applied to other groups of birds such as migratory waterfowl that are included in protected area designations (see Box 4.5 in Thaxter & Perrow, Chapter 4); although it should be borne in mind that at least some waterfowl show very high rates of avoidance (Plonczkier & Simms 2012).

In relation to seabirds in European waters (information for other parts of the world is scarce), most seabirds do not usually fly at blade height, with the exception of gulls (*Larus* spp.), skuas (*Stercorarius* spp.) and Northern Gannet *Morus bassanus* (Furness *et al.* 2013a). However, actual mortality estimates from OWFs are virtually absent (King 2019) and deaths are very difficult to measure accurately (Molis *et al.*, Chapter 6). Hence, there is a lack of knowledge on the real impact of offshore turbines, which could be higher than expected during unfavourable weather periods.

Shutdown on demand could also reduce impacts on migratory passerines and other nocturnal migrants. Most of these birds migrate on a broad front, at night and at high altitudes, but they frequently stop at promontories (capes) or islands. Adverse weather conditions associated with signalling lights could increase the attraction factor of OWFs (Hüppop & Hilgerloh 2012; Hüppop *et al.* 2019). These fluxes of migrants can be easily detected by radar, either locally, using surveillance radar, or on a wider scale by a network of weather radar sensors (Liechti *et al.* 2013). Acoustic detection and recording associated with radar could provide species identification. In either case, turbine shutdown, when a predefined threshold on the number of migrants is reached, shows some promise in being effective.

A preliminary evaluation of the use of shutdown at sea in Box 8.4 shows that it may be readily achieved from a technical perspective with minimal losses to production, and there is clear promise that it could be effective for a number of bird groups, including migratory soaring species in specific locations and, more generally, migratory waterfowl and passerines as well as seabirds.

Another alternative to shutdown for bats is to alter the 'cut-in' wind speeds of turbines by feathering the blades as described in Box 8.1. It is not known, however, whether this is also likely to be effective for birds as while this would also reduce the amount of time the turbines are operational, with potential subsequent reductions in collision risk, there are no known studies that illustrate that 'freewheeling' prior to cut-in is a specific threat to birds in the offshore environment. As yet, much remains to be learnt about the true extent of collision at sea (King 2019) and how this may be measured (Molis *et al.*, Chapter 6).

Compensation

If impacts cannot be avoided, minimised or reduced sufficiently, then compensation measures provide a means of offsetting the impacts so that there is at least no net loss (Langston & Pullan 2003; Gardner *et al.* 2013). However, it may be difficult to define what no net loss equates to in relation to a specific OWF, particularly against a backdrop of other OWFs and anthropogenic pressures. It remains very challenging to quantify levels of collision mortality (see Molis *et al.*, Chapter 6) and particularly in comparison to modelled estimates (see Cook & Masden, Chapter 5) at OWFs. The impacts of displacement, barrier and indirect effects also remain poorly understood, further complicating the definition of the levels of compensation that may be required.

An understanding of how impacts are allocated across metapopulations is also needed to identify appropriate compensation. However, the apportioning of impacts associated with a specific OWF is problematic, as birds using a site may originate from several different colonies or even regions, especially when incorporating all phenological periods including dispersal and wintering stages as well as the breeding period. In some cases, it may be straightforward to identify a particular population or colony of birds using the development area, for example using tracking technology (see Thaxter & Perrow, Chapter 4), or at least through apportioning methods based on theoretical foraging range during the breeding season. Where this is not possible, compensation should be directed to areas where measures would be likely to provide the most cost-effective results (Furness *et al.* 2013b).

In broad terms, compensation can be achieved by enhancing bird populations by manipulating limiting biological factors and minimising the impacts of other anthropogenic effects on populations (Marques *et al.* 2014). This is usually achieved by targeting breeding, foraging and resting areas for the affected species outside the wind farm. In the onshore arena, Marques *et al.* (2014) list some examples of compensatory measures which could be undertaken to enhance populations, including: (1) habitat creation, restoration or enhancement, (2) prey enhancement, (3) predator control, (4) removal of exotic/invasive species, (5) species reintroductions, and (6) supplementary feeding. These are briefly treated in turn in the following paragraphs in relation to the offshore environment.

The extension of protected areas around important seabird colonies into the offshore realm, to encompass foraging grounds (e.g. see Perrow *et al.* 2015 and other examples therein) is one possible way of enhancing the protection afforded to breeding seabirds and could improve or maintain colony productivity. There are many examples of habitat management and enhancement for seabirds (other than predator control; see below), such as those undertaken for Roseate Tern *Sterna dougallii* in the UK, through vegetation management, the provision of nest boxes on specially constructed terraces, and even the configuration of an island (Blue Circle Island) created from dredge spoil and quarry workings, to make it suitable for breeding terns (Cabot & Nisbet 2013). Habitat alterations are less feasible at sea, although other structures may be added, such as buoys or rafts that could, for example, provide social centres or resting sites between foraging bouts (see *Luring birds away from wind farms*, above). Specifically, in terms of compensatory habitat, Schippers *et al.* (2009) found that the loss of port-area habitat could result in a serious decline in the breeding population of Common Terns *Sterna hirundo,* but that compensation within the ports resulted in negligible declines, and the creation of extra breeding habitat outside the ports, in fish-rich waters, had a beneficial effect on the metapopulation.

The timing of implementation of compensatory measures needs to be carefully planned, as it may take a considerable amount of time to create and establish new habitat or put measures to control predators in place. Any benefits need to be realised at the same time as any negative effects begin to truly offset any impacts. This could involve implementing plans even before the OWF is constructed, such as in the case of the onshore site at Beinn an Tuirc, Scotland, where a pre-construction habitat management plan was implemented to reduce prey availability in the site and enhance foraging habitat for Golden Eagles *Aquila chrysaetos* away from it (Walker *et al.* 2005).

The enhancement of prey in the offshore environment is intuitively rather more difficult than onshore, although the serious, often negative interactions between seabirds and fisheries provides several possibilities, specifically as there is compelling evidence that many bird populations are limited by human overexploitation of shared prey species (e.g. Croxall *et al.* 2012; Cury *et al.* 2012). Thus, the closure or adaptation of fisheries may enhance prey stocks for seabirds (e.g. Daunt *et al.* 2008). Bycatch of seabirds from a range of commercial fisheries, known to be among most significant global threats to seabirds (Anderson *et al.* 2011; Croxall 2012; Žydelis *et al.* 2013), has also drawn attention in the UK (Bradbury *et al.* 2017). Measures to reduce bycatch in the North, Irish and Baltic Seas, which contain a significant proportion of the world's wind farms (Jameson *et al.* 2019), may thus be valuable as compensation for wind-farm losses.

Predator control is of huge importance in seabird conservation, from high-profile eradications of introduced and invasive rodents from entire islands (e.g. Martin 2015) to more mundane control of native predators at nature reserves (see examples in Cabot & Nisbet 2013). In addition, where predators are also birds of conservation interest, diversionary feeding may be successful in reducing losses of chicks of the desirable seabird species (Smart & Amar 2018). Directly feeding the seabird of interest does not appear to be undertaken in the manner of feeding scavenging raptors (González *et al.* 2006; García-Heras *et al.* 2013), although it is easy to see that adaptable scavengers such as gulls and skuas may respond favourably to hand-outs. The use of this strategy needs to be carefully monitored, however, as although it can be successful in raising productivity (González *et al.* 2006), poor-quality individuals may be disproportionately favoured (García-Heras *et al.* 2013). The removal of predators coupled with habitat management often leads to rapid recolonisation of previously affected seabird species and, as such, reintroduction is generally unnecessary.

In relation to all these possibilities, Furness *et al.* (2013b) provide a comprehensive review of potential compensatory and mitigation measures for OWF impacts on a variety

Figure 8.10 European Herring Gull *Larus argentatus* with Common Tern *Sterna hirundo* chick snatched from the Rockabill colony in Ireland. Large gulls may be significant predators of other seabirds and compensatory measures of wind-farm impacts on those species in some localities could benefit from increased control of gull populations. Ironically, in other locations it may be gull populations that require compensatory measures as a conservation priority. For example, European Herring Gull is red-listed in the UK (Eaton *et al.* 2015). (Martin Perrow)

of key seabirds species around the British Isles. The most cost-effective options were identified as: (1) provision of nest platforms for Red-throated Divers; (2) eradication of alien invasive mammalian predators on islands with Manx Shearwater *Puffinus puffinus* colonies; (3) supplementary feeding of breeding pairs of Arctic Skuas *Stercorarius parasiticus*; (4) culling of breeding Lesser Black-backed Gulls *Larus fuscus*, European Herring Gulls *L. argentatus* and Great Black-backed Gulls *L. marinus*, which may be significant predators of other seabirds (Figure 8.10), while introducing predator-proof fencing around mainland colonies of these gulls subject to fox predation; (5) closure of Lesser Sandeel *Ammodytes marinus* and European Sprat *Sprattus sprattus* fisheries in UK waters to increase productivity and survival of Black-legged Kittiwakes *Rissa tridactyla*, Common Guillemots *Uria aalge* and Razorbills *Alca torda*; (6) predator-proof fencing to exclude foxes from affected Sandwich Tern colonies and engineering to reduce the risk of tidal flooding where this is an issue; and (7) eradication of American Mink *Mustela vison* from islands with Common Tern colonies and deployment of predator-proof nesting rafts. It is of note, however, that wind-farm related impacts at a population scale have not yet been demonstrated for any of these species, although the level of displacement for Red-throated Diver (Mendel *et al.* 2019) points to such an effect.

Restoration

In theory, at the end of life of an OWF, decommissioning of all associated infrastructure, including the turbines, foundations, transition pieces, cabling, substations, meteorological masts, scour protection and any onshore components, should be undertaken to restore the area to as near to pre-development conditions as is reasonably practicable. The onus is on the owner to ensure this is achieved under relevant licence conditions. In the UK, Section 105 of the Energy Act 2004 details the requirement to prepare decommissioning programmes that are applicable to OWFs.

The Yttre Stengrund and Vindeby OWFs were decommissioned in 2016–2017, with around ten further sites predicted to be decommissioned (or repowered) in the coming decade (Topham & McMillan 2017). Experiences have shown that some infrastructure such as deep foundations may be particularly difficult to remove, and could even require destructive methods such as the use of explosives, with potential impacts upon fish and their bird predators. The most effective mitigation of such impacts may be the use of alternative construction methods such as GBF or floating foundations in the first place.

Moreover, in many cases repowering or refurbishment of sites may defer their decommissioning, and where pursued it will afford the opportunity to reassess mitigation requirements and implement any new measures that may be appropriate. Replacing many inefficient smaller turbines with fewer more efficient larger ones, which are carefully placed, could lead to significant decreases in collision risk (de Lucas *et al.* 2012; Smallwood 2017). Similarly, removing or relocating specific turbines that represent the greatest risk to birds could substantially reduce the overall impact (Drewitt & Langston 2006; Smallwood 2017). Whether an OWF is decommissioned, refurbished or repowered, monitoring should still maintain a key role in ensuring that any mitigation has been effective and understanding any residual impacts.

Where removal is ultimately required, it remains difficult to evaluate how long it may take habitats and ecosystems to recover to pre-development characteristics following a sustained perturbation (May 2017). Indeed, the bathymetry around an OWF may have been altered beyond practicable restoration, with associated changes in benthic communities and food webs. Extensive rehabilitation efforts may be required at this point to aid in the return to normal ecosystem functioning, assuming that any changes to the habitat and associated wildlife assemblage are undesirable. There is growing evidence (Dannheim *et al.* 2019; Gill & Wilhelmsson 2019) that positive reef effects will have occurred and a 'renewables-to-reefs' strategy may often be the most appropriate action from the perspective of biodiversity. In fact, many wind farms may have become so valuable that they would be considered as marine protected areas (Ashley *et al.* 2014). In this case, there will be an argument for only partial removal of the turbine foundations, perhaps above the sea surface. However, it remains unclear whether, on balance, impacts are positive for seabirds as a group, primarily as there is a general lack of monitoring of indirect and ecosystem effects, and understanding remains limited (Perrow 2019).

Concluding remarks

Plans for mitigation, targeting potential risks according to relevant assessments and using the precautionary approach, are required during planning and consenting. Introducing pre-emptive compensatory conservation measures as early as possible to reduce the risk of unforeseen impacts and provide protection for the most vulnerable birds (and bats) is clearly desirable. But despite the variety of possible measures to reduce impacts, few have been implemented at OWFs to date. This seems largely due to uncertainty associated with the need for mitigation, the efficacy of measures, a lack of incentive or pressure to adopt mitigation from the outset, and potentially prohibitive risks and costs for developers. Various reviews provide differing recommendations, illustrating the lack of evidence and changes in thinking in regard to risk and mitigation (Hüppop *et al.* 2006; Drewitt & Langston 2006; Edkins 2008; Cook *et al.* 2011; Marques *et al.* 2014; May *et al.* 2015). However, most studies agree that mitigation should be site and species specific.

To date, where mitigation has been applied, this has largely been in the planning phase, whereby important bird areas are identified and avoided at varying spatial scales. This offers a means of reducing risks by effectively decreasing the numbers of birds which may be subject to impacts within assessments. However, some of the other options discussed in this chapter may also be successful, as outlined in the following paragraphs.

Alternative construction methods, such as the use of non-piled foundations, or noise-mitigation systems including bubble curtains where piling is undertaken, offer scope to reduce indirect effects on the fish prey of seabirds. The timing of construction can also be modified or phased to avoid periods when species such as divers are most sensitive to displacement by vessels and/or construction.

Micro-siting turbines to increase or decrease the permeability of an array could aid in reducing displacement and barrier effects or collision risk. Furthermore, the use of fewer, larger and more productive turbines with the greatest possible air gap offers the potential to significantly reduce collision risk for seabirds, while maintaining site capacity, although the risk to migrating birds, at higher altitudes, could be increased. Such site-level planning must be balanced against the engineering and cost constraints for a productive and viable project.

Once the site has been constructed, adopting green, rather than red, lighting could reduce attraction to and therefore collision of migrating birds, although further research is needed on the effects of different lighting approaches on different species to make sure that there are no confounding impacts. Decoy structures could also be employed to lure sensitive species away from turbines, although again further research and trials are required to support their use. While auditory and visual bird deterrents could be effective for reducing densities of birds subject to risk, they are likely to be species-specific, birds could habituate, and some deterrents may be effective only in certain weather conditions.

Temporary turbine shutdown, as used at several terrestrial wind farms, also shows promise in being readily applicable at OWFs and could be highly effective for reducing collision mortalities for some species under certain scenarios, such as in poor weather during key migration periods at migration bottlenecks bringing large numbers of passerines into the range of turbines. Temporary shutdown and particularly increasing cut-in speed, as applied onshore, seem likely to be the most effective methods to reduce bat collisions at OWFs.

Of all the measures to reduce collision of birds, painting turbine blades with different patterns so that they be more easily perceived and thus avoided is perhaps the most attractive option with general applicability. Given the ease and low cost with which this could be implemented and the potential level of benefit that could accrue, there is an argument that painting turbines should be undertaken as a matter of course at all sites (many sites such at Alpha Ventus and other sites in German waters already appear to have done so) even if monitoring does not appear to have been undertaken to demonstrate the efficacy of the measure.

Measuring the success of any mitigation is difficult and ideally requires an understanding of the levels of impact in an unmitigated situation relative to one where mitigation has been applied to quantify efficacy. Of key importance is continued monitoring using appropriate survey methods to determine the response of birds (e.g. Harwood *et al.* 2017; Skov *et al.* 2018; Mendel *et al.* 2019) and bats to developments, while also providing information on any undesirable impacts, which may then be followed by the implementation of measures to address those impacts through an iterative process. The application of such adaptive management (Hanna *et al.* 2016) is variable between countries and projects and is difficult to put in place after construction. It is also debated whether monitoring has, in general, been sufficiently targeted and of sufficient rigour to be useful in the context of adaptive management (Perrow 2019). Monitoring and research need to be encouraged to provide sound evidence that the planned future expansion of the OWF industry is sustainable from the perspective of particular birds and bats.

If mitigation measures cannot be effectively applied, or even readily demonstrated, and the precautionary principle is adopted, then compensation may be required to offset both realised and potential residual impacts. Habitat creation or improvement, predator control, prey enhancement and supplementary feeding all show promise under different circumstances for different species, but few measures have yet to implemented as mitigation, partly because the evidence base for OWF impacts is only now becoming available (King 2019; Vanermen & Stienen 2019).

At the end of life of an existing OWF, repowering seems likely to be an attractive option, ideally incorporating any new technology and research to minimise risks to birds. Alternatively, where decommissioning is to be undertaken and the site restored to the baseline condition, using rehabilitation measures where necessary, there is likely to be an increasing argument for only the partial removal of turbine structures above the sea surface, as a result of the reef effect of turbine bases and scour protection benefiting biodiversity (Smyth *et al.* 2015; Dannheim *et al.* 2019; Gill & Wilhelmsson 2019). However, the specific value of this for many seabirds, perhaps apart from Great Cormorant and possibly some gulls, is uncertain (Perrow 2019).

Finally, there is an argument, as put forward by May (2017), that given the importance of renewables in reducing greenhouse gas emissions and the socio-economic factors at play, some level of unmitigated residual impact of OWFs on birds may be acceptable. An alternative argument is, however, that given the potential for cumulative and in-combination effects and the increasing prospect of population-scale impacts on some species as the offshore wind industry expands rapidly around the globe, and the fact that many birds (and bats) are already subject to a range of anthropogenic pressures, there is a case for the wind industry to enhance its 'green' reputation by ensuring that some funds from all OWFs are routinely put towards the conservation of key species. This could go some way towards offsetting any potential impacts and meeting the objective of the 'win–win' for wildlife and wind farms (Kiesecker *et al.* 2011).

Acknowledgements

We are indebted to Lothar Bach, Filipe Canário and Ricardo Tomé, and Roel May for their valuable contributions to this chapter in the form of information boxes. We are also grateful to Centrica Energy and Ørsted for permission to reproduce Figure 8.3 and Figure 8.4 that were originally part of assessment work for the Race Bank OWF and the Irish Sea Round 3 Zone, respectively, in UK waters.

References

Ahlén, I., Bach, L., Baagøe, H.J. & Pettersson, J. (2007) Bats and offshore wind turbines studied in southern Scandinavia. Report 5571 by Vindval. Retrieved April 8 2019 from https://www.naturvardsverket.se/Documents/publikationer/620-5571-2.pdf

Ahlén, I., Baagøe, H.J. & Bach, L. (2009) Behaviour of Scandinavian bats during migration and foraging at sea. *Journal of Mammalogy* 90: 1318–1323.

Alerstam, T., Rosén, M., Bäckman, J., Ericson, P.G. & Hellgren, O. (2007) Flight speeds among bird species: allometric and phylogenetic effects. *PLoS Biology* 5(8): e197. doi: 10.1371/journal.pbio.0050197.

American Bird Conservancy (2019) American Bird Conservancy wind risk assessment map. Retrieved 8 April 2019 from https://abcbirds.org/program/wind-energy-and-birds/wind-risk-assessment-map/

Amorim, F., Rebelo, H. & Rodrigues, L. (2012) Factors influencing bat activity and mortality at a wind farm in the Mediterranean region. *Acta Chiropterologica* 14: 439–457.

Anderson, O., Small, C. & French, G., Croxall, J.P., Dunn, E.K., Sullivan, B.J., Yates, O. & Black, A. (2011) Global seabird bycatch in longline fisheries. *Endangered Species Research* 14: 91–106.

Anderson, R., Morrison, M., Sinclair, K. & Strickland, D. (1999) Studying wind energy/bird interactions: a guidance document. Metrics and methods for determining or monitoring potential impacts on birds at existing and proposed wind energy sites. Washington, DC: National Wind Coordinating Committee. Retrieved 8 April 2019 from https://www.fwspubs.org/doi/suppl/10.3996/032012-JFWM-024/suppl_file/10.3996_032012-jfwm-024.s10.pdf

Arnett, E.B. (2017) Mitigating bat collision. In Perrow, M.R. (ed.) *Wildlife and Wind Farms, Conflicts and Solutions. Volume 2. Onshore: Monitoring and mitigation*. Exeter: Pelagic Publishing. pp. 167–184.

Arnett, E.B., Brown, W.K., Erickson, W.P., Fiedler, J.K., Hamilton, B.L., Henry, T.H., Jain, A., Johnson, G.D., Kerns, J., Koford, R.R., Nicholson, C.P., O'Connel, T.J., Piorkowski, M.D. & Tankersley, R.D. (2008) Patterns of bat fatalities at wind energy facilities in North America. *Journal of Wildlife Management* 72: 61–78.

Arnett, E.B., Huso, M.M.P., Hayes, J.P. & Schirmacher, M. (2010) Effectiveness of changing wind turbine cut-in speed to reduce bat fatalities at wind facilities. Final report submitted to the Bats and Wind Energy Cooperative. Austin, TX: Bat Conservation International. Retrieved 8 April 2019 from https://tethys.pnnl.gov/sites/default/files/publications/Arnett_and_Schirmacher_2010.pdf

Arnett, E.B., Huso, M.M.P., Schirmacher, M. & Hayes, J.P. (2011) Altering turbine speed reduces bat mortality at wind-energy facilities. *Frontiers in Ecology and the Environment* 9: 209–214.

Arnett, E.B., Hein, C.D., Schirmacher, M.R., Huso, M.M.P. & Szewczak, J.M. (2013a) Evaluating the effectiveness of an ultrasonic acoustic deterrent for reducing bat fatalities at wind turbines. *PLoS ONE* 8(6): e65794. doi: 10.1371/journal.pone.0065794.

Arnett, E.B., Barclay, R.M.R. & Hein, C.D. (2013b) Thresholds for bats killed by wind turbines. *Frontiers in Ecology and the Environment* 11: 171–171.

Ashley, M.C., Mangi, S.C. & Rodwell, L.D. (2014) The potential of offshore windfarms to act as marine protected areas – a systematic review of current evidence. *Marine Policy* 45: 301–309.

Bach, L. & Bach, P. (2009) Einfluss der Windgeschwindigkeit auf die Aktivität von Fledermäusen. *Nyctalus* 14: 3–13.

Bach, L., Bach, P., Pommeranz, H., Hill, R., Voigt, C., Göttsche, M., Göttsche, M., Matthes, H. & Seebens-Hoyer, A. (2017) Offshore bat migration in the German North and Baltic Sea in autumn 2016. Poster, European Bat Research Symposium, Donostia, 1–5 August 2017. Retrieved 8 April 2019 from https://www.researchgate.net/publication/318135846_Offshore_bat_migration_in_the_Germnan_North_and_Baltic_Sea_in_autumn_2016

Baerwald, E.F. & Barclay, R.M.R. (2009) Geographic variation in activity and fatality of migratory bats at wind energy facilities. *Journal of Mammalogy* 90: 1341–1349.

Band, W. (2012) Using a collision risk model to assess bird collision risks for offshore windfarms. Strategic Ornithological Support Services. Project SOSS-02. Thetford: British Trust for Ornithology. Retrieved 8 April 2019 from https://www.bto.org/sites/default/files/u28/downloads/Projects/Final_Report_SOSS02_Band1ModelGuidance.pdf

Behr, O., Brinkmann, R., Niermann, I. & Korner-Nievergelt, F. (2011) Akustische Erfassung der Fledermausaktivität an Windenergieanlagen. In Brinkmann, R., Behr, O., Niermann, I. & Reich, M. (eds) Entwicklung von Methoden zur Untersuchung und Reduktion des Kollisionsrisikos von Fledermäusen an Onshore- Windenergieanlagen. *Umwelt und Raum* 4: 177–286.

Behr, O., Brinkmann, R., Hochradel, K., Mages, J., Korner-Nievergelt, F., Reinhard, H., Simon, R., Stiller, F., Weber, N. & Nagy, M. (2018) Bestimmung des Kollisionsrisikos von Fledermäusen an Onshore-Windenergieanlagen in der Planungspraxis. Final report of the project financed by Federal Ministry for Economics and Energy (Förderkennzeichen 0327638E). Erlangen/Freiburg/Ettiswil. Retrieved 8 April 2019 from http://www.windbat.techfak.fau.de/Abschlussbericht/renebat-iii.pdf

Bradbury, G., Trinder, M., Furness, R., Banks, A.N., Caldow, R.W.G. & Hume, D. (2014) Mapping seabird sensitivity to offshore wind farms. *PLoS ONE* 9(9): e106366. doi: 10.1371/journal.pone.0106366.

Bradbury, G., Shackshaft, M., Scott-Hayward, L., Rexstad, E., Miller, D. & Edwards, D. (2017) Risk assessment of seabird bycatch in UK waters. Wildlife & Wetlands Trust (Consulting) Ltd Report to Department of Food and Rural Affairs (Defra). Slimbridge: Wildlife & Wetlands Trust (Consulting) Ltd. Retrieved 14 December 2018 from http://sciencesearch.defra.gov.uk/Document.aspx?Document=14236_MB0126RiskassessmentofseabirdbycatchinUKwaters.pdf

Bridges, J.M., Anderson, T.R., Shulund, D., Spiegel, L. & Chervick, T. (2004) Minimizing bird collisions: what works for the birds and what works for the utility? In Environment Concerns in Rights-of-Way Management 8th International Symposium, New York, 12–16 September 2004. New York: Elsevier. pp. 331–335. Retrieved 8 April 2019 from https://drive-electric.hu/driveelectricnet_files/Minimizing%20bird%20collisions.pdf

Bright, J., Langston, R.H.W., Bullman, R., Evans, E., Gardner, S. & Pearce-Higgins, J. (2008) Map of bird sensitivities to wind farms in Scotland: a tool to aid planning and conservation. *Biological Conservation* 141: 2342–2356.

Bright, J.A., Langstone, R.H.W. & Anthony, S. (2009) Mapped and written guidance in relation to birds and onshore wind energy development in England. RSPB Research Report No. 35. Sandy: Royal Society for the Protection of Birds. Retrieved 8 April 2019 from http://ww2.rspb.org.uk/Images/EnglishSensitivityMap_tcm9-237359.pdf

Brinkmann, R., Behr, O., Niermann, I. & Reich, M. (eds) (2011) *Entwicklung von Methoden zur Untersuchung und Reduktion des Kollisionsrisikos von Fledermäusen an Onshore-Windenergieanlagen. Ergebnisse eines Forschungsvorhabens.* Göttingen: Cuvillier.

Bundesamt für seeschifffahrt und Hydrographie (BSH) (2013) *Standard Investigation of the Impacts of Offshore Wind Turbines on the Marine Environment (2007).* Hamburg: Bundesamt für seeschifffahrt und Hydrographie.

Cabot, D. & Nisbet, I. (2013) *Terns. The New Naturalist Library.* London: HarperCollins.

Camphuysen, C.J. (1994) Scavenging seabirds at beam-trawlers in the southern North Sea: distribution, relative abundance, behaviour, prey selection, feeding efficiency, kleptoparasitism, and the possible effects of the establishment of 'protected areas'. BEON Report 1994-14. Texel: Netherlands Institute for Sea Research. Retrieved 8 April 2019 from www.vliz.be/imis-docs/publications/263122.pdf

Chaouachi, A., Covrig, C.F. & Ardelean, M. (2017) Multi-criteria selection of offshore wind farms: case study for the Baltic States. *Energy Policy* 103: 179–192.

Choong-Ki, K., Seonju, J. & Tae Yun, K. (2018) Site selection for offshore wind farms in the south-west coast of South Korea. *Renewable Energy* 120: 152–162.

Cook, A.S.C.P., Ross-Smith, V.H., Roos, S., Burton, N.H.K., Beale, N., Coleman, C., Daniel, H., Fitzpatrick, S., Rankin, E., Norman, K. & Martin, G. (2011) Identifying a range of options to prevent or reduce avian collision with offshore wind farms, using a UK-based case study. BTO Research Report No. 580. Thetford: British Trust for Ornithology. Retrieved 8 April 2019 from https://www.bto.org/sites/default/files/shared_documents/publications/research-reports/2011/rr580.pdf

Cook, A.S.C.P., Johnston, A., Wright, L.J. & Burton, N.H.K. (2012) A review of flight heights and avoidance rates of birds in relation to offshore wind farms. Report to The Crown Estate. Strategic Ornithological Support Services, Project SOSS-02. Thetford: British Trust for Ornithology. Retrieved 8 April 2019 from https://tethys.pnnl.gov/sites/default/files/publications/Cook-et-al-2012.pdf

Croxall, J.P., Butchart, S.H.M., Lascelles, B., Stattersfield, A.J., Sullivan, B., Symes, A. & Taylor, P. (2012) Seabird conservation status, threats and priority actions: a global assessment. *Birdlife Conservation International* 22: 1–34.

Cryan, P.M., Gorresen, P.M., Hein, C.D., Schirmacher, M.R., Diehl, R.H., Huso, M.M., Hayman, D.T.S., Fricker, P.D., Bonaccorso, F.J., Johnson, D.H., Heist, K. & Dalton, D.C. (2014) Behavior of bats at wind turbines. *Proceedings of the National Academy of Science of the United States of America* 111: 15126–15131.

Curry, R.C. & Kerlinger, P. (2000) Avian mitigation plan: Kenetech Model Wind Turbines, Altamont Pass WRA, California. Proceedings of National Avian-Wind Power Planning Meeting III, San Diego, California, May 1998 (PNAWPPM-III, Ed.). pp. 18–27. Ontario: LGL Ltd. Retrieved 8 April 2019 from http://www.windrush-energy.com/Update%2015-1-08/A%20WIND%20FOR%20OUR%20LIFE/Supplementary%20EA%20Reference%20Studies/Curry,%20R.C.%20&%20P.Kerlinger%20%282000%29.pdf

Cury, P.M., Boyd, I.L., Bonhommeau, S., Anker-Nilssen, T., Crawford, R.J. M., Furness, R.W., Mills, J.A., Murphy, E.J., Österblom, H., Paleczny, M., Piatt, J.F., Roux, J.-P., Shannon, L. & Sydeman, W.J. (2012) Global seabird response to forage fish depletion – one-third for the birds. *Science* 334: 1703–1706.

Dannheim, J., Degraer, S., Elliot, M., Smyth, K. & Wilson, J.C. (2019) Seabed communities. In Perrow, M.R. (ed.) *Wildlife and Wind Farms, Conflicts and Solutions. Volume 3. Offshore: Potential effects*. Exeter: Pelagic Publishing. pp. 64–85.

Daunt, F., Wanless, S., Greenstreet, S.P.R., Jensen, H., Hamer, K.C. & Harris, M.P. (2008) The impact of the sandeel fishery closure in the north-western North Sea on seabird food consumption, distribution and productivity. *Canadian Journal of Fisheries and Aquatic Sciences* 65: 362–381.

de Lucas, M. & Perrow, M.R. (2017) Birds: collision. In Perrow, M.R. (ed.) *Wildlife and Wind Farms, Conflicts and Solutions. Volume 1. Potential effects*. Exeter: Pelagic Publishing. pp. 155–190.

de Lucas, M., Ferrer, M., Bechard, M.J. & Muñoz, A.R. (2012) Griffon vulture mortality at wind farms in southern Spain: distribution of fatalities and active mitigation measures. *Biological Conservation* 147: 184–189.

Desholm, M. (2006) Wind farm related mortality among avian migrants – a remote sensing study and model analysis. PhD thesis, University of Copenhagen. Retrieved 8 April 2019 from www.vliz.be/imisdocs/publications/127432.pdf

Desholm, M. (2009) Avian sensitivity to mortality: prioritising migratory bird species for assessment at proposed wind farms. *Journal of Environmental Management* 90: 2672–2679.

Dierschke, V., Furness, R.W. & Garther, S. (2016) Seabirds and offshore wind farms in European waters: avoidance and attraction. *Biological Conservation* 202: 59–68.

Dooling, R. (2002) Avian hearing and the avoidance of wind turbines. NREL Report TP-500-30844. Golden, CO: National Renewable Energy Laboratory. Retrieved 8 April 2019 from http://www.nrel.gov/wind/pdfs/30844.pdf

Drewitt, A.L. & Langston, R.H.W. (2006) Assessing the impacts of wind farms on birds. *Ibis* 148: 29–42.

Drewitt, A.L. & Langston, R.H.W. (2008) Collision effects of wind-power generators and other obstacles on birds. *Annals of the New York Academy of Sciences* 1134: 233–266.

Eaton, M.A., Aebischer, N.J., Brown, A.F., Hearn, R., Lock, L., Musgrove, A.J., Noble, D.G., Stroud, D. & Gregory, R.D. (2015) Birds of conservation concern 4: the population status of birds in the United Kingdom, Channel Islands and Isle of Man. *British Birds* 108: 708–746.

Edkins, M.T. (2008) Impacts of wind energy development on birds and bats: looking into the problem. Juno Beach, FL: FPL Energy. Retrieved 8 April 2019 from https://tethys.pnnl.gov/sites/default/files/publications/Edkins-2014.pdf

EMODnet Bathymetry Consortium (2018) EMODnet Digital Bathymetry (DTM 2018). EMODnet Bathymetry Consortium. Retrieved 3 January 2019 from https://doi.org/10.12770/18ff0d48-b203-4a65-94a9-5fd8b0ec35f6

Erickson, W.P., Good, R.E., Johnson, G.D., Sernka, K.J., Strickland, M.D. & Young, D.P., Jr (2001) Avian collisions with wind turbines: a summary of existing studies and comparisons to other sources of avian collision mortality in the United States. National Wind Coordinating Committee. Retrieved 8 April 2019 from https://www.osti.gov/servlets/purl/822418

ESS Group Inc. (2012) Final Cape Wind avian and bat monitoring plan. Report to Cape Wind Associates, LLC. Retrieved 8 April 2019 from https://www.boem.gov/uploadedFiles/BOEM/Renewable_Energy_Program/Studies/Cape%20Wind%20ABMP.pdf

Everaert, J. & Stienen, E.W.M. (2007) Impact of wind turbines on birds in Zeebrugge (Belgium). *Biodiversity and Conservation* 16: 3345-3359.

Exo, K.-M., Hüppop, O. & Garthe, S. (2003) Birds and offshore wind farms: a hot topic in marine ecology. *Wader Study Group Bulletin* 100: 50–53.

Fijn, R.C., Krijsveld, K.L., Poot, M.J. & Dirksen, S. (2015) Bird movements at rotor heights measured continuously with vertical radar at a Dutch offshore wind farm. *Ibis* 157: 558-566.

Fox, A.D., Desholm, M., Kahlert, J., Christensen, T.K. & Petersen, I.B.K. (2006) Information needs to support environmental impact assessment of the effects of European marine offshore wind farms on birds. *Ibis* 148: 129–144.

Furness. R.W. & Wade, H.M. (2012) Vulnerability of Scottish Seabirds to offshore wind turbines – Report to Marine Scotland. Glasgow: MacArthur Green Ltd. Retrieved 8 April from https://www2.gov.scot/resource/0038/00389902.pdf

Furness, R.W., Wade, H.M. & Masden, E. (2013a) Assessing vulnerability of marine bird populations to offshore wind farms. *Journal of Environmental Management* 119: 56–66.

Furness, B., MacArthur, D., Trinder, M. & MacArthur, K. (2013b) Evidence review to support the identification of potential conservation measures for selected species of seabirds. MacArthur Green Technical Report for Cefas. September 2013. Retrieved 8 April 2019 from https://www.researchgate.net/publication/274931059_evidence_review_to_support_the_identification_of_potential_conservation_measures_for_selected_species_of_seabirds

García-Heras, M.-S., Cortés-Avizanda, A. & Donázar, J.-A. (2013) Who are we feeding? Asymmetric individual use of surplus food resources in an insular population of the endangered Egyptian vulture *Neophron percnopterus*. *PLoS ONE* 8(11): e80523. doi: 10.1371/journal.pone.0080523.

Gardner, T.A., Von Hase, A., Brownlie, S., Ekstrom, J.M.M., Pilgrim, J.D., Savy, C.E., Stephens, R.T.T., Treweek, J.O., Ussher, G.T., Ward, G. & Ten Kate, K. (2013) Biodiversity offsets and the challenge of achieving no net loss. *Conservation Biology* 27: 1254–1264.

Garthe, S. & Hüppop, O. (2004) Scaling possible adverse effects of marine wind farms on seabirds: developing and applying a vulnerability index. *Journal of Applied Ecology* 41: 724–734.

Gartman, V., Bulling, L., Dahmen, M., Geißler, G. & Köppel, J. (2016a) Mitigation measures for wildlife in wind energy development, consolidating the state of knowledge – Part 1: Planning and siting, construction. *Journal of Environmental Assessment Policy and Management* 18(3): 1650013-1-45.

Gartman, V., Bulling, L., Dahmen, M., Geißler, G. & Köppel, J. (2016b) Mitigation measures for wildlife in wind energy development, consolidating the state of knowledge – Part 2: Operation, decommissioning. *Journal of Environmental Assessment Policy and Management* 18(3):1650014-1-31.

Gill, A.B. & Wilhelmsson, D. (2019) Fish. In Perrow, M.R. (ed.) *Wildlife and Wind Farms, Conflicts and Solutions. Volume 1. Onshore: Potential Effects*. Exeter: Pelagic Publishing. pp. 86–111.

Gilsdorf, J.M., Hygnstrom, S.E. & VerCauteren, K.C. (2002) Use of frightening devices in wildlife damage management. Integrated Pest Management Reviews 7: 29–45.

González, L.M., Margalida, A., Sánchez, R. & Oria, J. (2006) Supplementary feeding as an effective tool for improving breeding success in the Spanish imperial eagle (*Aquila adalberti*). *Biological Conservation* 129: 477–486.

Gove, B., Langston, R.H.W., McCluskie, A.J., Pullan, D. & Scrase, I. (2013) Wind farms and birds: an updated analysis of the effects of wind farms on birds, and best practice guidance on integrated planning and impact assessment. In *Convention on the Conservation of European Wildlife and Natural Habitats 17 Sept 2013*. Report prepared by BirdLife International on behalf of the Bern Convention. Strasbourg: Council of Europe. Retrieved 3 June 2014 from http://www.birdlife.org/sites/default/files/attachments/201312_BernWindfarmsreport.pdf

Green, R.E., Langston, R.H.W., McCluskie, A., Sutherland, R. & Wilson, J.D. (2016) Lack of sound science in assessing wind farm impacts on seabirds. *Journal of Applied Ecology* 53: 1635–1641.

Hamer, K., Holt, N. & Wakefield, E. (2011) The distribution and behaviour of northern gannets in the Firth of Forth and Tay area. A review on behalf of Forth and Tay Offshore Wind Developers Group. Leeds: Institute of Integrative & Comparative Biology, University of Leeds.

Hanna, L., Copping, A., Geerlofs, S., Feinberg, L., Brown-Saracino, J., Gilman, P., Bennet, F., May, R., Köppel, J., Bulling, L. & Gartman, V. (2016) Assessing environmental effects (WREN): adaptive management white paper. Report by Bureau of Ocean Energy Management (BOEM), Marine Scotland Science, Norwegian Institute for Nature Research (NINA), Pacific Northwest National Laboratory (PNNL), Technische Universität Berlin and US Department of Energy (DOE). Retrieved 8 April 2019 from https://tethys.pnnl.gov/sites/default/files/publications/WREN-AM-White-Paper-2016.pdf

Harris, M.P., Bogdanova, M.I., Daunt, F. & Wanless, S. (2012) Using GPS technology to assess feeding areas of Atlantic puffins *Fratercula arctica*. *Ringing & Migration* 27: 43–49.

Harris, R.H. & Davis, R.A. (1998) Evaluation of the efficacy of products and techniques for airport bird control. Canada: LGL Ltd. Retrieved 8 April 2019 from http://nimby.ca/PDFs/TP13029B.pdf

Harwood, A.J.P., Perrow, M.R., Berridge, R.J., Tomlinson, M.L. & Skeate, E.R. (2017) Unforeseen responses of a breeding seabird to the construction of an offshore wind farm. In Köppel, J. (ed.) *Conference on Wind Energy and Wildlife Interactions Presentations from the CWW2015 conference*. Cham: Springer International Publishing. pp. 19–41.

Hatch, S.K., Connelly, E.E., Divoll, T.J., Stenhouse, I.J. & Williams, K.A. (2013) Offshore observations of eastern red bats (*Lasiurus borealis*) in the mid-Atlantic United States using multiple survey methods. *PLoS ONE* 8(12): e83803. doi: 10.1371/journal.pone.0083803.

Hein, C.D. (2017) Monitoring bats. In: Perrow, M.R. (ed.) *Wildlife and Wind Farms, Conflicts and Solutions. Volume 2. Onshore: Monitoring and Mitigation*. Exeter: Pelagic Publishing. pp. 31–57.

Ho, L.-W., Lie, T.-T., Leong, P.T.M. & Clear, T. (2018) Developing offshore wind farm siting criteria by using an international Delphi method. *Energy Policy* 113: 53–67.

Hodos, W. (2003) Minimization of motion smear: reducing avian collisions with wind turbines. Golden, CO: National Renewable Energy Laboratory. Retrieved 8 April 2019 from http://www.nrel.gov/docs/fy03osti/33249.pdf

Horn, J.W., Arnett, E.B. & Kunz, T.H. (2008) Behavioural responses of bats to operating wind turbines. *Journal of Wildlife Management* 72: 123–132.

Hötker, H., Thomsen, K.-M. & Jeromin, H. (2006) Impacts on biodiversity of exploitation of renewable energy sources: the example of birds and bats – facts, gaps in knowledge, demands for further research, and ornithological guidelines for the development of renewable energy exploitation. Bergenhusen, Germany: Michael-Otto-Institut im NABU. Retrieved 17 December 2018 from https://tethys.pnnl.gov/publications/impacts-biodiversity-exploitation-renewable-energy-sources-example-birds-and-bats

Hötker, H. (2017) Birds: displacement. In Perrow, M.R. (ed.) *Wildlife and Wind Farms, Conflicts and Solutions. Volume 1. Onshore: Potential Effects*. Exeter: Pelagic Publishing. pp. 119–154.

Houlsby, G.T., Kelly, R.B., Huxtable, J. & Byrne, B.W. (2005) Field trials of suction caissons in clay for offshore wind turbine foundations. *Géotechnique* 55: 287–296.

Hüppop, O. & Hilgerloh, G. (2012) Flight call rates of migrating thrushes: effects of wind conditions, humidity and time of day at an illuminated offshore platform. *Journal of Avian Biology* 43: 85–90.

Hüppop, O. & Hill, R. (2016) Migration phenology and behaviour of bats at a research platform in the south-eastern North Sea. *Lutra* 59: 5–22.

Hüppop, O., Dierschke, J., Exo, K.-M., Fredrich, E. & Hill, R. (2006) Bird migration studies and potential collision risk with offshore wind turbines. *Ibis* 148: 90–109.

Hüppop, O., Michalik, B., Bach, L., Hill, R. & Pelletier, S.K. (2019) Migratory birds and bats. In Perrow, M.R. (ed.) *Wildlife and Wind Farms, Conflicts and Solutions. Volume 3. Offshore: Potential Effects*. Exeter: Pelagic Publishing. pp. 142–173.

Jameson, H., Reeve, E., Laubek, B. & Sittel, H. (2019) The nature of offshore wind farms. In Perrow, M.R. (ed.) *Wildlife and Wind Farms, Conflicts and Solutions. Volume 3. Offshore: Potential effects*. Exeter: Pelagic Publishing. pp. 1–29.

Jameson, J.W. & Willis, C.K.R. (2014) Activity of tree bats at anthropogenic tall structures: implications for mortality of bats at wind turbines. *Animal Behaviour* 97: 145–152.

Johnson, G.D., Strickland, M.D., Erickson, W.P. & Yiung, D.P., Jr (2007) Use of data to develop mitigation measures for wind power development impacts to birds. In de Lucas, M., Janss, G. & Ferrer, M. (eds) *Birds and Wind Farms*. Madrid: Servicios Informativos Ambientales/Quercus. pp. 241–257.

Johnston, A., Cook, A.S.C.P, Wright, L.J., Humphreys, E.M. & Burton, N.H.K. (2014a) Modelling flight heights of marine birds to more accurately assess collision risk with offshore wind turbines. *Journal of Applied Ecology* 51: 31–41.

Johnston, A., Cook, A.S.C.P., Wright, L.J., Humphreys, E.M. & Burton, N.H.K. (2014b) Corrigendum. *Journal of Applied Ecology* 51: 1126–1130.

Joint Nature Conservation Committee (JNCC) (2010) Statutory nature conservation agency protocol for minimising the risk of injury to marine mammals from piling noise. Peterborough: Joint Nature Conservation Committee. Retrieved 8 April 2019 from http://jncc.defra.gov.uk/pdf/JNCC_Guidelines_Piling%20protocol_August%202010.pdf

Kelsey, E.C., Felis, J.J., Czapanskiy, M., Pereksta, D.M. & Adams, J. (2018) Collision and displacement vulnerability to offshore wind energy infrastructure among marine birds of the Pacific outer continental shelf. *Journal of Environmental Management* 227: 229–247.

Kiesecker, J.M., Evans, J.S., Fargione, J., Doherty, K., Foresman, K.R., Kunz, T.H., Naugle, D., Nibbelink, N.P. & Niemuth, N.D. (2011) Win-win for wind and wildlife: a vision to facilitate sustainable development. *PLoS ONE* 6(4): e17566. doi: 10.1371/journal.pone.0017566.

King, S. (2019) Seabirds: collision. In Perrow, M.R. (ed.) *Wildlife and Wind Farms, Conflicts and Solutions. Volume 3. Offshore: Potential effects*. Exeter: Pelagic Publishing. pp. 206–234.

Krijgsveld, K.L., Fijn, R.C., Japink, M., van Horssen, P.W., Heunks, C., Collier, M.P., Poot, M.J.M. & Dirken, S. (2011) Effect studies offshore wind farm Egmond aan Zee. Final report on fluxes, flight altitudes and behaviour of flying birds. Bureau Waardenburg Report No. 10-219. Commissioned by NordzeeWind. Retrieved 8 April 2019 from https://tethys.pnnl.gov/sites/default/files/publications/Krijgsveld%20et%20al.%202011.pdf

Kunz, T.H., Arnett, E.B., Erickson, W.P., Hoar, A.R., Johnson, G.D., Larkin, R.P., Strickand, M.D., Thresher, R.W. & Tuttle, M.D. (2007) Ecological impacts of wind development on bats: Questions, research needs, and hypotheses. *Frontiers in Ecology and the Environment* 5: 315–324.

Lagerveld, S., Jonge Poerink, B., Haselager, R. & Verdaat, H. (2014) Bats in Dutch offshore wind farms in autumn 2012. *Lutra* 57: 61–69.

Langston, R.H.W. (2013) Birds and wind projects across the pond: a UK perspective. *Wildlife Society Bulletin* 37: 5–18.

Langston, R.H.W. & Pullan, J.D. (2003) Wind farms and birds: an analysis of the effects of wind farms on birds, and guidance on Environmental Assessment criteria and site selection issues. Birdlife International report on behalf of the Bern Convention. Strasbourg: Council of Europe. Retrieved 17 December 2018 from https://www.rspb.org.uk/globalassets/downloads/documents/positions/climate-change/wind-power-publications/birdlife-international-report-to-the-bern-convention.pdf

Lapeña, B., Wijnberg, K., Hulscher, S. & Stein, A. (2010) Environmental impact assessment of offshore wind farms: a simulation-based approach. *Journal of Applied Ecology* 47: 1110–1118.

Larsen, J.K. & Guillemette, M. (2007) Effects of wind turbines on flight behaviour of wintering common eiders: implications for habitat use and collision risk. *Journal of Applied Ecology* 44: 516–522.

Lehnert, L.S., Kramer-Schadt, S., Schönborn, S., Lindecke, O., Niermann, I. & Voigt, C.C. (2014) Wind farm facilities in Germany kill noctule bats from near and far. *PLoS ONE* 9(8): e103106. doi: 10.1371/journal.pone.0103106.

Liechti, F., Guélat, J. & Komenda-Zehnder, S. (2013) Modelling the spatial concentrations of bird migration to assess conflicts with wind turbines. *Biological Conservation* 162: 24–32.

Lindeboom, H.J., Kouwenhoven, H.J., Bergman, M.J.N., Bouma, S., Brasseur, S., Daan, R., Fijn, R.C., de Haan, D., Dirksen, S., van Hal, R., Hille Ris Lambers, R., ter Hofstede, R., Krijgsveld, K.L., Leopold, M. & Scheidat, M. (2011) Short-term ecological effects of an offshore wind farm in the Dutch coastal zone; a compilation. *Environmental Research Letters* 6(3): 035101.

Long, C.V., Flint, J.A. & Lepper, P.A. (2011) Insect attraction to wind turbines: does colour play a role? *European Journal of Wildlife Research* 57: 323–331.

Loukogeorgaki, E., Vagiona, D.G. & Vasileiou, M. (2018) Site selection of hybrid offshore wind and wave energy systems in Greece incorporating environmental impact assessment. *Energies* 11: 2095. doi: 10.3390/en11082095.

Marques, A.T., Batalha, H., Rodrigues, S., Costa, H., Pereira, M.J.R., Fonseca, C., Mascarenhas, M. & Bernardino, J. (2014) Understanding bird collisions at wind farms: an updated review on the causes and possible mitigation strategies. *Biological Conservation* 179: 40–52.

Martin, G.R. (2011) Understanding bird collisions with man-made objects: a sensory ecology approach. *Ibis* 153: 239–254.

Martin, G.R. & Shaw, J.M. (2010) Bird collisions with power lines: failing to see the way ahead? *Biological Conservation* 143: 2695–2702.

Martin, T. (2015) *Reclaiming South Georgia – The defeat of furry invaders on a sub-Antarctic island*. Dundee: South Georgia Heritage Trust.

Masden, E.A., Haydon, D.T., Fox, A.D. & Furness, R.W. (2010) Barriers to movement: modelling energetic costs of avoiding marine windfarms amongst breeding seabirds. *Marine Pollution Bulletin* 60: 1085–1091.

Masden, E.A., Reeve, R., Desholm, M., Fox, A.D., Furness, R.W. & Haydon, D.T. (2012) Assessing the impact of marine wind farms of birds through movement modelling. *Journal of the Royal Society Interface* 9: 2120–2130.

May, R.F. (2015) A unifying framework for the underlying mechanisms of avian avoidance of wind turbines. *Biological Conservation* 190: 179–187.

May, R. (2017) Mitigation options for birds. In Perrow, M.R. (ed.) *Wildlife and Wind Farms, Conflicts and Solutions. Volume 2. Onshore: Monitoring and mitigation*. Exeter: Pelagic Publishing. pp. 124–145.

May, R., Reitan, O., Bevanger, K., Lorentsen, S.H. & Nygard, T. (2015) Mitigating wind-turbine induced avian mortality: sensory, aerodynamic and cognitive constraints and options. *Renewable and Sustainable Energy Reviews* 42: 170–181.

May, R., Gill, A.B., Köppel, J., Langston, R.H.W., Reichenbach, M., Scheidat, M., Smallwood, S., Voigt, C.C., Hüppop, O. & Portman, M. (2017) Future research directions to reconcile wind turbine–wildlife interactions. In Köppel, J. (ed.) *Conference on Wind Energy and Wildlife Interactions Presentations from the CWW2015 conference*. Cham: Springer International Publishing. pp. 255–276.

McGuinness, S., Mulddon, C., Tierney, N., Cummins, S., Murray, A., Egan, S. & Crowe, O. (2015) Bird sensitivity mapping for wind energy developments and associated infrastructure in the Republic of Ireland – guidance document. Kilcoole: BirdWatch Ireland. Retrieved 8 April 2019 from https://www.birdwatchireland.ie/portals/0/POLICY/Guidance_document.pdf

McGuire, L.P., Guglielmo, C.G., Mackenzie, S.A. & Taylor, P.D. (2012) Migratory stopover in the long-distance migrant silver-haired bat, *Lasionycteris noctivagans*. *Journal of Animal Ecology* 81: 377–385.

Mendel, B., Schwemmer, P., Peschko, V., Müller, S., Schwemmer, H., Mercker, M. & Garthe, S. (2019) Operational offshore wind farms and associated ship traffic cause profound changes in distribution patterns of Loons (*Gavia* spp.). *Journal of Environmental Management* 231: 429–438.

Nathan, R., Getz, W.M., Revilla, E., Holyoak, M., Kadmon, R., Saltz, D. & Smouse, P.E. (2008) A movement ecology paradigm for unifying organismal movement research. *Proceedings of the National Academy of Sciences of the United States of America* 105: 19052–19059.

Nehls, G., Betke, K., Eckelmann, S. & Ros. M. (2007) Assessment and costs of potential engineering solutions for the mitigation of the impacts of underwater noise arising from the construction of offshore windfarms. BioConsult SH Report, Husum, Germany, on behalf of COWRIE Ltd. COWRIE Ltd. Retrieved 8 April 2019 from https://tethys.pnnl.gov/sites/default/files/publications/Nehls%20et%20al%202007.pdf

Nehls, G., Harwood, A.J.P. & Perrow, M.R. (2019) Marine mammals. In Perrow, M.R. (ed.) *Wildlife and Wind Farms, Conflicts and Solutions. Volume 3. Offshore: Potential effects*. Exeter: Pelagic Publishing. pp. 112–141.

Nicholls, B. & Racey, P.A. (2007) Bats avoid radar installations: could electromagnetic fields deter bats from colliding with wind turbines? *PLoS ONE* 2(3): e297. doi: 10.1371/journal.pone.0000297.

Nicholls, B. & Racey, P.A. (2009) The aversive effect of electromagnetic radiation on foraging bats – a possible means of discouraging bats from approaching wind turbines. *PLoS ONE* 4(7): e6246. doi: 10.1371/journal.pone.0006246.

OSPAR (2008) OSPAR guidance on environmental considerations for offshore wind-farm development. No. 2008-3. Retrieved 8 April 2019 from www.vliz.be/imisdocs/publications/ocrd/224682.pdf

Pedersen, H.C. (2017) Box 11.1 Unexpected collision of willow ptarmigan with turbine towers at Smøla wind farm in Norway. In Perrow, M.R. (ed.) *Wildlife and Wind Farms, Conflicts and Solutions. Volume 1. Potential effects*. Exeter: Pelagic Publishing. pp. 256–259.

Peire, K., Nonneman, H. & Bosschem, E. (2009) Gravity base foundations for the Thornton Bank offshore wind farm. *Terra et Aqua* 115: 19–29.

Pelletier, S.K., Omland, K., Watrous, K.S. & Peterson, T.S. (2013) Information synthesis on the potential for bat interactions with offshore wind facilities – Final Report. OCS Study BOEM 2013-01163. Herndon, VA: US Department of the Interior, Bureau of Ocean Energy Management. Retrieved 8 April 2019 from https://www.boem.gov/ESPIS/5/5289.pdf

Perrow, M.R. (2019) A synthesis of effects and impacts. In Perrow, M.R. (ed.) *Wildlife and Wind Farms, Conflicts and Solutions. Volume 3. Offshore: Potential effects*. Exeter: Pelagic Publishing. pp. 235–277.

Perrow, M.R., Gilroy, J.J., Skeate, E.R. & Tomlinson, M.L. (2011) Effects of the construction of Scroby Sands offshore wind farm on the prey base of Little tern *Sternula albifrons* at its most important UK colony. *Marine Pollution Bulletin* 62: 1661–1670.

Perrow, M.R., Harwood, A.J.P., Skeate, E.R., Praca, E. & Eglington, S.M. (2015) Use of multiple data sources and analytical approaches to derive a marine protected area for a breeding seabird. *Biological Conservation* 191: 729–738.

Perrow, M.R., Harwood, A.J.P., Berridge, R. & Skeate, E.R. (2017) The foraging ecology of sandwich terns in north Norfolk. *British Birds* 110: 249–308.

Peste, F., Paula, A., da Silva, L.P., Bernardino, J., Pereira, P., Mascarenhas, M., Costa, H., Vieira, J., Bastos, C., Fonseca, C. & Ramos Pereira, J.M. (2015) How to mitigate impacts of wind farms on bats? A review of potential conservation measures in the European context. *Environmental Impact Assessment Review* 51: 10–22.

Petersen, T.S., Pelletier, S.K. & Giovanni, M. (2016) Long-term bat monitoring on islands, offshore structures, and coastal sites in the Gulf of Maine, mid-Altlantic, and Great Lakes – Final Report. Prepared for the US Department of Energy. Retrieved 8 April 2019 from https://tethys.pnnl.gov/sites/default/files/publications/Stantec-2016-Bat-Monitoring.pdf

Peterson, T.S., Pelletier, S.K., Boyden, S.A. & Watrous, K.S. (2014) Offshore acoustic monitoring of bats in the Gulf of Maine. *Northeastern Naturalist* 21: 86–107.

Plonczkier, P. & Simms, I.C. (2012) Radar monitoring of migrating pink-footed geese: behavioural responses to offshore wind farm development. *Journal of Applied Ecology* 49: 1187–1194.

Poerink, J.B. & Haselager, R. (2013) *Monitoring Migartie Vleermuizen Rottumeroog voor- en Najaar 2012*. Field work Company Report No. 20130102. Retrieved 8 April 2019 from https://www.boswachtersblog.nl/rottum/wp-content/uploads/sites/26/2013/03/vleermuismigratie-rottumeroog-2012-fieldwork-company.pdf

Poot, H., Ens, B.J., de Vries, H., Donners, M.A.H., Wernand, M.R. & Marquenie, J.M (2008) Green light for nocturnally migrating birds. *Ecology and Society* 13: 47.

PricewaterhouseCoopers (2010) *Biodiversity offsets and the mitigation hierarchy: a review of current application in the banking sector*. Washington, DC: Business and Biodiversity Offsets Programme & United Nations Environment Programme. Retrieved 8 April 2019 from https://www.unepfi.org/fileadmin/documents/biodiversity_offsets.pdf

Robinson, R.A. (2005) BirdFacts: profiles of birds occurring in Britain & Ireland. BTO Research Report 407. Thetford: British Trust for Ornithology. Retrieved 17 December 2018 from http://www.bto.org/birdfacts.

Rodrigues, L., Bach, L., Dubourg-Savage, M.-J., Karapandža, B., Kovač, D., Kervyn, T., Dekker, J., Kepel, A., Bach, P., Collins, J., Harbusch, C., Park, K., Micevski, B. & Mindermann, J. (2015) *Guidelines for consideration of bats in wind farm projects, Revision 2014*. Bonn: EUROBATS Publication Series No. 6. UNEP/EUROBATS Secretariat. Retrieved 8 April 2019 from https://www.

eurobats.org/sites/default/files/documents/publications/publication_series/pubseries_no6_english.pdf

Rydell, J., Bach, L., Dubourg-Savage, M.J., Green, M., Rodrigues, L. & Hedenström. A. (2010) Bat mortality at wind turbines in northwestern Europe. *Acta Chiropterologica* 12: 261–274.

Schippers, P., Snep, R.P.H., Schotman, A.G.M., Jochem, R., Stienen, E.W.M. & Slim, P.A. (2009) Seabird metapopulations: searching for alternative breeding habitats. *Population Ecology* 51: 459–470.

Schuster, E., Bulling, L. & Köppel, J. (2015) Consolidating the state of knowledge: a synoptical review of wind energy's wildlife effects. *Environmental Management* 56: 300–331.

Seagreen Wind Energy (2018) Ornithology technical report for Seagreen Alpha & Bravo Firth of Forth Offshore Wind Farm development 2018. Glasgow: Seagreen Wind Energy Ltd. Retrieved 4 January 2019 from http://marine.gov.scot/sites/default/files/lf009-env-ma-rpt-0023_eia_report_vol_3_app_8a_ornithology_technical_report.pdf

Seebens, A., Fuß, A., Allgeyer, P., Pommeranz, H., Mähler, M., Matthes, H., Göttsche, M., Göttsche, M., Bach, L. & Paatsch, C. (2013) Fledermauszug im Bereich der deutschen Ostseeküste. Unpublished report to Bundesamt für Seeschifffahrt und Hydrographie (BSH). Retrieved 8 April 2019 from http://www.bach-freilandforschung.de/images/download/Batmigration_German_Baltic_Sea.pdf

Sjollema, A.L., Gates, J.E., Hilderbrand, R.H. & Sherwell, J. (2014) Offshore activity of bats along the mid-Atlantic coast. *Northeastern Naturalist* 21: 154–163.

Skov, H., Desholm, M., Heinänen, S., Kahlert, J.A., Laubek, B., Jensen, N.E., Zydelis, R., Jensen, B.P. (2016) Patterns of migrating soaring migrants indicate attraction to marine wind farms. *Biology Letters* 12: 20160804.

Skov, H., Heinänen, S., Norman, T., Ward, R.M., Mendez-Roldan, S. & Ellis, I. (2018) ORJIP bird collision and avoidance study. London: The Carbon Trust. Retrieved 20 June 2018 from https://www.carbontrust.com/media/675793/orjip-bird-collision-avoidance-study_april-2018.pdf

Smallwood, K.S. (2017) The challenges of addressing wildlife impacts when repowering wind energy projects. In Köppel, J. (ed.) *Conference on Wind Energy and Wildlife Interactions Presentations from the CWW2015 conference.* Cham: Springer International Publishing. pp. 175–187.

Smallwood, K.S. & Karas, B. (2009) Avian and bat fatality rates at old-generation and repowered wind turbines in California. *Journal of Wildlife Management* 73: 1062–1071.

Smart, J. & Amar, A. (2018) Diversionary feeding as a means of reducing raptor predation at seabird breeding colonies. *Journal for Nature Conservation* 46: 48–55.

Smyth, K., Christie, N., Burdon, D., Atkins, J.P., Barnes, R. & Elliott, M. (2015) Renewables-to-reefs? – Decommissioning options for the offshore wind power industry. *Marine Pollution Bulletin* 90: 247–258.

Sonntag, N., Weichler, T., Weiel, S. & Meyer, B. (2006) Blinder Passagier – Zweifarbfledermaus (*Vespertilio murinus*) landet auf einem Forschungsschiff in der Pommerschen Bucht (südliche Ostsee). *Nyctalus* 11: 277–280.

Thaxter, C.B., Buchanan, G.M., Carr, J., Butchart, S.H.M., Newbold, T., Green, R.E., Tobias, J.A., Foden, W.B., O'Brien, S. & Pearce-Higgins, J.W. (2017) Bird and bat species' global vulnerability to collision mortality at wind farms revealed through a trait-based assessment. *Proceedings of the Royal Society of London B: Biological Sciences* 284(1862). doi: 10.1098/rspb.2017.0829.

Tomé, R., Leitão, A.H., Pires, N. & Canário, F. (2017a) Inter-and intra-specific variation in avoidance behaviour at different scales in migratory soaring birds. In: Book of abstracts. Conference on Wind Energy and Wildlife Impacts, Estoril, Portugal. Retrieved 8 April 2019 from http://cww2017.pt/images/Congresso/book-abstracts/BookOfAbstractCWW17_complete_4Set17.pdf

Tomé, R., Canário, F., Leitão, A.H., Pires, N. & Repas, M. (2017b) Radar assisted shutdown on demand ensures zero soaring bird mortality at a wind farm located in a migratory flyway. In Köppel, J. (ed.) *Conference on Wind Energy and Wildlife Interactions Presentations from the CWW2015 conference.* Cham: Springer International Publishing. pp. 119–133.

Topham, E. & McMillan, D. (2017) Sustainable decommissioning of an offshore wind farm. *Renewable Energy* 102: 470–480.

van de Laar, F.J.T. (2007) Green light to birds. Investigation into the effect of bird-friendly lighting. Report NAM locatie L15-FA-1. Assen: NAM. Retrieved 8 April 2019 from https://tethys.pnnl.gov/sites/default/files/publications/van-de-Laar-2007.pdf

Vanermen, N. & Stienen, E.W.M. (2019) Seabirds: displacement. In Perrow, M.R. (ed.) *Wildlife and Wind Farms, Conflicts and Solutions. Volume 3. Offshore: Potential effects*. Exeter: Pelagic Publishing. pp. 174–205.

Vanermen, N., Onkelinx, T., Courtens, W., Van de Walle, M., Verstraete, H. & Steinen, E.W.M. (2015) Seabird avoidance and attraction at an offshore wind farm in the Belgian part of the North Sea. *Hydrobiologia* 756: 51–61.

Verfuß, T. (2014) Noise mitigation systems and low-noise installation technologies. In BSH & BMU (2014) *Ecological Research at the Offshore Windfarm* Alpha Ventus*: Challenges, results and perspectives*. Federal Maritime and Hydrographic Society of Germany (BSH) and Federal Ministry for the Environment, Nature Conservation and Nuclear Safety (BMU). Weisbaden: Springer Spektrum. pp. 181–189.

Voigt, C.C., Popa-Lisseanu, A.G., Niermann, I. & Kramer-Schadt, S. (2012) The catchment area of wind farms for European bats: a plea for international regulations. *Biological Conservation* 153: 80–86.

Voigt, C.C., Lehnert, L.S., Petersons, G., Adorf, F. & Bach, L. (2015) Wildlife and renewable energy: German politics cross migratory bats. *European Journal of Wildlife Research* 61: 213–219.

Walker, D., McGrady, M., McCluskie, A., Madders, M. & McLeod, D.R.A. (2005) Resident Golden Eagle ranging behaviour before and after construction of a windfarm in Argyll. *Scottish Birds* 25: 24–40.

Walter, G., Matthes, H. & Joost, M. (2007) Fledermauszug über Nord- und Ostsee – Ergebnisse aus Offshore-Untersuchungen und deren Einordnung in das bisher bekannte Bild zum Zuggeschehen. *Nyctalus* 12: 221–233.

Wanless, S., Harris, M.P. & Greenstreet, S.P.R. (1998) Summer sandeel consumption by seabirds breeding in the Firth of Forth, south-east Scotland. *ICES Journal of Marine Science* 55: 1141–1151.

Willsteed, E.A., Jude, S., Gill, A.B. & Birchenough, S.N.R. (2018) Obligations and aspirations: a critical evaluation of offshore wind farm cumulative impact assessments. *Renewable and Sustainable Energy Reviews* 82: 2332–2345.

Wilson, J.C., Elliott, M., Cutts, N.D., Mander, L., Mendão, V., Perez-Dominguez, R. & Phelps, A. (2010) Coastal and offshore wind energy generation: is it environmentally benign? *Energies* 3: 1383–1422.

WindEurope (2018) Offshore wind in Europe: key trends and statistics 2017. Brussels: WindEurope. Retrieved 27 July 2018 from https://windeurope.org/wp-content/uploads/files/about-wind/statistics/WindEurope-Annual-Offshore-Statistics-2017.pdf

Winkelman, J.E. (1992) The impact of the Sep wind park near Oosterbierum (Fr.), The Netherlands, on birds, 3: Flight behaviour during daylight. RIN Report 92/4. Arnhem: DLO-Instituut voor Bos- en Natuuronderzoek.

Young, D.P., Erickson, W.P., Strickland, M.D., Good, R.E. & Sernka, K.J. (2003) Comparison of avian responses to UV-light-reflective paint on wind turbines. Golden, CO: National Renewable Energy Laboratory. Retrieved 8 April 2019 from https://www.nrel.gov/docs/fy03osti/32840.pdf

Žydelis, R., Small, C. & French, G. (2013) The incidental bycatch of seabirds in gillnet fisheries: a global review. *Biological Conservation* 162: 76–88.

CHAPTER 9

Perspectives on marine spatial planning

JOHANN KÖPPEL, JULIANE BIEHL, MARIE DAHMEN, GESA GEISSLER and MICHELLE E. PORTMAN

Summary

This planning-related chapter provides a review of approximately a decade's work on spatial planning approaches for marine ecosystems and seascapes, and relevant Strategic Environmental Assessments (SEAs). Attention is focused first on the initial motives and early pioneers of this young planning realm, recognising offshore wind farms (OWFs) as a crucial driver of marine spatial planning (MSP). Early approaches to MSP faced major challenges: a limited and rather dynamic marine database for planning purposes, with lessons of various marine uses still to be learned, especially considering actual impacts pertaining to OWFs. This chapter offers insights into pioneering MSP, primarily in the UK and Germany. The year 2021 marks the deadline for European member states to set up Marine Spatial Plans in accordance with the 2014 Maritime Spatial Planning Directive. In conjunction with the 2008 European Union Marine Strategy Framework Directive, this is to adopt an ambitious ecosystem-based management approach, both raising and addressing the question as to what degree this will modify future MSP and how SEA is conducted in relation to OWFs. In addition, challenges remain in preventing, assessing and mitigating intra- and inter-plan cumulative effects on the marine environment and tailoring impact assessment approaches accordingly, be they for national territorial waters or the Exclusive Economic Zone. At the same time, international cooperation is crucial, such as in the North and Baltic Seas, where multiple countries are involved in both the development and conservation approaches for shared marine ecosystems and their ecosystem services. In consequence, good practice and guidance for the making of Marine Spatial Plans and SEAs pertaining to OWFs continue to be refined and developed, yet much still remains to be implemented. The chapter concludes with a series of recommendations regarding future planning requirements and open questions concerning the continuously evolving field of MSP.

Introduction

Around the beginning of the new millennium, further activities and technological approaches increasingly fuelled traditional and novel uses of our seascapes and their management (Ehler *et al.* 2007; Jay 2010a). The United Nations Convention on the Law of the Sea (UNCLOS) had incontestably recognised the increasing economic importance of our globe's seas (UN 1982). In addition, UNCLOS shifted the focus beyond national territorial waters extending at most 12 nautical miles from shore, by setting a framework for the Exclusive Economic Zone (EEZ) up to 200 nautical miles from a nation's shores and the relevant rights and duties of the nation and its neighbours. Although Ehler *et al.* (2007) argued that marine spatial planning (MSP) is analogous to spatial or land-use planning in terrestrial environments, Jay (2010a) states that MSP was 'built at sea' and may be better handled as a marine adoption of terrestrial planning. Relevant environmental planning and impact assessment for oceans subsequently became an important issue (Jay 2010a; Portman 2016). In 2001, the European Union (EU) launched its Directive on Strategic Environmental Assessment, which targeted plans and programmes such as marine planning activities, including the development of infrastructure projects such as offshore wind farms (OWFs) (European Parliament & European Council 2001).

In general, planning addresses human-driven interactions at numerous levels, on different scales and for many purposes, including natural resource use and management (Portman 2016). Since natural resources are unequally distributed in space and time, planners and managers who understand how to work with the relevant temporal and spatial diversity are needed. MSP comes in various shapes and scales, yet it is comprised of a number of generic characteristics of spatial planning (Ehler *et al.* 2007; Jay 2010a; 2010b), including: (1) future orientation and (2) making use of strategies that seek to achieve agreed goals; and that it is (3) primarily a public-sector activity with different governance tiers, both (4) shaping and protecting the natural environment, (5) embodied in statutory procedures, and (6) concerned with the formation of entire patterns of activities across large territories.

Since the turn of the twenty-first century, what were initially informal, but have become yet more and more regulated spatial planning approaches have been developed all around the world. The development of renewable energies at sea added a new dimensions and urgency to managing ocean spaces (Jay 2010a; Boucquey *et al.* 2016). By 2018, around 70 countries had introduced MSP approaches, ranging from preparatory stages to planning revisions and adaptation (UNESCO 2018), in keeping with the fact that MSP evolved in parallel with the emergence of offshore renewable energy (Ehler *et al.* 2007; Johnson *et al.* 2012; Bradshaw *et al.* 2018; Ehler 2018; Yates *et al.* 2018). In a nutshell, MSP addresses the sustainable development of marine areas in that it shall "consider economic, social and environmental aspects to support [...] growth in the maritime sector, applying an ecosystem-based approach, and to promote the coexistence of relevant activities and uses", as paraphrased in the EU Maritime Spatial Planning Directive 2014/89/EC (European Parliament & European Council 2014, Art. 5) (Box 9.1). A standard approach in spatial planning applies zoning schemes to allow for the coexistence of different planned activities, including conservation (Stelzenmüller *et al.* 2013). A comprehensive overview on the emergence and current importance of MSP can be found in the introduction to *Environmental Planning for Oceans and Coasts* (Portman 2016: 97–113).

A myriad of ocean policies and planning efforts have been established or continue to be underway, be they on the initiative of individual countries, such as in the Netherlands, Belgium and the UK, or catalysed by European policy and legislation (Portman *et al.* 2009).

Box 9.1 Legislative background in the European Union: the advent of the Maritime Spatial Planning Directive

The EU Maritime Spatial Planning Directive was adopted in 2014, requiring member states to establish maritime spatial plans by 2021 (European Parliament & European Council 2014) (Figure 9.1). While aiming to achieve an integrated approach to marine

	1975	*Great Barrier Reef (GBR) Marine Park* (Australia) Spatial planning and zoning are considered as the cornerstones of the management strategy for protecting the GBR (one of the earliest marine zoning approaches)
UN Convention on the Law of the Sea (UNCLOS)	**1982**	Umbrella convention establishing a comprehensive regime for the law at sea in territorial waters, the contiguous zone, the continental shelf, and the high seas. It adopts a global approach and regulates conservation and maritime uses to generate a just and equitable international economic order.
OSPAR Convention Helsinki Convention	**1992**	Mechanism to protect the marine environment of the North-East Atlantic (OSPAR) and the Baltic Sea (Helsinki Convention, governed by the Helsinki Commission – HELCOM)
EU Water Framework Directive (WFD)	**2000**	Committing the European Union member states to achieve 'Good Environmental Status' (GES) of all water bodies (incl. marine waters up to 1 nautical miles from onshore)
	2005	*Integrated Management Plan for the North Sea* (Netherlands) MSP approach defines use zones for e.g. shipping routes, wind farms in operation/ under construction
EU Marine Strategy Framework Directive (MSFD)	**2008**	Establishing a framework for community action in the field of marine policy (Art. 1) to achieve GES of the EU's marine waters by 2020 (Art. 1) by applying an ecosystem-based approach to the management of human activities (EBM, Art. 3)
	2009	*Massachusetts Ocean Management Plan,* acc. to MA Oceans Act 2008 (USA, MA) Establishing management areas (e.g. prohibited / renewable energy / multi-use); Setting performance standards for the development within the plan area (e.g. standards for siting community-scale OWF and tidal energy facilities)
		MSPs for the German EEZ – MSP Baltic Sea and MSP North Sea (Germany) Zoning within the German EEZ by e.g. designating priority areas for OWF (i.e. wind energy production is granted priority over other / wind energy-incompatible uses); the German States (Länder) are responsible for planning in territorial waters
US National Ocean Policy	**2010**	MSP is one out of nine priority implementation objectives to address conservation, economic activity, user conflict, and sustainable use of the ocean, U.S. coasts, and the Great Lakes (foresees MSP to be produced by regional planning bodies)
	2011	*Blue Seas – Green Energy: A sectoral Marine Plan for Offshore Wind Energy in Scottish Territorial Waters* (Scotland) Selecting 9 short term sites by developers and The Crown Estate Commissioners and awarding Exclusivity Agreements; identification of further 25 areas (to be subjected to further assessment) for medium term development (2020–2030)
EU Maritime Spatial Planning Directive (MSP Directive)	**2014**	Establishing a coordinated and coherent framework for MSP across Europe Requiring the establishment of MSP for marine waters in EU member states by 2021 MSP shall aim at promoting sustainable growth of maritime economies, sustainable development of marine areas, and sustainable use of marine resources
		MSP for the Belgian part of the North Sea (Belgium) Zones for awarding domain concessions for the construction and exploitation of installations for generating electricity from water, tides or wind, since the Belgian EEZ was heavily overused (OWF, sand and gravel mining, marine conservation etc.)
	2015	*Offshore Renewable Energy Development Plan* (Ireland) 7 assessment areas and different technical options and combinations for wind, tidal, and wave energy – sectoral MSP
	2016	*National Offshore Wind Strategy* (USA) Aiming at 86 GW installed offshore wind energy capacity by 2050
	2017	*Draft MSP for Washington's Pacific Coast* (USA, WA) Guidance for new ocean uses (e.g. renewable energy projects, aquaculture), baseline data on coastal uses, recommendations to protect important and sensitive ecological areas and existing uses, e.g. fishing
	2018	*German EEZ MSPs revisions and area development plans 'Flächenentwicklungspläne' ongoing* (Germany) Pursuing an integrative approach concerning OWFs and grid development

Figure 9.1 Timeline of international treaties and principles affecting marine spatial planning (MSP), and MSP case study examples for offshore wind farm (OWF) development.

governance, it also attempts to secure and maintain the good environmental status of marine and coastal waters, in accordance with the Marine Strategy Framework Directive (Fernandes *et al.* 2017). Competent authorities need to be devolved, such as the Marine Management Organisation in England, Marine Scotland in Scottish territories, and the Federal Maritime and Hydrographic Agency for the EEZ in Germany. Marine spatial plans require the identification of spatial and temporal distributions of relevant existing and future activities and uses in marine waters (Art. 8, European Parliament & European Council 2014). These may include:

- Aquaculture and fishery areas
- Installations and infrastructure for the exploration of oil, gas and other energy resources, of minerals and aggregates, and for the production of energy from renewable sources
- Maritime transport routes and traffic flows
- Military training areas
- Nature and species conservation sites and protected areas
- Commodities extraction areas
- Scientific research
- Submarine cable and pipeline routes
- Tourism
- Underwater cultural heritage.

Moreover, the Maritime Spatial Planning Directive refers to the ecosystem-based approach. Even though experts and stakeholders have not agreed upon a generally accepted definition for ecosystem-based management, the concept is a core element to marine governance and thus central to the EU Maritime Spatial Planning Directive (Söderström & Kern 2017).

In the USA, coastal states bordering the Atlantic Ocean in the north-east, including Massachusetts, Rhode Island, Oregon, Washington and California, among others, mostly refer to MSP plans as 'ocean management plans' (Juda 2003; Boucquey *et al.* 2016) (see Box 9.2). In October 2017, the draft MSP plan for Washington State's coast was launched (State Ocean Caucus 2017), also targeting the Pacific Ocean for MSP. The Canadian Oceans Act was published in 1997, enabling legislation to promote the integrated management of Canada's oceans. Five Large Ocean Management Areas (LOMAs) were established on the Atlantic and Pacific coasts under the Oceans Act (Fisheries and Oceans Canada 2009).

In 2009, a comprehensive marine planning system was established in the UK, backed by the Marine and Coastal Access Act of 2009 (Parliament of the United Kingdom 2009). A Marine Policy Statement provided the context for marine plans and aimed at greater coherence in legislation throughout the UK (HM Government 2011). In 2018, the UK Government published a 25-year Environment Plan, which commits to having UK Marine Plans in place by 2021 (HM Government 2018). Among the different countries, 11 marine plan areas have been designated for the coastal (inshore) and EEZ (offshore) zones in England (MMO 2014), with the devolved administrations in Wales, Northern Ireland and Scotland issuing their own marine plans, including Scotland's National Marine Plan (Scottish Government 2015) and the Welsh National Marine Plan (Welsh Government 2018).

Box 9.2 Marine spatial planning as a driver for the offshore wind-farm markets in North America

Although offshore wind-farm (OWF) development in Europe was one of the major drivers for the marine spatial planning (MSP) process, the USA provides another narrative of marine planning being well established while OWF development is still in its early stages. Only recently has the "offshore industry [spread] beyond its northern European home to North America" (GWEC 2018: 54), with the first OWF in the USA, and North America in general, being commissioned off the coast of Rhode Island in 2016. Responsibility for Marine Spatial Plans is generally a prerogative of the individual federal states, although plans must be prepared in close cooperation with the federal authorities. State jurisdiction is traditionally limited to a zone of 3 nautical miles (Burger 2011), while the EEZ beyond this, to 12 nautical miles offshore, encompasses federal water (UN 1982). While OWF development is usually under the jurisdiction of the federal agencies within the EEZ, the grid connection and onshore facilities are administered by state authorities.

MSP in the USA was driven by other discourses over the use of the sea, such as conservation, fishing and relevant indigenous rights, aquaculture (State Ocean Caucus 2017) and transportation (EEA & CZM 2015). Marine Spatial Plans were first established in 2009 in Massachusetts (Ocean Management Plan) and Oregon (Territorial Sea Plan). With the National Offshore Wind Strategy issued in 2016, steps were undertaken at the strategic policy level to promote and encourage OWF development in the USA. Aiming to achieve a total of 86 GW of installed offshore wind energy capacity by 2050, this federal policy could boost OWF developments. Furthermore, some states have issued adjusted and often ambitious offshore wind energy targets, including New York State with 2.4 GW by 2030, New Jersey with 3.5 GW by 2030 and Massachusetts with 2.0 GW by 2020.

MSP in the USA aims at encouraging shared responsibilities and interagency cooperation among federal and state authorities during the planning and approval processes. While some states predefine sites for OWF development, which allow developers to propose OWF projects such as the Renewable Energy Zone in Rhode Island or four renewable energy suitability study areas in Oregon, other states identify OWF potentials and propose further studies at the consecutive planning tiers.

In Canada, the Pacific North Coast Integrated Management Area (PNCIMA) encompasses around two-thirds of the British Columbian coast. In a collaborative process, an integrated management plan for the PNCIMA was developed and published in 2007–2012. The process was led through an oceans governance agreement between the federal, provincial and First Nations governments, with contributions by various organisations, stakeholders and interested parties (PNCIMA Initiative 2017). Among other marine renewable approaches, the plan identifies wind energy to be of future interest in the plan area, with 79 wind-power tenures onshore and offshore already in place by 2010 and an additional 81 undergoing the application process (PNCIMA Initiative 2011).

In the rest of the world, areas such as the Antarctic and Arctic, which make an important contribution to biodiversity, are still in need of protection through conservation planning and Marine Protected Area (MPA) designation. Thus, the Antarctic and Arctic seas are perhaps the most relevant and emerging hotspots for MSP in the future, the latter of which is already facing pressure from fossil energy drilling exploration as well as novel fishery and shipping options under climate change conditions (Edwards & Evans 2017). Since the first MSP approaches, marine sanctuaries and MPAs have also increased dynamically (UNEP-WCM & IUCN 2018). In 2017, MPAs covered around 16% of national waters, which in turn represent around 39% of the global ocean. Thus, in total around 7% of the oceans are covered by 15,604 MPAs (UNEP-WCM & IUCN 2018). Moreover, when MSP first evolved, data were scarce and monitoring was scant. Geospatial databases have since become vital in collecting information on a variety of essential parameters, such as the distribution of species and a range of different uses of the sea, thereby assisting more informed decision making in the future (Boucquey *et al.* 2016).

At the same time, OWF development was set up rather dynamically in the early 2000s, with the first turbines installed in nearshore locations along the Dutch and Danish coasts (EWEA 2011). The first 'utility-scale' OWF, the Middelgrunden project, with 20 turbines and a total capacity of 40 MW, was installed in 2001 in Danish waters (Vølund & Hansen 2001). Ever since, the consent and development of OWFs has continued apace, with future objectives of installing turbines at increasingly greater distances out at sea (Simas 2017). By the end of 2017, there was 18.8 GW of installed capacity in 17 markets around the world, with 84% (15.7 GW) installed in European waters, mainly in the North and Baltic Seas (GWEC 2018). Moreover, the number of countries with pilot phases or industrial-scale OWF development projects is growing and, hence, "offshore wind is taking shape as a mainstream energy source" (GWEC 2018: 54).

Initially, MSP could not effectively follow up or even set an appropriate agenda on where to site (or exclude) OWFs. For example, the German offshore wind energy strategy was launched in 2002 while OWF project planning and permitting procedures started immediately. The first MSP plan was approved 7 years later, in 2009. Although including its own inherent Strategic Environmental Assessment (SEA), the first German MSP process was, to some degree, predefined since the planning and assessment processes drew on previous project-level OWF permitting and Environmental Impact Assessment (EIA) procedures, rather than identifying where these projects would be best placed (Köppel *et al.* 2017). Future pathways of MSP and delivery of relevant SEA may yet become more sophisticated, although key questions remain: Can MSP be applied in a timely manner? Can MSP actually become a powerful and proactive tool for OWF planning and assessment?

Scope

This chapter aims to illustrate how MSP emerged over time and how it has coincided with offshore wind energy development (see Figure 9.1 in Box 9.1). By highlighting issues in planning, such as environmental assessment, transboundary cooperation, co-location and stakeholder integration, the aim is not only to describe the *status quo* of MSP but also to pinpoint what is to be expected from marine plans in the future. This chapter draws on case study examples, mainly from Europe, the knowledge and expertise of the authors, and the body of scholarly work published on MSP.

In conducting a non-systematic literature review, the range of data gathered included peer-reviewed research articles, 'grey' literature, guidance material, and case study and policy documents. The literature search was conducted using various search engines, including Scopus, Science Direct, Tethys and Web of Knowledge, applying the following key words and combinations thereof: 'marine spatial planning', 'maritime spatial planning', 'offshore energy', 'offshore wind energy', 'strategic environmental assessment (SEA) for marine spatial plans', 'MSP for renewable energies', and 'exclusive economic zone'. Data collection for the chapter took place from November 2017 to September 2018.

The authors' prior experience with the emergence of MSP, especially in Germany and its early SEA approaches, may be summarised as follows. At the time of writing, Johann Köppel and Gesa Geißler serve as members of the scientific advisory board for the amendment of the MSP plans for the German EEZs in the North and Baltic Seas. Experiences of the initial phases of England's MSP and SEA for Rounds 1 and 2 of offshore wind development are drawn from Ines Kruppa's 2007 PhD thesis, as supervised by Johann Köppel, although more recent experiences from the UK have been included from the literature. In 2016, Michelle Portman published *Environmental Planning for Oceans and Coasts*, regarded as a seminal book in its field. She lead the initiative for Israel's first MSP plan and worked for the State of Massachusetts in coastal and marine management at the time the first Ocean Management Plan was developed for the state's waters. Marie Dahmen is a research assistant with the German Federal Maritime and Hydrographic Agency (BSH), where she coordinates the EU project 'Strategic Environmental Assessment on North Seas Energy' (SEANSE) and is part of the planning team for the revision of Germany's MSP plans.

Themes

Alternatives, zoning and Strategic Environmental Assessment

SEA for plans and programmes provides a tool to assess the likely environmental consequences of MSP plans and to include participatory procedures into decision-making processes. Although neither the Kyiv Protocol on SEA nor the European SEA Directive specifies MSP to be directly subject to SEA, MSP plans have to undergo an SEA since, for example, they "set a framework for future development consent of projects listed in Annexes I and II to Directive 85/337/EEC" (European Parliament & European Council 2001; UNECE 2003). The MSP Directive itself refers to the SEA Directive, in stating that if "maritime spatial plans are likely to have significant effects on the environment; they are subject to Directive 2001/42/EC", and should hence undergo an SEA.

In their frame-setting role, some MSP plans are guiding future OWF development through zoning approaches. This is the case in the MSP plan for the Belgian part of the North Sea (Belgian Federal Public Service Health, Food Chain Safety and Environment 2014b). The plan delineates a zone "for awarding domain concessions for the construction and exploitation of installations for generating electricity from water, tides or wind". The plan was subject to SEA and three alternatives were tested: (1) the draft MSP plan (Alternative 1), (2) a zero alternative, and (3) an additional Alternative 2, a variant with enhanced nature conservation among other use zones (Belgian Federal Public Service Health, Food Chain Safety and Environment 2014a).

The MSP plan for the German EEZ in the North Sea (BSH 2009b) and in the Baltic Sea (BSH 2009a) designated priority areas for wind energy. In these areas, the production of wind energy was given priority over other uses such as fisheries and exploitation of resources, while the alternatives assessed in the SEAs (BSH 2009c; 2009d) referred mostly to macro-siting (i.e. corridors and OWF areas) instead of technical (e.g. tidal energy vs

OWF) alternatives (cf. Rehhausen *et al.* 2018). The likely evolution of the environment without implementation of the plan, in other words a zero alternative analogous to that described above in relation to Belgium, was described in depth but was not considered to be a reasonable alternative, given Germany's energy and climate targets.

The Massachusetts Ocean Management Plan established three types of management areas: (1) prohibited areas (i.e. areas where certain uses, activities and facilities are prohibited, such as for electric power generation, transmission and distribution), (2) renewable energy areas, including both commercial and community scale, and (3) multi-use areas (EEA & CZM 2015). Among the potential sources of renewable energy, special attention was paid to wind energy. The plan proposed two wind energy areas, based on environmental screening that applied exclusionary criteria including high concentrations of marine avifauna and water-dependent marine uses such as ferry routes, and conducted a cumulative impact assessment.

In Ireland, a sectoral approach to MSP was pursued, addressing offshore renewable energy within an Offshore Renewable Energy Development Plan (OREDP) that assessed areas for technical (i.e. wind, wave and tidal energy, and a combination thereof) and macro-siting (i.e. different spatial distributions) alternatives (DCNER 2014). The OREDP was subject to SEA (Sustainable Energy Authority of Ireland *et al.* 2010), with the scope of the SEA determined early in the OREDP process, as both processes are closely interlinked. In the SEA, three scenarios for the development of offshore renewables to 2030, comprising low (800 MW OWF, 75 MW wave and tidal), medium (2,300 MW OWF, 500 MW wave and tidal) and high (4,500 MW OWF, 1,500 MW wave and tidal) development objectives, as well as a zero alternative, were used to assess the relevant environmental and socio-economic impacts. The SEA specified environmental effects that were to be dealt with at the subsequent project level through careful micro-siting for the different alternatives. Eventually, the zero alternative needed to be rejected, as the environmental and economic risks of not implementing the OREDP were found to be higher (e.g. continued reliance on imported electricity and fossil fuels, less economic investment in supply chains for offshore renewables, potential effects relating to not combating climate change), thus emphasising the necessity to implement the OREDP.

Early approaches to marine spatial planning in the UK

In England and Wales (Kruppa 2007), the Crown Estate was a crucial actor from the very beginning of OWF development, granting relevant leases in the territorial seas as well as licenses in the EEZ, based on tenders in subsequent rounds (Crown Estate 2016). A first round, Round 1, was launched in 2001, clearly serving as a pilot round, with no spatial presettings given and no zoning schemes set up at the time. The Crown Estate assessed bids on the basis of the financial standing of the candidates, their offshore development expertise and their wind turbine expertise (DTI 2002). North Hoyle, Scroby Sands (for results of intensive monitoring see Perrow *et al.* 2011; Skeate *et al.* 2012; Perrow 2019) and Kentish Flats OWFs, originating from Round 1, were all realised by 2006 (Crabtree *et al.* 2015). The dynamic pursuit of climate protection goals in the UK was a major driver in accelerating OWF development and thus Round 1 served in introducing wind projects and paving the way for the offshore wind industry.

In 2003, the Crown Estate launched the tender for a second round, which focused on three strategic development areas that had been identified in a prior SEA process: Thames, Greater Wash and Liverpool Bay to Solway (DTI 2002). The three strategic regions for the subsequent competitive bidding rounds identified in Round 2 drew on information such as water depth, transmission grid options, existing and planned uses, and environmental information (DTI 2002). However, more specific EIAs especially for individual Round 2 sites revealed sensitive siting conditions, especially for seabirds. The making of the SEA involved an interactive and participative approach, in which development scenarios in each of the strategic regions were tested. Applying a basic geographic information system (GIS)-based risk analysis, spatially explicit information on the abundance of marine mammals or bird migration corridors remained rather limited, however. Nevertheless, protected marine sites especially in the EEZ were largely missing in the early 2000s, since, for example, the Marine and Coastal Access Act was published only later in 2009, which formally introduced Marine Conservation Zones (Parliament of the United Kingdom 2009). As a result, in several cases the OWF sites needed to be modified because of the subsequent assessment on a project level. For instance, the London Array OWF in the Thames estuary needed to be altered because of internationally relevant populations of wintering Red-throated Diver (Loon) *Gavia stellata* (Figure 9.2). While initially planning for a phased development for the London Array OWF (development of phase 1 with a subsequent phase 2 development if no significant impacts were found during phase 1 monitoring), the second phase needed to be cancelled owing to uncertainty (Perrow 2019). Similarly, the Cirrus Shell Flat OWF in the North West region required modifications because of important wintering grounds for Common Scoter *Melanitta nigra* populations (Kaiser *et al.* 2002; 2006) (Figure 9.3). Furthermore, construction noise, possible cumulative effects and the grid connections could not be covered well enough at the time.

Thus, the strategic character of early SEA for the siting of OWF in English and Welsh waters was limited, particularly as its sole focus was on a few strategic regions, which were considered suitable for developing OWF. In other words, there was no systematic scrutiny of other regions as SEA was not conducted for the complete territorial and EEZ seascape. The assessment of alternatives was also limited to mere OWF development scenarios in

Figure 9.2 Red-throated Diver (Loon) *Gavia stellata* in winter plumage. Uncertainty of the effects on internationally important populations of this species led to the cancellation of the planned phase 2 of the London Array in the UK. The sensitivity of divers to displacement in particular (Mendel *et al.* 2019) is an important consideration for much wind-farm planning in the North Sea, especially in Germany. (Peter Massas/CC BY-SA 2.0)

Figure 9.3 A flock of Common Scoter *Melanitta nigra*, a species sensitive to disturbance from inshore wind farms, especially in relation to vessels and human presence, that became important in early consideration of wind-farm impacts in the UK, particularly in Liverpool Bay. (Martin Perrow)

relation to potential installed capacity and, critically, the SEA also adopted the approach of leaving any issues to be resolved at the site-based project EIA.

Taking on these experiences, siting decisions for Round 3 were made in two stages. The first of these was macro-siting, whereby national or 'strategic' selection of suitable seabed areas around the UK was conducted by Department of Energy and Climate Change (DECC) and the Crown Estate. A first Offshore Energy SEA was completed by DECC in 2009, followed by a second in 2011 and a third in 2016, with the conclusion that "at a strategic level there were no overriding environmental considerations to prevent achievement of deployment of up to 33 GW of offshore wind" (DECC 2009). In the second stage, the Crown Estate identified nine Round 3 zones, two within the Scottish EEZ and seven in England and Wales, comprising four within the EEZs of these countries and three within English territorial waters. In a separate zoning and project planning process, offshore wind developers, who were awarded Zone Development Agreements, held the right to locate OWFs within Round 3 zones, in principle with the flexibility of avoiding any particular limiting factors, including environmental issues such as important seabird foraging areas. Individual projects then required EIA, and, if necessary, Appropriate Assessments under the EU Habitats Directive (European Council 1992) in relation to European protected species or habitats, or under the EU Birds Directive (European Parliament & European Council 2009) in relation to Special Protection Areas for birds, which at that point were mostly seabird colonies connected as a result of breeding birds from specific colonies using proposed wind-farm areas.

It was, however, other considerations that ultimately meant that three of nine zones did not progress to development (Figure 9.4). Atlantic Array was abandoned as a result of challenging seabed conditions, water depth and difficult currents (Wind Power Offshore 2013). Challenging ground conditions were similarly cited as making the Irish Sea zone economically unviable (Wind Power Offshore 2014). The Navitus Bay site planned in the

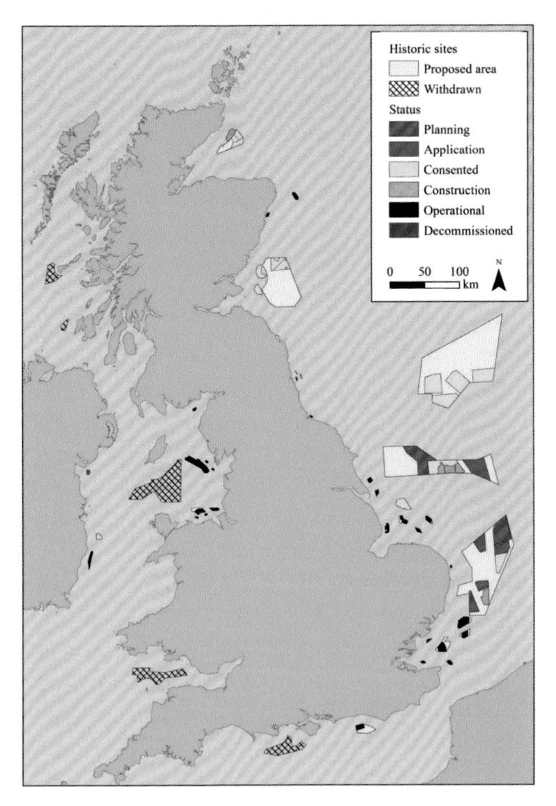

Figure 9.4 Map of the status of offshore wind-farm development (operational, under construction, consented, application submitted and in planning) in the UK at the beginning of 2019, incorporating Rounds 1, 2 and 3 of development. This includes three of the nine Round 3 zones (hatched areas in England & Wales) that were not developed. Sites have been developed within the other six Round 3 zones shown as proposed areas. Locations of five of the withdrawn Scottish territorial waters sites (see text) cannot be shown, as shapefiles are no longer available. (Information derived from OSPAR and the Crown Estate. Map of Europe from European Environment Agency)

West Isle of Wight zone was refused planning permission in September 2015 owing to the visual impact the development would have on the region, including the Jurassic Coast UNESCO World Heritage Site (Wind Power Offshore 2015).

In addition to the two Round 3 zones in Scotland, the Moray Firth and the Firth of Forth, the Scottish Government also released plans in 2010 to develop smaller OWF sites in Scottish territorial waters (STWs), that is, within 12 nautical miles of the Scottish coast (Marine Scotland 2010a) (five of these are shown in Figure 9.4). STWs are controlled directly by the Scottish Government, in contrast to planning applications outside the 12 nautical mile limit, which are handled by the UK Government. Of the ten STW sites, one site was withdrawn almost immediately by the developers because of interference with radar (*Courier* 2010), followed by three others in 2011. Issues with Kintyre included the proximity to the Campbeltown airport and the difficulties in dealing with local opposition groups (BBC News 2011). Developers withdrew from Solway Firth when the Scottish Government decided that the site was unsuitable for development based on problems with public acceptability and environmental and visual impacts (Marine Scotland 2014, Appendix 5). Wigtown Bay was deemed undevelopable by the Scottish Government following an economic review carried out in 2011 (Marine Scotland 2011) that identified that a wind farm would be likely to have adverse economic impacts on the local economy (Marine

Scotland 2014, Appendix 4). Developers also withdrew from a further three sites over a variety of issues, such as ground conditions and challenging wave conditions, as well as environmental considerations (populations of Basking Shark *Cetorhinus maximus*) in the case of Argyll Array (ScottishPower Renewables 2013), while the developers of Islay and Forth Array preferred to concentrate on other projects (*WindPower Monthly* 2010; *Telegraph* 2014). Although the SEA was carried out in parallel with the development of the plan for the STW sites (Marine Scotland 2010b), it did not predict many of the issues behind the loss of the majority (70%) of the STW sites initially proposed.

Of the remaining three projects, after initial consent, Neart na Gaoithe and Inch Cape were subject to judicial review linked to concern over significant impacts on internationally important populations of breeding seabirds in combination with sites in the nearby Round 3 zone. The judicial review was initially upheld and consents were overturned, although the Inner House later reinstated the consents (OffshoreWIND.biz 2017). Revised applications for fewer, larger turbines were submitted, and received consent in the case of Neart na Gaoithe (NnG Offshore Wind 2018), with the decision pending on Inch Cape (Inch Cape Wind 2018). The only STW site currently under construction at the time of writing, Beatrice, is is expected to be fully operational in 2019.

Moving into the future, the UK will be leasing tenders for Round 4 of offshore wind in 2019, "maintaining a pipeline of projects through to the late 2020s and beyond" (Crown Estate 2019). The proposed tender for Round 4 is likely to foresee capacity to be "increased from 6 GW to 7 GW, [...] available regions to be extended from 50 m to 60 m water depths [and] additional work [to be commissioned] on key issues of uncertainty, including further assessments on [MPA] sensitivities, ornithology and visibility analysis" (Crown Estate 2018). Following extensive stakeholder engagement, the Crown Estate has undertaken spatial analyses (including other sea uses, environmental sensitivities and technical resources), and proposes to include five regions in plans for a new leasing round, 'on the basis that they are technically feasible, include sufficiently large areas of available seabed for offshore wind development and have lower levels of development constraint' (Crown Estate 2018). In addition, four regions are proposed for further consideration, with an increased evidence base required to better understand seabed resources and constraints (Crown Estate 2019).

Germany's Exclusive Economic Zone Marine Spatial Plan 2009 Strategic Environmental Assessment reconsidered

In Germany, the coastal waters up to 12 nautical miles are governed by the states, whereas federal planning covers the EEZ. MSP plans are subject to SEA under the German EIA Act. A driving force for MSP in Germany's EEZ had been an ambitious policy to develop offshore wind power, to serve the iconic 'Energiewende' (Portman *et al.* 2009). Rehhausen *et al.* (2018) scrutinised the relevant SEA according to a set of criteria for sound SEA making (Table 9.1). While Germany was pioneering at the time in the early setting up of the MSP plan (Figure 9.5), some weaknesses were identified as well: the public was not involved in the scoping processes, and likely environmental consequences were displayed only for the preferred alternative, arguing that no other reasonable alternatives existed. The cumulative effects assessment remained limited to intra-plan effects, whereas at least state-level plans for the territorial sea could have been considered. The monitoring concept known as StUK 3, which was the standard investigation concept for impacts pertaining to OWFs until 2013 (BSH 2007), focused primarily on existing environmental monitoring and research on the effects of OWFs [since 2013, following BSH & BMU (2014) StUK 4 has replaced StUK 3 as the standard investigation concept for impacts pertaining to OWFs].

Table 9.1 A summary analysis of Strategic Environmental Assessment (SEA) on German Marine Spatial Plans showing selected features

Criteria	Indicators	Maritime Spatial Plans for the North and Baltic Seas
Well-timed integration of SEA into decision-making	Starting point of SEA	SEA began with the scoping process in 2005, when first meetings with stakeholders were set to start the planning process. Planning and SEA process ended in 2009.
	Well-timed integration of environmental objectives	Environmental objectives integrated into plan objectives as general principles ("Grundsätze") of spatial planning
	Well-timed integration of alternatives assessment	Due to limited alternatives assessment in the environmental report, the integration was also limited
Precise tiering and cooperation with other SEAs/EIAs in the process	Provision of tiering between SEAs and SEA-EIA	General recommendations for subsequent EIA for project approval of various possible project types were included in the environmental report
	Identification of mitigation and compensation opportunities for subsequent planning levels	Environmental report offers recommendations for mitigation and compensation opportunities
Scoping as prior structuring of SEA with public participation	Involved stakeholder and public	Federal and state agencies for health and environmental issues, environmental NGOs
	Scoping issues discussed	Status-quo analysis No-change alternative Methods for assessing the significance of impacts Avoidance and minimization measures Appropriate Assessment according to EU Habitats Directive Alternatives Monitoring
Selected alternatives and consideration of alternatives	Selection of alternatives	Limited to no-change alternative and some macro siting alternatives
	Environmental assessment for each alternative	No assessment for each alternative Alternatives were discussed but eliminated from further assessment
Cumulative effects assessment	Definition of cumulative effects	Intra-plan effects As the maritime spatial plan was the first maritime plan, no other federal plans could have been considered. But state level plans for coastal waters could have been considered.

Table 9.1 – *continued*

Criteria	Indicators	Maritime Spatial Plans for the North and Baltic Seas
Public participation	Timing of public consultation	No public participation during scoping Participation timeframes on the draft plan and the environmental report are not documented A hearing was held in October 2008
	Transparency of planning and decision-making process	The plan and the environmental report are permanently available on the Federal Maritime and Hydrographic Agency's website, but transparency is limited due to process documentation (e.g. public participation report) and the scoping report being unavailable
Monitoring concept	Clear description of monitoring concept within environmental report	Monitoring concept was described and relies mainly on existing monitoring programmes and research on the effects of offshore wind power plants
	Integration to planning cycle	The integration into the planning cycle was not defined in the monitoring concept

Reprinted from Rehhausen *et al.* (2018), with permission from Elsevier.
EIA: Environmental Impact Assessment; NGO: non-governmental organisation; EU: European Union.

The agency created a database to handle environmental monitoring information related to the North Sea and the Baltic Sea. However, it remains to be seen how effectively and transparently these data will be available for the amendment process of the German EEZ MSP 2019–2021.

Challenges of cumulative and in-combination effects

The assessment of both cumulative and in-combination effects constitutes a major task in conducting SEAs on MSP plans and involves the identification of impacts from multiple uses: those subject to the plan and those interacting with it. The consideration of cumulative effects proves most effective at higher planning levels, where developments might still avoid and mitigate unintended effects substantially. However, some aspects tend to be tiered to the licensing phase anyway, for example, the in-combination effects of pile driving and underwater noise, which remain uncertain until the exact piling position and construction period have been determined. For biota, the analysis may focus on the significance of affecting the most sensitive species and habitats. The leads to the challenging questions of acceptable levels of impact and the carrying capacity of ecosystems in terms of cumulative OWF development in combination with other anthropogenic stressors. In general, an absence of a common understanding about the nature of cumulative (and in-combination) effects prevails (Willsteed *et al.* 2018), causing a bottleneck for better practice. In a critical review of nine OWF Round 3 developments in England and Wales, Willsteed *et al.* (2018) found that spatial aspects and pressures were identified more comprehensively than temporal aspects.

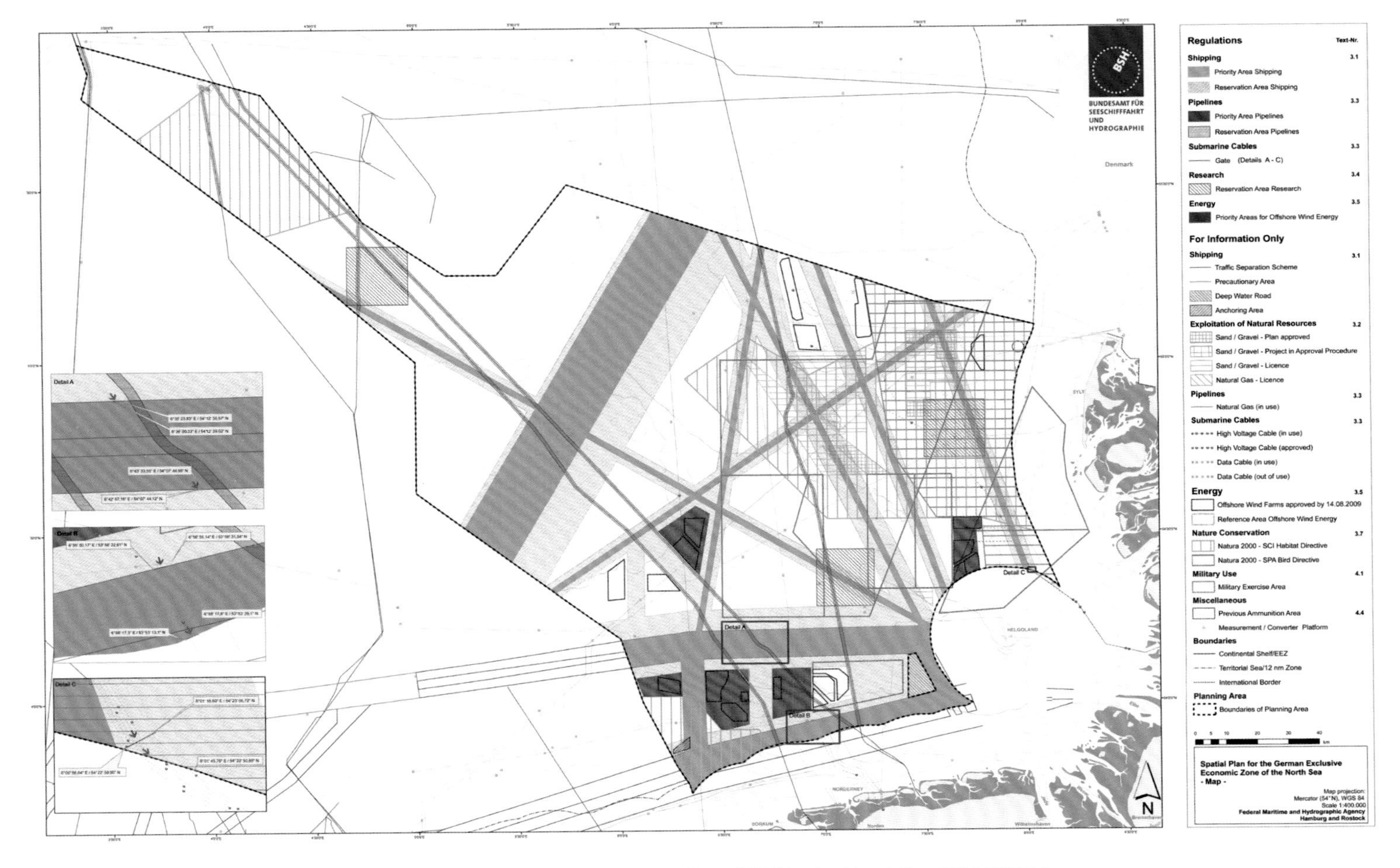

Figure 9.5 First Marine Spatial Plan map for the German Exclusive Economic Zone (EEZ) in the North Sea (BSH 2009b).

Figure 9.6 Total cumulative environmental effect in the Swedish Baltic Sea marine spatial planning area. The colour scale in the map shows the percentage of the maximal cumulative effect in the Baltic Sea: blue=0–10%; grey=10–20%; green=20–30%. (Swedish Agency for Marine and Water Management, www.havochvatten.se)

The Swedish Agency for Marine and Water Management uses the model-based tool 'Symphony' (SwAM 2017) in its MSP (Figure 9.6) to assess cumulative environmental impacts from different planning options (SwAM 2018). The ecosystem components of the Swedish territorial waters and EEZ are mapped in 250×250 m grids and overlaid by marine pressure maps. A sensitivity matrix describes how sensitive each ecosystem component is considered to be towards the multiple pressures, based on expert judgement. By repeating the calculation for different planning options, alternatives can be compared to support the decision process. Although model assumptions have their weaknesses and uncertainties, Symphony provides a worthwhile tool to analyse the cumulative impacts at a feasible spatial resolution. Mapping uses, pressures and sensitivities of ecosystems establish the first step in finding planning solutions for conflicting marine uses.

Yet, Willsteed *et al.* (2018) argue that the assessment of multiple stressors on receptors from both cumulative and in-combination effects of various projects is beset with challenges, particularly since knowledge gaps pertaining to effects of other marine uses exacerbate uncertainties. One method used "to filter and prioritise the complex interactions of sources, pressures, pathways and receptors" is the application of risk assessment criteria (Judd *et al.* 2015, p. 256). This means that all likely pressure–effect relations need to be identified, with their associated risk and the scale of impact, in order to define and implement suitable management actions.

Applying a drivers, pressures, state, impact and response (DPSIR) approach, Platteeuw *et al.* (2017) proposed a framework for cumulative impact assessment for OWF siting. This aims at criteria for the zonation, configuration and management of OWFs, and follows a six-step approach: (1) identification of pressures, (2) identification of potentially affected habitats and species, (3) description of other stressors and activities, (4) description of the nature and scale of cumulative effects, (5) evaluation of significance of effects, and (6) when necessary, adaptation and mitigation.

This approach has been further developed in the environmental working group of the Support Group on Maritime Spatial Planning (SG1), currently chaired by the Netherlands, which was established based on the political declaration on energy cooperation between the North Sea countries (European Commission 2016). The members conduct a Common Environmental Assessment Framework (CEAF), including a common knowledge basis, especially on cause–effect relations of offshore wind, an agreement on common species and habitats affected by wind-farm development, and on a common approach to how to assess cross-border cumulative effects. CEAF will be tested in SEANSE (Rijkswaterstaat Sea and Delta 2018) for different stages of wind energy development. In three case studies, the cumulative effects of existing and planned OWFs in the North Sea for the years 2023, 2030 and 2030 onwards will be assessed for the most sensitive marine species, namely Harbour Porpoise *Phocoena phocoena* (Figure 9.7), divers *Gavia* spp. (Figure 9.2), and seabirds. The assessment of future large-scale developments is based on observed effects from OWF monitoring, such as habitat use and displacement effects (Nehls *et al.* 2019; Vanermen & Stienen 2019), which serve as input parameters for population and collision risk models (Cook & Masden, Chapter 5). Within SEANSE, cumulative effects are considered not only in their spatial but also in their temporal extent, which is a promising approach for long planning horizons.

While the importance of assessing cumulative effects is widely accepted, actual approaches to handle the complex linkages and interactions in SEAs remain limited. A CEAF should eventually incorporate the ecosystem-based approach and principles of environmental risk assessments (Judd *et al.* 2015; Stelzenmüller *et al.* 2018).

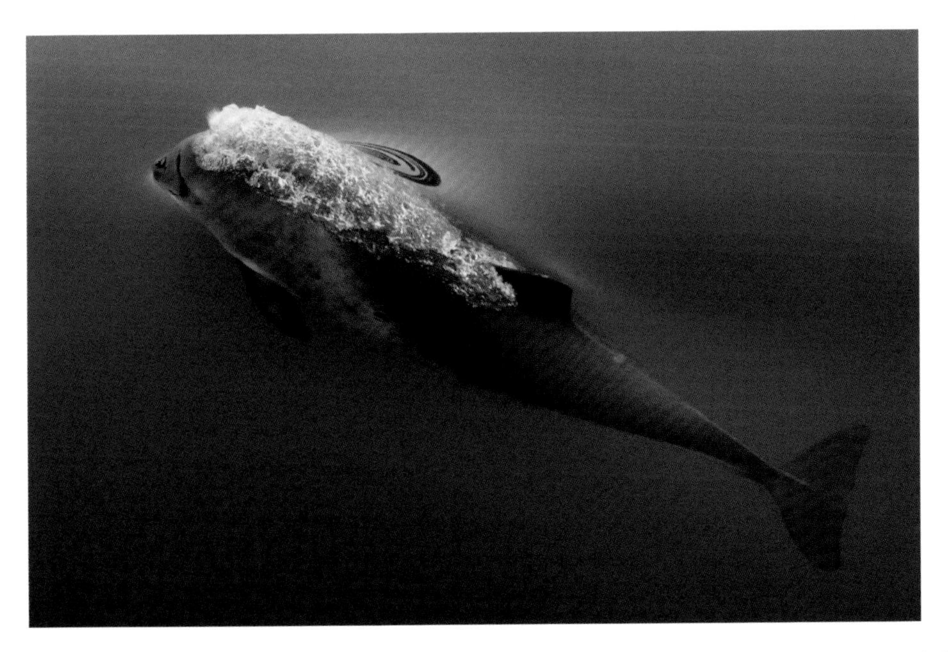

Figure 9.7 Surfacing Harbour Porpoise *Phocoena phocoena*; a small cetacean sensitive to wind-farm development, specifically pile driving during construction, that has led to the adoption of a noise-mitigation strategy to avoid population-level disturbance in the German part of the North Sea. (skeeze/Pixabay.com 752591)

Geospatial tools, baseline and timeline databases, and decision-support approaches

Knowledge on MSP has changed over the years and the first MSP amendments are underway; in Germany these cover a timespan from 2018 to at least 2021 (BBSR 2017). Globally, various baseline surveys and marine mapping results have become available and have been further elaborated (e.g. BfN 2018; Commonwealth of Massachusetts 2018; MARCO 2018). Ambitious tools and platforms have been established (Table 9.2), providing better planning baselines than a decade ago, but information is still limited and cumulative assessments remain beset by uncertainties (Willsteed *et al.* 2017; 2018). In Germany, at least the results from the responsible agency's database should be available; these draw on proponents' relevant obligations that have been set by the standard investigation programme StUK 4 (BSH 2013b). Overall, Willsteed *et al.* (2017, p. 19) argue for a shift "away from disparate [EIAs and limited SEAs], and [a] move towards establishing a common system of coordinated data and research to ecologically meaningful areas".

Another pending question is to what degree we might apply the full set of zoning categories in spatial planning as is undertaken in terrestrial planning, especially in terms of conservation. That is, might we not only identify possible 'no-go' areas for OWFs, but also use a tiered approach in terms of designated sites during decision making? In a similar way to onshore, can we set up MSP zoning categories that allow for OWFs to some degree, with the proviso of the requirement for specific mitigation measures, such as shutdown on demand in peak times of bird or even bat migration over the sea? (see Harwood & Perrow, Chapter 8).

With regard to the relevant scales of MSP and SEA, several sensitivity approaches have been proposed, especially for seabirds, such as SeaMaST (Bradbury *et al.* 2014), be they towards collision risks with OWFs (King 2019) or displacement (Vanermen & Stienen 2019). Furthermore, it remains to be seen whether multicriteria (decision) assessments (MCA or MCDA) can be applied widely in MSP, with its substantially overlaying marine use and conservation challenges. Will relevant MC(D)A (e.g. Hanssen *et al.* 2018), analytic hierarchy process or other decision support approaches and tools in scholarly work (e.g. for the selection of hybrid OWF and wave energy systems in Greece; Vasileiou *et al.* 2017) ever make it into practice? Different MC(D)A and analytic hierarchy processes have been combined with GIS applications primarily for siting onshore wind energy (Vasileiou *et al.* 2017). Stelzenmüller *et al.* (2010; 2011) combined GIS and a Bayesian belief network for marine planning purposes. They found it to be a useful tool to map and assess uncertainty-related changes in management measures as well as the risk of spatial management options. While the zoning challenges at hand seem quite tangible (Stelzenmüller *et al.* 2013, Figure 1), participatory planning processes (e.g. via public participation GIS) (Strickland-Munro *et al.* 2016) or power-driven governance deformations may substantially modify rationalism in the planning processes.

Cross-sectoral integration and transmission grid connections

It has become clear that OWF grid connections were not considered in the same intensity as was planned for the OWFs themselves, except for the US Mid-Atlantic MSP plan (Lüdeke *et al.* 2012), where (perhaps unsurprisingly) the transmission grid was considered as a planning backbone. In 2017, a paradigm shift in the German energy policy resulted from the enactment of the Offshore Wind Energy Act (WindSeeG 2016), which also introduced a tendering system for OWFs at sea. WindSeeG adopts an integrated approach by coupling grid development and OWF deployment at sea. This brought forward a more detailed

Table 9.2 Selected geospatial marine planning tools.

Provider	Link	Selected content
California Offshore Wind Energy Gateway, USA	https://caoffshorewind.databasin.org/	Geospatial information on ocean wind resources, ecological and natural resources, ocean commercial and recreational uses and community values
Massachusetts Ocean Resources, USA	http://maps.massgis.state.ma.us/map_ol/moris.php	Biology, fisheries, infrastructure, management and protected areas
Federal Maritime and Hydrographic Agency, Germany	https://www.geoseaportal.de/	Benthos, water pollution, marine use, marine mammals, vessel traffic density, seabed sediments
Federal Agency for Nature Conservation (BfN), Germany	Monitoring of seabirds: https://geodienste.bfn.de/seevogelmonitoring?lang=de	Seabird distribution, Marine Protected Areas, facilities and known sea uses, landscape designations
	Monitoring of Harbour Porpoises: https://geodienste.bfn.de/schweinswalmonitoring?lang=de	Harbour Porpoise distribution, Marine Protected Areas, facilities and known sea uses
Marine Management Organisation (MMO), England	http://defra.maps.arcgis.com/apps/webappviewer/index.html?id=3dc94e81a22e41a6ace0bd327af4f346	MMO marine plan areas, marine uses, conservation areas, landscape designations, recreation models
Natural England Open Data Geoportal	http://magic.defra.gov.uk/MagicMap.aspx	Marine uses, habitats and species, marine licensing, Marine Protected Areas, Water Framework Directive water bodies and sensitivities
Mid-Atlantic Ocean Data Portal (MARCO), USA	http://portal.midatlanticocean.org/visualize/#x=-73.24&y=38.93&z=7&logo=true&controls=true&basemap=Ocean&tab=data&legends=false&layers=true	Marine uses, marine life, recreation, renewable energy (lease areas, technology zones, wind speeds), oceanography (net primary productivity, sediments, etc.)
Northeast Ocean Data, USA	https://www.northeastoceandata.org/	Marine uses, culture and recreation, energy and infrastructure, marine fauna and habitats, restoration

planning scheme and scale, the so-called site development plan (*Flächenentwicklungsplan*), which supplements the MSP efforts (Figure 9.8). The funding rates for OWF development now follow competitive tendering, cutting costs considerably, with the average allowance at 0.44 ct/kWh (BNetzA 2017). By 2026, a central model is to be put in place, which invites bids once a year for an average of 700–900 MW annually (WindSeeG 2016).

Figure 9.8 Proposed areas for offshore wind farms in the German Exclusive Economic Zone in the North and Sea as sketched in the second draft of the site development plan *'Flächenentwicklungsplan'* (BSH 2019).

The German Offshore Wind Energy Act devolves new responsibilities within the German EEZ to the BSH. In addition to MSP and the licensing and monitoring of OWFs, BSH shall publish a sectoral development plan for offshore wind energy, the site development plan (Figure 9.9), in accordance with the Federal Network Agency (BNetzA), whose responsibilities lie with the grid development and tendering of renewables both onshore and offshore. The site development plan designates areas for the development of wind turbines, grid corridors and the location of converter platforms within the EEZ for the period 2026–2030. The site development plan determines specifically when BNetzA will invite tenders for which area, while aiming at a spatial and temporal coherence of commissioning OWFs and their connection to the grid. Besides efficiency criteria, such as the use of existing grid connections and the expected capacity in an area, which depends on wind conditions and area geometry, preference is given to areas where fewer conflicts, such as with conservation interests, are expected (BSH 2018). With the new legislation, the BSH provides information for the different areas, which are invited for tenders starting in 2021. Information comprises data on the marine environment, seabed, wind and oceanographic conditions and is crucial for developers to place bids (Breuch-Moritz 2018).

Policy coherence between the Marine Strategy Framework and Marine Spatial Planning Directives

A major challenge for MSP lies in Europe's MSP plans and the Marine Strategy Framework Directive's (MSFD) (European Parliament & European Council 2008) reference to the 'ecosystem approach'. In essence, MSFD 'enhanced' the terrestrial, generally freshwater,

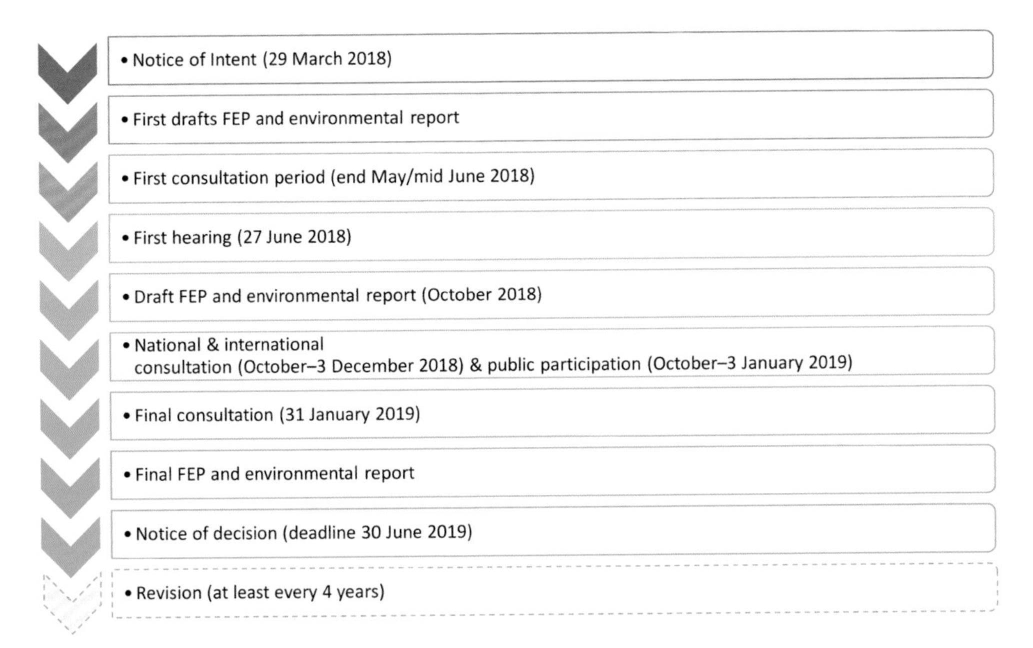

Figure 9.9 Scheduled procedural milestones for the site development plan 2019 *'Flächenentwicklungsplan 2019'* (FEP) (BSH 2018), including Strategic Environmental Assessment (SEA). (Translated by the authors)

EU Water Framework Directive (WFD) (European Commission & European Parliament 2000) and applied it to the sea. As such, it provides a list of 11 qualitative descriptors of good environmental status (Annex I WFD, European Commission & European Parliament 2000). In 2017, Annex III of the Directive (European Commission 2017b) was amended to link ecosystem components and anthropogenic pressures and impacts with the descriptors in accordance with 2017's Decision on Good Environmental Status (European Commission 2017a). The operationalisation of the objectives may become an even more ambitious assignment than was faced when applied to WFD. At least in a medium-term perspective, the question will no longer be limited to whether wind farms might harm the marine environment, but rather ask: Does the wind farm comply with the management goals of the MSFD? MSP is already an inherently complex endeavour, considering the various environmental regulations for our seas, as Boyes & Elliott (2016) demonstrated for the international and EU marine governance system.

Good marine environmental status focuses on pressure and impacts such as nutrient enrichment and contamination by hazardous substances that are less directly influenced by the deployment of offshore wind energy. However, Annex III also lists to some degree more OWF-prone pressures and impacts, such as physical loss of habitat (i.e. the minor fractions of original habitat lost under permanent construction) and disturbances as a result of underwater noise, or biological disturbance following colonisation or exclusion. The EU's Decision on Good Environmental Status (European Commission 2017a) adds criteria and methodological standards for the monitoring and assessment of predominant pressures and impacts, as for underwater noise in descriptor 11: "Member States shall establish threshold values for […] the spatial distribution, temporal extent, and levels of anthropogenic impulsive sound sources [to] not exceed levels that adversely affect populations of marine animals" (D11C1 – Primary criteria; European Commission 2017a). Furthermore, "the population abundance of [relevant] species" (covering also OWF-relevant species, e.g. cetaceans and divers) shall not "become adversely affected due to anthropogenic pressures, such that its long-term viability is ensured" (D1C2 – Primary criteria; European Commission 2017a).

Far-reaching noise reduction strategies to protect marine mammals, especially the Harbour Porpoise (Figure 9.7) have been pursued, specifically in terms of OWF pile-driving effects (see Thomsen & Verfuβ, Chapter 7). The Federal Ministry for the Environment, Nature Conservation and Nuclear Safety (BMU) has published a noise-mitigation strategy for the German North Sea, which foresees the establishment of a noise-mitigation zone from September to April. To avoid population-level disturbance of Harbour Porpoise, the noise-protection strategy sets a threshold for the extent of pile-driving affected areas in the EEZ. This states that at any one time, a maximum 10% of the North Sea EEZ can be affected by defined noise levels radiating from OWF under construction (BMU 2013; Bellmann *et al.* 2017). Standard threshold values for noise during OWF construction were introduced in 2003 on advice from the Federal Environment Agency (UBA). Based on new research, the recommended thresholds (dual criteria) have been part of OWF licensing since 2008 (UBA 2008). The incidental provision number 14 of the BSH for German OWFs thus requires a noise-mitigation concept, which is that it must be ensured that the sound exposure level (SEL) does not exceed 160 dB re 1 μPa^2 s and the peak level (L_{peak}) does not exceed 190 dB re 1 µPa at a distance of 750 m from the construction site (BSH 2013a). Given the common European framework, further similar approaches ensuring uniformity across member states may be an upcoming issue. Merchant *et al.* (2018) propose a methodology for marine noise management to map and quantify risk at the population level. By defining risk-based noise-exposure indicators, the noise pollution exposure to populations can be quantified and managed (Merchant *et al.* 2018). Merchant *et al.* (2018) argue that noise could be integrated in assessments of cumulative effects as a stressor, as the proposed methodology is compatible with risk-mapping approaches.

Another challenge is revealed by the ambitious approach to come up with any comparable and standardised spatially explicit surveys of marine mammals. The Surveys for Small Cetacean Abundance in the North Sea (SCANS) data are an important resource to track the distribution of populations over years and to estimate population dynamics. The last survey, SCANS-III (Hammond *et al.* 2017), was conducted in summer 2016 (Kaymaz 2018; see Scheidat & Porter, Chapter 2). However, owing to the large time gaps, spatial resolution and some inconsistencies in the data set, the usability of the data for the purposes of EIA in individual sites remains rather limited. The undeniable dynamics in (protected) species distribution patterns in the marine realm, which may follow patchily distributed prey resources, which in turn may shift under climate change scenarios, also needs to be taken into account. The main question revolves around long-term effects, especially when developing less sensitive, and thus less suitable, OWF sites, for instance, once initiating OWF repowering. Owing to the unbounded and dynamic nature of the sea and human activities within it (Jay *et al.* 2016; Gill *et al.* 2018), uncertainties tend to be high, thus requiring flexible and adaptive approaches to both MSP and baseline survey and monitoring approaches. In other words: "[…] marine spatial planning can evolve to account for this dilemma by focussing not on static environmental baselines, but by placing emphasis instead on the resilience of the ecosystem and the ecosystem services they provide" (Yates *et al.* 2018, p. 1). Given the great mosaic of baseline conditions and survey methods, over time only informal and technical guidance for MSP for offshore energy can lead to coherence and comparability, although the competence for the provision of guidelines lies mostly with the individual states and national authorities. A supranational example has been the OSPAR Guidance on Environmental Considerations for Offshore Wind Farm Development (OSPAR Commission 2008, amendment ongoing).

Transboundary cooperation

Transboundary cooperation lies "at the heart of marine spatial planning [...] as marine and coastal ecosystems transcend administrative boundaries and steer planning towards wider regional or sea basin considerations" (Jay *et al.* 2016, p. 85). In adopting an ecosystem approach, the MSF Directive calls for a close coordination and coordinated monitoring programmes within a marine region or subregion (Article 11) to facilitate coherence of action across the member states (European Parliament & European Council 2008). However, Jay *et al.* (2016) argue for the existence of an apparent dichotomy, which becomes apparent in the MSP Directive as national authorities retain competency for MSP in their waters. For the development and implementation of transnationally coherent and consistent MSP in the North and Baltic Sea region, the EU Commissions Department on Maritime Affairs, 'DG Mare', supports, *inter alia*, the cooperation between national maritime planning agencies.

With the MSP Directive (2014/89/EU), all EU coastal states are obliged to draw up maritime spatial plans by 2021. In their planning procedures and management processes, the member states must ensure that the national plans are coordinated with each other and, in particular, take account of transnational issues. Whereas in the North Sea region almost all countries have MSP plans in place, the Baltic Sea countries are in different stages of the planning process, with the only enforced plans in Lithuania and Germany at the time of writing. The member states are involved in several research projects to support the planning process and exchange ideas and information in MSP expert groups such as the EU Member States Expert Group on MSP and the HELCOM-VASAB Maritime Spatial Planning Working Group (2019).

With the Baltic Marine Environment Protection Commission, otherwise known as the Helsinki Commission (HELCOM), founded in 1974, and the Oslo–Paris Convention (OSPAR 1992), there are intergovernmental structures that pursue objectives for the entire Baltic Sea and North Sea and that require cross-border coordination. HELCOM aims to expand cross-sectoral cooperation in reaching a good environmental status of the Baltic Sea, regulating marine activities, reducing the discharge of hazardous substances and implementing the ecosystem approach in MSP (HELCOM 2018). In the North Sea, OSPAR is leading international cooperation for marine environmental protection in the North East Atlantic. MSP is seen as an instrument for controlling human activities, especially regarding wind energy. The guidance on environmental considerations for the development of OWFs recommends minimising negative effects by careful site selection, for example, by avoiding important bird habitats and migration routes (OSPAR Commission 2008). Such guidance from intergovernmental institutions can lead to better practice and to harmonised approaches for MSP and environmental assessments.

Ecosystem services and co-location

The question of co-location of OWF with other uses, including MPAs, has entered the arena (Christie *et al.* 2014; Yates *et al.* 2015; Astariz & Iglesias 2016). A further discourse links the OWF planning and permitting processes to the ecosystem services concept (Hooper *et al.* 2017), covering also the well-known land-based 'green vs green' dilemma in which climate protection via renewable energy is pitched against the well-being of biodiversity. Hooper *et al.* (2017) published a review of 78 publications, which exemplarily compiled ecosystem services provided by OWFs. This scholarly work shows that negative effects, such as the role of wind farms in introducing invasive species and the decrease in the appeal of the seascape to tourists, as well as positive effects, such as the proliferation of filter-feeding mussels and the artificial reef effect (Dannheim *et al.* 2019), temporary carbon capture and the potential for attraction of tourists, should be accounted for. Gee & Burkhard (2018) further discuss coastal–marine cultural ecosystem services as sense of place or regional image.

Trade-offs between industries, for instance fisheries and marine renewables, should also be evaluated, which could, in turn, identify potential for co-location opportunities (Kite-Powell 2017). Kite-Powell (2017) identify transitory and non-transitory and production functions of OWFs in the form of regulating and supporting, cultural and provisioning. For example, Stelzenmüller *et al.* (2016) show that in 2010 and 2011 up to 90% of Danish (range 10–90%) and 40% of German (in total 20–80%) gill-net fleet landings of European Plaice *Pleuronectes platessa* (a flatfish) overlapped with OWF areas. Yates *et al.* (2015) propose a mitigation strategy for minimising the total cost of planning solutions by facilitating and maximising co-location opportunities between compatible uses that are competing for space. Often, OWFs become exclusion zones for fishing, especially in relation to trawling where there are safety issues of snagging wind-farm infrastructure, among others; yet Dutch approaches explore co-location options (Jansen *et al.* 2016). Dutch multi-use debates have emphasised co-location of either bivalve or seaweed production within OWFs, with Blue Mussel *Mytilus edulis* cultivation in Dutch waters appearing to be the most feasible option from a commercial perspective (Kite-Powell 2017).

Jansen *et al.* (2016) also explore the potential risks and synergies of co-location of aquaculture and other marine offshore activities in the Dutch North Sea in terms of ecological, economic and governance aspects. Besides calling for pilot projects with mussel cultures, Jansen *et al.* (2016) argue for the development of remote management and monitoring systems to obtain data and monitor culture performance without *in situ* visits, which would otherwise be time consuming and expensive. In the UK, Christie *et al.* (2014) propose an adaptive management approach for assessing co-location options for offshore renewable energy, particularly within MPAs and with aquaculture projects.

Thus, even OWFs and MPAs do not necessarily represent opposing spatial uses (Ashley *et al.* 2018) (see Box 9.3). Christie *et al.* (2014) argue that a 'deploy and monitor' process based on pilot projects can determine whether co-location is practicable. Effectively, the question is whether and how MSP can assist in overcoming potential trade-offs, for example, by fostering both socio-economic and ecological goals and thus facilitating the co-evolution of Sustainable Development Goals.

Box 9.3 Positive synergies through co-location of wind farms and Marine Protected Areas

In marine spatial planning (MSP), both energy infrastructure and marine conservation are likely to be considered as activities that must be allocated space in the marine environment. A traditional view of the in-combination effect of one use compared to another is that this is likely to lead to risks and negative consequences for the environment (e.g. Portman 2014; Stelzenmüller *et al.* 2018). However, there could be some advantages to considering Marine Protected Areas (MPAs) together with OWFs. For example, research that suggests that energy infrastructure can contribute in some ways to marine conservation by providing additional areas that are closed to damaging activities such as destructive trawling by fisheries (Christie *et al.* 2014; Ashley *et al.* 2018; Dannheim *et al.* 2019). If this can be true even for coastal coal-fired power plants, as suggested by Shabtay *et al.* (2017, 2018), then it seems that an even better case can be made for OWFs that both are less intrusive to the marine environment and have clear benefits for marine biodiversity, including desirable predatory fish (Dannheim *et al.* 2019; Gill & Wilhelmsson 2019). In addition, whereas many power plants are situated close to shore, OWFs are often purposely situated farther from shore to exploit good wind-for-power conditions. This means that OWFs farther offshore can serve as closed (or semi-closed) areas that act as buffers, or as refugia for a variety of marine life (Ban *et al.* 2016); promote the extension of MPAs farther out at sea; and, in particular, enhance the diversity, range and geographic spread of the MPA network.

Moreover, there should be a strong connection between MPAs and OWFs because: (1) MPAs are limited in their ability to counteract consequences of climate change, and (2) OWFs mitigate climate change. This means that in terms of progressing towards the Sustainable Development Goals 13 ('Climate Action') and 14 ('Life Below Water'), potential positive synergies should also be considered (Simard *et al.* 2016). This is relatively new thinking, but provides an approach that can be used readily within the framework of MSP, which already provides the means of looking at myriad uses together, especially when compatibility matrices are used in the process. Such a tool allows for consideration of compatibility as well as conflicts. As for wildlife impacts, there needs to be much more integrated ecosystem-based research (Perrow 2019), whereby ecological functioning and various species and their multiple conservation goals, in relation to the protection afforded by both MPAs and OWFs, are all considered simultaneously. Once there are findings from such research, the challenge will be to bring these to bear on MSP in an effective and useful way.

Concluding remarks

While around 65 countries have introduced MSP approaches, implementation stages vary substantially. For instance, African countries, including Angola, South Africa and the Seychelles, are pursuing initial stages, while Asian-Pacific countries, such as China, Vietnam and the Philippines, are implementing plans. Revisions and adaptations to plans were issued or are underway for the Great Barrier Reef Marine Park and northern European countries, including Belgium, Germany and the Netherlands (UNESCO 2018).

With offshore wind energy operating in more and more markets, and with more ambitious climate protection targets in the EU for example, the planning and commissioning of OWF is likely to become a priority for an increasing number of coastal states in the years to come (Gill *et al.* 2018; Ehler 2018). The expansion of offshore renewable energy, and wind energy in particular, continues to compete for space with other maritime uses in "an already busy seascape" (Yates *et al.* 2018, p. 1).

With this challenge in mind, and as far as guidance and good practice in MSP planning and assessment processes is concerned, the EU-funded Baltic SCOPE Project developed a checklist toolbox for the ecosystem approach in MSP (Baltic SCOPE Project 2017b). The guidance draws on HELCOM-VASAB's 'Guidelines for the implementation of the ecosystem-based approach in MSP in the Baltic Sea area' (HELCOM-VASAB Maritime Spatial Planning Working Group 2015), thus addressing key ecosystem approach elements for marine spatial planners and for those who prepare SEAs. These projects provide the backbone for the good practice principles for MSP recommended and future challenges identified by the authors of this chapter, as follows:

- By focusing on the long-term and major strategic pathways and actions, MSP can provide visions and directions for development and conservation (Ehler *et al.* 2007; Ehler 2018). One should pursue 'optimacy', that is the extent to which an MSP and SEA process follows best practice according to international standards (Bond *et al.* 2018; Rehhausen *et al.* 2018), rather than the basic requirements that all too often are the primary focus.
- The precautionary principle matters offshore, since impacts are often not predictable and beset with substantial uncertainties (HELCOM-VASAB Maritime Spatial Planning Working Group 2015). In managing uncertainty, risk assessments that adopt the ALARP principle (as low as reasonably possible), focusing on realistic risks and not simply worst case scenarios (Gill *et al.* 2018), could be applied in MSP.
- By covering various types of impacts, including cumulative, in-combination, synergistic, short-term and long-term, as well as adverse and beneficial, planning can pursue an integrative approach, identifying potential conflicts but also synergies between maritime uses (HELCOM-VASAB Maritime Spatial Planning Working Group 2015).
- Addressing the accurate planning level is of importance to cope successfully with the issues and problems at hand (HELCOM-VASAB Maritime Spatial Planning Working Group 2015). One size does not necessarily fit all MSP tiers. Whereas the overarching MSP plan needs to integrate and balance all sectors and uses, sectoral marine plans have specific targets and often focus on wind energy, leaving less room for the consideration of alternative developments (Dahmen 2017).
- Planning human use is to be based on the latest state of knowledge of ecosystems, thus 'safeguarding the components of the marine ecosystem in the best possible way' (HELCOM-VASAB Maritime Spatial Planning Working Group 2015). In adopting common approaches and methods to evidence-based planning and by sharing data widely, knowledge on the marine environment and MSP can grow.
- However, the crucial question is: Will the new data be used if there is more development of OWFs and other marine uses? This question refers to the requirement to establish and apply the 'best available science' mandate in planning and impact assessment (Wolters *et al.* 2016; Weber *et al.* 2019). With an increasing body of research and generated knowledge, best practice approaches will have to be updated regularly (Gill *et al.* 2018).
- Reasonable alternatives need to avoid or reduce environmental impacts (HELCOM-VASAB Maritime Spatial Planning Working Group 2015). A major responsibility to pursue a sound discussion of alternatives early on in the process lies with the making of SEA of MSP plans. However, Rehhausen *et al.* (2018) demonstrated how much still needs to be done in terms of that paramount requirement. It remains to be seen, for example, whether the amendment (until 2021) of previous plans, for example in the German case, will prove to be more 'strategic' than its forerunners.
- Transboundary dimensions in MSP can be more important than in terrestrial prototypes of spatial planning, such as in the case of the North and Baltic Seas, where various EEZs cut through a contiguous seascape. MSP should also reach across sectors and governmental responsibilities, and among different levels of government, as horizontal and vertical tiering, respectively. Moreover, data can be harmonised and the joint workforce and knowledge of different stakeholders can be utilised to develop solutions to any issues (Baltic SCOPE Project 2017a; Ehler 2018).
- To facilitate evidence-based planning and decision making, an ever increasing variety of decision support tools are available to be applied in the construction of MSP plans

(Pınarbaşı *et al.* 2017; Stelzenmüller *et al.* 2010; 2011; 2013). In adopting a reflective and iterative angle, more attention is being paid to monitoring, evaluation and reporting approaches for greater evidence-based management of marine resources (Addison *et al.* 2018).

- Addison *et al.* (2018, pp. 949–950) pinpointed key solutions for the latter: '(i) integrating models into marine management systems to help understand, interpret, and manage the environmental and socio-economic dimensions of uncertain and complex marine systems; (ii) utilising big data sources and new technologies to collect, process, store, and analyze data; and (iii) applying approaches to evaluate, account for, and report on the multiple sources and types of uncertainty'.
- Carefully designed participatory processes not only increase transparency of decision making, but can allocate local knowledge and expertise and allow planners to consider various interests early on in the planning process. Continuous information and participation throughout the planning process may generate ownership and acceptance of the spatial planning process (HELCOM-VASAB Maritime Spatial Planning Working Group 2015).
- Planning processes should be accompanied by a resonating advisory board, as was undertaken in the Washington MSP process (State Ocean Caucus 2017), and for the amendment of the German EEZ MSP (see *Germany's Exclusive Economic Zone Marine Spatial Plan 2009 Strategic Environmental Assessment reconsidered*, above). By involving expertise and know-how from different disciplines and stakeholders, an advisory board can enable quality assurance, iterative feedback loops and critical reflections of the entire planning process.
- Mitigation measures to prevent, reduce and, as far as possible, offset adverse effects on the environment should be considered early in the planning process, implemented, monitored and evaluated carefully (HELCOM-VASAB Maritime Spatial Planning Working Group 2015).
- Knowledge about ocean seascapes, ecosystems and species has increased over the past few decades. Yet planning faces uncertainties, as do effects and impacts, particularly at the ecosystem and population levels. In applying adaptive and iterative processes, through monitoring and follow-up, uncertainties can be managed and feedback loops established. Organisations and planning entities entrusted with MSP plans should be capable of learning while doing, and reflective of the process and its performance.
- MSP is required to balance ecological, economic, social and cultural goals and objectives to facilitate sustainable development (Ehler 2018). As required by various international treaties, management and planning approaches should be transnational and centred on the ecosystem in question (UN 1982; European Parliament & European Council 2008; 2014).
- SEA has been found to promote and foster learning among stakeholders involved in the planning and SEA process (e.g. Jha-Thakur *et al.* 2009; Jones & Morrison-Saunders 2017; Sánchez & Mitchell 2017). With manifold MSP processes being amended or under review, the SEAs may provide the chances for learning from previous activities and adopting improvements in future MSP plans.
- At the same time, the proficiency of MSP should not be overestimated. The obvious and tremendously complex requirements to be considered for the marine realm leading to a potential 'horrendogram', as espoused by Boyes & Elliot (2016), might otherwise easily paralyse MSP's core assignments.

In conclusion, spatial planning both onshore and offshore, has always, perhaps invariably, been somewhat delayed, meaning it has a 'latecomer' stigma in terms of its actual governing capabilities. For instance, Hurlimann & March (2012) identified spatial (land-based) planning for climate change adaptation as often being passive instead of proactive in shaping change and modifying markets. Proactive adaptation therefore remains a key challenge for spatial planning (Hurlimann & March 2012). A major challenge for early and to another degree contemporary MSP has been its emergence into a highly dynamic, but in many ways rather unexplored environment (Ehler *et al.* 2007). Jentoft & Knol (2014, p. 4) argue that "MSP is applied to social and ecological systems [...] that are inherently diverse, complex and dynamic, and work at multitude scales"; MSP deals with "wicked problems", finding users "in a situation where more space for one means less for other users".

Integrated and participatory planning processes therefore need their time. On one hand, marine ecosystems first have to be mapped in a suitable manner for spatial planning, but on the other hand understanding of the environmental consequences of marine exploitation or the use of new technologies, such as OWFs, needs to be informed by empirical and monitoring results in order to undertake meaningful mapping and assessment. This introduces the classic 'chicken or egg' syndrome. Data collection and monitoring at sea are also costly and time intensive (Breuch-Moritz 2018). Ultimately, much will depend on the involved agencies' and stakeholders' capacities and willingness to establish institutional learning processes, with ongoing and future amendments of initial MSP plans as obvious test beds. As MSP is a rather novel planning task for most agencies, the risks of path dependency might be less prevalent, meaning that learning may be more likely to happen.

Acknowledgements

We particularly thank our masters' students Nora F. Sprondel and Maria Kuznetsova for their contributions of topical details. Marie Dahmen's contributions represent her own opinions and do not necessarily reflect the BSH's views. We want to express our gratitude for valuable feedback from Martin Perrow, particularly in relation to early and iterative commenting on previous versions of the manuscript. Eleanor Skeate of ECON Ecological Consultancy Ltd provided additional information in relation to planning in the UK, while Andrew Harwood of the same kindly drew Figure 9.4. We also thank the Swedish Agency for Marine and Water Management (www.havochvatten.se) for providing Figure 9.6, and Elsevier for permission to reproduce Table 9.1 (reprinted from *Environmental Impact Assessment Review*, 73, Rehhausen *et al.*, Quality of federal level strategic environmental assessment – A case study analysis for transport, transmission grid and maritime spatial planning in Germany, 41–59, 2018).

References

Addison, P.F.E., Collins, D.J., Trebilco, R., Howe, S., Bax, N., Hedge, P., Jones, G., Miloslavich, P., Roelfsema, C., Sams, M., Stuart-Smith, R.D., Scanes, P., von Baumgarten, P. & McQuatters-Gollop, A. (2018) A new wave of marine evidence-based management: emerging challenges and solutions to transform monitoring, evaluating, and reporting. *ICES Journal of Marine Science* 75: 941–952.

Ashley, M., Austen, M., Rodwell, L. & Mangi, S. (2018) Co-locating offshore wind farms and marine protected areas: a United Kingdom perspective. In Yates, K.L. & Bradshaw, C.J.A.

(eds) *Offshore Energy and Marine Spatial Planning*. London: Routledge. pp. 246–259.

Astariz, S. & Iglesias, G. (2016) Co-located wind and wave energy farms: uniformly distributed arrays. *Energy* 113: 497–508.

Baltic SCOPE Project (2017a) Recommendations on maritime spatial planning across borders. Gothenburg: Baltic SCOPE. Retrieved 10 September 2018 from www.balticscope.eu/content/uploads/2015/07/BalticScope_OverallRecomendations_EN_WWW.pdf

Baltic SCOPE Project (2017b) The ecosystem approach in maritime spatial planning – a checklist toolbox. Gothenburg: Baltic SCOPE. Retrieved 16 December 2018 from www.balticscope.eu/content/uploads/2015/07/BalticScope_Ecosystem_Checklist_WWW.pdf

Ban, S.S., Alidina, H.M., Okey, T.A., Gregg, R.M. & Ban, N.C. (2016) Identifying potential marine climate change refugia: a case study in Canada's Pacific marine ecosystems. *Global Ecology and Conservation* 8: 41–54.

BBC News (2011) SSE ditches plans for Kintyre offshore wind farm. Retrieved 5 February 2019 from https://www.bbc.co.uk/news/uk-scotland-business-12619091.

Belgian Federal Public Service Health, Food Chain Safety and Environment (2014a) Environmental impact report (EIR-plan) for the draft Marine Spatial Plan. Brussels: Belgian Federal Public Service Health, Food Chain Safety and Environment. Retrieved 18 April 2018 from www.health.belgium.be/sites/default/files/uploads/fields/fpshealth_theme_file/19086905/Plan-EIA%20MSP_ENGLISH.pdf%20.pdf

Belgian Federal Public Service Health, Food Chain Safety and Environment (2014b) Summary of the marine spatial plan for the Belgian part of the North Sea. Brussels: Belgian Federal Public Service Health, Food Chain Safety and Environment. Retrieved 18 April 2018 from www.health.belgium.be/sites/default/files/uploads/fields/fpshealth_theme_file/19094275/Summary%20Marine%20Spatial%20Plan.pdf

Bellmann, M., Schuckenbrock, J., Gündert, S., Müller, M., Holst, H. & Remmers, P. (2017) Is there a state-of-the-art to reduce pile-driving noise? In Köppel, J. (ed.) *Wind Energy and Wildlife Interactions. Presentations from the CWW2015 Conference*. Cham: Springer International Publishing. pp. 161–172.

Bond, A., Retief, F., Cave, B., Fundingsland, M., Duinker, P.N., Verheem, R. & Brown, A.L. (2018) A contribution to the conceptualisation of quality in impact assessment. *Environmental Impact Assessment Review* 68: 49–58.

Boucquey, N., Fairbanks, L., St. Martin, K., Campbell, L.M. & McCay, B. (2016) The ontological politics of marine spatial planning: assembling the ocean and shaping the capacities of 'community' and 'environment'. *Geoforum* 75: 1–11.

Boyes, S.J. & Elliott, M. (2016) Brexit: the marine governance horrendogram just got more horrendous! *Marine Pollution Bulletin* 111: 41–44.

Bradbury, G., Trinder, M., Furness, B., Banks, A.N., Caldow, R.W.G. & Hume, D. (2014) Mapping seabird sensitivity to offshore wind farms. *PLoS ONE* 9 (9): e106366. doi: 10.1371/journal.pone.0106366.

Bradshaw, C.J.A., Greenhill, L. & Yates, K.L. (2018) The future of marine spatial planning. In Yates, K.L. & Bradshaw, C.J.A. (eds) *Offshore Energy and Marine Spatial Planning*. London: Routledge. pp. 284–293.

Breuch-Moritz, M. (2018) Bilanz-Pressekonferenz 2017.Hamburg, Rostock: Bundesamt für Seeschifffahrt und Hydrographie. Retrieved 27 March 2019 from https://www.bsh.de/DE/PRESSE/Pressemitteilungen/_Anlagen/Downloads/Bilanz-Pressekonferenz-2017-Langfassung.pdf?__blob=publicationFile&v=1

Bundesamt für Naturschutz (BfN) (2018) BfN: Karten. Bonn: Bundesamt für Naturschutz. Retrieved 12 April 2018 from www.bfn.de/infothek/karten.html

Bundesinstitut für Bau-, Stadt- und Raumforschung (BBSR) (2017) Koordinierte maritime Raumordnung in der Nordsee: InterregB. Bonn: Bundesinstitut für Bau-, Stadt- und Raumforschung. Retrieved 12 April 2018 from www.interreg.de/INTERREG2014/DE/Projekte/Gute-Beispiele/RaumentwicklungGovernance/DL/dl_northsee.pdf?__blob=publicationFile&v=2

Bundesministerium für Umwelt, Naturschutz, Bau und Reaktorsicherheit (BMU) (2013) Konzept für den Schutz der Schweinswale vor Schallbelastungen bei der Errichtung von Offshore-Windparks in der deutschen Nordsee (Schallschutzkonzept). Bonn: Bundesministerium für Umwelt, Naturschutz und Reaktorsicherheit. Retrieved 4 April 2018 from www.bfn.de/fileadmin/BfN/awz/Dokumente/schallschutzkonzept_BMU.pdf

Bundesnetzagentur (BNetzA) (2017) Ergebnisse der 1. Ausschreibung vom 01.04.2017: Bekanntgabe der Zuschläge. Bonn: Bundesnetzagentur. Retrieved 27 March 2019 from https://www.bundesnetzagentur.de/DE/

Service-Funktionen/Beschlusskammern/1_GZ/BK6-GZ/2017/2017_0001bis0999/BK6-17-001/Ergebnisse_erste_Ausschreibung.pdf?__blob=publicationFile&v=3

Burger, M. (2011) Consistency Conflicts and Federalism Choice: Marine Spatial Planning Beyond the States' Territorial Seas. *Environmental Law Reporter*, 41 (7). Retrieved 06 September 2018 from https://law.pace.edu/school-of-law/sites/pace.edu.school-of-law/files/PELR/Consistency_Conflicts_and_Federalism_Choice.pdf

Christie, N., Smyth, K., Barnes, R. & Elliott, M. (2014) Co-location of activities and designations: a means of solving or creating problems in marine spatial planning?. *Marine Policy* 43: 254–261.

Commonwealth of Massachusetts (2018) Massachusetts Ocean Resource Information System (MORIS). Boston: Commonwealth of Massachusetts. Retrieved 12 April 2018 from www.mass.gov/service-details/massachusetts-ocean-resource-information-system-moris

The Courier (2010) Bell Rock wind farm plan off the radar. Retrieved 5 February 2019 from https://www.thecourier.co.uk/news/local/angus-mearns/116831/bell-rock-wind-farm-plan-off-the-radar/

Crabtree, C.J., Zappalá, D. & Hogg, S.I. (2015) Wind energy: UK experiences and offshore operational challenges. *Proceedings of the Institution of Mechanical Engineers, Part A: Journal of Power and Energy* 229: 727–746.

Crown Estate (2016) The Crown Estate's role in the development of offshore renewable energy. London: The Crown Estate. Retrieved 5 April 2018 from www.thecrownestate.co.uk/media/5411/ei-the-crown-estate-role-in-offshore-renewable-energy.pdf

Crown Estate (2018) The Crown Estate shares further detail on plans for Round 4, including proposed locations to be offered for new seabed rights. London: The Crown Estate. Retrieved 19 January 2019 from www.thecrownestate.co.uk/en-gb/media-and-insights/news/2018-the-crown-estate-shares-further-detail-on-plans-for-round-4-including-proposed-locations-to-be-offered-for-new-seabed-rights/

Crown Estate (2019) Offshore wind potential new leasing. London: The Crown Estate. Retrieved 19 January 2019 from www.thecrownestate.co.uk/en-gb/what-we-do/on-the-seabed/energy/offshore-wind-potential-new-leasing/

Dahmen, M. (2017) Alternatives in energy SEAs. Unpublished master's thesis, Technische Universität Berlin.

Dannheim, J., Degraer, S., Elliott, M., Smyth, K. & Wilson, J.C. (2019) Seabed communities. In Perrow, M.R. (ed.) *Wildlife and Wind Farms, Conflicts and Solutions. Volume 3. Offshore: Potential effects*. Exeter: Pelagic Publishing. pp. 64–85.

Department of Communications, Energy and Natural Resources (DCNER) (2014) Offshore renewable energy development plan: a framework for the sustainable development of Ireland's offshore renewable energy resource. Dublin: DCNER. Retrieved 19 April 2018 from www.dccae.gov.ie/documents/20140204%20DCENR%20-%20Offshore%20Renewable%20Energy%20Development%20Plan.pdf

Department of Energy and Climate Change (DECC) (2009) UK Offshore Energy Strategic Environmental Assessment: future leasing for offshore wind farms and licensing for offshore oil & gas and gas storage. London: DECC. Retrieved 10 April 2018 from assets.publishing.service.gov.uk/government/uploads/system/uploads/attachment_data/file/194328/OES_Environmental_Report.pdf

Department of Trade and Industry (DTI) (2002) Future offshore – a strategic framework for the offshore wind industry. London: DTI. Retrieved 9 April 2018 from tethys.pnnl.gov/sites/default/files/publications/A_Strategic_Framework_for_the_Offshore_Wind_Industry.pdf

Edwards, R. & Evans, A. (2017) The challenges of marine spatial planning in the Arctic: results from the ACCESS programme. *Ambio* 46 (Suppl. 3): 486–496.

Ehler, C. (2018) Marine spatial planning. In Yates, K.L. & Bradshaw, C.J.A. (eds) *Offshore Energy and Marine Spatial Planning*. London: Routledge. pp. 6–17.

Ehler, C., Douvere, F. & Intergovernmental Oceanographic Commission (IOC) (2007) Visions for a sea change: Report of the first International Workshop on Marine Spatial Planning. Paris: UNESCO. Retrieved 10 December 2018 from unesdoc.unesco.org/ark:/48223/pf0000153465

European Commission (2016) Political declaration on energy cooperation between the North Seas countries. Retrieved 29 October 2018 from ec.europa.eu/energy/sites/ener/files/documents/Political%20Declaration%20on%20Energy%20Cooperation%20between%20the%20North%20Seas%20Countries%20FINAL.pdf

European Commission (2017a) Commission Decision (EU) 2017/848 of 17 May 2017 – laying down criteria and methodological standards on good environmental status of marine waters and specifications and standardised methods for monitoring and assessment, and repealing Decision 2010/477/EU. Retrieved 19 January 2019 from eur-lex.europa.eu/legal-content/EN/TXT/?qid=1495097018132&uri=CELEX:32017D0848

European Commission (2017b) Commission Directive (EU) 2017/845 of 17 May 2017 amending Directive 2008/56/EC of the European Parliament and of the Council as regards the indicative lists of elements to be taken into account for the preparation of marine strategies. Retrieved 19 January 2019 from eur-lex.europa.eu/legal-content/EN/TXT/?uri=CELEX%3A32017L0845

European Commission & European Parliament (2000) Directive 2000/60/EC of the European Parliament and of the Council of 23 October 2000 establishing a framework for Community action in the field of water policy: Water Framework Directive. Official Joural of the European Communities: 32000L0060. Retrieved 19 January 2019 from eur-lex.europa.eu/legal-content/EN/TXT/?uri=celex%3A32000L0060

European Council (1992) Council Directive 92/43/EEC of 21 May 1992 on the conservation of natural habitats and wild fauna and flora. Official Journal of the European Communities: 31992L0043. Retrieved 18 January 2019 from eur-lex.europa.eu/legal-content/EN/TXT/?uri=CELEX:31992L0043

European Parliament & European Council (2001) Directive 2001/42/EC of the European Parliament and of the Council on the assessment of effects of certain plans and programmes on the environment. Official Journal of the European Communities: 32001L0042. Retrieved 28 March 2019 from https://eur-lex.europa.eu/legal-content/EN/ALL/?uri=CELEX%3A32001L0042

European Parliament & European Council (2008) Directive 2008/56/EC of the European Parliament and of the Council of 17 June 2008 establishing a framework for community action in the field of marine environmental policy (Marine Strategy Framework Directive). Official Journal of the European Communities: 32008L0056. Retrieved 19 January 2019 from eur-lex.europa.eu/legal-content/EN/TXT/?uri=CELEX:32008L0056

European Parliament & European Council (2009) Directive 2009/147/EC of the European Parliament and of the Council of 30 November 2009 on the conservation of wild birds. Official Journal of the European Communities: 32009L0147. Retrieved 18 January 2019 from eur-lex.europa.eu/legal-content/EN/TXT/?uri=CELEX:32009L0147

European Parliament & European Council (2014) Directive 2014/89/EU of the European Parliament and of the Council of 23 July 2014 establishing a framework for maritime spatial planning. Official Journal of the European Communities: 32014L0089. Retrieved 19 December 2018 from eur-lex.europa.eu/legal-content/EN/TXT/?uri=uriserv:OJ.L_.2014.257.01.0135.01.ENG%20

European Wind Energy Association (EWEA) (2011) Wind in our sails: the coming of Europe's offshore wind energy industry. Brussels: European Wind Energy Association. Retrieved 5 April 2018 from www.ewea.org/fileadmin/files/library/publications/reports/Offshore_Report.pdf

Executive Office of Energy and Environmental Affairs (EEA) and the Massachusetts Office of Coastal Zone Management (CZM) (2015) 2015 Massachusetts Ocean Management Plan: Volume 1- management and administration. Boston: Commonwealth of Massachusetts. Retrieved 19 January 2019 from www.mass.gov/service-details/2015-massachusetts-ocean-management-plan

Federal Maritime and Hydrographic Agency (BSH) (2007) Standard – investigation of the impacts of offshore wind turbines on the marine environment (StUK 3). Hamburg: Bundesamt für Seeschifffahrt und Hydrographie. Retrieved 5 December 2018 from tethys.pnnl.gov/sites/default/files/publications/Standard_Investigation_of_the_Impacts_of_Offshore_Wind_Turbines_on_the_Marine_Environment.pdf

Federal Maritime and Hydrographic Agency (BSH) (2009a) Spatial Plan for the German Exclusive Economic Zone in the Baltic Sea: Text section: unofficial translation. Hamburg: Bundesamt für Seeschifffahrt und Hydrographie. Retrieved 18 April 2018 from www.bsh.de/en/Marine_uses/Spatial_Planning_in_the_German_EEZ/documents2/Spatial_Plan_Baltic_Sea.pdf

Federal Maritime and Hydrographic Agency (BSH) (2009b) Spatial Plan for the German Exclusive Economic Zone in the North Sea: Text section: Unofficial translation. Hamburg: Bundesamt für Seeschifffahrt und Hydrographie. Retrieved 18 April 2018 from www.bsh.de/en/Marine_uses/Spatial_Planning_in_the_German_EEZ/documents2/Spatial_Plan_North_Sea.pdf

Federal Maritime and Hydrographic Agency (BSH) (2009c) Umweltbericht zum Raumordnungsplan für die deutsche ausschließliche Wirtschaftszone (AWZ) in der Nordsee. Hamburg: Bundesamt für Seeschifffahrt und Hydrographie. Retrieved 19 April 2018 from www.bsh.de/de/Meeresnutzung/Raumordnung_in_der_AWZ/Dokumente_05_01_2010/Umweltbericht_Nordsee.pdf

Federal Maritime and Hydrographic Agency (BSH) (2009d) Umweltbericht zum Raumordnungsplan für die deutsche ausschließliche Wirtschaftszone (AWZ) in der Ostsee. Hamburg: Bundesamt für Seeschifffahrt und Hydrographie. Retrieved 19 April 2018 from www.bsh.de/de/Meeresnutzung/Raumordnung_in_der_AWZ/Dokumente_05_01_2010/Umweltbericht_Ostsee.pdf

Federal Maritime and Hydrographic Agency (BSH) (2013a) Offshore wind farms prediction of underwater sound: minimum requirements on documentation. Hamburg: Bundesamt für Seeschifffahrt und Hydrographie. Retrieved 9 April 2018 from www.bsh.de/en/Products/Books/Standard/Prediction_of_Underwater.pdf

Federal Maritime and Hydrographic Agency (BSH) (2013b) Standard: Untersuchung der Auswirkungen von Offshore-Windenergieanlagen auf die Meeresumwelt (StUK4). Hamburg: Bundesamt für Seeschifffahrt und Hydrographie. Retrieved 19 January 2019 from www.bsh.de/DE/PUBLIKATIONEN/_Anlagen/Downloads/Offshore/Standards-EN/Standard-Investigation-impacts-offshore-wind-turbines-marine-environment.pdf?__blob=publicationFile&v=4

Federal Maritime and Hydrographic Agency (BSH) (2018) Entwurf Flächenentwicklungsplan 2019 für die deutsche Nord- und Ostsee. Hamburg: Bundesamt für Seeschifffahrt und Hydrographie. Retrieved 5 December 2018 from www.bsh.de/DE/THEMEN/Offshore/Meeresfachplanung/_Anlagen/Downloads/Aktuelles_FEP_Entwurf_FEP2.pdf?__blob=publicationFile&v=3

Federal Maritime and Hydrographic Agency (BSH) (2019) Karte zweiter Entwurf Flächenentwicklungsplan 2019 Nordsee. Hamburg: Bundesamt für Seeschifffahrt und Hydrographie. Retrieved 12 June 2019 from https://www.bsh.de/DE/THEMEN/Offshore/Meeresfachplanung/meeresfachplanung.html?nn=1653366

Federal Maritime and Hydrographic Agency (BSH) & Federal Ministry for the Environment, Nature Conservation and Nuclear Safety (BMU) (eds) (2014) *Ecological Research at the Offshore Windfarm Alpha Ventus: Challenges, results and perspectives.* Wiesbaden: Springer Spektrum.

Fernandes, M.d.L., Esteves, T.C., Oliveira, E.R. & Alves, F.L. (2017) How does the cumulative impacts approach support maritime spatial planning? *Ecological Indicators* 73: 189–202.

Fisheries and Oceans Canada (2009) The role of the Canadian Government in the oceans sector. Ottowa, Ontario: Oceans Directorate Fisheries and Oceans Canada. Retrieved 10 September 2018 from waves-vagues.dfo-mpo.gc.ca/Library/337909.pdf

Gee, K. & Burkhard, B. (2018) Tracing regime shifts in the provision of coastal-marine cultural ecosystem services. In Yates, K.L. & Bradshaw, C.J.A. (eds) *Offshore Energy and Marine Spatial Planning*. London: Routledge. pp. 113–131.

Gill, A.B. & Wilhelmsson, D. (2019) Fish. In Perrow, M.R. (ed.) *Wildlife and Wind Farms, Conflicts and Solutions. Volume 3. Offshore: Potential effects.* Exeter: Pelagic Publishing. pp. 86–111.

Gill, A.B., Birchenough, S.N.R., Jones, A.R., Judd, A., Jude, S., Payo-Payo, A. & Wilson, B. (2018) Environmental implications of offshore energy. In Yates, K.L. & Bradshaw, C.J.A. (eds) *Offshore Energy and Marine Spatial Planning*. London: Routledge. pp. 132–168.

Global Wind Energy Council (GWEC) (2018) Global Wind Report. Annual market update 2017. Brussels: Global Wind Energy Council. Retrieved 8 May 2018 from files.gwec.net/register?file=/files/GWR2017.pdf

Hammond, P., Lacey, C., Gilles, A., Viquerat, S., Börjesson, P., Herr, M., Macleod, K., Ridoux, V., Santos, M. B., Scheidat, M., Teilmann, J., Vingada, J. & Øien, N. (2017) Estimates of cetacean abundance in European Atlantic waters in summer 2016 from the SCANS-III aerial and shipboard surveys. St Andrews: University of St Andrews. Retrieved 10 September 2018 from https://synergy.st-andrews.ac.uk/scans3/files/2017/04/SCANS-III-design

Hanssen, F., May, R., van Dijk, J. & Rød, J.K. (2018) Spatial multi-criteria decision analysis tool suite for consensus-based siting of renewable energy structures. *Journal of Environmental Assessment Policy and Management* 20(03): 1840003.

HELCOM (2018) Implementation of the Baltic Sea Action Plan 2018: background document to the 2018 HELCOM Ministerial Meeting. Helsiniki: HELCOM. Retrieved 13 December 2018 from www.helcom.fi/Lists/Publications/Implementation%20of%20the%20BSAP%202018.pdf

HELCOM-VASAB Maritime Spatial Planning Working Group (2015) Guideline for the implementation of ecosystem-based approach in maritime spatial planning (MSO) in the Baltic Sea area. Riga: VASAB Secretariat. Retrieved 27 March 2019 from http://vasab.org/wp-content/uploads/2018/06/Guideline-for-the-implementation-of-ecosystem-based-approach-in-MSP-in-the-Baltic-Sea-area-1.pdf

HELCOM-VASAB Maritime Spatial Planning Working Group (2019) Joint HELCOM-VASAB maritime spatial planning working group. Helsiniki: HELCOM. Retrieved 19 January 2019 from www.helcom.fi/helcom-at-work/groups/helcom-vasab-maritime-spatial-planning-working-group

HM Government (2011) UK Marine Policy Statement. London: The Stationery Office Limited. Retrieved 19 January 2019 from assets.publishing.service.gov.uk/government/uploads/system/uploads/attachment_data/file/69322/pb3654-marine-policy-statement-110316.pdf

HM Government (2018) A Green Future: Our 25 year plan to improve the environment. Kew, London: The National Archives. Retrieved 4 April 2018 from assets.publishing.service.gov.uk/government/uploads/system/uploads/attachment_data/file/693158/25-year-environment-plan.pdf

Hooper, T., Beaumont, N. & Hattam, C. (2017) The implications of energy systems for ecosystem services: a detailed case study of offshore wind. *Renewable and Sustainable Energy Reviews* 70: 230–241.

Hurlimann, A.C. & March, A.P. (2012) The role of spatial planning in adapting to climate change. *Wiley Interdisciplinary Reviews: Climate Change* 3: 477–488.

Inch Cape Wind (2018) Inch Cape Offshore Wind Development announces supply chain commitments. Edinburgh: ICOL (Inch Cape Offshore Limited). Retrieved 6 February 2019 from www.inchcapewind.com/news/Inch_Cape_Offshore_Wind_Development_announces_supply_chain_commitments

Jansen, H.M., van den Burg, S., Bolman, B., Jak, R.G., Kamermans, P., Poelman, M. & Stuiver, M. (2016) The feasibility of offshore aquaculture and its potential for multi-use in the North Sea. *Aquaculture International* 24: 735–756.

Jay, S. (2010a) Built at sea: marine management and the construction of marine spatial planning. *Town Planning Review* 81: 173–192.

Jay, S. (2010b) Planners to the rescue: spatial planning facilitating the development of offshore wind energy. *Marine Pollution Bulletin*. 2010: 493–499.

Jay, S., Alves, F.L., O'Mahony, C., Gomez, M., Rooney, A., Almodovar, M., Gee, K., Vivero, J.L.S. de, Gonçalves, J.M.S., da Luz Fernandes, M., Tello, O., Twomey, S., Prado, I., Fonseca, C., Bentes, L., Henriques, G. & Campos, A. (2016) Transboundary dimensions of marine spatial planning: fostering inter-jurisdictional relations and governance. *Marine Policy* 65: 85–96.

Jentoft, S. & Knol, M. (2014) Marine spatial planning: risk or opportunity for fisheries in the North Sea? *Maritime Studies* 13: 1. doi: 10.1186/2212-9790-13-1.

Jha-Thakur, U., Gazzola, P., Peel, D., Fischer, T.B. & Kidd, S. (2009) Effectiveness of strategic environmental assessment – the significance of learning. *Impact Assessment and Project Appraisal* 27: 133–144.

Johnson, K., Kerr, S. & Side, J. (2012) Pentland Firth and Orkney Waters Scotland: second run. Orkney; Heriot-Watt University. Retrieved 11 September 2018 from www.mesma.org/FILE_DIR/06-10-2013_13-16-55_46_7_MESMA-FW-Case-Study2-Scotland.pdf

Jones, M. & Morrison-Saunders, A. (2017) Understanding the long-term influence of EIA on organisational learning and transformation. *Environmental Impact Assessment Review* 64: 131–138.

Juda, L. (2003) Changing national approaches to ocean governance: the United States, Canada, and Australia. *Ocean Development & International Law* 34: 161–187.

Judd, A.D., Backhaus, T. & Goodsir, F. (2015) An effective set of principles for practical implementation of marine cumulative effects assessment. *Environmental Science & Policy* 54: 254–262.

Kaiser, M., Elliott, A., Galanidi, M., Rees, E.I.S., Caldow, R.W.G., Stillmann, R., Sutherland, W.J. & Showler, D. (2002) Predicting the displacement of common scoter *Melanitta nigra* from benthic feeding areas due to offshore windfarms: COWRIE-BEN-03-2002.Anglesey: Bangor University and Natural Environment Research Council (NERC). Retrieved 28 March 2019 from https://tethys.pnnl.gov/sites/default/files/publications/Kaiser%20et%20al.%202002.pdf

Kaiser, M.J., Galandini, M., Showler, D., Elliott, A.J., Caldow, R.W.G., Rees, E.I.S., Stillmann, R. & Sutherland, W.J. (2006) Distribution and behaviour of common scoter *Melanitta nigra* relative to prey resources and environmental parameters. *Ibis* 148: 110–128.

Kaymaz, I. (2018) Monitoring the possible impacts of offshore wind farms on Cetacean species in the North Sea region: Developing alternative SDG14 indicators. Master's thesis, Technische Universität Berlin.

King, S. (2019) Seabirds: collision. In Perrow, M.R. (ed.) *Wildlife and Wind Farms, Conflicts and Solu-*

tions. Volume 3. Offshore: Potential effects. Exeter: Pelagic Publishing. pp. 206–234.

Kite-Powell, H.L. (2017) Economics of multi-use and co-location. In Buck, B.H. & Langan, R. (eds) *Aquaculture Perspective of Multi-use Sites in the Open Ocean: The untapped potential for marine resources in the Anthropocene*. Cham: Springer International Publishing. pp. 233–249.

Köppel, J., Geißler, G., Rehhausen, A., Wende, W., Albrecht, J., Syrbe, R.-U., Magel, I., Scholles, F., Putschky, M., Hoppenstedt, A. & Stemmer, B. (2017) Strategische Umweltprüfung und (neuartige) Pläne und Programme auf Bundesebene – Methoden, Verfahren und Rechtsgrundlagen: im Auftrag des Umweltbundesamtes. Dessau-Roßlau: Umweltbundesamt. Retrieved 19 January 2019 from www.umweltbundesamt.de/sites/default/files/medien/1410/publikationen/2018-10-18_texte_81-2018_supbundesplanung.pdf

Kruppa, I. (2007) Steuerung der Offshore-Windenergienutzung vor dem Hintergrund der Umweltziele Klima- und Meeresumweltschutz. PhD thesis, Technische Universität Berlin.

Lüdeke, J., Geißler, G. & Köppel, J. (2012) Der neue Offshore-Netzplan zur Regelung der Anbindung von Offshore Windparks. Analyse und Diskussion der Prüfung seiner Umweltauswirkungen. *UVP-Report* 26(3+4): 183–190.

Marine Management Organisation (MMO) (2014) Marine plan areas in England. Newcastle upon Tyne: MMO. Retrieved 19 June 2018 from www.gov.uk/government/uploads/system/uploads/attachment_data/file/325688/marine_plan_areas.pdf

Marine Scotland (2010a) Draft Plan for Offshore Wind Energy in Scottish Territorial Waters. Edinburgh: Marine Scotland. Retrieved 28 March 2019 from https://www2.gov.scot/Resource/Doc/336236/0110013.pdf

Marine Scotland (2010b) Strategic Environmental Assessment (SEA) of Draft Plan for Offshore Wind Energy in Scottish Territorial Waters: Volume 1: Environmental Report. Edinburgh: Marine Scotland. Retrieved 28 March 2019 from https://www2.gov.scot/Resource/Doc/312161/0098588.pdf

Marine Scotland (2011) Blue Seas – Green Energy. A Sectoral Marine Plan for Offshore Wind Energy in Scottish Territorial Waters. Part B: Post Adoption Statement. Edinburgh: Marine Scotland. Retrieved 28 March 2019 from https://tethys.pnnl.gov/sites/default/files/publications/Blue_Seas_Green_Energy_Part_B.pdf

Marine Scotland (2014) Planning Scotland's Seas: Sectoral Marine Plans for Offshore Wind, Wave and Tidal Energy in Scottish Waters. Consultation Analysis Report. Appendix 4 Campaign Text Regarding OWSW1/TWS1, Appendix 5 Campaign Text Regarding OWSW2. Edinburgh: Marine Scotland. Retrieved 28 March 2019 from https://www.gov.scot/binaries/content/documents/govscot/publications/consultation-responses/2014/05/planning-scotlands-seas-sectoral-marine-plans-offshore-wind-wave-tidal/documents/00448870-pdf/00448870-pdf/govscot%3Adocument

Mendel, B., Schwemmer, P., Peschko, V., Müller, S., Schwemmer, H., Mercker, M. & Garthe, S. (2019) Operational offshore wind farms and associated ship traffic cause profound changes in distribution patterns of Loons (*Gavia* spp.). *Journal of Environmental Management* 231: 429–438.

Merchant, N.D., Faulkner, R.C. & Martinez, R. (2018) Marine noise budgets in practice. *Conservation Letters* 11(3): e12420.

Mid-Atlantic Regional Council on the Ocean (MARCO) (2018) Mid-Atlantic Ocean Data Portal. Retrieved 12 April 2018 from portal.midatlanticocean.org/

Nehls, G., Harwood, A.J.P. & Perrow, M.R. (2019) Marine mammals. In Perrow, M.R. (ed.) *Wildlife and Wind Farms, Conflicts and Solutions. Volume 3. Offshore: Potential effects*. Exeter: Pelagic Publishing. pp. 112–141.

NnG Offshore Wind (2018) Planning and consenting. Edinburgh: Neart na Gaoithe. Retrieved 6 February 2019. From http://nngoffshorewind.com/about/planning-and-consenting

OffshoreWIND.biz (2017) Neart na Gaoithe moves forward as court dismisses RSPB legal action. Schiedam: Navingo BV. Retrieved 28 March 2019 from https://www.offshorewind.biz/2017/07/19/neart-na-gaoithe-moves-forward-as-court-dismisses-rspb-legal-action/

OSPAR Commission (2008) OSPAR guidance on environmental considerations for offshore wind farm development. Reference No. 2008-3. London: OSPAR. Retrieved 5 December 2018 from ospar-archive.s3-eu-west-1.amazonaws.com/DECRECS/AGREEMENTS/08-03e_agreement_consolidated_guidance_for_offshore_windfarms.doc?response-content-disposition=attachment%3B%20filename%3D%2208-03e_agreement_consolidated_guidance_for_offshore_windfarms.doc%22&AWSAccessKeyId=AKIAJIACMW2T5USCSU5A&Expires=1544021175&Signature=0t9bJ2RUVUy9k8CQyojgerLgbSY%3D

Pacific North Coast Integrated Management Area (PNCIMA) Initiative (2011) Atlas of the Pacific North Coast Integrated Management Area. British Columbia: Pacific North Coast Integrated Management Area Initiative. Retrieved 11 September 2018 from www.pncima.org/media/documents/atlas/pncima-atlas_print_online.pdf

Pacific North Coast Integrated Management Area (PNCIMA) Initiative (2017) Pacific North Coast Integrated Management Area plan. British Columbia: Pacific North Coast Integrated Management Area Initiative. Retrieved 11 September 2018 from www.pncima.org/media/documents/2016-plan/2316-dfo-pncima-report-v17-optimized.pdf

Parliament of the United Kingdom (2009) Marine and Coastal Access Act 2009: c 23. London: Her Majesty's Stationery Office. Retrieved 10 November 2018 from www.legislation.gov.uk/ukpga/2009/23/pdfs/ukpga_20090023_en.pdf

Perrow, M.R. (2019) A synthesis of effects and impacts. In Perrow, M.R. (ed.) *Wildlife and Wind Farms, Conflicts and Solutions. Volume 3. Offshore: Potential effects*. Exeter: Pelagic Publishing. pp. 235–277.

Perrow, M.R., Gilroy, J.J., Skeate, E.R. & Tomlinson, M.L. (2011) Effects of the construction of Scroby Sands offshore wind farm on the prey base of Little tern *Sternula albifrons* at its most important UK colony. *Marine Pollution Bulletin* 62:1661–70.

Pınarbaşı, K., Galparsoro, I., Borja, Á., Stelzenmüller, V., Ehler, C.N. & Gimpel, A. (2017) Decision support tools in marine spatial planning: present applications, gaps and future perspectives. *Marine Policy* 83: 83–91.

Platteeuw, M., Bakker, J., van den Bosch, I., Erkmann, A., Graafland, M., Lubbe, S. & Warnas, M. (2017) A framework for assessing ecological and cumulatve effects (FAECE) of offshore wind farms on birds, bats, and marine mammals in the southern North Sea. In Köppel, J. (ed.) *Wind Energy and Wildlife Interactions. Presentations from the CWW2015 Conference*. Cham: Springer International Publishing. pp. 219–237.

Portman, M.E. (2014) Regulatory capture by default: offshore exploratory drilling for oil and gas. *Energy Policy* 65: 37–47.

Portman, M. (2016) *Environmental Planning for Oceans and Coasts: Methods, tools and technologies*. Cham: Springer International Publishing.

Portman, M.E., Duff, J.A., Köppel, J., Reisert, J. & Higgins, M.E. (2009) Offshore wind energy development in the exclusive economic zone: legal and policy supports and impediments in Germany and the US. *Energy Policy* 37: 3596–3607.

Rehhausen, A., Köppel, J., Scholles, F., Stemmer, B., Syrbe, R.-U., Magel, I., Geißler, G. & Wende, W. (2018) Quality of federal level strategic environmental assessment – a case study analysis for transport, transmission grid and maritime spatial planning in Germany. *Environmental Impact Assessment Review* 73: 41–59.

Rijkswaterstaat Sea and Delta (2018) Strategic Environmental Assessment on North Seas Energy. Rjikswijk: Rijkswaterstaat Sea and Delta. Retrieved 19 January 2019 from northseaportal.eu/

Sánchez, L.E. & Mitchell, R. (2017) Conceptualizing impact assessment as a learning process. *Environmental Impact Assessment Review* 62: 195–204.

Scottish Government (2015) Scotland's National Marine Plan: a single framework for managing our seas. Edinburgh: The Scottish Government. Retrieved 4 April 2018 from www.gov.scot/Resource/0047/00475466.pdf

ScottishPower Renewables (2013) ScottishPower renewable update on Argyll Array offshore windfarm. Glasgow: Scottish Power Ltd. Retrieved 5 February 2019 from https://www.scottishpowerrenewables.com/news/pages/scottishpower_renewables_update_on_argyll_array_offshore_windfarm.aspx

Shabtay, A., Portman, M.E. & Carmel, Y. (2017) Incorporating principles of reconciliation ecology to achieve ecosystem-based marine spatial planning. *Ecological Engineering* 120: 595–600.

Shabtay, A., Portman, M.E. & Carmel, Y. (2018) Contributions of marine infrastructures to marine planning and protected area networking. *Aquatic Conservation: Marine and Freshwater Ecosystems* 28: 830–839.

Simard, F., Laffoley, D. & Baxter, J.M. (eds) (2016) Marine Protected Areas and climate change: Adaptation and mitigation synergies, opportunities and challenges. Gland: International Union for Conservation of Nature. Retrieved 27 March 2019 from http://dx.doi.org/10.2305/IUCN.CH.2016.14.en

Simas, T. (2017) Offshore wind: current challenges for sustainable development. In *Conference on Wind Energy and Wildlife Impacts 2017 – Book of abstracts:* 15. Linda-a-Velha: Conference on Wind Energy and Wildlife Impacts. Retrieved 28 March 2019 from http://cww2017.pt/images/Congresso/book-abstracts/BookOfAbstractCWW17_complete_4Set17.pdf

Skeate, E.R., Perrow, M.R. & Gilroy, J.J. (2012) Likely effects of construction of Scroby Sands offshore wind farm on a mixed population of harbour *Phoca vitulina* and grey *Halichoerus grypus* seals. *Marine Pollution Bulletin* 64: 872–81.

Söderström, S. & Kern, K. (2017) The ecosystem approach to management in marine environmental governance: institutional interplay in the Baltic Sea region. *Environmental Policy and Governance* 27: 619–631.

State Ocean Caucus (2017) Washington marine spatial planning. Olympia: Washington State Governor's Office. Retrieved 5 April 2018 from msp.wa.gov/

Stelzenmüller, V., Lee, J., Garnacho, E. & Rogers, S.I. (2010) Assessment of a Bayesian belief network-GIS framework as a practical tool to support marine planning. *Marine Pollution Bulletin* 60: 1743–1754.

Stelzenmüller, V., Schulze, T., Fock, H.O. & Berkenhagen, J. (2011) Integrated modelling tools to support risk-based decision-making in marine spatial management. *Marine Ecology Progress Series* 2011: 197–212.

Stelzenmüller, V., Lee, J., South, A., Foden, J. & Rogers, S.I. (2013) Practical tools to support marine spatial planning: a review and some prototype tools. *Marine Policy* 38: 214–227.

Stelzenmüller, V., Diekmann, R., Bastardie, F., Schulze, T., Berkenhagen, J., Kloppmann, M., Krause, G., Pogoda, B., Buck, B.H. & Kraus, G. (2016) Co-location of passive gear fisheries in offshore wind farms in the German EEZ of the North Sea: a first socio-economic scoping. *Journal of Environmental Management* 183: 794–805.

Stelzenmüller, V., Coll, M., Mazaris, A.D., Giakoumi, S., Katsanevakis, S., Portman, M.E., Degen, R., Mackelworth, P., Gimpel, A., Albano, P.G., Almpanidou, V., Claudet, J., Essl, F., Evagelopoulos, T., Heymans, J.J., Genov, T., Kark, S., Micheli, F., Pennino, M.G., Rilov, G., Rumes, B., Steenbeek, J. & Ojaveer, H. (2018) A risk based approach to cumulative effect assessments for marine management. *Science of the Total environment* 612: 1132–1140.

Strickland-Munro, J., Kobryn, H., Brown, G. & Moore, S.A. (2016) Marine spatial planning for the future: using Public Participation GIS (PPGIS) to inform the human dimension for large marine parks. *Marine Policy* 73: 15–26.

Sustainable Energy Authority of Ireland, AECOM Envrionment, METOC & CMRC (2010) Strategic Environmental Assessment (SEA) of Offshore Renewable Energy Development Plan (OREDP) in the Republic of Ireland. Dublin: Sustainable Energy Authorithy of Ireland. Retrieved 19 April 2018 from www.dccae.gov.ie/documents/SEA%20Environmental%20Report%20Compressed%205MB.pdf

Swedish Agency for Marine and Water Management (SwAM) (2017) Symphony – a tool for ecosystem-based marine spatial planning.Gothenberg: Swedish Agency for Marine and Water Mangement. Retrieved 5 December 2018 from www.havochvatten.se/en/swam/eu--international/marine-spatial-planning/symphony---a-tool-for-ecosystem-based-marine-spatial-planning.html

Swedish Agency for Marine and Water Management (SwAM) (2018) Strategic Environmental Assessment of the Marine Spatial Plan proposal for the Baltic Sea. Gothenburg: Björn Sjöberg. Retrieved 5 December 2018 from https://www.havochvatten.se/download/18.246c08de16436c8f9886dda2/1530277197844/sea-baltic-sea-swedish-consultation-msp.pdf

The Telegraph (2014, 26 March) SSE scraps £20bn offshore wind farm plan and questions viability of sector. Retrieved 5 February 2019 from https://www.telegraph.co.uk/finance/newsbysector/energy/10725596/SSE-scraps-20bn-offshore-wind-farm-plan-and-questions-viability-of-sector.html

Umweltbundesamt (UBA) (2008) Beurteilung von Umweltauswirkungen bei der Genehmigung von Offshore Windenergieanlagen. Dessau-Roßlau: Umweltbundesamt. Retrieved 9 April 2018 from www.umweltbundesamt.de/sites/default/files/medien/pdfs/offshore-windenergie.pdf

UN Environment World Conservation Monitoring Centre (UNEP-WCM) & International Union for Conservation of Nature and Natural Resources (IUCN) (2018) Marine protected planet. Retrieved 5 April 2018 from protectedplanet.net/marine

United Nations (UN) (1982) United Nations Convention on the Law of the Sea (UNCLOS). New York: UN Office of Legal Affairs. Retrieved 10 November 2018 from www.un.org/depts/los/convention_agreements/texts/unclos/unclos_e.pdf

United Nations Economic Commission for Europe (UNECE) (2003) Protocol on Strategic Environmental Assessment to the convention on the Environmental Impact Assessment in a transboundary context: Kyiv (SEA) Protocol. New York: UN Office of Legal Affairs. Retrieved 12 April 2018 from www.unece.org/fileadmin/

DAM/env/eia/documents/legaltexts/protocolenglish.pdf

United Nations Educational, Scientific and Cultural Organization (UNESCO) (2018) Marine Spatial Planning Programme – World Application.Paris: Intergovernmental Oceanographic Commission (IOC) of UNESCO. Retrieved 17 October 2018 from msp.ioc-unesco.org/world-applications/overview/

Vanermen, N. & Stienen, E.W.M. (2019) Seabird displacement. In Perrow, M.R. (ed.) *Wildlife and Wind Farms, Conflicts and Solutions. Volume 3. Offshore: Potential effects*. Exeter: Pelagic Publishing. pp. 174–205.

Vasileiou, M., Loukogeorgaki, E. & Vagiona, D.G. (2017) GIS-based multi-criteria decision analysis for site selection of hybrid offshore wind and wave energy systems in Greece. *Renewable and Sustainable Energy Reviews* 73: 745–757.

Vølund, P. & Hansen, J. (2001) Middelgrunden 40 MW offshore wind farm near Copenhagen, Denmark, installed year 2000. World Sustainable Energy Day, 2001 (Wels, Austria).Retrieved 28 March 2019 from http://www.middelgrunden.dk/middelgrunden/sites/default/files/public/file/Artikel_Middelgrunden_40_MW_offshore_wind_farm_near_Copenhagen.pdf

Weber, J., Biehl, J. & Köppel, J. (2019) Lost in bias? Multifaceted discourses framing the communication of wind and wildlife research results – the PROGRESS case. In Bispo, R., Bernardino, J., Coelho, H. & Lino Costa, J. (eds) *Wind Energy and Wildlife Impacts: Balancing energy sustainability with wildlife conservation*. Basel: Springer Nature. pp. 179-204.

Welsh Government (2018) Draft Welsh National Marine Plan. Cardiff: Welsh Government. Retrieved 5 April 2018 from https://gov.wales/sites/default/files/consultations/2018-02/draft-plan-en.pdf

Willsteed, E., Gill, A.B., Birchenough, S.N.R. & Jude, S. (2017) Assessing the cumulative environmental effects of marine renewable energy developments: establishing common ground. *Science of the Total Environment* 577: 19–32.

Willsteed, E.A., Jude, S., Gill, A.B. & Birchenough, S.N.R. (2018) Obligations and aspirations: a critical evaluation of offshore wind farm cumulative impact assessments. *Renewable and Sustainable Energy Reviews* 82: 2332–2345.

WindPower Monthly (2010, 22 November) FOR pulls out of Forth Array offshore project. Retrieved 5 February 2019 from https://www.windpowermonthly.com/article/1042327/pulls-forth-array-offshore-project

Wind Power Offshore (2013) Analysis: Why did RWE decide to drop Atlantic Array? Twickenham: Haymarket Media Group Ltd. Retrieved 5 February 2019 from https://www.windpoweroffshore.com/article/1223085/analysis-why-rwe-decide-drop-atlantic-array

Wind Power Offshore (2014) Dong and Centrica abandoned Irish Sea Round 3 Zone. Twickenham: Haymarket Media Group Ltd. Retrieved 5 February 2019 from https://www.windpoweroffshore.com/article/1305992/dong-centrica-abandond-irish-round-3-zone

Wind Power Offshore (2015) Navitus Bay rejected. Twickenham: Haymarket Media Group Ltd. Retrieved 5 February 2019 from https://windpoweroffshore.com/article/1363660/navitus-bay-rejected

Windenergie-auf-See-Gesetz (WindSeeG) (2016) Windenergie-auf-See-Gesetz (2016) vom 13. Oktober 2016 (BGBl. I S. 2258, 2310), das zuletzt durch Artikel 2 Absatz 19 des Gesetzes vom 20. Juli 2017 (BGBl. I S. 2808) geändert worden ist. Retrieved 28 March 2019 from http://www.gesetze-im-internet.de/windseeg/WindSeeG.pdf.

Wolters, E.A., Steel, B.S., Lach, D. & Kloepfer, D. (2016) What is the best available science? A comparison of marine scientists, managers, and interest groups in the United States. *Ocean & Coastal Management* 122: 95–102.

Yates, K.L., Schoeman, D.S. & Klein, C.J. (2015) Ocean zoning for conservation, fisheries and marine renewable energy: assessing trade-offs and co-location opportunities. *Journal of Environmental Management* 152: 201–209.

Yates, K.L., Polsenberg, J., Kafas, A. & Bradshaw, C.J.A. (2018) Introduction: Marine spatial planning in the age of offshore energy. In Yates, K.L. & Bradshaw, C.J.A. (eds) *Offshore Energy and Marine Spatial Planning*. London: Routledge. pp. 1–5.

Index

Page numbers in *italics* denote figures and in **bold** denote tables.

VATTENFALL